EXPLOSIVE PHENOMENA IN ASTROPHYSICAL COMPACT OBJECTS

Related Titles from the AIP Conference Proceedings Subseries on Astronomy and Astrophysics

555 Cosmology and Particle Physics: CAPP 2000
Edited by Ruth Durrer, Juan Garcia-Bellido, and Mikhail Shaposhnikov, March 2001, 1-56396-986-6

540 Particle Physics and Cosmology: Second Tropical Workshop
Edited by José F. Nieves, October 2000, 1-56396-965-3

537 Waves in Dusty, Solar, and Space Plasmas
Edited by F. Verheest, M. Goossens, M. A. Hellberg, and R. Bharuthram, October 2000, 1-56396-962-9

528 Acceleration and Transport of Energetic Particles Observed in the Heliosphere: ACE 2000 Symposium
Edited by Richard A. Mewaldt, J. R. Jokipii, Martin A. Lee, Eberhard Möbius, and Thomas H. Zurbuchen, July 2000, 1-56396-951-3

526 Gamma-Ray Bursts: 5th Huntsville Symposium
Edited by R. Marc Kippen, Robert S. Mallozzi, and Gerald J. Fishman, June 2000, CD-ROM included, 1-56396-947-5

523 Gravitational Waves: Third Edoardo Amaldi Conference
Edited by Sydney Meshkov, June 2000, 1-56396-944-0

522 Cosmic Explosions: Tenth Astrophysics Conference
Edited by Stephen S. Holt and William W. Zhang, June 2000, 1-56396-943-2

516 26th International Cosmic Ray Conference: ICRC XXVI, Invited, Rapporteur, and Highlight Papers
Edited by Brenda L. Dingus, David B. Kieda, and Michael H. Salamon
May 2000, 1-56396-939-4

515 GeV-TeV Gamma Ray Astrophysics Workshop: Towards a Major Atmospheric Cherenkov Detector VI
Edited by Brenda L. Dingus, Michael H. Salamon, and David B. Kieda, May 2000, 1-56396-938-6

510 The Fifth Compton Symposium
Edited by Mark L. McConnell and James M. Ryan, March 2000, 1-56396-932-7

To learn more about these titles, or the AIP Conference Proceedings Series, please visit the webpage **http://www.aip.org/catalog/aboutconf.html**

EXPLOSIVE PHENOMENA IN ASTROPHYSICAL COMPACT OBJECTS

First KIAS Astrophysics Workshop

Seoul, Korea   24–27 May 2000

EDITORS
Heon-Young Chang
KIAS, Seoul, Korea
Chang-Hwan Lee
SUNY Stony Brook, New York and KIAS, Seoul, Korea
Mannque Rho
CEA, Saclay, France
Insu Yi
KIAS, Seoul, Korea

Melville, New York, 2001
AIP CONFERENCE PROCEEDINGS ■ VOLUME 556

Editors:

Heon-Young Chang
Korea Institute for Advanced Study (KIAS)
207-43 Cheongryangri-dong, Dongdaemun-gu
Seoul 130-012
KOREA

E-mail: hyc@kias.re.kr

Chang-Hwan Lee
Department of Physics and Astronomy   and   Korea Institute for Advanced Study (KIAS)
State University of New York at Stony Brook   207-43 Cheongryangri-dong, Dongdaemun-gu
Stony Brook, NY 11794-3800   Seoul 130-012
USA   KOREA

E-mail: chlee@tonic.physics.sunysb.edu   E-mail: chlee@kias.re.kr

Mannque Rho
Service de Physique Théorique
Commisariat à l'Énergie Atomique (CEA)
91191 Gif-sur-Yvette cedex
FRANCE

E-mail: rho@spht.saclay.cea.fr

Insu Yi
Korea Institute for Advanced Study (KIAS)
207-43 Cheongryangri-dong, Dongdaemun-gu
Seoul 130-012
KOREA

E-mail: iyi@kias.re.kr

Authorization to photocopy items for internal or personal use, beyond the free copying permitted under the 1978 U.S. Copyright Law (see statement below), is granted by the American Institute of Physics for users registered with the Copyright Clearance Center (CCC) Transactional Reporting Service, provided that the base fee of $18.00 per copy is paid directly to CCC, 222 Rosewood Drive, Danvers, MA 01923. For those organizations that have been granted a photocopy license by CCC, a separate system of payment has been arranged. The fee code for users of the Transactional Reporting Service is: 1-56396-987-4/01/$18.00.

© 2001 American Institute of Physics

Individual readers of this volume and nonprofit libraries, acting for them, are permitted to make fair use of the material in it, such as copying an article for use in teaching or research. Permission is granted to quote from this volume in scientific work with the customary acknowledgment of the source. To reprint a figure, table, or other excerpt requires the consent of one of the original authors and notification to AIP. Republication or systematic or multiple reproduction of any material in this volume is permitted only under license from AIP. Address inquiries to Office of Rights and Permissions, Suite 1NO1, 2 Huntington Quadrangle, Melville, N.Y. 11747-4502; phone: 516-576-2268; fax: 516-576-2450; e-mail: rights@aip.org.

L.C. Catalog Card No. 2001086988
ISBN 1-56396-987-4
ISSN 0094-243X
Printed in the United States of America

CONTENTS

Preface ix
List of Participants xi
Group Photo xv

RELATIVISTIC ASTROPHYSICS AND BLACK HOLES

The Force-Free Magnetosphere of a Black Hole 3
    C. H. Lee and H. K. Lee
Magnetic Extraction of Spin Energy from a Black Hole 10
    J. H. Krolik
Magnetic Extraction of Black Hole Rotational Energy 23
    H. K. Lee
Electron-Positron Outflow from Black Holes and the Formation of Winds ... 30
    H. P. M. M. van Putten

ACCRETION, MAGNETOHYDRODYNAMICS, AND DYNAMICS

The Evolution of Compact Young Star Clusters
and Their Relation to Hypernovae 39
    S. Portegies Zwart
Neutron Star Mergers in and out of Globular Clusters 48
    H. M. Lee
Recent Progress in the Research of X-ray Pulsars 56
    F. Nagase
Stellar Evolution and Black Hole Formation 68
    G. E. Brown
Iron Line Diagnostics of X-ray Binaries 78
    T. Dotani
Simulations of Radio Galaxies: Synthetic Observations 85
    I. L. Tregillis, D. Ryu, and T. W. Jones
The Role of the Outer Boundary Condition in Accretion Disk Models 93
    F. Yuan, Q. Peng, J. Lu, and J. Wang
Some Consequences of Magnetic Fields in High Energy Sources 100
    E. G. Blackman
Interaction of Radiation with Matter in Accretion Flow 112
    M.-G. Park
Simulations of Time-Dependent High Energy Emission from Blazars 119
    M. Kusunose, F. Takahara, and H. Li
Galactic Center ADAF Ruled out by Polarization 125
    E. Agol
Models for Type Ia Supernovae and Evolutionary Effects with Redshift ... 132
    P. Höflich

ASTRO-HADRON PHYSICS AND PARTICLE ASTROPHYSICS

Little Bang at Big Accelerators: Heavy Ion Physics from AGS to LHC 147
    J. Schukraft
Physics of Dense and Superdense Matter 160
    M. Rho
Chirally Protected Pion Mass 172
    T.-S. Park, H. Jung, and D.-P. Min
Kaon Production in Heavy Ion Collisions and Kaon Condensation in Neutron Stars 178
    C.-H. Lee
Neutrino Interactions in Color-Flavor-Locked Dense Matter 184
    D. K. Hong, H. K. Lee, M. A. Nowak, and M. Rho
Magnetoelastic Pulsations of Neutron Stars 197
    S. Bastrukov, J. Yang, D. Podgainy, and F. Weber
Neutron Star Equation of State 205
    J. M. Lattimer

GAMMA-RAY BURSTS

Very High Energy Phenomena in GRBs 221
    T. Totani
Crustal Shear Oscillations, Magnetar Spindown, and the 1998 August 27 Flare 228
    R. C. Duncan
The Afterglows of Gamma-Ray Bursts 240
    S. R. Kulkarni, E. Berger, J. S. Bloom, F. Chaffee, A. Diercks,
    S. G. Djorgovski, D. A. Frail, T. J. Galama, R. W. Goodrich,
    F. A. Harrison, R. Sari, and S. A. Yost
Prompt GRB Optical Follow-Up Experiments 261
    H. S. Park, G. Williams, E. Ables, D. Band, S. Barthelmy,
    R. Bionta, T. Cline, N. Gehrels, D. Hartmann, K. Hurley,
    M. Kippen, R. Nemiroff, W. Pereira, and R. Porrata
The GRB-Afterglow Transition: Black Holes, Bullets, and Beams 268
    R. A. M. J. Wijers
Effects of Environment and Energy Injection on Gamma-Ray Burst Afterglows 276
    Z. G. Dai
The Emission Features of GRB Afterglows 284
    D. M. Wei and T. Lu
On the Physics of Gamma-Ray Burst Engines and Its Observational Consequences 291
    H.-Y. Chang, I. Yi, C. Kim, and K. Kwak

SUPERNOVAE AND JETS

Jet Induced Supernovae: Hydrodynamics and Observational Consequences 301
    A. Khokhlov and P. Höflich

The Collapsar Model for Gamma-Ray Bursts 313
    A. I. MacFadyen

Bi-polar Supernova Explosions 324
    L. Wang, P. Höflich, and J. C. Wheeler

Cosmic Ray Acceleration at Supernova Remnants 331
    H. Kang

Pulsars, Magnetars, and Asymmetric Supernovae 338
    J. C. Wheeler, I. Yi, P. Höflich, and L. Wang

Modern Supernova Search 350
    M. G. Lee

ASTROPHYSICS

Is the Cas A Point Source Discovered by Chandra a Black Hole or a Neutron Star? 359
    H. Umeda, K. Nomoto, S. Tsuruta, and S. Mineshige

Accretion Disk Emission and Raman Scattering in Symbiotic Stars 368
    H.-W. Lee

Flares in X-ray Transients 375
    S.-W. Kim

Submillimeter-wave and Millimeter-wave Observations of the Interaction of Supernova Remnants with Molecular Clouds 382
    K. Tatematsu, Y. Arikawa, Y. Sekimoto,
    and the Mt. Fuji Submillimeter-wave Telescope Team

Intense Laser Astrophysics 389
    H. Takabe

Author Index 403

PREFACE

Three things distinguish this KIAS Workshop from other meetings in Korea. First, such a meeting *was* held in Korea and in particular in KIAS. Second, it had a major theme that encompasses a variety of domains of physics, namely, astro-particle physics, astro-hadron physics and relativistic astrophysics. Third, it purported to promote active collaboration between Korea and the rest of the world.

A workshop of this nature with a principal focus on highly explosive phenomena in astrophysical systems explored from a wide-ranging vista has not been held before in Korea. The topic dealing with direct confrontation with observations in Nature that characterizes this domain is somewhat foreign to much of the Korean physics community. To some, particularly to the young participants, it was a novel and inspiring experience to witness both theory *and* observations put on the same footing and presented by actively working experts. It suggested a fruitful direction that a new and ambitious institute like KIAS could possibly take.

Physics under extreme conditions addresses energetic astrophysical processes as well as those simulated in high-energy laboratories. This meeting made what we think is the first, albeit exploratory, attempt to bring the two together. This meant that various other fields of physics than astrophysics, such as condensed matter physics, particle physics and hadron physics, had to converge onto one common theme so as to relate the "Little Bang" in the big terrestrial laboratories already built or to be built to the energetic collapse phenomena involving black holes and other compact stars taking place in the Universe. If nothing else at present, the proper common languages had to be introduced.

The participation in the meeting of the active researchers from all over the world was combined with the initiation of new projects and the continuation of on-going ones with the Korean physicists, both established and upcoming. This objective has been met with success, with numerous collaborations generated from the gathering.

The workshop owes its success to the generous support of KIAS as well as to the efficient help of the KIAS' staff. The secretaries of the workshop, Heon-Young Chang and Chang-Hwan Lee, did a remarkably efficient job in both helping organize the meeting and editing this volume. We acknowledge with gratitude the active participation of Ralph Wijers in all stages of organization with his effective "arm-twisting" persuasion. The enthusiasm of the participants and their high-quality contributions – even those that failed to be included in this volume – are highly appreciated.

Mannque Rho, Insu Yi

LIST OF PARTICIPANTS

Eric. Agol	Johns Hopkins, USA	agol@pha.jhu.edu
Changrim Ahn	Ewha, Korea	ahn@nm.ewha.ac.kr
Sang-Hyeon Ahn	Seoul, Korea	sha@astro.snu.ac.kr
Sergey.I. Bastrukov	JINR, Dubna, Russia	bast@mm.ewha.ac.kr
Richard Bionta	LLNL, USA	bionta1@llnl.gov
Eric. G. Blackman	Rochester, USA	blackman@pas.rochester.edu
Gerry. E. Brown	Stony Brook, USA	popenoe@nuclear.physics.sunysb.edu
Yong-Ik Byun	Yonsei, Korea	byun@darksky.yonsei.ac.kr
Heon-Young Chang	KIAS, Korea	hyc@kias.re.kr
Inyong Cho	Emory, USA	cho@physics.emory.edu
Kyunghwa Cho	Ewha, Korea	clarity@hanmail.net
Chul-Sung Choi	KAO, Korea	cschoi@hanul.issa.re.kr
Yun-Young Choi	Ewha, Korea	yychoi@mm.ewha.ac.kr
Zigao Dai	Nanjing, China	daizigao@public1.ptt.js.cn
Tadayasu Dotani	ISAS, Japan	dotani@astro.isas.ac.jp
Robert C. Duncan	Austin, Texas, USA	duncan@astro.as.utexas.edu
Young S. Hahn	KAIST	
Peter A. Höflich	Austin, Texas, USA	pah@astro.as.utexas.edu
Deog Ki Hong	Pusan, Korea	dkhong@hyowon.cc.pusan.ac.kr
Soon-Tae Hong	Sogang, Korea	sthong@ccs.sogang.ac.kr
Jai-Chan Hwang	Kyungpook, Korea	jchan@bh.knu.ac.kr
Hyesung Kang	Pusan, Korea	kang@uju.es.pusan.ac.kr
Young-Gwang Kang	Osaka, Japan	ygkang@ile.osaka-u.ac.jp
Chung Wook Kim	KIAS, Korea	cwkim@kias.re.kr
Chunglee Kim	KIAS	ciel@newton.kias.re.kr
Do-Won Kim	Kangnung, Korea	do.won.kim@cern.ch
Hongsu Kim	Hanyang, Korea	
Hui Kyung Kim	Hanyang, Korea	black072@hepth.hanyang.ac.kr
Hyun Hee Kim	Ewha, Korea	meclien@hanmail.net
Joon-Il Kim	Seoul, Korea	jikim@phya.snu.ac.kr
Soo Bong Kim	Seoul, Korea	sbkim@phya.snu.ac.kr
Sang Chul Kim	Seoul, Korea	sckim@astro.snu.ac.kr
Soon-Wook Kim	Seoul, Korea	skim@astro6.snu.ac.kr
Yoonho Kim	KIST, Korea	yhkim@kist.re.kr
Youngman Kim	Seoul, Korea	youngman@mulli.snu.ac.kr
Seoktae Koh	Hanyang, Korea	kst@hepth.hanyang.ac.kr
Yoon Suk Koh	Seoul, Korea	
Bon-Chul Koo	Seoul, Korea	koo@astrohi.snu.ac.kr

Julian H. Krolik	Johns Hopkins, USA	jhk@phya.jhu.edu
Shri R. Kulkarni	Caltech, USA	srk@phobos.caltech.edu
Pawan Kumar	IAS, USA	pk@ias.edu
Masaaki Kusunose	Kwansei, Japan	kusunose@kwansei.ac.jp
James M. Lattimer	Stony Brook, USA	lattimer@astro.sunysb.edu
Bum-Hoon Lee	Sogang, Korea	bhl@ccs.sogang.ac.kr
Byung-Joo Lee	Soongsil, Korea	bjlee@physics,soongsil.ac.kr
Chang-Hwan Lee	Stony Brook & KIAS	chlee@kias.re.kr
Choonkyu Lee	Seoul, Korea	cklee@phya.snu.ac.kr
Chul Hoon Lee	Hanyang, Korea	chlee@hepth.hanyang.ac.kr
Eunyoung Lee	Ewha, Korea	sponky@hanmail.net
Hee Won Lee	Yonsei, Korea	hwlee@galaxy.yonsei.ac.kr
Hyun Kyu Lee	Hanyang, Korea	hklee@hepth.hanyang.ac.kr
Hyung Mok Lee	Seoul, Korea	hmlee@astro.snu.ac.kr
Kimyung Lee	KIAS	
Myung Gyoon Lee	Seoul, Korea	mglee@astrog.snu.ac.kr
Taejin Lee	Kangwon, Korea	taejin@cc.kangwon.ac.kr
Won Woo Lee	Hanyang, Korea	wwlee@astro.hanyang.ac.kr
Andrew I. MacFadyen	Santa Cruz, USA	andrew@ucolick.org
Grant Mathews	Notre Dame, USA	mathews@yso.mtk.nao.ac.jp
Dong-Pil Min	Seoul, Korea	dpmin@phya.snu.ac.kr
Young Joo Moon	Ewha, Korea	yjmoon@mm.ewha.ac.kr
Fumiaki Nagase	ISAS, Japan	nagase@astro.isas.ac.jp
Soonkeon Nam	Kyung Hee, Korea	nam@khu.ac.kr
Ramesh Narayan	Harvard, USA	rnarayan@cfa.edu
Hye-Sook Park	LLNL, USA	park@ursa.llnl.gov
Myeong-Gu Park	Kyungpook, Korea	mgp@knu.ac.kr
Myunghee Park	Ewha, Korea	
Soon-mi Park	Ewha, Korea	smda@bcline.com
Yung Kyung Park	Ewha, Korea	teddy113@hanmail.net
Sang-Jin Sin	Hanyang, Korea	
Young-Ho Song	Seoul, Korea	singer@phya.snu.ac.kr
Mijoo Syn	Ewha, Korea	mijoosyn@hanmail.net
Simon Portegies Zwart	Boston, USA	spz@komodo.bu.edu
Maurice van Putten	MIT, USA	mvp@math.mit.edu
Sun Hong Rhie	Notre Dame, USA	srhie@nd.edu
Mannque Rho	Saclay, France	rho@spht.saclay.cea.fr
Dongsu Ryu	Chungnam, Korea	ryu@canopus.chungnam.ac.kr
Jurgen Schukraft	CERN, Swiss	Jurgen.Schukraft@cern.ch
Hideaki Takabe	Osaka, Japan	takabe@ile.osaka-u.ac.jp

Ken'ichi Tatematsu	NAO, Japan	tatematsu@nro.nao.ac.jp
Tomonori Totani	NAO, Japan	totani@yso.mtk.nao.ac.jp
Hideyuki Umeda	Tokyo, Japan	umeda@astron.s.u-tokyo.ac.jp
Jian-Min Wang	IHEP, China	wangjm@astrosv1.ihep.ac.cn
Lifan Wang	Austin, Texas, USA	lifan@astro.as.utexas.edu
Daming Wei	PMO, China	dmwei@pmo.ac.cn
John C. Wheeler	Austin, Texas, USA	wheel@astro.as.utexas.edu
Ralph A.M.J. Wijers	Stony Brook, USA	rwijers@mail.astro.sunysb.edu
Jongmann Yang	Ewha, Korea	jyang@mm.ewha.ac.kr
Insu Yi	KIAS, Korea	iyi@kias.re.kr
Jee-eun Yi	Ewha, Korea	yi203@yahoo.com
Feng Yuan	Nanjing, China	fyuan@nju.edu.cn

First KIAS Astrophysics Workshop *Explosive Phenomena in Astrophysical Compact Objects*
May 24 - 27 (Wed - Sat), 2000 KIAS International Conference Hall

RELATIVISTIC ASTROPHYSICS
AND
BLACK HOLES

The Force-Free Magnetosphere of a Black Hole

Chul H. Lee and Hyun Kyu Lee

Department of Physics, Hanyang University, Seoul 133-791, Korea

Abstract. The magnetic field plays an essential role in extracting energy from a rotating black hole. We consider the structure of stationary axisymmetric force-free black hole magnetospheres using the "3 + 1" formalism. The structure can be characterized by three scalar functions - the stream function, the current potential and the angular velocity of the magnetic field lines. We discuss the possible analytic solutions to the equations governing those scalar functions. From one such solution that can represent the electromagnetic field configuration in the space with an accretion disk on the equatorial plane of the black hole, we estimate the amount of the electric charge induced on the stretched horizon of the black hole.

INTRODUCTION

Black holes have been considered as a candidate of a powerhouse of active galactic nuclei [1] and, more recently, of gamma-ray bursts [2]. In the so-called Blandford-Znajek process the rotational energy of the black hole is supposed to be extracted to power those astrophysical phenomena. There is also an attempt to explain the gamma-ray burst as an outburst of the energy originally stored in the electric field around a charged black hole [3].

In the process of extracting rotational energy from the black hole a strong magnetic field plays an essential role. When the magnetic field is sufficiently strong, the inertia of the plasma around the black hole can be ignored, and the force-free condition

$$\rho_e \mathbf{E} + \mathbf{j} \times \mathbf{B} = 0 \quad (1)$$

can be a good approximation. The region around a black hole where the force-free condition holds is called a force-free magnetosphere.

The structure of a stationary axisymmetric force-free magnetosphere was studied by Blandford and Znajek [1] in the seventies. Macdonald and Thorne [4] extended the study later using the "3 + 1" formalism. In order to be able to use the terms electric field and magnetic field it is natural to use the 3 + 1 formalism. The decomposition of the electromagnetic field tensor $F_{\mu\nu}$ into the electric and magnetic field vectors is an observer dependent process, and one has to specify an observer at each spacetime point. A fiducial observer in the 3 + 1 formalism can naturally play this role.

The structure of a force-free magnetosphere can be characterized by three scalar functions - the stream function, the current potential and the angular velocity of the magnetic field lines. These functions have to satisfy the so-called stream equation. We discuss the possible analytic solutions to the equation.

It is interesting to note that a rotating black hole immersed in a magnetic field may accrete electric charge. Wald derived the electromagnetic field occurring when a stationary axisymmetric black hole is placed in an originally uniform magnetic field, and showed that the black hole can accrete charges until its charge becomes $Q = 2B_0 J$ where B_0 is the strength of the magnetic field and J is the angular momentum of the black hole [5]. We can ask the question whether the existence of a magnetosphere induces electric charge on a black hole. From one solution of the stream equation that can represent the electromagnetic field configuration in the space with an accretion disk on the equatorial plane of the black hole, we estimate the amount of the electric charge induced on the stretched horizon of the black hole.

THE 3 + 1 FORMALISM

In the "3 + 1" formalism, the four dimensional spacetime is decomposed into a succession of three dimensional spacelike hypersurfaces with the direction normal to them taken to be the "universal time" direction. Then the metric of the spacetime can be written in the following form;

$$ds^2 = -\alpha^2 dt^2 + \gamma_{ij}(dx^i + \beta^i dt)(dx^j + \beta^j dt) \tag{2}$$

where α is the lapse function, β^i is the shift function and γ_{ij} is the metric of the three dimensional hypersurface. This hypersurface is called an "absolute" space when γ_{ij} is independent of t. At each spacetime point a fiducial observer is chosen to have the 4-velocity

$$U^\mu = \frac{1}{\alpha}(1, -\beta^i). \quad (U_\mu = (-\alpha, 0, 0, 0)) \tag{3}$$

The the electric and magnetic fields are defined by

$$E^\mu = F^{\mu\nu} U_\nu \tag{4}$$

$$B^\mu = -\frac{1}{2}\epsilon^{\mu\nu\rho\sigma} U_\nu F_{\rho\sigma} \tag{5}$$

and, in terms of E^μ and B^μ, the electromagnetic field tensor $F^{\mu\nu}$ is decomposed as follows;

$$F^{\mu\nu} = U^\mu E^\nu - U^\nu E^\mu \epsilon^{\mu\nu\rho\sigma} U_\nu B_{\rho\sigma} \tag{6}$$

It is easy to see that the zeroth components of E^μ and B^μ are zero,

$$E^\mu = (0, \mathbf{E}), \quad B^\mu = (0, \mathbf{B}) \tag{7}$$

so that $\mathbf{E}$ and $\mathbf{E}$ can be treated as 3-vectors in spacelike hypersurfaces. Likewise, the current 4-vector J^μ can be decomposed as

$$J^\mu = \rho_e U^\mu + j^\mu \tag{8}$$

where ρ_e and j^μ are defined by

$$\rho_e = -J^\mu U_\mu \tag{9}$$
$$j^\mu = J^\mu + J^\nu U_\nu U^\mu. \tag{10}$$

The zeroth component of j^μ is also zero,

$$j^\mu = (0, \mathbf{j}) \tag{11}$$

and $\mathbf{j}$ can be treated as a 3-vector in spacelike hypersurfaces.

In terms of these 3-vectors, the Maxwell's equations are written by [6]

$$\nabla \cdot \mathbf{E} = 4\pi \rho_e \tag{12}$$
$$\nabla \cdot \mathbf{B} = 0 \tag{13}$$
$$\nabla \times \alpha \mathbf{E} = -\partial_0 \mathbf{B} + (\beta \cdot \nabla)\mathbf{B} - (\mathbf{B} \cdot \nabla)\beta \tag{14}$$
$$\nabla \times \alpha \mathbf{B} = 4\pi \alpha \mathbf{j} + \partial_0 \mathbf{E} - (\beta \cdot \nabla)\mathbf{E} + (\mathbf{E} \cdot \nabla)\beta \tag{15}$$

The derivatives in these equations are covariant derivatives with respect to the metric of the absolute space γ_{ij}.

In this work, the background metric of the spacetime is taken to be the Kerr metric for which the lapse and shift functions are given by

$$\alpha = \frac{\rho\sqrt{\Delta}}{\Sigma}, \quad \beta^r = \beta^\theta = 0, \quad \beta^\phi = -\frac{2Mra}{\Sigma^2} \tag{16}$$

where $\rho^2 = r^2 + a^2 \cos^2\theta$, $\Delta = r^2 - 2Mr + a^2$ and $\Sigma^2 = (r^2 + a^2)^2 - \Delta a^2 \sin^2\theta$. The metric of the absolute space is given by

$$\gamma_{ij} = \begin{pmatrix} \frac{\rho^2}{\Delta} & 0 & 0 \\ 0 & \rho^2 & 0 \\ 0 & 0 & \frac{\Sigma^2 \sin^2\theta}{\rho^2} \end{pmatrix} \tag{17}$$

THE STRUCTURE OF A FORCE-FREE MAGNETOSPHERE

The stationary axisymmetric magnetosphere can be characterized by the following three scalar functions; the stream function $\Psi(=\int d\mathbf{S}\cdot\mathbf{B})$, the current potential $I(=\int d\mathbf{S}\cdot\alpha\mathbf{j})$ and the angular velocity of the magnetic field lines Ω_F [4].

Taking a single magnetic field line and rotating it 360° around the rotation axis of the black hole will produce a 2-dimensional surface which is called a "magnetic surface". Given the values of the coordinates r and θ, a magnetic surface is specified. The stream function $\Psi(r,\theta)$ represents the total magnetic flux contained inside it. Then the poloidal part(r and θ components) of the magnetic field is given by

$$\mathbf{B}^P = \frac{\nabla\Psi \times \mathbf{e}_{\hat{\phi}}}{2\pi\tilde{\omega}} \qquad (18)$$

¿From the integral version of Eq.(14), one can see that the toroidal part(ϕ component) of the electric field is zero;

$$\mathbf{E}^T = 0 \qquad (19)$$

Then, with the assumption of $\mathbf{E}\cdot\mathbf{B}=0$, the poloidal part of the electric field $\mathbf{E}^P$ can be written by

$$\mathbf{E}^P = -\mathbf{v}_F \times \mathbf{B}^P \qquad (20)$$

Here $-\mathbf{v}_F$ can be interpreted as the toroidal velocity of fiducial observers through the magnetic field. That is to say that the electric field is entirely induced by the motion of fiducial observers with a local rest frame of the magnetic field defined as that frame in which the electric field vanishes. $\mathbf{v}_F$ is then the velocity of magnetic field lines with respect to fiducial observers and can be written by

$$\mathbf{v}_F = \frac{1}{\alpha}(\Omega_F - \omega)\tilde{\omega}\mathbf{e}_{\hat{\phi}} \quad (\tilde{\omega} = \frac{\Sigma}{\rho}\sin\theta,\ \omega = -\beta^\phi) \qquad (21)$$

where Ω_F is the angular velocity of the magnetic field lines mentioned above. Ω_F turns out to be a constant on each magnetic surface [1], and therefore is a function of Ψ.

¿From Eq's (1) and (19) one can see that the poloidal part of the electric current density is parallel to the poloidal part of the magnetic field. Therefore, the current flows along a magnetic surface, and the total current toward the black hole through a surface whose boundary is specified by the values of the coordinates r and θ is a function of the magnetic flux; $I = I(\Psi)$. From the integral version of Eq.(15), the poloidal component of the magnetic field is derived to be

$$B^T = -\frac{2I}{\alpha\tilde{\omega}}. \qquad (22)$$

Finally, making use of Eq.(22) and the force-free condition, one can derive the expression for the electric current density **j**;

$$\mathbf{j} = \rho_e \mathbf{v}_F - \frac{1}{\alpha}\frac{dI}{d\Psi}\mathbf{B}. \tag{23}$$

We have seen that **E**, **B** and **j** can be calculated from the three scalar functions, Ψ, I and Ω_F. Under the force-free condition, the equation that these three scalar functions have to satisfy turns out to be [4]

$$\nabla \cdot [\frac{\alpha}{\tilde{\omega}^2}(1 - \frac{(\Omega_F - \omega)^2 \tilde{\omega}^2}{\alpha^2})\nabla\Psi] + \frac{\Omega_F - \omega}{\alpha}\frac{d\Omega_F}{d\Psi}\nabla\Psi \cdot \nabla\Psi + \frac{16\pi^2}{\alpha\tilde{\omega}^2}I\frac{dI}{d\Psi} = 0. \tag{24}$$

This equation, called the stream equation, is quite complicated and one cannot expect to be able to obtain analytic solutions in a general situation. However, in the restricted case of a non-rotating black hole with $\Omega_F = \frac{dI}{d\Psi} = 0$, Eq.(24) reduces to the following simple form;

$$\nabla \cdot (\frac{\alpha}{\tilde{\omega}^2}\nabla\Psi) = 0 \tag{25}$$

or explicitly

$$r^2 \partial_r[(1 - \frac{2M}{r})\partial_r\Psi] + \sin\theta \partial_\theta[\frac{1}{\sin\theta}\partial_\theta\Psi] = 0. \tag{26}$$

The following exact solutions to Eq.(26) have been discussed in Ref's [1], [7] and [8];

$$\Psi_1 = \cos\theta \tag{27}$$
$$\Psi_2 = (r - 2M)(1 \mp \cos\theta) - 2M(1 \pm \cos\theta)\ln(1 \pm \cos\theta) \tag{28}$$

Now we take a linear combination of the above solutions

$$\Psi = \Psi_0[(r - 2M)(1 - \cos\theta) + r_0\cos\theta - 2M(1 + \cos\theta)\ln(1 + \cos\theta) \tag{29}$$

for the northern hemisphere($0 \le \theta < \frac{\pi}{2}$), and

$$\Psi = \Psi_0[(r - 2M)(1 + \cos\theta) + r_0\cos\theta - 2M(1 - \cos\theta)\ln(1 - \cos\theta) \tag{30}$$

for the southern hemisphere($\frac{\pi}{2} < \theta \le \pi$). Then the stream function has a discontinuity at the equatorial plane($\theta = \frac{\pi}{2}$) which gives rise to the discontinuity of the radial component of the magnetic field. This can naturally represent the situation where an accretion disk exists on the equatorial plane with the toroidal component of the current density

$$j^{\hat{\phi}} = \frac{\Psi_0(r + r_0)}{4\pi^2 r \Sigma}\delta(\theta - \frac{\pi}{2}). \tag{31}$$

The magnetic flux leaving the northern hemisphere of the black hole is calculated to be

$$\Phi_B = \Psi(\frac{\pi}{2}) - \Psi(0)$$
$$= \Psi_0(4M\ln 2 + r_0). \tag{32}$$

ELECTRIC CHARGE INDUCED ON THE BLACK HOLE

Coulomb's law of electrodynamics, Eq.(12), suggests that the net electric charge of a system can be obtained by the surface integration

$$Q = \frac{1}{4\pi} \oint d\Sigma \cdot \mathbf{E}. \tag{33}$$

over a closed surface surrounding the system. Applied to the black hole case, Eq.(33) becomes

$$Q = \frac{1}{2} \int_{horizon} d\theta \sin\theta \Sigma E_n \tag{34}$$

where $E_n(= E^{\hat{r}})$ is obtained by, from Eq's (20), (18) and (21),

$$E_n = -\frac{\Omega_F - \omega}{2\pi\alpha} \frac{\sqrt{\Delta}}{\rho} \partial_r \Psi \tag{35}$$

We do not have the solution for Ψ to use in Eq.(35) in the general rotating black hole case. However, in the numerical study by Macdonald [7], it is demonstrated that the effect of the black hole rotation to the poloidal magnetic field structure is small even at a high angular momentum, $a \leq 0.75M$. Therefore, we use the solution for Ψ shown in the previous section to calculate the order of magnitude of the electric charge induced on a rotating black hole. Also we assume the optimal case where $\Omega_F \approx \frac{\Omega_H}{2}$ (Ω_H is the value of ω at the horizon) [9]. Then Eq.(34) becomes

$$Q = \frac{1}{2\pi} \Psi_0 \Omega_H \frac{(r_H^2 + a^2)^2}{r_H^2} \int_0^{\pi/2} d\theta \frac{\sin\theta(1-\cos\theta)}{1 + \frac{a^2}{r_H^2}} \cos^2\theta$$

$$\approx \frac{\Phi_B}{4M} \frac{2Ma}{r_H} \frac{1}{4}$$

$$\approx \frac{aMB}{4} \tag{36}$$

The following relations are used in the second and the last steps respectively in the above equation; $\Omega_H = \frac{2Mar_H}{(r_H^2+a^2)^2}$ and $\Phi_B \approx 4\pi B(r_H^2 + a^2) = 8\pi BMr_H$.

We see that the amount of electric charge induced on a black hole is roughly proportional to the angular momentum($L = aM$) of the black hole times the strength of the magnetic field at the horizon. For the case of a rapidly rotating black hole of $M = 3M_\odot$ and $a = 0.5M$ with a strong magnetic field of $B = 10^{15}G$ at the horizon, for example, the charge Q is calculated to be about $10^{15}C$. It turns out to be much smaller than the extremal charge $Q_{extremal} \approx 10^{20} \frac{M}{M_\odot} C$, and using Kerr metric as the background geometry for the magnetosphere is justified.

ACKNOWLEDGMENTS

This work was supported by the Interdisiplinary Research program of the KOSEF(grant No. 1999-2-003-5).

REFERENCES

1. Blandford, R. and Znajek, R., *MNRAS* **179**, 433 (1977).
2. Lee, H., Wijers, R. and Brown, G., *Phys. Rep.* **325**, 83 (2000).
3. Preparata, G., Ruffini, R. and Xue, S., *Astron. Astrophys.* **338**, L87 (1998).
4. Macdonald, D. and Thorne, K., *MNRAS* **198**, 345 (1982).
5. Wald, R., *Phys. Rev.* **D10**, 1680 (1974).
6. Thorne, K. and Macdonald, D., *MNRAS* **198**, 339 (1982).
7. Macdonald, D., *MNRAS* **211**, 313 (1984).
8. Ghosh, P. and Abramowicz, M., *MNRAS* **292**, 887 (1997).
9. Thorne, K., Price, R. and Macdonald, D., *Black Holes: The Membrane Paradigm*, New Haven and London: Yale University Press, 1986, ch. III, pp. 145.

Magnetic Extraction of Spin Energy from a Black Hole

J.H. Krolik*

Department of Physics and Astronomy, Johns Hopkins University, 3300 N. Charles St., Baltimore, MD 21218; jhk@pha.jhu.edu

Abstract. Numerous variations have been proposed on the original suggestion by Blandford and Znajek that magnetic fields could be used to extract rotational energy from black holes. A new categorization of these variations is proposed so that they may be considered in a systematic way. "Black hole spindown" is defined more precisely, distinguishing decrease in the spin parameter a/M from decreases in angular momentum a and rotational kinetic energy, $M - M_i$. Several key physical questions are raised: Can the "stretched horizon" of a black hole communicate with the outside world? Do accretion disks bring any net magnetic flux to the black holes at their centers? Is the magnetic field adjacent to a black hole force-free everywhere?

BACKGROUND

Adding angular momentum to a black hole while keeping its mass fixed decreases its surface area:

$$A = 8\pi M^2 \left[1 + \sqrt{1 - a_*^2}\right], \tag{1}$$

where, as usual, M is the black hole mass, $a_* \equiv a/M$ is its dimensionless spin parameter with a the specific angular momentum, and all quantities are in relativistic units (i.e., $G = c = 1$, so that, for example, the unit of length is $r_g = GM/c^2$). Hawking and Ellis [12] showed that A cannot decrease; consequently, no matter what happens, a black hole of surface area A cannot ever have a mass less than $(A/16\pi)^{1/2}$, its "irreducible mass" M_i. However, if $a_* > 0$, $M > M_i$. Therefore, reducing a_* can make M smaller; that is, braking a black hole's spin can make it lose energy. In principle, this energy can be delivered in usable form to the outside world. If $a_* = 1$, the energy potentially tappable is $(1 - 1/\sqrt{2})M \simeq 0.293M$. Comparing this quantity to the amount of energy released in the course of accretion (generally estimated as $\sim 0.1M$), we see that there is potentially as much energy stored in black hole spin as can be released in ordinary accretion.

The first proposal of a specific mechanism to extract this energy was due to Penrose [26]. His scheme made use of the fact that particles inside the ergosphere can achieve *negative* total energy (i.e., including their rest-mass energy) if their orbital frequencies are small enough ($\Omega < a^{-1}(1 - r/2M)$). However, Bardeen et al. [4] pointed out that it would be very difficult to accomplish this task by particle-particle events because there is an extremely large velocity difference between positive and negative energy orbits.

Since Penrose's initial suggestion, a number of other ideas have been proposed. The unifying theme of all these proposals is to couple the black hole's spin to external matter by magnetic forces. In some sense, they are non-local realizations of the Penrose process that make use of electromagnetic effects to convey negative energy into the hole, whether by injecting it with negative energy wave modes or negative energy particles.

Because the first such scheme was proposed by Blandford & Znajek [7], there has been a tendency in the literature to refer to them all generically as the "Blandford-Znajek" mechanism. Although all of them share the fundamental idea of magnetic couplings, and so in that sense are different versions of the same basic idea, there are now enough variations on the theme to make a clearer categorization useful. It is the object of this paper to present such a categorization, along with some comments about these variations' relative standing.

THE RATE OF ROTATIONAL ENERGY CHANGE

Before embarking on this discussion it is worthwhile to pause a moment to discuss several points of principle in order to clarify our language. First of all, let us list the ways by which black holes can gain or lose angular momentum (we will omit mechanisms involving singular events such as the initial creation of the black hole or a merger with a comparable mass object). One way is through the action of its own gravitational field: matter orbiting with an angular momentum vector oriented obliquely to the angular momentum of a black hole feels a torque. Here we will ignore this effect, assuming that any nearby matter orbits in the plane normal to the black hole spin. A second way is through matter crossing the event horizon bearing angular momentum. The third and final mechanism is through capture of photons with orbital angular momentum.

Next, let us consider what is required in order to identify the source of energy for a given event. It is occasionally assumed (e.g., in [19]) that all energy lost from stably-orbiting matter in an accretion disk is drawn from gravitational potential energy lost by accreting matter, with none coming from the spin of the black hole. However, this distinction—made locally—can be problematic because the shape of the gravitational potential depends on the black hole spin. To take an extreme example, when $a_* = a/M > 0.943$, the marginally stable orbit falls inside the ergosphere; in that case, part of the orbital kinetic energy of the disk material is due to the rotation of the black hole. If there are other ways of coupling the rotation of the black hole to the disk (see below), the identity of the source of energy at any particular location can become even fuzzier.

Although the local energy source can be ill-defined, it is always possible to make this distinction globally. All that is necessary is to compute the fluxes of angular momentum and energy across the black hole's event horizon.

However, even for this distinction one must be careful about definitions. For example, if matter accretes with exactly the specific angular momentum $\hat{L}_{ms}M$ of the marginally stable orbit and the angular momentum of captured photons is neglected, the spin parameter a_* always increases toward unity; when captured radiation is included in the accounting, a_* reaches equilibrium at $\simeq 0.998$ [33]. If the accreting matter arrives at the hole with a fraction $1 - \mathcal{L}_{ms}$ of $\hat{L}_{ms}M$, we may regard the spin as being tapped because the hole's rotational energy is less than it would have been if only the usual amount of energy (i.e. the binding energy at the marginally stable orbit) had been released.

One might also choose to impose the stronger condition that the hole is being "spun down" if a_* actually decreases. Dividing the angular momentum fluxes into those depending on accretion of rest-mass and those depending on photon capture (photon capture also includes capture of electromagnetic waves), we find that the rate of change of a_* is

$$\frac{da_*}{dt} = \frac{\dot{M}_o}{M}\left[(1 - \mathcal{L}_{ms})\hat{L}_{ms} - 2a_*(1-\eta)\right] + \frac{\dot{M}_\gamma}{M}\left[\mathcal{J}_\gamma - 2a_*\right], \quad (2)$$

where $\dot{M}_o$ is the rate of rest-mass accretion, η is the fraction of rest-mass energy lost before matter arrives at the black hole, $\dot{M}_\gamma$ is the rate at which the black hole mass changes due to photon accretion, and $\mathcal{J}_\gamma M$ is the mean angular momentum per photon. Note that $\dot{M}_\gamma$ depends on the energy of photons as measured at infinity; that can be quite different from the locally-measured photon energy. If photons are unimportant, whether a_* increases or decreases depends on the balance between the angular momentum brought into the black hole $(1 - \mathcal{L}_{ms})\hat{L}_{ms}$ and the angular momentum required to maintain the same ratio of spin to mass, $2a_*(1-\eta)$.

It is apparent from equation 2 that a_* can fall even while the angular momentum a increases: the rate of change of a is the same as for a_* but without the two terms $\propto a_*$. Moreover, as equation 2 also shows, a_* might remain constant even while M increases. If that is so, the total rotational energy still increases. This fact suggests a still stronger definition–that the actual rotational energy $M - M_i$ decreases. Writing this criterion in terms of its component quantities, we have

$$\frac{d(M-M_i)}{dt} = \dot{M}_o\left\{(1-\eta)\left[1 - m_i - \frac{a_*^2}{2m_i(1-a_*^2)^{1/2}}\right] + \frac{a_*\hat{L}_{ms}(1-\mathcal{L}_{ms})}{4m_i(1-a_*^2)^{1/2}}\right\}$$
$$+ \dot{M}_\gamma\left[1 - m_i - \frac{a_*^2}{2m_i(1-a_*^2)^{1/2}} + \frac{a_*\mathcal{J}_\gamma}{4m_i(1-a_*^2)^{1/2}}\right], \quad (3)$$

where $m_i \equiv M_i/M$.

Thus, we see that the vague term "spindown" can connote any of several distinguishable results: diminishing a_*, a, or $M - M_i$. Because rotation contributes only a small amount to the total energy of the black hole when a_* is small, the different criteria are very similar for $a_* \lesssim 0.8$; for larger values of a_*, the difference becomes more significant. When considering stored-energy reservoirs, the last definition is the most precise.

THE ORIGINAL BLANDFORD-ZNAJEK MECHANISM

As remarked above, the first plausible mechanism for removing black hole spin energy was proposed by Blandford & Znajek [7]. They imagined that as a black hole accretes, it would inevitably trap some net magnetic flux due to accreted field lines with connections to infinity. There would then be an approximately time-steady magnetic field configuration with field lines embedded in the hole's event horizon, even while their far ends close at very large distance from the black hole. The enforced rotation of space-time due to the black hole spin would then drive an MHD wind. In their initial formulation, the field structure was supposed to be force-free everywhere, i.e. there was so little plasma attached to these field lines that $B^2/4\pi \gg \rho c^2$ everywhere. Phinney [27] extended this picture to include a small, but finite, plasma rest-mass density so that MHD wave group speeds would remain (slightly) less than c. Even in this modification, there is so little inertia outside the event horizon of the black hole that the orbital motion of nearby plasma has little effect on the rotation rate of magnetic field lines; only the distant region where the energy is delivered is considered to have any significant inertia.

Similarly, because the accretion rate is taken to be negligible, the change in the black hole's rotational energy is due solely to the term proportional to M_γ in equation 3; i.e., negative angular momentum and energy are brought to the black hole by zero rest-mass electromagnetic waves, not by ordinary matter.

These processes may also be visualized in terms of their associated electric fields. Suppose that the magnetic field is stationary with respect to a distant observer. Matter just outside the black hole is compelled to rotate with the hole, so in the matter's frame there is an electric field perpendicular to both the field and the rotational velocity whose magnitude is proportional to the velocity mismatch between the field frame and the matter frame. This field drives a current from one field-line to another through the resistance of the horizon; the current then flows out along the new field-line until somewhere far from the black hole it crosses back to its original field-line and closes the circuit. When the resistance at infinity R_l is comparable to the resistance in the horizon R_h, energy is dissipated in the distant "resistor" at a rate $\sim cr_g^2 B_h^2$, where B_h is the magnitude of the field at the event horizon. This impedance matching is equivalent to the condition that all the work done by the horizon forcing the field to rotate reaches infinity.

In the usual formulation, the magnetic field is taken to be time-independent and force-free everywhere between the black hole and a distant, but localized, "load". A corollary to the assumption that the load is localized is that ideal MHD and flux-freezing apply everywhere between the black hole and the load. The force-free assumption then assures that wherever the field and the plasma move in lock-step the field controls the rotation rate. Given those assumptions, matching the current through the horizon to the current through the load determines the rotation rate of the flux-lines linked by the current loop:

$$\Omega_F = \frac{\Omega_h R_l + \Omega_l R_h}{R_h + R_l}, \quad (4)$$

where Ω_h is the rotation rate of the black hole and Ω_l the rotation rate of the load [5]. The load's rotation rate is determined by a balance between whatever forces (gravitational, magnetic, ...) act upon it.

When thinking in terms of the equivalent circuit, it is important to distinguish true dissipative resistance from effective resistance (this distinction is closely related to the distinction between ordinary and radiation resistance). For example, if the field-lines at infinity pass through plasma with high conductivity, the major part of the energy transmitted to infinity can be in the form of bulk work; i.e., the distant matter can be accelerated coherently. By contrast, ordinary resistance produces heat. The impedance matching referred to earlier refers to the total load at infinity, not solely the dissipative part.

Because B_h depends on the history of past accretion onto the black hole, specifically the total amount of net magnetic flux accreted, it is unclear how large B_h should be. Because there is resistance in the horizon, the field must be supported by currents somewhere in the vicinity of the black hole. Rees et al. [31] suggested that they might be located in a "fossil" accretion disk so that they might not be any farther from the horizon than a distance $\sim r_g$. Rees et al. then argued that the field could not be any larger than $\sim (r_{ms}/r_g)^2$ times the pressure in the fossil disk, or there would be no dynamical equilibrium. Another way to set the scale is to suppose that there is a small amount of accretion, and estimate B_h as comparable to the field strength in the nearby accretion disk; if, as seems likely from recent numerical simulations [8,32], $B^2 < 8\pi p$, this estimate yields a result somewhat smaller than the previous one [10]. Combining the more conservative estimate with the assumption of good impedance matching, the expected luminosity is

$$L_{BZ} \sim \dot{M}_o c^2 (v_{orb,ms}/c)(r_{ms}/h_{ms})(r_g/r_{ms})^2, \quad (5)$$

where everything with subscript ms is evaluated at the marginally stable orbit.

There are several issues regarding this mechanism that remain incompletely understood. Punsly & Coroniti [29] raised the question of whether the black hole event horizon and plasma far from the black hole could be in causal contact. They argued that, although the plasma density is taken to be negligible in the usual formulation of the Blandford-Znajek mechanism, it cannot be literally zero. There would then be some accretion onto the black hole, and the fast magnetosonic speed would be (slightly) less than c. If so, there would inevitably be a surface surrounding the event horizon within which the inward velocity of the plasma would be greater than the fast magnetosonic speed, and no signal carrying energy or information could propagate outward across that surface. Punsly [28] further argued that, because of severe gravitational redshifting, electromagnetic waves (in this case, the fast magnetosonic mode) can carry only tiny amounts of energy away from the event horizon. The significance of these criticisms is still unclear.

Further questions may also be raised about this form of the Blandford-Znajek scenario: Must the black hole necessarily accumulate a net flux? Perhaps all field lines suffer enough reconnection in the course of accretion that only closed loops are brought to the black hole. If this is the case, the field loops would close near the black hole, not far away. Would the mechanism still work if the plasma is not magnetically-dominated everywhere between the black hole and the load? Field lines passing through the accretion disk, for example, will surely have substantial inertia attached. Similarly, the load region (presumably near the Alfvenic surface of the wind where $B^2/4\pi \sim \rho v^2$) might not be too far from the black hole even if the field lines close at much greater distance. One immediate consequence of non-force-free behavior would be to change equation 4 to involve a mean value of Ω_l weighted inversely by the local resistance between the magnetic flux surfaces. Two other consequences would be more serious, however. Significant plasma inertia would cause the magnetosonic speed to fall well below c, so that the inner magnetosonic surface would likely move well outside the event horizon. Plasma inertia would also slow the approach to stationarity and possibly disrupt it altogether. In that case, the whole time-steady picture—a fixed magnetosonic surface, well-defined field rotation rates, a steady-state lumped-parameter electrical circuit analog—would be undercut.

Differing answers to these questions may be regarded as the basis for the variations on this scheme that have been suggested. Alternatively, they may be used as the structure of a classification scheme. We will organize the remainder of this paper along those lines, dividing schemes according to their answers to three questions:

• Does the innermost part of the magnetic field run through the event horizon or through plasma in the ergosphere?
• Do the field loops close nearby (e.g., in the disk) or far away?
• Is there anywhere in the system other than the horizon and a localized load where plasma inertia matters?

Labelled by this scheme, the original Blandford-Znajek mechanism is one in which the field lines run through the event horizon and out to infinity, and the field is force-free everywhere except in the event horizon and the load.

FIELD LINES ANCHORED OUTSIDE THE HORIZON, CLOSED AT INFINITY, AND FORCE-FREE

The first alternative was invented by Punsly and Coroniti [29,30] and elaborated by Punsly (1996). It has also been recently discussed by Li [18]. They suggested that a better means to extract the rotational energy of the black hole might be a magneto-centrifugal wind anchored in plasma orbiting in the ergosphere, but well away from the event horizon (see also [24] for a similar idea). This device would finesse the possible causality problem of field lines tied to the event horizon, but would be otherwise quite similar to the original Blandford-Znajek idea: the field lines extend to infinity, the structure is supposed at least roughly time-steady, and the plasma inertia is required to be low enough as to satisfy the condition of magnetic domination everywhere but possibly in the distant load.

Preliminary numerical simulations of a version of this idea exist [15]. In these simulations, the initial magnetic field was taken to have uniform intensity and to be directed parallel to the rotation axis. Although the Alfven speed is relatively small in the equatorial plane, it is $\simeq c$ everywhere outside the disk. As a result, magnetic braking of orbiting material is very strong, leading to dramatic infall, shocks, and a (transient?) jet that is pressure-driven along the axis but magnetically-propelled at larger radius. Just as for the classic Blandford-Znajek mechanism, the characteristic luminosity scale is $\sim |\vec{B}|^2 r_g^2 c$, but the coefficient could be considerably less than unity. Based on this point of view, Meier et al. [21] and Meier [20] have also suggested that the relative strength of the magnetic field at the base of the jet could explain the morphological contrast between FR1 and FR2 radio galaxies.

FIELD LINES ANCHORED IN THE HORIZON, CLOSED IN THE DISK, AND FORCE-FREE

In the original Blandford-Znajek scheme, the field lines are anchored in the event horizon, but extend to infinity. It is also possible that some or all the field lines threading the horizon could instead close by passing through the disk. Because the material in the disk will arrive at the black hole in an infall time, the details of such a structure must be transitory. However, by the same token, if the loops close well outside the inner edge of the disk, the characteristic duration of this structure would be $\sim \alpha^{-1}(r/h)^2$ dynamical times, where α is the usual ratio of r–ϕ stress to local pressure and h is the disk thickness. Particularly in a thin disk, this could be a relatively long time.

For this reason, some discussions of this scheme have assumed a time-steady situation and its corollary, field lines with a fixed rotation frequency (e.g., [17]; but see [11] for an intrinsically non-steady version). In this case, because the resistance of disk plasma is so much less than R_h, Ω_F for all field-lines must be very nearly the rotational frequency Ω_d of the disk matter they thread. To estimate the offset, we observe that the $\vec{J} \times \vec{B}$ force can be written in two equivalent ways. Most often, this force is evaluated by using Ampère's Law (neglecting the displacement current) to write $\vec{J} = (c/4\pi)\nabla \times \vec{B}$. However, one can also use Ohm's Law to compute the current, i.e. set $\vec{J} = \sigma(\vec{E} + \vec{v}/c \times \vec{B})$. Identifying the field-line velocity with the $\vec{E} \times \vec{B}$ drift speed [14], we can then equate the two forms and find

$$|\Omega_F - \Omega_d| \sim \frac{c^2}{8\pi hr\sigma} \tag{6}$$

where h is the disk thickness and r is the radial coordinate of the place the field-lines pass through. If resistivity in accretion disks is due to electron-ion Coulomb collisions, $\sigma \sim 10^{14}$ s^{-1}; even if particle-wave scattering or turbulence magnifies the effective collision rate, $|\Omega_F - \Omega_d| \ll \Omega_d$.

To compute the rate at which black hole rotational energy is tapped, one must next estimate the total resistive load due to the disk. The energy per unit volume dissipated by true electrical resistance is the characteristic rate of dissipation for turbulent magnetic field in the disk, $\sim (B^2/8\pi)\Omega$, times a very small factor $\sim c^2/(v_{orb}\sigma h)$. On the other hand, if $\langle B_r B_\phi \rangle \neq 0$, the rotational work done by the field is likely to be much larger. The problem is that the field structure within the disk is very much *not* force-free, so the usual simplifications invoked to estimate power output in the Blandford-Znajek mechanism don't apply. As a result, detailed calculations are necessary to evaluate the true load—and therefore the power that is actually generated—in this model.

FIELD LINES ANCHORED IN PLUNGING PLASMA, CLOSED IN THE DISK, AND NOT FORCE-FREE

In contrast to the previous two pictures, let us now consider the consequences of a disk whose material is actually accreting. Within it, the growth of the magneto-rotational instability to nonlinear amplitude creates strong MHD turbulence. When plasma follows circular orbits whose frequency declines outward, the shear automatically biases the turbulence so that $\langle B_r B_\phi \rangle < 0$ and angular momentum is carried outward, thereby permitting accretion [3].

When material leaves the marginally stable orbit and plunges inward, its high electrical conductivity assures that it carries magnetic flux along. Whether or not there is any net magnetic flux threading the black hole's event horizon, most magnetic field lines in the plunging plasma are closed relatively close by, for the MHD turbulence in the disk should lead to much reconnection. The question now arises as to whether these field lines, threading plasma plunging along unstable orbits through the ergosphere, can exert enough stress to tap the black hole's rotational energy at an interesting rate.

Novikov & Thorne [23] argued that in a time-steady, axi-symmetric disk the stress $T_{r\phi}$ should approach zero at the marginally stable orbit because the inertia of material in the plunging region inside r_{ms} would be too small to exert any significant stress. If this were the case, this version of the Blandford-Znajek mechanism (and the previous one!) would be unimportant. However, as remarked in [33,25], it is possible for magnetic forces to be significant even in the absence of substantial inertia. If so, $T_{r\phi}$ need not go to zero at r_{ms}.

In fact, the same MHD processes that account for angular momentum transport in the disk proper are likely to apply in the neighborhood of r_{ms}. The existence of the magneto-rotational instability depends only on the disk shear; it is therefore just as strong in the vicinity of r_{ms} as it is farther out in the disk. There is also no particular reason why MHD turbulence should damp more quickly in the inner disk than the outer. We may therefore expect the amplitude of MHD turbulence (normalized to the disk pressure) to remain roughly constant as $r \to r_{ms}$. Because the shear is still strong, we can likewise expect $T_{r\phi}/Tr(T)$ to not change dramatically in this region. In fact, if anything, the ratio of magnetic stress to pressure is likely to rise near r_{ms} because the pressure begins to decline as the inflow velocity rises, whereas we can expect the magnetic field strength to vary more slowly. It therefore appears that there is no local mechanism to enforce the decline in $T_{r\phi}$ demanded by the Novikov-Thorne model.

If this is the case, flux-freezing through the plunging region can lead to magnetic forces becoming comparable to gravity in a roughly time-steady and axi-symmetric accretion flow when $v_r \sim v_\phi$ [16]. $T_{r\phi}$ at r_{ms} large enough to significantly alter the accretion efficiency follows as a corollary [9]; in a time-steady, axi-symmetric state, the additional dissipation per unit area in the disk falls roughly as $r^{-7/2}$ at $r \geq r_{ms}$ [1].

Recent simulations by Hawley & Krolik [13] (3-d non-relativistic MHD in the pseudo-Newtonian Pacyński-Wiita potential) support these arguments (but see also [2]). They find that the inner regions of disks become highly turbulent and non-steady, but that the azimuthally-averaged and vertically-integrated magnetic stress is essentially flat from the innermost part of the disk proper through the marginally stable orbit and well into the plunging region. One might therefore expect the accretion efficiency to be somewhat greater than the one predicted on the basis of zero-stress at r_{ms} even without black hole rotation.

Black hole rotation (and its attendant frame-dragging) may enhance this supplemental energy release because it is mediated by magnetic torques [9]. In other words, when the black hole rotates, this mechanism is very similar to the previous one, but transfers the inner field "anchor" to plunging plasma and does not require either force-free field structure or time-steadiness.

As for any mechanism proposed to tap black hole spin, "spindown" must be defined carefully. If the weakest of the three criteria is used, any increase in the accretion efficiency above the nominal is associated with a diminution of the black hole's rotational energy, for it means that the accreting matter enters the hole with smaller angular momentum than it might otherwise have. If the constant a_* criterion is used and captured radiation is ignored, spindown occurs when the efficiency is greater than

$$\eta_{eq} = 1 - \frac{\sqrt{1 - 3/r_{ms} + 2a_*/r_{ms}^{3/2}}}{1 - a_*/r_{ms}^{3/2}}, \qquad (7)$$

[1]. For $0 \leq a_* \lesssim 0.95$, $\eta_{eq} \simeq 0.3 - 0.35$. Captured radiation diminishes η_{eq} by $\simeq 0.1$ for $a_* \gtrsim 0.99$. The contrast between this criterion and the strongest one (in which the rotational energy actually declines) is illustrated in Figure 1. Although a genuine evaluation of how large this effect might be requires numerical

MHD simulations, simplified (simplistic?) analytic estimates indicate that even this criterion might be met when $a_* \gtrsim 0.6$ [9].

Because the spindown luminosity in this mechanism is intimately tied to accretion, its natural luminosity scale is the accretion luminosity scale, $\sim \dot{M}_o c^2$. However, effects such as mass clumping or magnetic reconnection could diminish it, possibly by sizable factors.

FIELD LOOPS ANCHORED IN THE ERGOSPHERE, CLOSED INSIDE THE MARGINALLY STABLE ORBIT, AND NOT FORCE-FREE

If, like the magnetic field on the horizon, the field strength in the plunging region is comparable to its value in the disk, it could in principle drive a substantial magneto-centrifugal wind. Unlike most MHD wind models, this one would be very

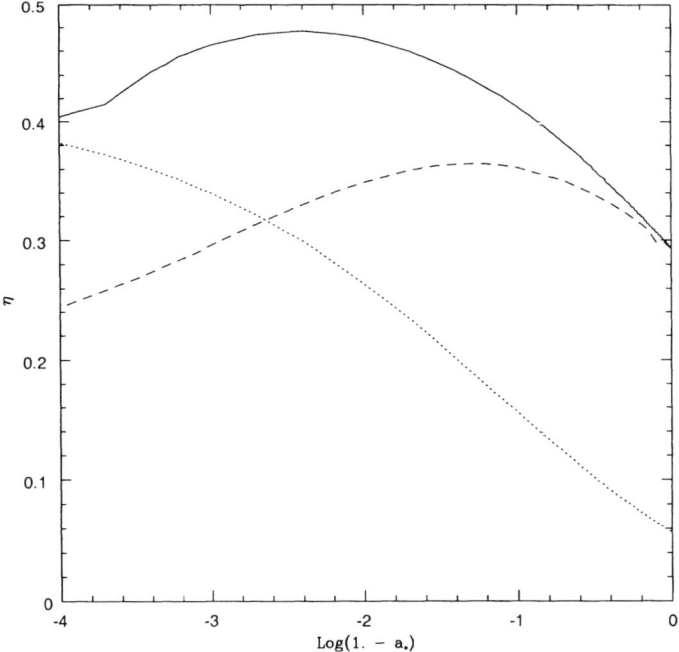

FIGURE 1. The efficiency η above which the rotational energy is diminished is shown by a solid curve. For comparison, the dotted curve shows the Novikov-Thorne efficiency; the dashed curve shows the efficiency at which a_* does not change [1]. Both the solid and dashed curves are specific to the model of this section.

far from time-steady, for the character of the field in this region changes in a free-fall time. In a time-steady wind, field lines must extend to infinity, for fluid elements have been travelling outward, carrying their magnetic flux with them, for effectively forever. By contrast, when the wind is highly variable, there is no requirement for any continuity in field structure between fluid elements expelled long ago and those just entering the wind today.

Nonetheless, the same dynamics that drive time-steady winds should also apply in this context. In fact, circumstances might be more propitious toward wind creation in this region than in the disk, for the Alfven speed should be rather higher due to the similar strength magnetic field and the much smaller inertia (cf. the simulations reported in [20]). Local fluctuations in the angular momentum of plunging fluid elements could lead to shocks in this region, and strong local heating. Just as for the Blandford-Znajek mechanism, the characteristic luminosity scale is $\sim |\vec{B}|^2 r_g^2 c$, but the coefficient could be considerably less than unity.

DISK WINDS

Finally, it is useful as a standard of comparison to contrast these variations on the Blandford-Znajek mechanism with magneto-centrifugal winds launched from an accretion disk. Indeed, Livio et al. [19] claimed that the black hole spin energy reservoir would *always* be relatively unimportant because electromagnetic energy loss from the accretion disk would always exceed that from the black hole itself through the Blandford-Znajek mechanism. Their argument (an elaboration of one in [7]) hinged on a simple order-of-magnitude comparison: They estimated that the luminosity of a disk-driven wind would be generically $\sim O(10) c r_g^2 B_{pd} B_{td}$, where $B_{(p,t)d}$ are the poloidal and toroidal magnetic field components in the disk and the factor $O(10)$ accounts for the fact that the surface area of the disk is rather larger than the surface area of the black hole. Supposing that all currents supporting magnetic fields are found in the disk, they then estimated $B_h \lesssim B_{pd}$, so that the disk luminosity would always be greater than the Blandford-Znajek luminosity.

However, this luminosity estimate is based on the implicit assumption that the r-ϕ component of the Maxwell stress in the disk wind has a vertical scale height comparable to the disk radius, not the (gas density-determined) disk thickness. This is a very strong assumption. At the moment, there is no way to rigorously evaluate its quality, but there are a few suggestive results in hand already from simulations [22,13]. Both of these indicate that the magnetic stresses decline vertically more slowly than does the gas density, but the scale height ratio is only a factor of two or three. If later work confirms these tentative results, the estimated magnetic wind luminosity of the disk would be greatly reduced. Mechanisms that derive their energy from the spin of the black hole, but do not put the energy into winds, would then rise in relative importance.

CONCLUSIONS

Virtually every idea anyone has suggested about how to extract spin energy from black holes involves magnetic fields anchored in the black hole's ergosphere in one way or another. At the same time, the variations possible among these ideas (whether the field lines close at infinity or in the disk, whether they are anchored in the horizon or farther away, ...) raise different questions and different opportunities. In discussions of these ideas, it is therefore important to distinguish carefully between them, for questions and criticisms raised about one version do not necessarily apply to the others. It has been the aim of this short paper to provide a convenient framework for making these distinctions, and to assess (in a necessarily subjective fashion) their current standing.

In order for further progress to be made, it is apparent that (at least) three key questions must be answered:

1.) Is it possible for energy and angular momentum to be carried away from a black hole when the field lines are anchored only in the horizon?

2.) Do black holes accumulate net magnetic flux (so that field loops close far away from the black hole and the structure changes on very long timescales) or does enough reconnection accompany accretion that field loops all close nearby, and the structure changes on the infall timescale?

3.) Can one think of the field as being force-free everywhere but in a small load, or is the plasma inertia qualitatively important?

ACKNOWLEDGMENTS

I am happy to acknowledge stimulating and instructive conversations with Eric Agol, Roger Blandford, Doug Eardley, and Ethan Vishniac. This work was partially supported by NSF Grant AST-9616922.

REFERENCES

1. Agol, E. & Krolik, J.H. 2000 Ap.J. 528, 161
2. Armitage, P., Reynolds, C. & Chiang, J. 2000, astro-ph/0007042
3. Balbus, S.A. & Hawley, J.F. 1998, Revs. Mod. Phys. 70, 1
4. Bardeen, J., Press, W.H., and Teukolsky, S.A. 1972, Ap.J. 178, 347
5. Blandford, R.D. 1990, in Active Galactic Nuclei: 1990 Saas-Fee Lectures, eds. T.J.-L. Courvoisier and M. Mayor (Berlin: Springer-Verlag)
6. Blandford, R.D. & Payne, D.G. 1982, M.N.R.A.S. 199, 883
7. Blandford, R.D. & Znajek, R.L. 1977, M.N.R.A.S. 179, 433
8. Brandenburg, A., Nordlund, A., Stein, R.F. & Torkelsson, U. 1996, Ap.J. Lett. 458, L45
9. Gammie, C.F. 1999, Ap.J. Lett. 522, L57
10. Ghosh, P. & Abramowicz, M. 1997, M.N.R.A.S. 292, 887

11. Gruzinov, A. 2000, astro-ph/9908101
12. Hawking, S.W. & Ellis, G.F.R. 1973, The Large Scale Structure of Space-Time (Cambridge: Cambridge University Press)
13. Hawley, J.F. & Krolik, J.H. 2000, astro-ph/0006456, submitted to Ap.J.
14. Jackson, J.D. 1975, Classical Electrodynamics, 2d ed. (New York: Wiley)
15. Koide, S., Shibata, K. & Kudoh, T. 1998, Ap.J. Lett. 495, L63 Ap.J.
16. Krolik, J.H. 1999, Ap.J. Lett. 515, L73
17. Li, L.-X. 2000a, Ap.J. Lett. 533, L115
18. Li, L.-X. 2000b, astro-ph/0007353
19. Livio, M., Ogilvie, G. & Pringle, J.E. 1998, Ap.J. Lett. 512, L100
20. Meier, D.L. 1999, Ap.J. 522, 753
21. Meier, D.L., Edgington, S., Godon, P., Payne, D.G. & Lind, K.R. 1997, Nature 388, 350
22. Miller, K. & Stone, J. 2000, Ap.J. 534, 398
23. Novikov, I.D. & Thorne, K.S. 1973, in Black Holes, eds. C. De Witt and B. De Witt (New York: Gordon & Breach), p. 343
24. Okamoto, I. 1992, M.N.R.A.S. 254, 192
25. Page, D. & Thorne, K.S. 1974, Ap.J. 191, 499
26. Penrose, R. 1969, Riv. Nuovo Cimento 1, 252
27. Phinney, E.S. 1983, unpublished Cambridge University Ph.D. thesis
28. Punsly, B. 1996, Ap.J. 467, 105
29. Punsly, B. & Coroniti, F.V. 1990a, Ap.J. 350, 518
30. Punsly, B. & Coroniti, F.V. 1990b, Ap.J. 354, 583
31. Rees, M.J., Phinney, E.S., Begelman, M.C. & Blandford, R.D. 1982, Nature 295, 17
32. Stone, J.M., Hawley, J.F., Gammie, C.F. & Balbus, S.A. 1996, Ap.J. 463, 656
33. Thorne, K.S. 1974, Ap.J. 191, 507

Magnetic Extraction of Black Hole Rotational Energy

Hyun Kyu Lee

Department of Physics, Hanyang University, Seoul 133-791, Korea
and
Service de Physique Theorique, CE Saclay, 91191 Gif-sur-Yvette, France

Abstract. As a possible inner engine of gamma-ray bursts(GRB's), a black hole with an external magnetic field supported by a surrounding disk or torus is discussed. The Blandford-Znajek process where the rotational energy of the black hole can be extracted as a Poynting flux is reviewed. It is demonstrated that using a simple circuit analysis the disk power increases the effective power of the black hole-accretion disk system, although a part of disk power is dissipated into black hole entropy. We discuss also the evolutions of the black hole mass and angular momentum, taking into account the flow of accreting matter from a strongly magnetized disk using a simplified model.

INTRODUCTION

The estimated energy of each gamma ray burst(GRB) is known to be up to 2×10^{54} ergs comparable to the rest mass energy of our sun. Although this could be reduced if considerable beaming is involved, it is likely that the central engine for the GRB must be able to extract a substantial fraction of the rest mass energy of a compact object, neutron star or black hole, and convert it into energy of GRB. Recently a black hole with an external magnetic field supported by a surrounding disk or torus has been considered as a viable model for an inner engine of GRB's [1] [2] [3]

To get an idea on the numerical values relevant to this system, it is useful to give some order-of-magnitude estimations of physical quantities involved in the discussion of a compact object(black hole) under strong magnetic field.

The energy E, size R and angular velocity Ω can be expressed in terms of the mass of a black hole, M. For a stellar mass black hole,

$$E \sim M_\odot [\frac{M}{M_\odot}](c^2) = 1.8 \times 10^{54} [\frac{M}{M_\odot}] erg \tag{1}$$

$$R \sim M_\odot [\frac{M}{M_\odot}](\frac{G}{c^2}) = 1.5 \times 10^5 [\frac{M}{M_\odot}] cm \tag{2}$$

$$\Omega \sim \frac{1}{M_\odot}[\frac{M_\odot}{M}](\frac{1}{Gc^3}) = 2 \times 10^5[\frac{M_\odot}{M}]s^{-1} \qquad (3)$$

With strong magnetic field of $B \sim B_{15} = 10^{15}$ G, energy density u_B, Poynting flux s_P and power P are given by

$$u_B \sim B_{15}^2 = 10^{30} erg/cm^3 \qquad (4)$$

$$s_P \sim B_{15}^2(c) = 3 \times 10^{40} \frac{erg}{cm^2 \, s} \qquad (5)$$

$$P \sim B_{15}^2 M_\odot^2 [\frac{M}{M_\odot}]^2 (G^2/c^3) = 6.7 \times 10^{50} [\frac{M}{M_\odot}]^2 erg/s \qquad (6)$$

Throughout the paper, the units with G=c =1 will be adopted.

Suppose a fraction of black hole's energy is extracted out by Poynting power, then the time scale for energy extraction can be estimated by

$$\tau \sim \frac{M_\odot}{B_{15}^2 M_\odot^2}[\frac{M_\odot}{M}] = 2.7 \times 10^3 [\frac{M_\odot}{M}]s \qquad (7)$$

which is comparable to $\tau_{GRB} \lesssim 10^3$s.

BLANDFORD-ZNAJEK PROCESS

Blandford and Znajek [4] showed that a magnetic braking of a rotating black hole with angular momentum $J = \tilde{a}M^2$(angular velocity Ω_H) is possible to extract out a substantial part of its rotational energy. For a steady state which can be obtained with the force-free environment, they obtained the extraction rate of the energy and the angular momentum out of a black hole using the relativistic formulation. The Blandford-Znajek power dP_{BZ} and the angular momentum loss rate $dP_J i$ along the magnetic field lines encompassed by two magnetic surfaces with the magnetic flux $d\Psi$ is given by

$$dP_{BZ} = \frac{1}{2\pi} \Omega_F \, I \, d\Psi, \quad dP_J = \frac{1}{2\pi} I \, d\Psi \qquad (8)$$

where Ψ, I and Ω_F denote the magnetic flux, current and the angular velocity of field line respectively.

For the optimal process, $\Omega_F \sim \Omega_H/2$ [5], the total Blandford-Znajek power [2] is calculated to be

$$P_{BZ} \sim 10^{50} \tilde{a}^2 (\frac{M}{M_\odot})^2 (\frac{<B_H>}{10^{15} gauss})^2 \, erg/s. \qquad (9)$$

The maximum available energy is the rotational energy of a black hole, $E_{rot} = (1 - \sqrt{[1 + \sqrt{1 - \tilde{a}^2/2}]})M$. For $\tilde{a} \sim 1$, the available rotational energy is $0.29M$. However the portion which can be extracted out by the Blandford-Znajek process depends on how a black hole evolves. In fact, the total loss rate of rotational energy, P_H, is greater than Blandford-Znajek power(for $\Omega_H > \Omega_F$):$dP_H = \frac{1}{2\pi}\Omega_H I\, d\Psi$. The rest of rotational energy is dissipated into black hole to increase its entropy(or irreducible mass): $dP_{dissipation} = dP_H - dP_{BZ}$.

For the open field lines to infinity, the evolution of black hole mass with optimal power [7] is given by

$$M(t) = M_0 e^{\frac{1}{2H_0} - \frac{1}{2H}} \sqrt{H/H_0} \qquad (10)$$

where $H_{\tilde{a}=1} = 1$ and $H_{\tilde{a}=0} = 2$. As $\tilde{a} \to 0$, the irreducible mass is increasing by $0.2M$ while the extracted energy into infinity along the open field lines is $0.09M$ [2].

One can also consider the field lines from a black hole anchored onto the inner edge of an accretion disk. In this case, it is natural to identify the angular velocity of field line as that of accretion disk, $\Omega_F \sim \Omega_D$. Assuming the whole magnetic flux goes onto the acctretion disk, the energy extracted during the process in which angular momentum is deceasing from $\tilde{a} \sim 1$ to $\tilde{a} 0.36$ is calculated to be $0.15M$ [8]. However, for $\tilde{a} < 0.36$, $\Omega_H < \Omega_D$ and the black hole is spun up rather than slowed down.

The realistic situation is rather complicated than these two ideal extreme cases and it might be a combination of the open field lines to infinity and the closed field lines to the accretion disk.

A SYSTEM OF BLACK HOLE AND MAGNETIZED ACCRETION DISK

The magnetic field on a black hole cannot be supported by itself alone. An accretion disk surrounding the black hole is the natural candidate for the supporting system for the strong magnetic field on the black hole. Therefore the disk can put some constraint on the Blandford-Znajek power.

One of the simple ways of analyzing the effect of the disk is to adopt a circuit analogy [5] [9] [10]. The currents out of the black hole in the equatorial plane may pass through the accretion disk to make the closed circuit of currents together with the inflowing currents onto the horizon from the loading region at infinity.

Then the current conservation implies [2] $B_\phi^H > B_\phi^{disk}$. From the boundary conditions on the horizon in the optimal case and on the accretion disk with angular velocity Ω_D [11] respectively, one can show that the magnetic field on the horizon cannot be smaller than that on the inner edge of the accretion disk, $B_H > B_z(r_{in})$. Similar observation has been discussed by Ghosh and Abramowicz [12] using the field configuration motivated by Macdonald [13]. It implies that the Blandford-Znajek power is not dominated by the disk power due to the magnetic braking [14] [9].

TOTAL POWER

The Poynting flux out of the horizon is carried out to the load region along the poloidal magnetic field lines. Consider the Poynting flux through the funnel between the magnetic surfaces with magnetic flux $\Delta\Psi$. One can replace the complexity of the load region by the equivalent resistance ΔR_L which can be formally defined by

$$\frac{1}{2\pi}\Omega_F \Delta\Psi = I \Delta R_L, \tag{11}$$

where ΔR_L is the equivalent resistance across the load region at infinity encompassed by two magnetic surfaces. On the horizon the magnetic braking induces the electromotive force $\mathcal{E}_H$: $\Delta\mathcal{E}_H = \frac{1}{2\pi}\Omega_H \Delta\Psi$, between the magnetic surfaces,

For a simple analysis we will consider the equivalent circuit for the Blandford-Znajek power, where the electromotive forces are summed to be an electromotive force $\mathcal{E}_H$: The current I (total current flowing into the black hole), the total horizon surface resistance R_H and the equivalent load resistance R_L. We get the power of energy extraction out of the rotating black hole: $P_{BZ}^H = I_H^2 R_L = (\frac{\mathcal{E}_H}{R_L+R_H})^2 R_L$.

A similar analysis can be applied to the Poynting power from the magnetic braking of the magnetized accretion disk. Now considering the magnetic fields both on the horizon of the rotating black hole and on the accretion disk, there are two electromotive forces $\mathcal{E}_H$ and $\mathcal{E}_D$. Since The magnetic field lines do not cross each other, we can sum the powers from each region to get the total power. Assuming the high conductivity of the accretion disk, the power, P_{BZ}^{HD}, at the loading region powered by the system of the rotating black hole and the magnetized accretion disk can be obtained by replacing $\mathcal{E}_H$ by $\mathcal{E}_H + \mathcal{E}_D$. The power is enhanced by the addition of the accretion disk:

$$\frac{P_{BZ}^{HD}}{P_{BZ}^H} = (1+(\mathcal{E}_D/\mathcal{E}_H))^2 > 1, \tag{12}$$

Since the electromotive forces are induced by the magnetic braking, the ratio $\mathcal{E}_H/\mathcal{E}_D$ depends on how to estimate the magnetic fields on the disk and the horizon. The comparison of these two powers is possible because the magnetic fields on the black hole and the disk are supposed to be related. For example, using the relation [2] obtained from the boundary conditions on the horizon and on the disk, we get for $\tilde{a} \sim .5$, $\mathcal{E}_H/\mathcal{E}_D \sim 3$, $\frac{P_{BZ}^{HD}}{P_{BZ}^H} \sim 1.7$.

EVOLUTION

From the loss rate of angular momentum of a disk by magnetic braking, the accretion rate can be calculated in the standard method as

$$\dot{M}_+ = 2r B_\phi B_z / \Omega_D = 4\, r^2 B_z^2. \tag{13}$$

where the axial symmetric solution suggested by Blandford [11], $B_\phi = 2r\Omega_D B_z$, is used. Therefore the strong magnetic field($\sim 10^{15}$G) on the disk implies a high accretion rate [2], $\dot{M} \sim 10^{-4} M_\odot s^{-1} \gg \dot{M}_E$.

The evolution of a black hole now is determined not only by the Blandford-Znajek process but also by the accretion from the disk. The evolution of mass and angular momentum [16] can be written by

$$\dot{M}_H = -P_{BZ} + \dot{M}_+ \tilde{E}, \quad \dot{J}_H = -\frac{P_{BZ}}{\Omega_H} + \dot{M}_+ \tilde{l} \qquad (14)$$

iwhere $\tilde{E}$ and $\tilde{l}$ are the specific energy and angular momentum of accreting matters [15].

The presence of the accretion disk with appreciable pressure or magnetic field is essential for the magnetic field on the horizon, the evolution of the accretion disk might be responsible also to the evolution of the magnetic field on the disk and B_H.

Since the system of a black hole-accretion disk we are considering for the central engine of the gamma ray bursts is supposed to emerge in a final stage of the binary merging processes [17] or of the hypernovae [18], the mass of the accretion disk should not be much greater than the solar mass. Inferred from the large accretion rate due to the strong magnetic field the life time is supposed to be less than few thousand seconds [2]. Therefore the evolution of the disk during the period in which Blandford-Znajek process is effective is also important to calculate the time dependent accretion rate. Since the presence of the accretion disk with appreciable pressure or magnetic field is essential for the magnetic field on the horizon, the evolution of the accretion disk might be responsible also to the evolution of the magnetic field on the disk and B_H.

For numerical calculations [15] we take a simple ansatz on decreasing mass and magnetic field of the disk, in the following form:

$$B_H^2 = B_H^2(0) D(t), \qquad (15)$$

We take $D(t)$ to be vanishing as the total accreting mass becomes $\sim$ solar mass, $D(t) = 1 - (\int_0^t \dot{M}_+)/M_\odot$.

As an example, the evolution of the black hole is calculated taking $M(0) = 3 M_\odot$ and $B_H(0) = 10^{15}$G. When we take the initial angular momentum parameter to be $\tilde{a}(0) = 1$, the mass of the black hole increases up to $\sim 3.6 M_\odot$. Compared to the initial mass sum of $4 M_\odot$, $\sim 10\%$ of the rest - mass energy of the system is extracted by the Blandford-Znajek process, which is about the same fraction of energy that can be taken from the black hole without considering the accretion from the disk [2]. Or compared to disk mass $M_D = M_\odot$, it can be understood as 40% of the accreted mass energy is extracted out [19].

DISCUSSION

In this work, the magnetic extraction of the rotational energy of a black hole is discussed in the frame work of the Blandford-Znajek process. The effects of the accretion disk are discussed using a circuit analysis and using a simple accretion model due to the magnetic braking.

For the Blandford-Znajek process, non vanishing magnetic flux on the horizon is essential. There have been several works [20] which seem to imply that magnetic flux repulsion can suppress the efficiency of the Blandford-Znajek process substantially. The effect of the flux expulsion is substantial only for the extreme Kerr black hole. The effect gets reduced rapidly as $\tilde{a}$ decreases. For example for $\tilde{a} = 0.5$, the effect is reduced to 6%. Hence for the practical purpose of explaining the gamma-ray burst power, where the black holes in the center are rapidly rotating but not necessarily at the extreme rotation, it is not a severe restriction. More discussions with a different point of view for the force-free environment can be found in ref [2].

The current I onto the black hole(or poloidal component of magnetic field) is also essential for the magnetic braking. Since the current consists of flow of particles, it is important to see whether the fluid flows along the magnetic field lines can reach the horizon. This condition gives nontrivial constraints on the energy flow into the horizon: if negative energy flow is possible, then the energy of a black hole can be extracted out by the Poynting flux as discussed in the previous sections.

Takahashi et al. [21] showed that a flow into horizon with negative energy flux is possible for $0 < \Omega_F < \Omega_H$, which is likely the case particularly for the optimal condition. For a magnetically dominated flow, it corresponds to the Blandford-Znajek process off equatorial plane.

Around equatorial plane, it is expected that there are accereting matter flows and the interaction between accreting matter and the magnetic field becomes important. Although the situation is much more complicated than off-equatorial plane, it has been studied [22] to demonstrate that the negative energy flow or equivalently energy extraction out of black hole is also possible provided a strong magnetic field coupled to the accreting matter.

In summary, it is likely that the Blandford-Znajek process in extracting the rotational energy of a black hole is at least as effective as in the original formulation, which has enough efficiency for powering gamma-ray bursts, provided that there is a strongly magnetized accretion disk. However since we have considered simplified situations, there are several issues to be discussed in the future, particularly more detailed analysis on the magnetic field and the related properties of disk.

ACKNOWLEDGMENTS

I would like to thank G.E. Brown, H.-K.Kim, C.-H. Lee, L.-X. Li, M. Takahashi, R.A.M.J. Wijers, and I. Yi for discussions and collaborations on which this paper is based. I am also very grateful to Mannque Rho for the hospitality at Saclay, where this paper was completed. This work is supported partially by KOSEF Grant No. 1999-2-112-003-5 and by BK21 program.

REFERENCES

1. P. Meszaros and M.J. Rees, ApJ 482, L29(1997); B. Paczynski, ApJ 494, L45(1998)
2. H.K. Lee, R.A.M.J. Wijers and G.E. Brown, Phys. Rep. 325, 83(2000)
3. S.F. Portegies-Zwart, C.-H. Lee , H.K. Lee, ApJ 520, 666(1999)
4. R.D. Blandford and R.L. Znajek, MNRAS 179, 433(1977)
5. K.S. Thorne, R.H. Price and D.A. Macdonald, *Black Holes; The Membrane Paradigm* (Yale University Press, New Haven and London, 1986)
6. R.L. Znajek, MNRAS 179, 457(1977)
7. I. Okamoto, MNRAS 294, 192(1992)
8. L.-X. Li, ApJ 533, L115(2000)
9. L.-X. Li, Phys. Rev. D 61, 084016(2000)
10. H.K. Lee, G.E. Brown and R.A.M.J. Wijers, ApJ 536, 416(2000)
11. R.D. Blandford, MNRAS, 176, 465(1976)
12. P. Ghosh and M.A. Abramowicz, MNRAS 292, 887(1997)
13. D. Macdonald, MNRAS 211, 313(1984)
14. M. Livio, G.I. Ogilvie and J.E. Pringle, ApJ. 512, 100(1999)
15. H.K. Lee and H.-K Kim, J. Korean Phys. Soc. 36, 188(2000)
16. S.J. Park and E.T. Vishniac, ApJ 332, 135(1988)
17. H.-Th, Janka, M. Ruffert and T. Erbel, astro-ph/9801005
18. G.E. Brown, R.A.M.J. Wijers, C.-H. Lee, H.K. Lee, and H.A. Bethe, astro-ph/9905337; G.E. Brown, R.A.M.J. Wijers, C.-H. Lee, H.K. Lee, G. Islaelian, and H.A. Bethe, astro-ph/0003361, New Astronomy in press.
19. L.-X. Li and B. Paczynski, ApJ 534, L197(2000)
20. J. Bicak, V. Janis, MNRAS 212, 899(1985) and references there in.
21. M. Takahashi, S. Nitta, Y. Tatematsu, A. Tomimatsu, ApJ 363, 206(1990)
22. K. Hirotani, M. Takahashi, S. Nitta, A. Tomimatsu, ApJ 386, 455(1992); B. Punsly, ApJ 506, 790(1998); J.H. Krolik, ApJ. 515, L73(1999); C.F. Gammie, ApJ. 522, L57(1999)

Electron-Positron Outflow from Black Holes and the Formation of Winds

Maurice H.P.M. van Putten

MIT, Cambridge, MA 02139

Abstract. The collapse of young massive stars or the coalescence of a black hole-neutron star binary is expected to give rise to a black hole-torus system. When the torus is strongly magnetized, the black hole produces electron-positron outflow along open magnetic field-lines. Through curvature radiation in gaps, this outflow rapidly develops into a $e^{\pm}\gamma$-wind, which is ultra-relativistic and of low comoving density, proposed here as a possible input to GRB fireball models.

Here, I discuss some aspects of black holes when exposed to external magnetic fields. For example, black hole-torus systems are a probable outcome of the collapse of young massive stars [28,17] and the coalescence of black hole-neutron star binaries [18], both of which are possible progenitors of cosmological gamma-ray bursts (GRBs). If all black holes are produced by stellar collapse, they should be nearly maximally rotating [1,2]. A surrounding torus or accretion disk is expected to be magnetized by conservation of magnetic flux and linear amplification (cf. [17,15]).

A black hole-torus system will have open magnetic field-lines from the horizon to infinity and closed magnetic field-lines between the black hole and the torus [23]. The closed magnetic field-lines mediate Maxwell stresses [24]. This may be seen by way of similarity to pulsar magnetospheres [10]. In a poloidal cross-section, the torus can be identified with a pulsar which rotates at an angular velocity $\Omega_P \sim \Omega_H - \Omega_T$, wherein the black hole horizon corresponds to infinity. Then, the inner light-surface [29] corresponds to the pulsar light-cylinder, and a 'bag' attached to the torus to the last closed field-line. Here, Ω_H and Ω_T denote the angular velocities of the black hole and the torus, respectively. The work performed by the Maxwell stresses is commonly attributed to an outgoing Poynting flux emanating from the horizon [3,21]. These Maxwell stresses are likely to be important to the evolution of the torus, and tend to delay accretion onto the black hole. The open magnetic field-lines, on the other hand, enable the black hole to produce an outflow to infinity. Such outflows generate emissions by deceleration against the interstellar medium and through internal shocks.

Here, the outflow along open magnetic field-lines is studied, and found to produce a pair-dominated $e^{\pm}\gamma$–wind in combination with curvature radiation.

Open field-lines from the horizon to infinity have radiative ingoing boundary conditions at the horizon as seen by zero-angular momentum observers (ZAMOs), and outgoing boundary conditions at infinity. It is well-known that for an outflow to exist, there must be regions in which pairs are created (gaps), somewhere on these open field-lines [3,19,20,4]. The gaps are powered by an electric current I along the field-lines, which is limited by a horizon surface resistivity of 4π, in the presence of a certain potential drop across them. The net particle flow is limited by the black hole luminosity into the gap. The magnetosphere within the gaps is differentially rotating, beyond which the magnetosphere may be force-free and in rigid rotation. Note that, in contrast, the currents along closed magnetic field-lines are fixed by the angular velocity Ω_T of the surrounding matter, where the gaps are most likely residing between the horizon and the inner light surface. Of interest here is the location of the gaps on the open magnetic field-lines and the power dissipated within, as sites of linear acceleration of charged particles and their curvature radiation.

A rotating black hole tends to produce electrons and positrons by spontaneous emission along open magnetic field-lines in an effort to evolve to a lower energy state by shedding off its angular momentum. Indeed, in the adiabatic limit, the radiated particles possess a specific angular momentum of at least $2M$, whereas the specific angular momentum of the black hole, a, is at most M. In the approximation of an asymptotically uniform magnetic field, e.g., in a Wald-field [27], the emissions at infinity to satisfy a Fermi-Dirac distribution of radiative Landau states, neglecting curvature radiation and magnetic mirror effects. This results from a modification to the Hawking radiation process [25]. This is a highly idealized picture derived in the perturbative limit of small particle densities, which will be modified significantly by curvature radiation and the formation of force-free regions. The spontaneous emission process concerns particles with energy-at-infinity ω below the Fermi-level V_F. Here, V_F is the energy-at-infinity associated with the particles as seen on a null-generator of the horizon, such as the ZAMO-derivative $\xi^a \partial_a = \partial_t - \beta \partial_\phi$, where β denotes the angular velocity of the sky as seen by ZAMOs. That is,

$$V_F \psi = [\xi^a D_a]_\infty^H \psi = (\nu \Omega_H - eV)\psi, \tag{1}$$

where Ω_H is the angular velocity of the horizon, using the sign-convention $\psi \propto e^{-i\omega t} e^{i\nu \phi}$. The energy-at-infinity ω and the azimuthal quantum number ν are associated with the asymptotically time-like Killing vector ∂_t and azimuthal Killing vector ∂_ϕ, whereas $D_a = i^{-1}\partial_a + eA_a$ denotes the gauge-covariant derivative in the presence of an electromagnetic vector potential A_a. In calculating V_F, it is relevant to identify the ground state of the black hole-magnetic field configuration. It has been shown that the lowest energy state, in the process of an angular momentum exchange between the black hole and the surrounding electromagnetic field by variations of the horizon charge q, assumes when $q = 2BJ$ [27,7], where J is the angular momentum of the black hole. Rotation of the equilibrium charge $q = 2BJ$ on the horizon recovers $4\pi BM^2$ as the maximal horizon flux of the magnetic field from the uncharged flux $4\pi BM^2 \cos\lambda$, where $\sin\lambda = J/M^2$ [7,25]. With the sign convention that B is parallel to Ω_H, we then have $V_F = \nu\Omega_H$ with $\nu = eA_\phi$ (for e^-) and $A_a = B(\partial_\phi)_a/2$ in the Wald electrostatic equilibrium state. Note that the spontaneous emission process is anti-symmetric under pair-conjugation.

The rate of spontaneous emission is given by a certain barrier transmission coefficient in the level-crossing picture of electrons and positrons [6]. This follows from frame-dragging by β, and the resulting shift between the energy-at-infinity ω and the energy ω_Z as seen by ZAMOs:

$$\omega_Z = \pm\sqrt{m_e^2 + |eB|(2n+1\pm\alpha)} = \omega + \nu\beta = \begin{cases} \omega - \nu\Omega_H & \text{on the horizon} \\ \omega & \text{at infinity.} \end{cases} \qquad (2)$$

Here, it is the quantum number ν which gives rise to different energies between ZAMOs and Boyer-Linquist observers. Figure 1 (a) shows an equivalent classical picture, where the frame-dragging β induces a potential energy $V_{BL} = e\beta A_\phi$ on a flux surface A_ϕ=const. with respect to the axis of rotation, itself at zero potential in the $q = 2BJ$ state. Since β describes a differentially rotating space-time, it varies with distance to the black hole and V_{BL} introduces a potential energy drop along the magnetic field-lines. When sufficiently strong, a Schwinger-type process is set in place, which locally produces pairs at a certain rate per unit volume. Formally, the rate of pair-production follows from a scattering calculation in the WKB approximation [14,9,6] (cf. also [13]). The pair-production rate is found to be given by a barrier transmission coefficient $\Gamma \sim e^{-\pi B_c/B\theta^2}$, where $B_c = m_e^2/e = 4.4 \times 10^{13}$G is the QED value of the magnetic field-strength and θ is the poloidal angle in Boyer-Linquist coordinates. More precisely, the gradient $\eta = -\nabla V_{BL}$ parallel to B drives a pair-production rate per unit volume by a Schwinger-process $d^2N/dtdV \sim (e^2\eta B/4\pi^2)e^{-\pi B_c/\eta\theta^2}$ ($B \gg B_c$, $a \sim M$). This pair-production process will be in place, whenever the charge-density is low so that the magnetosphere remains in differential rotation.

The magnetosphere on open field-lines away from a gap assumes a force-free, rigidly rotating state with a Goldreich-Julian charge density [3,21,22]. This is similar to the analogous case in pulsar magnetospheres [11,12]. In view of the horizon boundary conditions below, I shall assume that the gap is attached to the horizon.

To a first approximation, the local structure of a gap follows from the ingoing radiative horizon-boundary conditions. The flow in a gap is described by a charge-density ρ_e, a pair-density n_w and a Lorentz factor Γ. This flow is powered by an electric current I along the open field-lines to infinity through a polar cap of area A_p at the cost of a certain potential drop across. The ingoing radiative boundary condition at the horizon applies to electrons and positrons alike: in the limit as we approach the horizon, I and the electric charge density ρ_e are no longer independent, but become proportional to one another (cf. [20]):

$$I \longrightarrow -\rho_e A_p, \qquad (3)$$

since all particles fall into the black hole with the velocity of light. The sign of ρ_e in (3) is that seen by ZAMOs. Here, ρ_e (and n_w) are normalized by factoring in the redshift factor. I saturates against the horizon surface resistivity of 4π: $4\pi I \sim \nu\Omega_H$, up to a logarithmic factor on the left hand-side. Hence, $\rho_e \sim \rho_{GJ}/2$, where $\rho_{GJ} = B\Omega_H/2\pi$ is the Goldreich-Julian charge density near the horizon. With curvature $R_B \sim \sqrt{2M}/\theta^2$ of the Wald-field, curvature radiation produces $n_w \gg \rho_e/e$ in

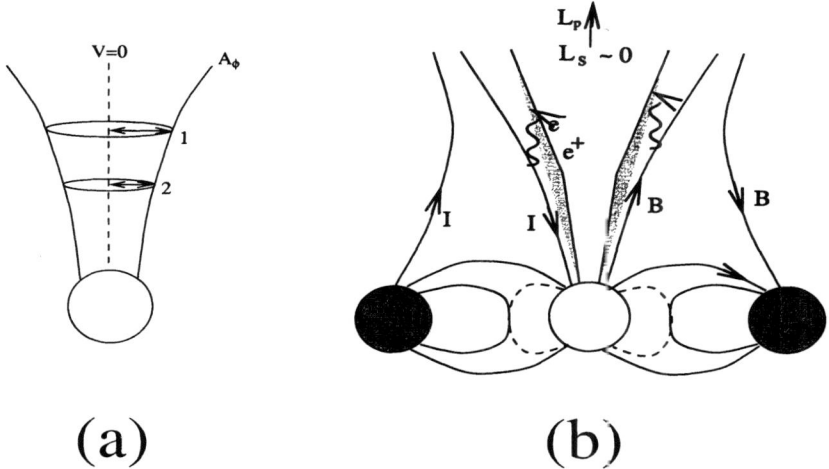

FIGURE 1. (a) A classical picture of the potential energy V_{BL} as seen in Boyer-Lindquist coordinates along surfaces of constant magnetic flux A_ϕ. Note that the axis of rotation has zero potential $V=0$ in electrostatic equilibrium $q=2BJ$. Hence, $V_{BL}=e\beta A_\phi$ in the presence of frame-dragging β. Since β describes differential rotation in the surrounding space-time, a potential drop emerges along A_ϕ =const.: $\Delta V_{BL} = (\beta_2 - \beta_1) A_\phi$. When the potential drop is steep, a Schwinger-type process generates electron–positron pairs. (b) Cartoon of the formation of a black hole-wind in a black hole-torus system. There is a minimum opening angle $\theta_{min} \sim \sqrt{B_c/3B}$, beyond which spontaneous emission along open field-lines by the black hole is effective. Flux surfaces with $\theta \sim \theta_{min}$ have a gap length of order M, which decreases for $\theta > \theta_{min}$. These gaps, indicated in grey, create pairs, which are subject to linear acceleration and produce curvature radiation. The net outflow L_p in particles is a combination of an inner, current-free outflow with vanishingly small Poynting flux L_S inside $\theta < \theta_{min}$ and an outer, current-carrying outflow with $\theta > \theta_{min}$. Both derive most of their particles from pair-cascade through curvature radiation, and flow along open field-lines to infinity.

momentum balance: $n_w 2e^2 \Gamma^4/3R_B^2 \sim \rho_e E_\perp$, where $E_\perp \sim \nu \Omega_H/Me$ is the equivalent electric field normal to the horizon as seen in Boyer-Lindquist coordinates. Note, however, that the magnetic field of the torus will have larger curvature than that of the Wald-field. Given energy balance of the outflow $n_w \Gamma m_e A_{cap}$ with the total power $IE_\perp L$ dissipated in the gap, where L is the linear gap size, the solutions are governed by the unknown L.

The gap size L determines the degree to which the black hole luminosity is put to work in accelerating particles. The gap produces a radiation pressure $\propto L$, which acts on the interface with the force-free magnetosphere above. The interface is probably Raleigh-Taylor unstable against this radiation pressure. Moreover, the gap itself may well widen due to this pressure. The arguments given above are intended as a first sketch towards the structure of the gaps, and it appears to be of interest to consider the gap size L in the context of a detailed stability analysis.

A continuous outflow establishes with appropriate current closure. Note that closure over the polar axis introduces Poynting flux with negative helicity, whereas current closure over a gap across the equator of the black hole and the bag of the torus (corresponding to the last field-line in pulsar magnetospheres) introduces Poynting with positive helicity - indicative of positive energy and angular momentum transport outwards. As the latter is energetically favorable over the former, thereby leaving negligible Poynting flux over the axis of symmetry. A similar conclusion has been found in the case of current closure around neutron stars [11,12]. It follows that the black hole-wind is pair-dominated with the property that $\sigma = L_S/L_p \sim 0$ within $\theta < \theta_{min} = \sqrt{B_c/3B}$, where L_S and L_p are the luminosities in Poynting flux and pairs, respectively [26]. Figure 1 (b) sketches this wind–formation process, assuming a L to be large on the flux surfaces with $\theta \sim \theta_{min}$.

It will be of interest to look for observational evidence in GRB–afterglow emissions for the presented ultra–relativistic, low density pair-dominated wind.

ACKNOWLEDGEMENTS

This work is partially supported by NASA Grant 5-7012 and an MIT Reed Award. The author thanks the hospitality of Theoretical Astrophysics, Caltech, and the Korean Institute of Advanced Study (KIAS), where some of this work was performed, and gratefully acknowledges stimulating discussions with P. Goldreich, E.S. Phinney, R.D. Blandford and K.S. Thorne.

REFERENCES

1. Bardeen J.M., *Nature*, **226**, 64 (1970).

2. Bethe H.A. & Brown G.E., astro-ph/9805355.
3. Blandford R.D. & Znajek R.L. *Mon. Not. R. Astron. Soc.* **179**:433–456 (1977).
4. Beskin V.S. & Kuznetsova I.V., *Il Nuovo Cimento* (to appear, 2000).
5. Chevalier R.A. and Li Z.-Y., *astro-ph/9908272* (1999).
6. Damour T. *in* Proc. 1st Marcel Grossmann Meeting on Gen. Rel., ed.R. Ruffini (North Holland, Amsterdam, 1977), p459-482.
7. Dokuchaev V.I. *Sov. Phys. JETP*, **65**(6):1079-1086 (1987).
8. Gavrilov S.P. & Gitman D.M., *Phys. Rev. D*, **53**:7162 (1996).
9. Gibbons G.W. *MNRAS*, **177**:37P (1976).
10. Goldreich P. & Julian W.H., *ApJ.*, **157**:869 (1969).
11. Goldreich P., *in* Publ Astron Soc Pacific, **83**(495):599 (1971).
12. Goldreich P., *in* Accademia Nazionale Dei Lincei, N. 162, p151 (1971).
13. Goldreich P. & Tremaine S., *ApJ.*, **222**:850 (1978).
14. Hawking S.W. *Commun. Math. Phys.*, **43**:199 (1975).
15. Kluzniak W. & Ruderman M., *ApJ*, **505**:L113 (1998).
16. Lin Q., hep-th/9810037 (1998).
17. Paczyński B. *ApJ.*, **494**:45 (1998).
18. Paczyński B. *Acta Astron.*, **41**: 257-267 (1991).
19. Phinney E.S., *in* Astrophysical Jets (Reidel, Dordrecht 1983), p201.
20. Punsly B. & Coroniti F.V., *Phys. Rev. D.*, **40**:3834 (1989); *ibid. ApJ.*, **350**:518 (1990).
21. Thorne K.S., Price R.H. and Macdonald D.A. *Black holes: the membrane paradigm* (Yale Univ. Press, 1986).
22. Treves A., Turolla R. & Popov S.B., astro-ph/0005508 (2000).
23. van Putten M.H.P.M. and Wilson A. *in* ITP Conference on black holes: Theory confronts reality, http://www.itp.ucsb.edu/online/bhole−c99 (1999).
24. van Putten M.H.P.M. *Science*, **284**:115-118 (1999).
25. van Putten M.H.P.M., *Phys. Rev. Lett.*, **84**(17):3752 (2000).
26. van Putten M.H.P.M., http://online.itp.ucsb.edu/online/astro99.
27. Wald R.M., *Phys. Rev. D.*, **10**:1680–1684 (1974).
28. Woosley S., *ApJ.*, **405**:273-277 (1993).
29. Znajek R.L., *Mon. Not. R. Astron. Soc.*, **179**:457–472 (1977).

ACCRETION, MAGNETOHYDRODYNAMICS, AND DYNAMICS

The Evolution of Compact Young Star Clusters and Thier Relation to Hypernovae

Simon Portegies Zwart

Massachusetts Institute of Technology, Cambridge, MA 02139, USA
Hubble Fellow

Abstract. We study the evolution and observability of young and compact star clusters near the Galactic center, such as the Arches cluster and the Quintuplet. The star clusters are modeled with a combination of techniques; using direct N-body integration to calculate the motions of all stars and detailed stellar and binary evolution to follow the evolution of the stars. The modeled star clusters dissolve within 10 to 60 million years in the tidal field of the Galaxy. The projected stellar density in the modeled clusters drops within 5% to 70% of the lifetime to a level comparable to the projected background density towards the Galactic center. We therefore conclude that between 20 and 50 clusters with characteristics similar to the Arches and the Quintuplet cluster exist. At young age these clusters are highly collisional. This leads to the build-up of a massive collision runaway in each of these clusters. The mass of the collision runaway easily exceeds several hundred $M_\odot$, and the star will be rotating close to break-up. The collision runaway may well explode in a hypernova. Based on our model calculations and estimates of the birthrate of such stars we conclude that the hypernova rate in our Galaxy is about 10^{-6} per year.

INTRODUCTION

A number of compact and young star clusters have been observed within the inner few ten parsec from the Galactic center. Most noticeable are the Arches cluster (Object 17, Nagata et al. 1995) and the Quintuplet cluster (AFGL 2004, Nagata et al. 1990; Okuda et al. 1990). But it is not excluded that others exist as these clusters are well hidden behind thick layers of absorbing material. The Arches and the Quintuplet clusters form the galactic counterparts of NGC 2070 (or R 136); the central star cluster in the 30 Doradus region in the Large Magellanic Could (Massey & Hunter 1998). The structural parameters of these clusters, size, mass and density profile are quite similar as are their ages. The Arches and the Quintuplet clusters are at a projected distance of $\lesssim 50$ pc from the Galactic center. Their evolution is therefore dramatically affected by the presence of the tidal field of the Galactic bulge and inner disc.

In this paper we report the results of N-body simulations of young and compact star clusters, such as the Arches cluster and the Quintuplet cluster in the vicinity of the Galactic bulge. These cluster are particularly interesting because a strong coupling between stellar evolution, stellar dynamics and the tidal field of the Galaxy may exist. In addition to this, excellent observational data are available. Many unusually bright and massive stars are present in both clusters which, due to the high central density of 10^5 to 10^6 stars pc^{-3} are likely to interact strongly with each other.

A number of intriguing question about these clusters makes it worth to model them in great detail, these are 1.) are these clusters the progenitors of globular star clusters, 2.) what is their contribution to the star formation rate in the Galaxy, 3.) are their mass functions intrinsically flat as has been suggested by observations, 4.) how far are these clusters really from the Galactic center and 5.) how many are still hidden, waiting to be discovered. I will address the latter conundrum in this paper and a more detailed paper is in preparation (Portegies Zwart, Makino, McMillan & Hut, 2000b).

THE MODEL

We study the evolution of the Arches and Quintuplet cluster by integrating the equations of motion of all stars and at the same time we account for the evolution of the stars and binaries. The adopted N-body integration algorithm, evolution of stars and binaries, and the interface between the dynamical calculations and the stellar evolution are described extensively by Portegies Zwart et al. (2000a, see also http://www.sns.ias.edu/~starlab).

The N-body portion of the simulations is carried out using kira, operating within the Starlab software environment (Portegies Zwart et al. 1998). Time integration of stellar orbits is accomplished using a fourth-order Hermite scheme (Makino & Aarseth 1992). Kira also incorporates block time-steps (McMillan 1986a; 1986b; Makino 1991) special treatment of close two-body and multiple encounters of arbitrary complexity, and a robust treatment of stellar and binary evolution and stellar collisions. The special-purpose GRAPE-4 (Makino et al. 1997) system is used to accelerate the computation of gravitational forces between stars.

The evolution of stars and binaries are carried out by SeBa (see Portegies Zwart & Verbunt, 1996, Sect. 2.1) the binary evolution package which is combined in the starlab software toolset. The treatment of collisions and mass loss in the main-sequence stage for massive stars are described by Portegies Zwart et al. (1998; 1999; 2000a).

INITIAL CONDITIONS

The observed parameters for the Arches and the Quintuplet clusters are presented in Tab. 1. These clusters have masses of about $\sim 10^4\,M_\odot$ and are extremely

TABLE 1. Observed parameters for the Arches and the Quintuplet clusters. Columns list cluster name, reference, age, mass, projected distance to the Galactic center, tidal radius ($r_{\rm tide}$), and half mass radius ($r_{\rm hm}$). The final column presents an estimate of the density within the half mass radius.

Name	ref	Age [Myr]	M [10^3 M$_\odot$]	$r_{\rm GC}$	$r_{\rm tide}$ [pc]	$r_{\rm hm}$	$\rho_{\rm hm}$ [10^5 M$_\odot$/pc^2]
Arches	a	1–2	12–50	30	1	0.2	0.6 – 2.6
Quintuplet	b	3–5	10–16	50	1	~ 0.5	0.08 – 0.13

References: a) Brandl et al. (1996); Campbell et al. (1992); Massey & Hunter (1998). b) Figer et al. (1999);

TABLE 2. Overview of initial conditions for our model calculations. From left to right the columns list the model name, the distance to the Galactic center, the initial King parameter W_0, the initial tidal– and half mass relaxation times, half-mass crossing time, core radius, half-mass radius and distance to the first Lagrangian point in the tidal field of the Galaxy and the time at which the cluster mass drops below 5% of its initial value. The last column gives the number of stars which experience a collision in a runaway collision merger.

Model	$R_{\rm gc}$ [pc]	W_0	$t_{\rm rxt}$ [Myr]	$t_{\rm rxh}$ [Myr]	$t_{\rm hm}$ [kyr]	$r_{\rm core}$ [pc]	$r_{\rm hm}$	r_{L1}	$t_{\rm end}$ [Myr]	$N_{\rm coll}$
R34W4	34	4	46	3.3	25	0.05	0.12	0.7	12	6
R90W4	90	4	130	8.2	65	0.09	0.22	1.4	30	12
R150W4	150	4	206	13	100	0.14	0.30	1.9	59	12

compact $r_{\rm hm} \lesssim 1$ pc (Figer, McLean & Morris 1999). The projected distance from the Arches cluster to the Galactic center is about 34 pc, the Quintuplet cluster is somewhat farther away. If the third component of the projected distance to the Galactic center is zero then the observed distance equals the real distance.

Our calculations start with 12k (12288) stars at zero age. We assign masses to stars between 0.1 M$_\odot$ and 100 M$_\odot$ from the mass function suggested for the Solar neighborhood by Scalo (1986). The median mass of this mass function is about 0.3 M$_\odot$, and the mean mass $\langle m \rangle \simeq 0.6$ M$_\odot$. For the models with 12k stars this results in a total cluster mass of ~ 7500 M$_\odot$. Initially all stars are single, but binaries may form via three body encounters in which one star carries away the excess energy and angular momentum. We adopt three distances from the Galactic center, 34 pc, 90 pc and 150 pc. The initial density profile and velocity dispersion for the models are taken from a Heggie & Ramamani (1995) model with $W_0 = 4$. At birth the clusters are assumed to perfectly fill the zero velocity surface in the tidal field of the Galaxy. An overview of the initial conditions for the computed models is summarized in Tab. 2.

The tidal field is characterized by the Oort (1927) constants A and B and the local stellar density. The mass within the clusters' orbit at a distant $r_{\rm GC}$ from the

Galactic center is calculated with (Mezger et al. 1999)

$$M_{\text{Gal}}(r_{\text{GC}}) = 4.25 \times 10^6 \left(\frac{r_{\text{GC}}}{[\text{pc}]}\right)^{1.2} \ [M_\odot]. \quad (1)$$

And with this we can derive the appropriate Oort constants.

Once the tidal field, the mass of the cluster and its density profile are selected the N-body system is fully determined. The total mass of the stellar system determines the unit of mass in the N-body system, the distance to the first Lagrangian point r_{L1} in the tidal field of the Galaxy sets the distance unit and the velocity dispersion together with the size of the stellar system sets the time scale. The evolution of the cluster is subsequently followed using the direct N-body integration including stellar and binary evolution and the tidal field of the Galaxy (see Portegies Zwart et al. 2000b).

For economic reasons not all stars are kept in the N-body system, but stars are removed when they are $3r_{\text{L1}}$ from the center of the star cluster.

RESULTS

Figure 1 shows the evolution of the mass and number of stars of the models from Table. 2. Star clusters which are located further away from the galactic center live considerably longer than the closer clusters. Estimating the cluster lifetime naively via the initial relaxation time would lead to an age of 48 Myear ($\equiv$ 12.4Myear × 12.9/3.3) for model R150W4 (see Portegies Zwart et al 2000b for a discussion).

The collision rates in our models are more than 10 times higher than simple estimates based on cross sections. In the computations 9.6 ± 2 collisions occurred in a timespan of about 5–10 Myr whereas the cross section argument predicts only ~ 1 collision.

In our simulations high-mass stars predominantly participated in encounters. The most massive star participates in several collisions with other stars. Typically, the mass of such a runaway grows to exceed 150 $M_\odot$ (see also Portegies Zwart et al. 1999). The rejuvenation of the runaway merger delays its collapse to a compact object following a supernova.

DISCUSSION

How many of these clusters do exist

Although the Arches and Quintuplet clusters are very compact, it may still be hard to notice them near the Galactic center. The local stellar density is high and we can only see these clusters in projection onto the background.

The projected density near the galactic center can be calculated by differentiating Eq. 1 with respect to the distance along line of sight. Portegies Zwart et al. (2000b) perform this calculation numerically and arrive at a projected density of about 3000 $M_\odot$ pc^{-3}.

Figure 2 shows the evolution of the density within the projected half mass radius for models R34W4 (dotes), R90W4 (dashes) and R150W4 (solid). The two error bars give the projected half mass densities for the observed clusters: Arches (left) and Quintuplet (right). The horizontal dotted line gives the surface density at a projected distance of 34 pc from the Galactic center.

The projected densities of the observed clusters are about an order of magnitude higher than the background. Clusters with a lower background density may remain unnoticed among the background stars, as observers may have difficulty to distinguish the cluster from the background.

The projected density of model R150W4 only exceeds the background density in the first few million years. The two models which started more compact R34W4 and R90W4 have projected densities well above the background for a larger fraction of their lifetime. The cluster farther away from the Galactic center has a lower density because it is more extended; its tidal radius is larger. After the first few million years these cluster may be hard to notice among the dense stellar background. We

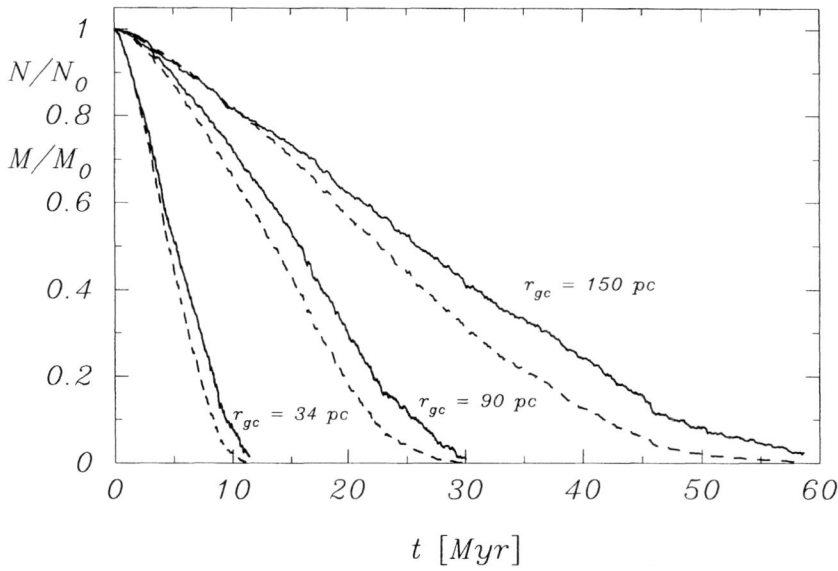

FIGURE 1. Evolution of the total mass M (solid) number of stars N (dashes) for the models at $r_{GC} = 34$ pc (left), $r_{GC} = 90$ pc (middle) and $r_{GC} = 150$ pc (right) Cluster farther away from the Galactic center live longer.

adopt a minimum contrast about three times the projected background density, i.e.: $10^4 \, M_\odot \, pc^{-2}$ required for distinguishing a star cluster among the background. In that case the cluster at a distance of 150 pc would be visible for only about 3 million years ($\sim$ 5% of its lifetime), and almost 9 million years ($\sim$ 70% of its lifetime) for the cluster at a distance of 34 pc. Although the cluster far away from the Galactic center (model R150W4) lives much longer, it only remains visible for a small fraction of its lifetime.

Expressed as fraction of their lifetime the two models are observable for about 5% to 70% of the time for the clusters at a distance of 150 pc and 34 pc, respectively. We therefore expect that a large population of clusters with characteristics similar to the Arches and Quintuplet may still be hidden in the direction of the Galactic center. Based on the presented calculations we estimate that more than 50 clusters remain to be found within a projected distance of 200 pc from the Galactic center. Most of these will be older than the Arches and Quintuplet clusters but not exceeding 60 million years, as these clusters dissolve on a shorter time scale.

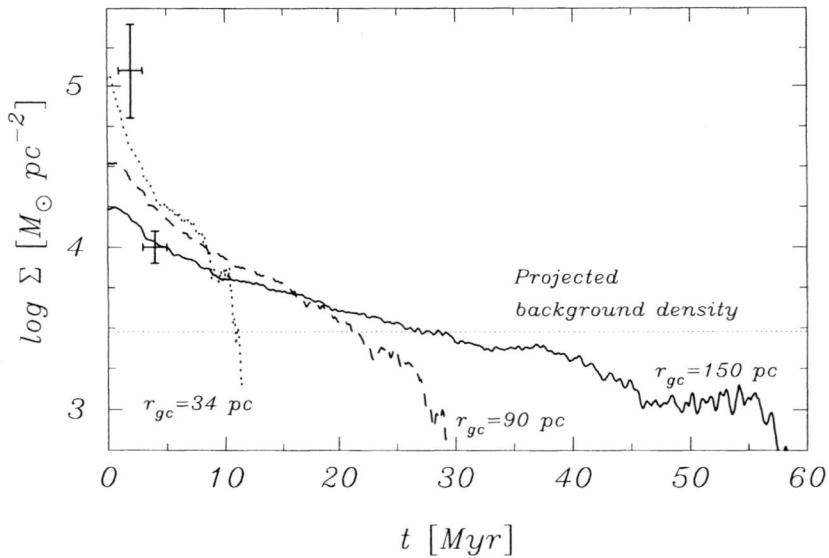

FIGURE 2. Evolution of the surface density within the projected half-mass radius for the models at $r_{GC} = 34$ pc (dotted line), at 90 pc (dashed line) and at $r_{GC} = 150$ pc (solid line). The horizontal dotted line gives the integrated background density at a projected distance of 34 pc from the Galactic center. The two error bars give the observed surface densities of the Arches and the Quintuplet clusters.

The relation to hypernovae

Recently Portegies Zwart et al. (1999) studied the collision history in clusters with similar characteristics as the calculations performed here. They find that physical collisions between stars in these models are frequent and that the evolution of the most massive stars and the dynamical evolution of the cluster are closely coupled. In all cases, a single star grows steadily in mass through mergers with other stars, forming a very massive ($\gtrsim 100 M_\odot$) star in less than 3–4 Myr. The growth rate of such a runaway merger is much larger than estimates based on simple cross-section arguments, mainly because the star is typically found in the core and tends to form binaries with other massive stars there (see also Portegies Zwart et al 1999) The runaway is "rejuvenated" by each new collision, and its lifetime is extended considerably as a consequence. Observationally, such a star will appear in the Hertzsprung-Russell diagram as a blue straggler. When the runaway forms a black hole, the binary in which it is found is usually dissociated.

Also in the models carried out for this paper multiple collisions are very frequent. The last column in Tab. 1 gives the number of collisions which happened in each of the calculations. All the runaway collision products experience a supernova explosion at the end of their lifetime. The exploding star is typically more massive than 100–$200\,M_\odot$. Due to the impact of several massive stars the runaway collision product will be rotating near break-up. The supernovae produced by these stars may have characteristics similar to what Paczynski (1998) calls a hypernova. Iwamoto et al. (2000) argue that a hypernova can be produced by a massive star which is rotating rapidly. Following their model such rapid rotation can originate from the merger in a binary system. In this case, however, the rapid rotation may originate from the impacts of other stars.

If, indeed, each of these clusters produce one hypernova we can derive the hypernova rate from our model calculations. We estimated that about 50 young and compact star clusters populate the inner 200 pc of the Galaxy. Each of which lives for less than 6×10^7 years, resulting in a cluster birthrate of 10^{-6} per year in the Galaxy. Each cluster produces one hypernova leading to a hypernova rate of 10^{-6} per year in the Galaxy.

CONCLUSION

We studied the evolution of the two young and compact star clusters near the Galactic center, the Arches cluster and the Quintuplet. Based on the background stellar density towards the Galactic center and the projected density of the modeled and observed clusters we conclude that the number of such clusters must be far greater than observed. Most of these clusters remain hidden in the stellar background and they are observable only in the first few million years of their existence, when they are still very compact. Within a projected distance of 200 pc from the Galactic center between 20 and 50 clusters with characteristics similar to the Arches cluster and the Quintuplet may be hidden. These hidden clusters are likely to be somewhat older and less compact than those already found.

The collision rate in these clusters is very high and typically one star experiences multiple collisions, causing it to grow well beyond $150\,M_\odot$. This runaway collision product will be rotating rapidly and its explosion may be associated with a hypernova. Each of these clusters can produce one hypernova. The hypernova rate then equals the formation rate of young and compact star clusters, which is about 10^{-6} per year in the Galaxy.

ACKNOWLEDGEMENTS

I thank Piet Hut, Jun Makino and Steve McMillan for discussions. I am grateful to Drexel University and Tokyo University for the use of their GRAPE hardware and to the Korean Institute for Advanced Study for their hospitality. This work was supported by NASA through Hubble Fellowship grant HF-01112.01-98A awarded by the Space Telescope Science Institute, which is operated by the Association of Universities for Research in Astronomy.

REFERENCES

1. Brandl, B., Sams, B. J., Bertoldi, F., Eckart, A., Genzel, R., Drapatz, S., Hofmann, R., Loewe, M., Quirrenbach, A. 1996, ApJ, 466, 254
2. Campbell, B., Hunter, D. A., Holtzman, J. A., Lauer, T. R., Shayer, E. J., Code, A., Faber, S. M., Groth, E. J., Light, R. M., Lynds, R., O'Neil, E. J., J., Westphal, J. A. 1992, AJ, 104, 1721
3. Figer, D. F., Kim, S. S., Morris, M., Serabyn, E., Rich, R. M., McLean, I. S. 1999a, ApJ, 525, 750
4. Figer, D. F., McLean, I. S., Morris, M. 1999b, ApJ, 514, 202
5. Heggie, D. C., Ramamani, N. 1995, MNRAS, 272, 317
6. Iwamoto, K., Nakamura, T., Nomoto, K., Mazzali, P. A., Danziger, I. J., Garnavich, P., Kirshner, R., Jha, S., Balam, D., Thorstensen, J., 2000, ApJ 534, 660
7. Makino, J. 1991, ApJ, 369, 200
8. Makino, J., Aarseth, S. J. 1992, PASJ 44, 141
9. Makino, J., Taiji, M., Ebisuzaki, T., Sugimoto, D. 1997, ApJ, 480, 432
10. Massey, P., Hunter, D. A. 1998, ApJ, 493, 180
11. Mezger, P. G., Zylka, R., Philipp, S., Launhardt, R. 1999, A&A, 348, 457
12. McMillan, S. L. W. 1986a, ApJ, 306, 552
13. McMillan, S. L. W. 1986b, ApJ, 307, 126
14. Nagata, T., Woodward, C. E., Shure, M., Pipher, J. L., Okuda, H. 1990, ApJ, 351, 83
15. Nagata, T., Woodward, C. E., Shure, M., Kobayashi, N. 1995, AJ, 109, 1676
16. Okuda, H., Shibai, H., Nakagawa, T., Matsuhara, H., Kobayashi, Y., Kaifu, N., Nagata, T., Gatley, I., Geballe, T. R. 1990, ApJ, 351, 89

17. Oort, J. 1927, ban, 3, 275
18. Paczynski, B., 1998, in 4th Huntsville Symposium, (Eds C. A. Meegan, R. D. Preece & T. M. Koshut. Woodbury, New York, p.783
19. Portegies Zwart, S. F., Verbunt, F. 1996, A&A, 309, 179
20. Portegies Zwart, S. F., Hut, P., Makino, J., McMillan, S. L. W. 1998, A&A, 337, 363
21. Portegies Zwart, S. F., Makino, J., McMillan, S. L. W., Hut, P. 1999, A&A, 348, 117
22. Portegies Zwart, S. F., McMillan, F., Makino, J., Hut, P. 2000a, MNRAS in press (astro-ph/0005248)
23. Portegies Zwart, S. F., Makino, J., McMillan, F., Hut, P. 2000b *in preparation*
24. Scalo, J. M., 1986, Fund. of Cosm. Phys., 11,

Neutron Star Mergers in and out of Globular Clusters

Hyung Mok Lee

Astronomy Program, SEES, Seoul National University, Seoul 151-742, Korea

Abstract. Neutron stars become the most massive component in globular clusters in less than 2 Gyr. The dynamical friction brings large fraction of neutron stars into the central parts in relative short time scale. Neutron star binaries can be formed by three-body processes in dense nuclei of globular clusters, and they are eventually pumped out of the cluster through the interactions with single stars. Some of the ejected binaries will undergo mergers well within the Hubble time. Since the merger of neutron star pair could release large amount of radiation in short time scale, they can be observed at large distances. We estimate that more than 10^4 neutron star merger may have taken place during last Hubble time in our Galaxy.

INTRODUCTION

Neutron star mergers can release a large amount of energy in short time scale so that a flash can be observed at large distances. It was considered to be one of the possibilities for the gamma-ray bursts, although very energetic explosions such as GRB 990123 would require very efficient conversion of gravitational energy into gamma-ray photons. More likely outcome of merger is a supernova-like event that emits bulk of radiation in optical and UV wavelengths [1]. Then they may be detected by the extragalactic supernova survey.

The merger of neutron stars is likely to take place in binaries because the direct collision between isolated neutron stars would be too rare. The evolution of a binary system of massive stars could lead to binary neutron stars through stellar evolution. Neutron star binaries lose energy and angular momentum by gravitational radiation if orbital separation is very small.

The estimate of the rate of neutron star merger requires detailed knowledge of the evolution of binary systems as a function of binary parameters, as well as the distribution of binary parameters themselves, but the rate could be as high as one in every 10^5 years [2] in our Galaxy. Note that the cosmological origin of gamma-ray bursts also requires the event rate of about 10^{-6} per year per galaxy (e.g, [3]).

Neutron star merger can also happen through purely dynamical processes in dense stellar systems such as globular clusters. There are ample evidences for the presence of neutron stars in globular clusters. Large number of X-ray sources, which are presumed to be binaries containing neutron stars, are found from globular clusters [4]. The surface brightness distribution of M15 can be best fitted by a mixture of normal stars and dark objects with mass much greater than the normal stars. A large number of short period pulsars has been found from a number of globular clusters (see [6] for a review).

In this paper, we present the process that could lead to the merger of neutron stars through the formation and dynamical evolution of binaries in globular clusters.

DYNAMICAL EVOLUTION OF CLUSTERS

The globular clusters we now observe are mostly composed of stars with mass less than about $0.8 M_\odot$ whose main-sequence lifetime is comparable to the age of the universe. Any stars with higher mass have already evolved off the main-sequence and most of them have already disappeared.

The dynamical evolution of globular clusters initially containing massive stars and binaries is a difficult problem to study. The general tendency of the core collapse via gravothermal instability would be delayed by the evolution of stars: the stellar mass loss provides indirect heating effect to the cluster. The binaries also play important role in preventing the core collapse. However, the mass spectrum and even a small amount of rotation accelerate the core collapse (e.g., [5]). Such a complex type of evolution can be best studied by the direct integration of N-body equation of motion, but the current computing power is not enough to do that. Here we briefly describe the course of evolution of globular clusters based on our understanding of simple models for non-rotating, spherical star clusters.

Since the stellar evolution time scale is a steep function of mass, we expect that a cluster can be approximated as a two component system: low mass stars whose life time is very long, and the remnant stars resulting from the evolution of massive stars. The remnant stars include black holes (around 10 $M_\odot$), neutron stars (around 1.4 $M_\odot$), and white dwarfs (around 0.8 $M_\odot$). The white dwarfs are expected to be very abundant, but their individual masses are similar to the normal stars, and would not play any interesting dynamical roles. Black holes are rare, but would quickly settle down toward the central parts of the cluster because of large mass ratio with normal stars. The black holes then form binaries via three-body processes. Since binaries can easily ejected through interactions with single black holes. Thus, in early phase of the evolution, most of the black holes should have disappeared (e.g., [7], [8]).

Neutron stars begin to play important role after the ejection of black holes. The central part becomes dominated by neutron stars and the binaries between neutron stars are formed by three-body processes. The tidal capture also could take place, most predominantly between a normal star and a neutron star. However, there are considerable uncertainties in the outcome of tidal capture: some fraction will lead to direct collisions. In some cases, the normal star can be completely disrupted.

If some fraction of tidally captured systems survive as close binaries, they are subject to the dynamical interactions with other stars or binaries. The interactions between a neutron star and a binary composed of a neutron star and a normal star could result in an exchange: a binary composed of two neutron stars can form through this process. Such a binary has very tight orbit, and could be ejected just after the exchange interactions (see §3). However, it is difficult to estimate the expected number of such events. We mainly concentrate on binary neutron stars formed via three-body processes in this paper.

NEUTRON STAR BINARIES

The physical conditions of the central parts of globular clusters composed of neutron stars and normal stars are favorable for the formation of neutron star binaries via three-body processes because of high stellar density.

The dynamic nature of binaries can be measured by the relative binding energy to the average kinetic energy of stars. The hardness parameter x is defined as

$$x \equiv \frac{E_B}{\frac{1}{2}mv^2}, \qquad (1)$$

where E_B is the binding energy (defined as a positive value) of a binary, m is the mass of individual background stars, and v is the three-dimensional velocity dispersion. If x is greater than about 3, the binaries are called 'hard'. Hard binaries become harder as a result of close encounters with other stars. If x is smaller than ~ 3, the binaries are called 'soft' and they are disrupted through interactions.

Therefore, we need to consider only binaries with $x > 3$. The rate of formation of such binaries have been obtained as a function of stellar density and velocity dispersion (e.g., [9]):

$$\frac{dn_B}{dt} \approx 105 \frac{G^3 m^5}{v^9}, \qquad (2)$$

where n_B is the number density of binaries.

The binary evolves in (x, e) space in such a way that x always increases. The eccentricity (e) distribution becomes a thermal distribution of $f(e)de \approx 2ede$ where $f(e)de$ is the fraction of binaries in the range $e \sim e + de$. The interactions between binaries and singles provides kick velocity to both components, and the amount of energy released by a single close encounter is proportional to the binding energy itself. Since clusters have rather shallow potential, the interacting stars eventually get ejected when the binary star is very hard.

Consider close interactions between a single star of mass m and an equal-mass binary of mass $2m$. If the initial total momentum of binary-single system is negligible, $mv_1 = 2mv_2$, where v_1 and v_2 are velocity of single and a binary after the interaction, respectively. Since the average amount of energy release per close interaction is $0.4Gm^2/2a$ (e.g., [10]), the condition for the ejection of the binary (i.e., $v_2 > v_{esc}$) becomes

$$a < a_{crit} = \frac{Gm}{15v_{esc}^2}. \qquad (3)$$

For a cluster of $v_{esc} = 40$ km/sec, the escaping binaries have $a < 0.052$ AU. The corresponding orbital period is about 2.6 days.

The gravitational radiation eventually leads to the merger of two neutron stars. The time scale for gravitational radiation merger is very sensitive function of the semi-major axis and eccentricity of the binary orbit. Assuming that the mass of the neutron star is 1.4 $M_\odot$, we have shown the fraction of escaping binaries whose gravitational merger time scale is shorter than 5×10^9 years in Fig. 1, as a function of the escape velocity of the cluster. The solid line assumes that the the escaping binaries satisfy Eq. 3, while the dotted line assumes somewhat tighter orbits for ejected binaries (see below). The binaries ejected from clusters with low velocity dispersion are wide, and the probability of merger is very small.

The binding energy of escaping binaries should be larger than the estimates given above, because binaries just before the ejection criteria will become much

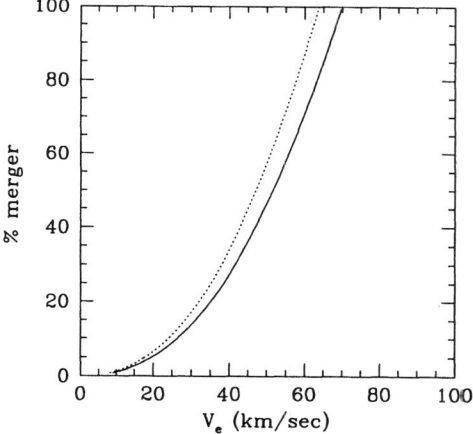

FIGURE 1. Fraction of ejected binaries whose gravitational merger time scale is shorter than 5×10^9 years. The solid line assumes that the the binaries have the critical semi-major axis given in Eq. 3, while the dotted line assumes that the average binding energy of ejected binaries are about 20 % higher than the critical value (see text).

more tightly bound as a result of interaction, but still remains in the cluster. Only those with $a < a_{crit}$ are ejected. To see such an effect, we have carried out simple monte-carlo simulations: starting from a relative soft binary, we traced the evolution of (a, e) taking into account the close interactions of binaries with singles. We have confirmed the eccentricity distribution of binaries closely follow the thermal distribution of $f(e)de \propto e de$. The distribution of binding energies of ejecting binaries is shown in Fig. 2. On average, the binding energy of escaping binaries are larger than the critical value by about 30 %, and the median value is about 16% larger. Since the gravitational merger time scale is proportional to a^4, the time scale for gravitational merger will decrease substantially. The dotted line in Fig. 1 assumes that the the semi-major-axes of ejected binaries are 20 % smaller than the critical value estimated by Eq. (3).

In Fig. 3, we have shown the distribution of escape velocities of Galactic globular clusters based on the data compiled by Webbink [12]. Majority of clusters have small escape velocity. However, escape velocity of massive clusters are larger. Although the number of clusters with large escape velocity is small, most of globular clusters is contained in these clusters. Using the merger probability shown in Fig. 1 as a function of escape velocity, we have estimated the fraction of neutron star binaries whose gravitational radiation merger time scale is shorter than 5×10^9 years

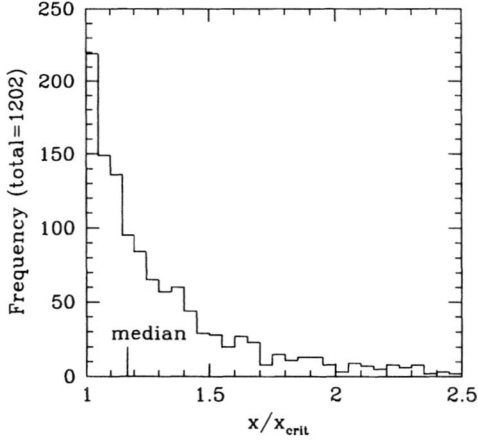

FIGURE 2. The distribution of binding energies of escaping binaries based on monte-carlo simulation of three-body interactions. Most binaries have binding energies much greater than the critical binding energy.

for the Galactic globular clusters. Assuming constant mass-to-luminosity ratio of M/L_v, where L_v is the luminosity in visual band, we found that about 30 % of the escaping neutron stars will undergo merger in 5×10^9 years after being ejected when we used the solid line of Fig. 1. The fraction becomes about 35 % when if we use the dotted line.

Number of observed globular clusters in the Galaxy is about 150. Since we cannot observe some clusters opposite to the Galactic center, the actual number could be around 200. The globular clusters continuously lose their masses due to the tidal force of the Galaxy. The tidal shock should have been very important in early evolution of globular clusters [11]. We expect some clusters already have been completely disrupted during the last Hubble time.

Suppose globular clusters have about 1% in mass in the form of neutron stars. The total mass of the globular clusters in the Galaxy could be about $10^8 M_\odot$, and the total number of neutron stars would be about 7×10^5. It is not clear how many globular clusters have been destroyed during the last Hubble time. The fraction of 'collapsed' globular clusters is estimated to be about 20 %, and similar fraction may have been disrupted. Suppose that about 10 % of the neutron stars in globular clusters have been ejected in the form of close binaries. Assuming that 30% of the ejected binaries undergo gravitational radiation merger in 5×10^9 years, we estimate that about 10^4 neutron star mergers have taken place during last Hubble

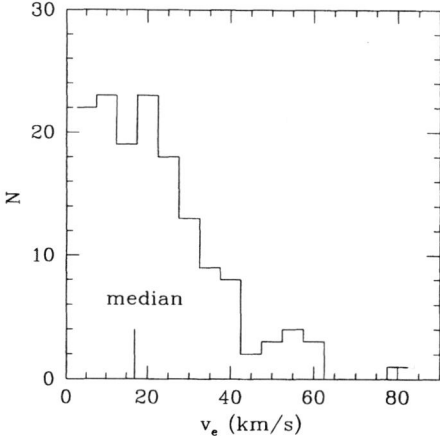

FIGURE 3. The distribution of escape velocities of Galactic globular clusters. The median value is about 17 km/sec.

time. This translates into one merger per 10^6 years per galaxy.

DISCUSSION

We have presented the process for the merger of neutron star binaries formed in globular clusters. The binaries are formed by three-body processes and undergoes orbital evolution through interactions with other stars. The mergers take place outside of the clusters because the binaries are ejected before they become mergers. We estimate that about 10^4 mergers may have taken place in our Galaxy during the last Hubble time. This is about an order of magnitude smaller than the estimated upper limit of 10^{-5} [2] for the disk populations. Because of different environments, the neutron star mergers of globular cluster origin and of the disk population may produce different observational signatures.

Giant elliptical galaxies have more globular clusters per mass than spiral galaxies. If we assume that the properties of globular clusters are similar to our Galaxy, the event rate originating from globular clusters for these galaxies will be much larger than that of our Galaxy.

The neutron star merger described in this paper would not be uniformly distributed over time. The evolution of globular clusters requires rather long time scale. Thus the neutron star ejection rate will probably increase in time and many events should be taking place in galaxies of $z < 1$.

ACKNOWLEDGEMENTS

This research was supported by the KOSEF grant no. 1999-2-113-001-5. I also wish to thank BK21 program for its general support.

REFERENCES

1. Li, L.-X., & Paczynski, B., *Astrophys. J. Lett.*, 507, L59, (1998)
2. Kalogera, V. & Lorimer, D. R., *Astrophys. J.*, 530, 890 (2000)
3. Fishman, G. J., & Meegan, C. A., *Ann. Rev. Astron. Astrophys.*, 33, 415 (1995)
4. Hertz, P. & Grindlay, J. E., *Astrophys. J.*, 275, 105 (1983)
5. Kim, E., Lee, H. M., & Spurzem, R., *in preparation*
6. Phinney, E. S., in *Structure and Dynamics of Globular Clusters, ASP Conference Series*, Vol. 50, eds. S. G. Djorgovski and G. Meylan, p141 (1993)
7. Kulkarni, S., Hut, P., & McMillan, S. L. W., *Nature*, 364, 421 (1993)
8. Lee, H. M., *Mon. Not. Royal Ast. Soc*, **272**, 605 (1995)
9. Goodman, J., & Hut, P., *Astrophys. J.*, 403, 271 (1993)
10. Spitzer, L. Jr., *Dynamical Evolution of Globular Clusters*, Princeton University Press (1987)
11. Gnedin, O., Lee, H. M., & Ostriker, J. P., *Astrophys. J.* **522**, 935 (1999)

12. Webbink, R. E., in *Dynamics of Star Clusters, IAU Symp. No. 113.* eds., J. Goodmann and P. Hut, p541 (1985)

Recent Progress in the Research of X-Ray Pulsars

Fumiaki Nagase*

*The Institute of Space and Astronautical Science
3-1-1 Yoshinodai, Sagamihara, Kanagawa 229-8510, Japan

Abstract. Significant progress has been achieved in the last few years in X-ray pulsar research, including both the "rotation-powered" X-ray pulsars and the "accretion-powered" X-ray pulsars. In this paper I will review some recent topics, such as detection of X-ray emission and pulsations from known radio pulsars, magnetar interpretation of soft gamma repeaters (SGRs) and anomalous X-ray pulsars (AXPs), discovery of millisecond pulsation in a low mass X-ray binary (LMXB) system, a rush of pulsar discoveries in the Small Magellanic Cloud (SMC), the doubling of the number of accreting X-ray pulsars, and a soft-excess feature that is seen commonly in the spectra of all sub-classes of accreting X-ray pulsars.

INTRODUCTION

Fast-rotating neutron stars with strong surface magnetic fields often radiate X-rays with a coherent intensity modulation along the spin of the neutron star, providing a class of X-ray source called "X-ray pulsars." Currently more than a hundred X-ray pulsars are known including both the "rotation-powered" X-ray pulsars which are isolated neutron stars in which X-ray pulsations are powered by the rotation energy of the neutron star and the "accretion-powered" X-ray pulsars which are neutron stars in a binary system in which mass accretion onto the neutron star from a companion star powers the pulsating X-ray emission. Although the population of X-ray pulsars is only about one tenth of that of radio pulsars, X-ray pulsars exhibit a lot of variety in the system parameters and X-ray emission mechanism. Neutron stars with pulsating X-ray emission will be classified as follows roughly in the order from rotation-powered X-ray pulsars to accretion-powered X-ray pulsars:

- Isolated cooling neutron stars (ICNSs; 4)

- Recycled millisecond pulsars (ms-PSRs; 5)

- Young pulsars associated with supernova remnants (Crab-like PSRs; 7)

- Soft gamma-ray repeaters (SGRs; 2)

- Anomalous X-ray pulsars (AXPs; 6)
- Millisecond X-ray pulsar in a low-mass X-ray binary (ms-LMXB; 1)
- Transient/Be-star binary X-ray pulsars (BeXBs; 65)
- Low-mass X-ray binary pulsars (LMXBs; 5)
- High-mass X-ray binary pulsars (HMXBs; 13)

Approximate numbers of pulsating X-ray sources in each subclass are shown in parentheses, although they are changing rapidly with new discoveries. Clearly, binary systems consisting of a neutron star and a Be-star companion are the majority of X-ray pulsars. AXPs were once considered to be LMXBs; however they are now interpreted as magnetars, together with SGRs. Details are described later in individual sections.

ROTATION-POWERED X-RAY PULSARS

Before observations with ROSAT, X-ray emission and pulsations were detected from only a few young Crab-like pulsars, despite the theoretical scenario of neutron star production in the supernova explosion of massive stars. In the last decade, however, the number of X-ray emitting rotation-powered pulsars has increased to 27 pulsars from observations with ROSAT and ASCA, as reviewed by Becker & Trümper [5] (see left panel of Fig. 1). X-ray pulsations were detected from 11 pulsars among the 27 X-ray detected pulsars.

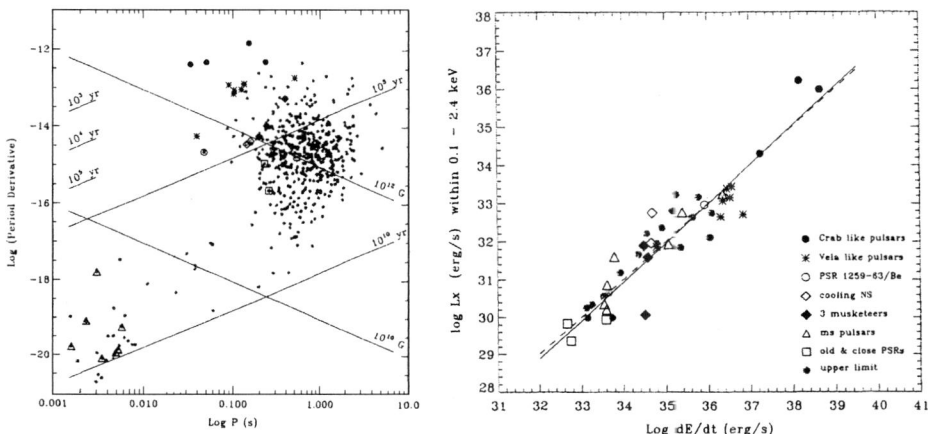

FIGURE 1. Left panel: Period vs. period derivative diagram of rotation-powered pulsars, in which those with larger symbols indicate X-ray emission detected with ROSAT. Right panel: X-ray luminosity vs. spin-down energy for pulsars detected with ROSAT (from [5], © Springer-Verlag).

TABLE 1. Parameters of five millisecond pulsars with X-ray pulsations are compared with those of the Crab pulsar (after [28]).

PSR Name	P ms	$\dot{P}$ s s^{-1}	T_c yr	$\dot{E}$ erg s^{-1}	B_s G	B_L G
J0437−4715	5.76	$(8 \pm 7) \times 10^{-21}$	1.1×10^{10}	1.7×10^{33}	2.2×10^{8}	1.0×10^{4}
J2124−3358	4.93	1.1×10^{-20}	7.2×10^{9}	3.6×10^{33}	2.3×10^{8}	1.8×10^{4}
B1821−24	3.05	1.6×10^{-18}	3.0×10	2.2×10^{36}	2.3×10^{8}	7.2×10^{5}
J0218+4232	2.32	8.0×10^{-20}	4.6×10^{8}	2.5×10^{35}	4.4×10^{8}	3.2×10^{5}
B1937+21	1.56	1.1×10^{-19}	2.4×10^{8}	1.1×10^{36}	4.1×10^{8}	9.9×10^{5}
B0531+21	33.40	4.2×10^{-13}	1.3×10^{3}	4.5×10^{38}	3.9×10^{12}	9.5×10^{5}

From the results of ROSAT and ASCA observations of pulsars, they revealed that the observed X-ray luminosity is closely correlated with the rotational energy loss as $L_x(\dot{E}) = 10^{-3} \times \dot{E}$, despite large dispersions of their ages and magnetic field strengths (see right panel of Fig. 1). This is firm evidence that the observed X-rays are produced by magnetospheric emission originating from the co-rotating magnetosphere.

X-ray Pulsations from Isolated Cooling Neutron Stars

X-ray pulsations were detected with ROSAT from two cooling neutron stars, PSR B0656+14 [13,5], and PSR B1055−52 [39,5] with blackbody temperatures of 70-130 eV, suggesting the existence of a hot polar cap on the neutron star surface. Although there is still no evidence of radio emission and supernova association, RX J0720.4−3125 [18], and RX J0420.0−5022 [19] are candidates for X-ray pulsating ICNSs with pulse periods 8.4 s and 22.7 s and temperatures 80 and 60 keV. Alternative X-ray emission mechanisms, such as accretion of interstellar matter or Thorne-Żytkow objects [54], and magnetars [48] are discussed for these sources (see [18,37]), since corresponding radio emission is not detected yet.

X-ray pulsations of millisecond pulsars

X-ray pulsations have been so far detected with ROSAT and ASCA from five millisecond pulsars as listed in Table 1 [6,28], in which pulse perid (P) and its derivative ($\dot{P}$), characteristic age (T_c), rotational energy loss ($\dot{E}$), magnetic field strengths at the neutron star surface (B_s) and at the light cylinder (B_L) are listed. Among these, PSR B1821−24, PSR J0218+4232, and PSR B1937+21 exhibit hard power law spectra and sharp double-peaked pulse profiles, whereas the other two show soft spectra with broad sinusoidal single-peaked profiles. The values of B_L in the former three pulsars are comparable with that of the Crab pulsar as shown in

TABLE 2. Young Crab-like X-ray pulsars in association with SNRs

Pulsar Name	P ms	$\dot{P}$ s s^{-1}	SNR association	Ref.
PSR B0531+21	33.40	4.21×10^{-13}	Crab Nebula	[46,5]
PSR B0833−45	89.29	1.25×10^{-13}	Vela SNR	[46,5]
PSR B1509−58	150.23	1.54×10^{-12}	MSH 15-52	[46,5]
PSR B0540−69	50.37	4.79×10^{-13}	N158A in LMC	[46,5]
AX J1811.5−1926	64.67	4.38×10^{-14}	G11.2-0.3	[49,52]
PSR J1617−5055	69.34	1.35×10^{-14}	RCW 103	[50,26]
PSR J0537−6910	16.12	5.13×10^{-14}	N157B in LMC	[34,35]
RX J0822−4300	75.3	1.49×10^{-13}	Pup A	[41]

Table 1, suggesting the correlation of pulse emission properties with the magnetic field strength at the light cylinder.

Young Crab-like pulsars in supernova remnant

The Crab pulsar, Vela pulsar, PSR B1509−58 and PSR B0540−69 are the only rotation-powered pulsars so far known to be associated with supernova remnants [46]. Recently three new young pulsars associated with supernova remnants, Kes 73, RCW 103, and G11.2−0.3 have been discovered.

Torii et al. [49] discovered a source pulsating with a period of 65 ms in the supernova remnant, G11.2−0.3 using ASCA archival data. The rate of spin-down for this pulsar was measured later to be $\dot{P} = 4.4 \times 10^{-14}$ s s^{-1} [52]. In ASCA archival data including the supernova remnant RCW 103, they have also found a 69 ms pulsar. The pulsar is at a position 7 arcminutes away from the center of the supernova remnant, the angular diameter of which is 9 arcminutes [50]. In spite of the seperation of the pulsar from the center of the supernova remnant, their association seems plausible after the discovery of a 69-ms radio pulsar counterpart, PSR J1617−5055 [26]. A giant glitch is suspected to have occurred in between 1993 August and 1997 September [53]. Marshall et al. [34] discovered a 16 ms pulsar associated with N157B in the Large Magellanic Cloud near SN1987A and PSR B0540-69, and derived a spin-down rate of $\dot{P} = 5.126 \times 10^{-14}$ s s^{-1} using RXTE and ASCA. Follow-up monitoring by RXTE revealed a large glitch in this Crab-like pulsar [35].

Pavlov et al. [41] recently reported the detection of X-ray pulsations at 75.3 ms from the central source in Puppis A, RX J0822−4300 using the ROSAT archival PSPC and HRI data with a period derivative $\dot{P} = 1.49 \times 10^{-13}$ s s^{-1} from the two observations of 4.56 yr separation. Since the X-ray radiation of this pulsar has a thermal-like spectrum, they suspect that its pulsations may be due to a nonuniform temperature distribution over the neutron star surface. Thus RX J0822−4300 is a young pulsar in a supernova remnant; its pulse emission mechanism is different

TABLE 3. List of anomalous X-ray pulsars and soft gamma-ray repeaters

Source Name	P sec	$\dot{P}$ yr	Γ	kT keV	Ref.
1E 1048.1−5937	6.44	$(1.4 - 4) \times 10^{-11}$	3	0.64	[33,54,15]
1E 2259+586	6.98	$\sim 5 \times 10^{-13}$	4	0.41	[33,54,15]
4U 0142+61	8.69	$\sim 2 \times 10^{-12}$	4	0.39	[33,54,15]
RX J170849−4009	11.0	2×10^{-11}	3.5	—	[47,27]
1E 1841−045	11.76	4×10^{-11}	3.2	0.55	[55,16]
AX J1845−0300	6.97	—	5	0.64	[51]
SGR 1806−20	7.48	8×10^{-11}	~ 2.2	—	[31]
SGR 1900+14	5.16	$(5 - 14) \times 10^{-11}$	1.1	~ 0.5	[23,32,59]

from other Crab-like pulsars which show a power law spectrum. The properties of newly discovered Crab-like pulsars and the pulsar in Pup A are listed in Table 2 together with the previously known ones.

SGRs and AXPs

Recently, Kouveliotou et al. [31] discovered coherent pulsations with a period of 7.47 s in the soft gamma-ray burst repeater, SGR 1806−20 from the RXTE observation, and derived a spindown rate of 8.3×10^{-11} s s^{-1} combining the RXTE data with the ASCA archival data. Assuming that the spin-down is due to the magnetic field decay, these results yield a pulsar age and magnetic field strength of about 1.5×10^3 yr and 8×10^{14} G, respectively. This result suggests the existence of "magnetars", neutron stars with magnetic fields about 100 times stronger than those of young radio pulsars.

From an ASCA observation, pulsations were also discovered from another soft gamma-ray burst repeater, SGR 1900+14 with a period of 5.16 s [23]. The derivative of the period $\dot{P} = 1.1 \times 10^{-10}$ s s^{-1} was measured from RXTE follow-up observations [32]. From these results, the characteristic age and magnetic field are estimated as about 1.5×10^3 yr and 5×10^{14} G, respectively.

These results of pulsations in SGR 1806−20 and SGR 1900+14 confirm the theoretical prediction by Thompson and Duncan [48] that the soft gamma repeaters are possible candidates for magnetars. Baring and Harding [4] suggested that these magnetars should be radio quiet pulsars, because electron-positron pair production in the neutron star magnetosphere is efficiently supressed in the ultra-strong magnetic field by the action of two-photon splitting in quantum electrodynamical processes [22,3].

Mereghetti & Stella [33] suggested that there is a group of LMXB pulsars that show common features of: (1) pulse periods in the range 5-10 s, (2) very soft X-ray spectra with a power law of photon index $\Gamma \geq 3$ and a blackbody component of $kt \simeq 0.5$ keV, (3) relatively low ($\sim 10^{35}$ erg s^{-1}) luminosity, (4) stable trend of spin down, (5) little intensity variability on a wide time scale, and (6) no identification of an optical counterpart. They categolized these pulsars as a new sub-class of pulsars called "anomalous 6-s X-ray pulsars (AXPs)". They discussed the LMXB origin of these AXPs, whereas van Paradijs et al. [54] proposed Thorne-Żtkow Objects (TZO) where the neutron star has been spiraling into an evolved massive companion star.

Properties of currently accepted candidates of AXPs are listed in Table 3 together with those of SGRs. The new AXPs are AX J1708.8-4009 (= 1RXS J170849.0 − 400910) detected during the ASCA Galactic plane survey [47], 1E 1841−045 in supernova remnant Kes 73 [55], and AX J1845.0−0300 [51,15].

Among these the pulsar 1E 1841−045 is of special interest. From the period of 11.8 s and its derivative of 4×10^{-11} s s^{-1}, this pulsar in the SNR Kes 73 is suggested to young ($\sim 2 \times 10^3$ yr) and to have an extremely strong dipole magnetic field, $\sim 8 \times 10^{14}$ G. At the same time this pulsar has properies common to the AXPs mentioned above. Hence, this pulsar will provide a hint on the link between SGRs and AXPs. And this will support the interpretation that AXPs are a subclass of magnetars with extremely strong magnetic field [48,4].

Assuming the spin-down in these AXPs is due to the magnetic field decay of isolated neutron stars, the spin-down rate yields the magnetic field strengths of $10^{14}-10^{15}$ G. The positions of these AXPs and SGRs in the period-period derivative diagram are shown in Figure 2, showing the clear separation of these sources from ordinary radio pulsars.

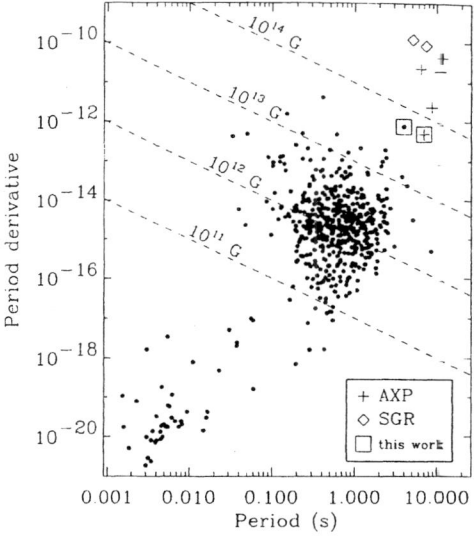

FIGURE 2. Pulsar period vs. perid derivative diagram, in which data for SGRs and AXPs are plotted together showing their strong magnetic field strengths (from [43]).

ACCRETION-POWERED X-RAY PULSARS

X-ray binaries containing neutron stars are classified into two types, the High Mass X-ray Binaries (HMXB) or Low Mass X-ray Binaries (LMXB) depending on the mass of the companion star. Most of the HMXBs are young binary systems and show X-ray pulsations due to a strong dipole magnetic field and mass accretion onto the magnetic poles at the neutron star surface. Bildsten et al. [7] tabulated 44 X-ray pulsars in their review article. However, within a few years after their review, a dramatic rush of pulsar discoveries has been continuing (see e.g., [38]), and the total number of accreting X-ray pulsars has doubled so far. A dozen HMXBs among them are classified as wind-fed high mass X-ray binaries. In addition, four LMXBs are known to be X-ray pulsars. Although six anomalous X-ray pulsars were once considered to be members of LMXBs [33], they are currently considered to be magnetars and are discussed in the previous section.

The majority (remaining three quadrants) of HMXBs show a transient nature and most of them are identified to be massive Be star binaries (see e.g., [9] for review of optical observations of Be star systems). Thus the Be star X-ray binary pulsars (BeXBs) represent the largest sub-class of massive X-ray binaries. Thus it seems a good assumption that all the transient X-ray pulsars are Be star binary systems. The pulse periods of BeXB pulsars, including other classes of X-ray pulsars, distribute uniformly in the range of a few seconds to a thousand seconds with a few exceptions as shown in Figure 3. This is in good contrast to the radio pulsars, the pulse periods of which are more narrowly distributed below a few seconds.

It is worth noting that more than two dozen BeXB pulsars were discovered in the Magellanic Clouds in the last few years, in contrast to the fact that only three accreting pulsars were known for the past two decades. This rush of pulsar discoveries in LMC/SMC is due to the ideal line-up of X-ray observatories, ROSAT, ASCA, BeppoSAX, RXTE, and CGRO/BATSE. The discovery of a LMXB, 1SAX J1808.4−3658 with coherent millisecond pulsations is unique, since it provides firm evidence for the evolution scenario of the recycled millisecond pulsars [1].

1SAX J1808.4−3658

1SAX J1808.4−3658 is a unique transient X-ray source discovered with BeppoSAX [24]. During the observation, two bright type-I X-ray bursts were detected, demonstrating that the compact object is a neutron star. Subsequently, coherent pulsations of 2.49 ms [56] and a 2.0-hr intensity modulation due to orbital motion [8] were detected from an RXTE/PCA observation giving firm evidence of a LMXB. It is a widely accepted theoretical model that millisecond radio pulsars started as ordinary pulsars which lost most of their magnetic field and were 'spun up' to millisecond spin periods by accretion of matter from a companion star in a LMXB system [1]. The millisecond X-ray pulsar, 1SAX J1808.4−3658 will become a millisecond radio pulsar when the accretion turns off, thus providing direct evidence that supports the evolutionary scenario of recycled millisecond pulsars.

Be star binary systems in the Magellanic Clouds

Extensive surveys of the LMC and SMC regions were performed with ROSAT and more than 500 X-ray sources in the LMC [42] and 248 sources in the SMC [25] were catalogued. Variabilities of 27 LMC X-ray sources were studied using the ROSAT observations between 1990 and 1994 [19]. Complete compilation and optical identification for the X-ray sources in the LMC/SMC regions were performed in a series of papers [44,10,45].

Recently more than 20 BeXB pulsars were discovered in the LMC/SMC regions (see e.g., [38]) from ROSAT, ASCA, BeppoSAX and RXTE observations. Yokogawa et al. [60] conducted a systematic analysis of the ASCA survey data of the SMC region and found coherent pulsations from 12 sources among 39 sources detected with ASCA. Most of them were found to exhibit long-term flux variability, suggesting that they are BeXBs. Harberl and Sasaki [21] have newly identified 25 BeXBs from the ROSAT SMC catalogue; thus, a total 47 BeXBs are listed at present including identification of all ASCA detected pulsars to BeXBs. Yokogawa et al. [60] also found that these BeXB pulsars in the SMC can be clearly separated from SNRs and other classes of sources using the hardness ratio analysis (see Fig. 4). The figure gives a hint to classify dim sources from which ASCA data was unable to find pulsations or line features characteristic of the thermal emission.

It is remarkable that very few LMXBs were discovered so far from the SMC in contrast to the recent tremendous discoveries of BeXBs. Thus the number ratio of HMXBs to LMXBs in the SMC is strikingly different to our Galaxy where

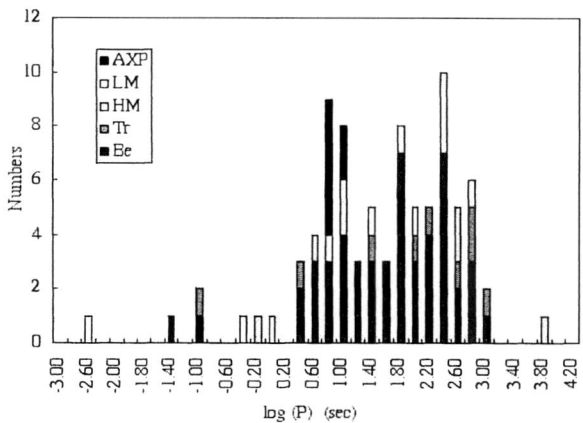

FIGURE 3. Period distribution of accreting X-ray pulsars.

the population of LMXBs is larger than that of HMXBs as suggested in [45,60]. This implies that the SMC has been more active than our Galaxy in massive star formation. Since HMXBs are relatively young it is suspected that strong star formation activity in the SMC took place in the recent past, several million years ago.

Soft-excess feature in spectra of accreting pulsars

A feature of "soft-excess" spectrum, that is the intensity excess at energies below 1 keV in the spectrum over the extrapolation of a power low spectrum fitted to the higher energy, has been reported from various sub-classes of X-ray pulsars except for BeXB pulsars. The soft excess feature is considered to be a common property in the spectra of anomalous X-ray pulsars and the excess emission is usually fitted by a blackbody of temperature $kT = 0.4 - 0.7$ keV. Some LMXBs, such as Her X-1 and 4U 1626−67, show a soft excess feature in their spectra (e.g., [36,11,12] for Her X-1 and [2,40] for 4U 1626−67). Evidence of such a soft excess feature has also been reported from HMXBs, LMC X-4 [58] and SMC X-1 [57] from the combined analyses of ROSAT and Ginga data.

Such a clear soft excess feature, however, has not been reported so far from BeXBs, because most of Galactic BeXBs are subjected to heavy soft X-ray absorption. Recently, clear examples of such a soft excess feature were obtained from ASCA observations of BeXB pulsars in the SMC, such as RX J0059.2−7138 [14,29], and XTE J0111.2−7317 [61].

The energy spectrum with soft excess feature observed with ASCA from a BeXB pulsar, XTE J0111.2−7317 is compared in Figure 5 with that observed with ASCA from a HMXB pulsar SMC X-1. Both can be fitted by a model of power law plus

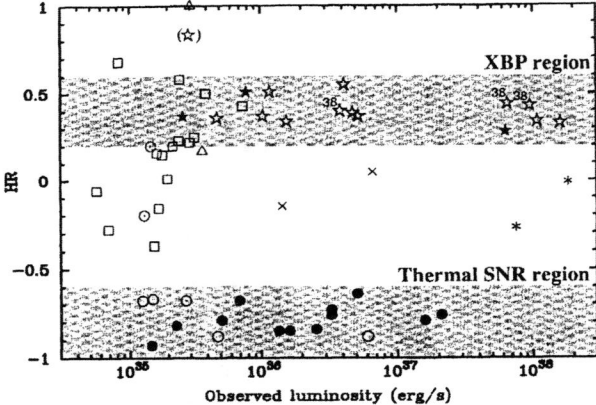

FIGURE 4. Hardness ratios of massive X-ray binary pulsars and X-ray emitting supernova remnants in the LMC/SMC observed with ASCA (from [60]).

blackbody emission. The best fit parameters are $\Gamma = 0.8$, $kT = 0.15$ keV, and $N_H = 2.7 \times 10^{21}$ cm^{-2}, and $\Gamma = 0.8$, $kT = 0.18$ keV, and $N_H = 2.2 \times 10^{21}$ cm^{-2}, respectively for XTE J0111.2−7317 and SMC X-1. Thus, the spectral model that involves a power law and a soft blackbody emission, which is widely adopted to fit the soft-excess spectra observed from X-ray pulsars, can be adopted also to interpret the soft-excess feature seen in the spectra of some BeXBPs.

However, the total X-ray luminosities of these sources are 1.8×10^{38} erg s^{-1} and 2.4×10^{38} erg s^{-1}, respectively at the distance of SMC. If the soft excess emission is really a blackbody, the luminosity fraction of the blackbody emission is about one tenth of the total luminosity. This requires blackbody radii of several hundred

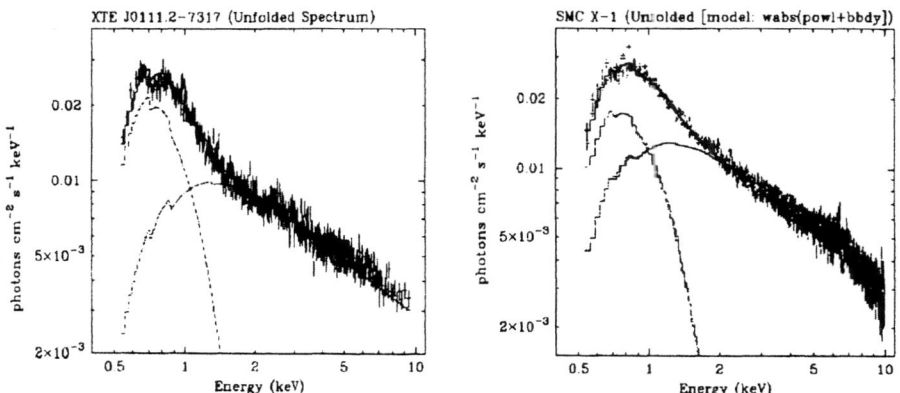

FIGURE 5. Soft excess features in energy spectra observed with ASCA from XTE J0111.2-7313 and SMC X-1 (from [38]).

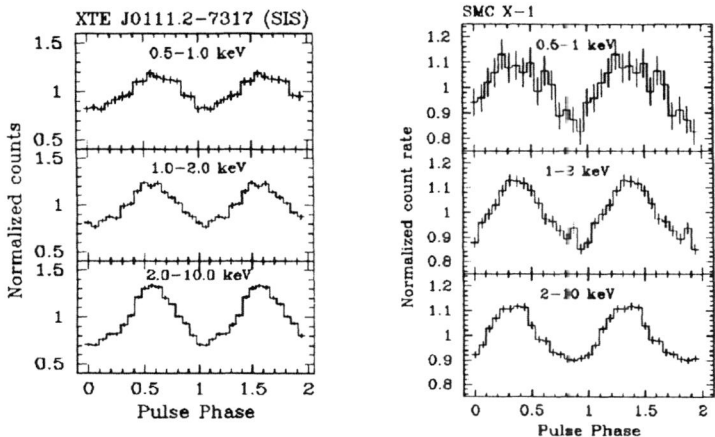

FIGURE 6. Energy dependent pulse profiles of XTE J0111.2-7313 and SMC X-1 (from [38]).

km if the sources are really located in the SMC.

The energy dependent pulse profiles of XTE J0111.2−7317 and SMC X-1 indicates that the pulse shape and pulse fraction do not change drastically with energies in these pulsars (see Fig. 6). This means that the soft excess component is also pulsating in the same manner as the power law component. Thus the blackbody interpretation seems implausible, because the corresponding size of blackbody emission is several tens of times the neutron star radius at the distance of SMC. Hence the blackbody component with such a large size is unlikely to explain the soft excess since the soft excess component should be pulsating in the same manner as the hard power law componenet.

Hence, it is likely that all the emission in the 0.5-10 keV band, including the soft excess emission, originates from the same site, for instance, from accretion columns on the neutron star surface at the magnetic poles. The soft excess feature in spectra would be a feature intrisically common to X-ray pulsar emission from neutron star accretion columns, because many pulsars in all sub-class of X-ray pulsars exhibit such a feature. Further investigations are needed to establish the interpretation of the soft excess feature seen in the energy spectra of X-ray pulsars.

REFERENCES

1. Alpar, M.A., et al., *Nature* **300**, 728 (1982).
2. Angelini, L. et al., *Astrophys. J. Letters* **449**, L41 (1995).
3. Baring M.G., and Harding A.K., *Astrophys. J.* **482**, 372 (1997).
4. Baring M.G., and Harding A.K., *Astrophys. J. Letters* **507**, L55 (1998).
5. Becker W., and Trümper J., *Astron. Astrophys.* **326**, 682 (1997).
6. Becker W., and Trümper J., *Astron. Astrophys.* **341**, 803 (1999).
7. Bildsten L., et al., *Astrophys. J. Suppl.* **113**, 367 (1997).
8. Chakrabarty D., Morgan E.H., *Nature* **394**, 346 (1998).
9. Coe, M.J., in *The Evolution of Galaxies on Cosmological Timescales*, ASP Conf. Series, (Astro-ph/9911272) (1999).
10. Cowley, A.P., et al., *Publ. Astron. Soc. Pacific* **109**, 21 (1997).
11. Dal Fiume, D., et al., *Astron. Astrophys.* **329**, L41 (1998).
12. Endo, T., Nagase, F., and Mihara, T., *Publ. Astron. Soc. Japan* **52**, 223 (2000).
13. Finley, J.P., Ögelman, H., & Kiziloglu, Ü., *Astrophys. J. Letters* **394**, L21 (1992).
14. Hughes, J.P., *Astrophys. J.* **427**, L25 (1994).
15. Gotthelf E.V., and Vasisht G., *New Astron.* **3**, 293 (1998).
16. Gotthelf E.V., and Vasisht G., in *Pulsar Astronomy − 2000 and Beyond*, AIP Conf. Series, Vol 202, 699 (2000).
17. Haberl F., et al., *Astron. Astrophys.* **318**, 490 (1997a).
18. Haberl F., et al., *Astron. Astrophys.* **326**, 662 (1997b).
19. Haberl F., Pietch, W., and Motch, C., *Astron. Astrophys.* **351**, L53 (1999).
20. Haberl, F., and Pietsch, W., *Astron. Astrophys.* **344**, 521 (1999).
21. Haberl, F., and Sasaki, M., *Astron. Astrophys.* **359**, 573 (2000).
22. Harding A.K., Baring, M.G., and Gontheir, P.L., *Astrophys. J.* **476**, 246 (1997).

23. Hurley K., et al., *Astrophys. J. Letters* **510**, L111 (1999).
24. in't Zand, J.J.M., et al., *Astron. Astrophys.* **331**, L25 (1998).
25. Kahabka, P., et al., *Astron. Astrophys. Suppl.* **136**, 81 (1999).
26. Kaspi V.M., et al., *Astrophys. J. Letters* **503**, L161 (1998).
27. Kaspi V.M., et al., *Astrophys. J. Letters* **503**, L161 (2000).
28. Kawai, N., and Saito, Y., *Astrophys. Lett. and Comm.* **38**, 1 (1999).
29. Kohno, M. Yokogawa, J.,and Koyama, K., *Publ. Astron. Soc. Japan* **52**, 299 (2000).
30. Kouveliotou C., et al., *Nature* **379**, 799 (1996).
31. Kouveliotou C., et al., *Nature* **393**, 235 (1998).
32. Kouveliotou C., et al., *Astrophys. J. Letters* **510**, L115 (1999).
33. Mereghetti S., & Stella L., *Astrophys. J. Letters* **442**, L17 (1995).
34. Marshall F.E., et al., *Astrophys. J. Letters* **499**, L179 (1998).
35. Marshall F.E., et al., in *Pulsar Astronomy − 2000 and Beyond*, AIP Conf. Series, Vol 202, 335 (2000).
36. McCray, R.A., et al., *Astrophys. J.* **262**, 301 (1982).
37. Motch, C., in Proceedings of "X-ray Astronomy '999: Stellar Endpoints, AGN, and the Diffuse Background," to be published in *Astrophys. Lett. and Comm.* (2000).
38. Nagase, F., in Proceedings of "X-ray Astronomy '999: Stellar Endpoints, AGN, and the Diffuse Background," to be published in *Astrophys. Lett. and Comm.* (2000).
39. Ögelman, H., and Finley, J.P., *Astrophys. J. Letters* **413**, L31 (1993).
40. Orlandini, M., et al., *Astrophys. J. Letters* **500**, L163 (1998).
41. Pavlov G.G., Zavlin V.E., and Trümper J., *Astrophys. J. Letters* **511**, L45 (1998).
42. Pietsch, W., and Kahabka P., in *Lecture Notes in Physics 416: New Aspects of Magellanic Cloud Research* eds. B. Baschek, G. Klare, K. Beuermann, p. 59 (1992).
43. Pivovaroff, M.J., Kaspi, V.M., and Camilo, F., *Astrophys. J.* **535**, 379 (2000).
44. Schmidtke, P.C., et al., *Publ. Astron. Soc. Pacific* **106**, 843 (1994).
45. Schmidtke, P.C., et al., *Astron. J.* **117**, 927 (1999).
46. Seward F.D., et al., *Astrophys. J. Letters* **287**, L19 (1984).
47. Sugizaki M., et al., *Publ. Astron. Soc. Japan* **49**, L25 ().
48. Thompson C., and Dancan R.C., *Astrophys. J.* **473**, 322 (1996).
49. Torii K., et al., *Astrophys. J. Letters* **489**, L145 (1997).
50. Torii K., et al. *Astrophys. J. Letters* **494**, L207 (1998a).
51. Torii K., et al., *Astrophys. J.* **508**, 854 (1998b).
52. Torii K., et al., *Astrophys. J. Letters* **523**, L69 (1999).
53. Torii K., et al., *Astrophys. J. Letters* **534**, L71 (2000).
54. van Paradijs J., Taam, R.E., and van den Heuvel, E.P.J., *Astron. Astrophys.* **299**, L41 (1995).
55. Vasisht, G., and Gotthelf E.V., *Astrophys. J. Letters* **486**, L129 (1997).
56. Wijnands R., and van der Klis, M., *Nature* **394**, 344 (1998).
57. Wojdowski, P., et al., *Astrophys. J.* **502**, 253 (1998).
58. Woo, J.W., et al., *Astrophys. J.* **467**, 811 (1996).
59. Woods, P.M., et al., *Astrophys. J. Letters* **535**, L55 (2000).
60. Yokogawa, J., et al., *Astrophys. J. Suppl.* in press (2000a).
61. Yokogawa, J., et al., *Astrophys. J.* **539**, 191 (2000b).

Stellar Evolution and Black Hole Formation

Gerald E. Brown

Department of Physics & Astronomy, State University of New York Stony Brook, NY 11794, USA

Abstract. The evolution of the black hole transient sources is sketched. These contain a black hole of mass $\sim 7M_\odot$ with a low-mass companion star. In five of the six binaries in which the companion is in main sequence, it is a K-star; in the other case, an M-star. In three other cases the companion is a more massive star in subgiant phase. The near equality in main sequence masses of the companions shows that they originate at a well defined initial separation a_i, also that there cannot be a wide variation in main sequence mass of the black hole progenitor.

We make the case that these transient sources are relics of gamma ray bursters.

INTRODUCTION

Ralph Wijers [1] conducted a workshop in Cambridge, England, in which he gave a discussion of the black-hole transient sources. He estimated that more than 3000 of these are presently "in operation" in the Galaxy; i.e., we could observe them if we were in the right place at the right time. They are observable only up to a few kiloparsecs, so we can see only a small proportion of those in the Galaxy from our vantage point. These binaries have black hole masses of $M_{BH} \sim 7M_\odot$. Most amazingly, in all of the binaries in which the companion is in main sequence, its mass is $\sim 1M_\odot$ (within a factor of two). A rough estimate gives a progenitor black hole mass of $\sim 25M_\odot$, so this means a progenitor ratio of about $q \sim 1/25$. Observationally we have seen no binaries with $q < 1/4$, so the evolution of the black hole transient sources clearly involves new concepts.

Brown, Lee, & Bethe [2] developed the scenario that high-mass black holes in the range of ZAMS mass stars $20 - 35M_\odot$ could be made only if they burned helium while "clothed" with a hydrogen envelope. This has been confirmed by Wellstein & Langer [3]. Developing this theme, we believe we can account for the regularity found in the main sequence companions in the black hole transient sources.

In Figure 1 we show the evolutionary tracks of ZAMS $20M_\odot$ and $25M_\odot$ stars by Schaller et al. [4]. No stars between 25 and $40M_\odot$ were evolved by Schaller et al., so we do not have their tracks. But, in any case, these authors used wind loss rates for the He winds which were a factor 2 − 3 too large [3] and these large wind loss rates severely effect stars in this range of masses.

Because of the large wind losses employed, the ZAMS $40M_\odot$ star of Schaller et al. [4] has blown away by the end of He core burning. We expect some of it to be left in Wolf Rayet form after evolution with factor 2 − 3 lower He wind losses and we expect the $30-35M_\odot$ stars to be not too different from the $25M_\odot$ one shown in Figure 1 except they might not reach a radius as large as $800R_\odot$ and we would not expect them to increase in radius during He shell burning, but to actually decrease because of the He winds. It is not ruled out that transient sources can involve a ZAMS $40M_\odot$ star, once the latter is evolved with lower He wind losses. The matters discussed in this paragraph cannot be made more precise until a better evolution in the ZAMS mass region $25-40M_\odot$ is carried out.

Our prototype $25M_\odot$ star of Schaller et al. is particularly interesting because of its very rapid rise in radius, up to $\sim 400R_\odot$ towards the end of H shell burning

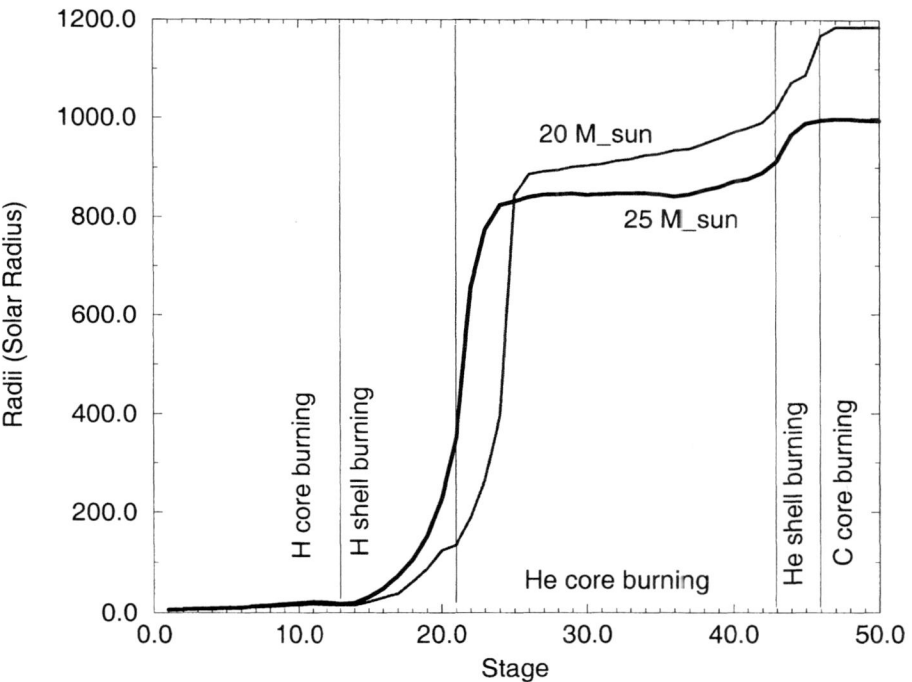

FIGURE 1. Evolutionary tracks of $20M_\odot$ and $25M_\odot$ stars from [4].

and from 400 to $800R_\odot$ in the beginning of He burning (which begins before the star has reached its red giant tip). from the Schaller et al. tables, we see that just at $25M_\odot$ their central carbon abundance at the end of He core burning is $\sim 15\%$. This is the limit at which Weaver & Woosley [5] find below which convective core carbon burning is skipped. The processes $C + C \rightarrow{}^{24}Mg$, ${}^{20}Ne + \alpha$, etc. which take place at a high temperature of $T \sim 80 - 100$ keV are then also skipped. In lower mass stars, where they take place, they carry off a lot of entropy through neutrino pair emission which goes as T^9. Thus, the Schaller et al. range of star above $25M_\odot$ have much more entropy than those below, and the former will end up with much larger Fe cores. The only way to accommodate larger entropy in the final core, which has entropy $S/k \sim 1$ per nucleon [6] is to have a larger number of nucleons. Schaller et al. use a ${}^{12}C(\alpha,\gamma){}^{16}O$ capture rate of S(300 keV)=100 keV barns, whereas Woosley and collaborators use 170 keV barns. Their calculations show that convective core carbon burning is skipped at main sequence $19M_\odot$. if we choose a rate ~ 150 keV barns, our best bet, this will happen at $21M_\odot$. With the sudden increase in Fe core mass, we pinpoint this mass as the value for which the (single) star will evolve into a black hole.

Schaller et al. have used He wind losses of a factor of $2-3$ too high and a ${}^{12}C(\alpha,\gamma){}^{16}O$ rate which we believe to be somewhat too low. With a lower He wind loss rate and a 150 keV value for ${}^{12}C(\alpha,\gamma){}^{16}O$ we can only guess that all stars in the ZAMS mass range of $\sim 20 - 35 M_\odot$ have the evolutionary track we show in Figure 1. Assuming this, we shall play the game which follows.

The radius of the $25M_\odot$ star increases rapidly at the onset of He core burning up to $\sim 800R_\odot$ and then levels off. We shall use Case C mass transfer; i.e., mass transfer following core helium burning. The massive star has only $\sim 10^4$ yrs before it goes supernova, whereas its thermal time scale $\tau_{th} = 3 \times 10^7/(M/M_\odot)^2$ yrs is ~ 5 times this. Thus, as in Brown et al. [2] we require that the massive star actually expand to meet the low mass companion (rather than beginning mass transfer at the first-reached Roche lobe, which would be at a radius $R_L \sim 2/3 a_i$ for the large mass ratio considered here). Since the evolutionary track is flat, it does not matter whether our requirement is that the star complete all, or only $\sim$ half, of its He core burning while clothed. If it completes some fraction, it will complete all. This greatly simplifies matters.

We label the initial mass of the massive star by $M_{B,i} = 25M_\odot$, the final He star mass as $M_{B,f} \simeq 0.3 M_{B,i}$ as in [7] and the mass of the main sequence star as M_A, neglecting any possible accretion onto it during common envelope evolution. Then, the energy equation for common envelope evolution is

$$\frac{GM_A M_{B,f}}{a_f} = \frac{0.6 GM_{B,i}^2}{\alpha_{ce} a_i} \qquad (1)$$

We can solve for a_i;

$$M_A a_i = \left[0.6\left(\frac{M_{B,i}}{M_{B,f}}\right) a_f\right]\left(\frac{M_{B,i}}{\alpha_{ce}}\right). \qquad (2)$$

The quantity in square brackets is assumed known. Bethe & Brown [7] took $M_{B,f}/M_{B,i} = 0.3$, and $a_f = 5R_\odot$ is the Roche Lobe of the main sequence companion. Eq. (2) then reduces to

$$M_A = \frac{10\,R_\odot}{\alpha_{ce}} \frac{M_{B,i}}{a_i}. \tag{3}$$

Our assumed range of $M_{B,i}$ is

$$20\,M_\odot < M_{B,i} < 35\,M_\odot \tag{4}$$

and that of a_i is

$$800\,R_\odot < a_i < 1000\,R_\odot \tag{5}$$

taken from Fig. 1 as the interval between He core burning and the final radius of the $25R_\odot$ star just before explosion. This then gives us a possible interval in M_A of

$$\frac{0.2\,M_\odot}{\alpha_{ce}} < M_A < \frac{0.44\,M_\odot}{\alpha_{ce}}. \tag{6}$$

In the Appendix we derive the approximate mass-period relation

$$P_{orb} = 9\text{ hr}\left(\frac{M_A}{M_\odot}\right)^{0.82}. \tag{7}$$

The observed periods in Table 1 are in the interval

$$5.9\text{ hr} < P_{orb} < 11.2\text{ hr}. \tag{8}$$

In eq. (6) we choose $\alpha_{ce} = 1/3$ so as to obtain

$$6.0\text{ hr} < P_{orb} < 11.3\text{ hr}. \tag{9}$$

A priori we would have expected $\alpha_{ce} \sim 0.5$ so that the envelope is expelled with the same kinetic energy as it originally possessed, but $\alpha_{ce} = 1/3$ is not far from this.

The main sequence companion is restricted to the narrow range of $a_i = 800 - 1000\,R_\odot$ by our present considerations. This means a logarithmic interval of $\ln(1.25)$, so that with the assumed total logarithmic interval assumed by Brown et al. [2], the fractional favorable interval is only $\ln(1.25)/7 = 0.03$, substantially smaller than 0.11 found by these authors. We believe this smaller favorable interval to be required by the small variation in masses of the main sequence companions. Cutting this interval down from 0.11 to 0.03 would give the estimated rate of black hole transient source production to be 2.4×10^{-6} yr^{-1} in the Galaxy, and

TABLE 1. Parameters of suspected black hole binaries with measured mass functions [2]. N means nova, XN means X-ray nova. Numbers in parenthesis indicate errors in the last digits. Reprinted from [2], Copyright 1999, with permission from Elsevier Science.

X-ray names	other name(s)	compan. type / q (M_{opt}/M_X)	P_{orb} (d) / K_{opt} (km s^{-1})	$f(M_X)$ ($M_\odot$) / i (degree)	M_{opt} ($M_\odot$) / M_X ($M_\odot$)	(l,b) / d (kpc)
Cyg X-1 1956+350	V1357 Cyg HDE 226868	O9.7Iab	5.5996 74.7(10)	0.25(1)	33(9) 16(5)	(73.1,+3.1) 2.5
LMC X-3 0538−641		B3Ve	1.70 235(11)	2.3(3)	5.6−7.8	(273.6,−32.1) 55
LMC X-1 0540−697		O7−9III	4.22 68(8)	0.14(5)		(280.2,−31.5) 55
XN Mon 75 A 0620−003	V616 Mon N Mon 1917	K4 V 0.057−0.077	0.3230 443(4)	2.83−2.99 37−44	0.53−1.22 9.4−15.9	(210.0,−6.5) 0.66−1.45
XN Oph 77 H 1705−250	V2107 Oph	K3 V	0.5213 420(30)	4.44−4.86 60−80	0.3−0.6 5.2−8.6	(358.6,+9.1) 5.5:
XN Vul 88 GS 2000+251	QZ Vul	K5 V 0.030−0.054	0.3441 520(16)	4.89−5.13 43−74	0.17−0.97 5.8−18.0	(63.4,−3.1) 2
XN Cyg 89 GS 2023+338	V404 Cyg N Cyg 1938, 1959	K0 IV 0.055−0.065	6.4714 208.5(7)	6.02−6.12 52−60	0.57−0.92 10.3−14.2	(73.2,−2.2) 2.2−3.7
XN Mus 91 GS 1124−683		K5 V 0.09−0.17	0.4326 406(7)	2.86−3.16 54−65	0.41−1.4 4.6−8.2	(295.0,−6.1) 3.0
XN Per 92 GRO J0422+32		M0 V 0.029−0.069	0.2127(7) 380.6(65)	1.15−1.27 28−45	0.10−0.97 3.4−14.0	(197.3,−11.9)
XN Sco 94 GRO J1655−40		F5-G2 0.33−0.37	2.6127(8) 227(2)	2.64−2.82 67−71	1.8−2.5 5.5−6.8	(345.0,+2.2) 3.2
XN 4U 1543−47	MX 1543-475	A2 V	1.123(8) 124(4)	0.20−0.24 20−40	1.3−2.6 2.0−9.7	(330.9,+5.4) 9.1(11)
XN Vel 93		K6-M0 0.137± 0.015	0.2852 475.4(59)	3.05−3.29 ∼ 78	0.50−0.65 3.64−4.74	

an estimated number of 2400, to be compared with a lower limit of 3000 found by Wijers [1] from observational arguments.

The three subgiants GS 2023+338, GRO J1655−40 and 4U1543−47 have longer periods, from 1 − 6 days. Before filling their Roche Lobes they were obviously "silent". Any more massive companions will end up outside of their Roche Lobes while in main sequence, and fill them only as they evolve. This basically removes the factor of Δq in the population synthesis, so we will have about 25 times more "silent partners" than binaries with main sequence companion; therefore, ∼ 50,000 high-mass black-hole, main-sequence companion binaries in the Galaxy.

---

[1]) Following, otherwise, the arguments of [2].

TRANSIENT BLACK HOLE SOURCES AS RELICS OF GAMMA RAY BURSTERS

Brown et al. [8] have made the case that the progenitors of the transient black hole sources are promising progenitors of gamma ray bursters. In particular, the binary Nova Scorpii 1994 (GRO J1655−40) has the characteristics of being the relic of a GRB.

The general idea of what is necessary for a GRB follows Woosley's Collapsar model [9]. Namely, the center of a rapidly rotating He star falls into a black hole, the outer material being supported by centrifugal force. Jets of matter are driven. Brown et al. [8] supplement this model by magnetohydrodynamical effects; namely they power the GRB and hypernova explosion by extracting the rotational energy of the black hole. We now describe some of this paper. In detail, the low-mass companion star in the transient black hole source brings the He core of the massive star into corotation as it expels the hydrogen envelope of the massive star. The center of the He envelope burns to an Fe core. Once the Chandrasekhar mass for the electron fraction in the core is reached, collapse into a neutron star ensues. At the same time, the remaining He envelope tranforms into an accretion disc, and the neutron star is brought into corotation with the inner part of the disc. The inner disc is thin, essentially Keplerian, because of neutrino cooling. At this point the Kerr parameter

$$\tilde{a} = \frac{Jc}{GM^2} \tag{10}$$

of the neutron star can be calculated. The neutron star is sufficiently large that a nonrelativistic calculation will suffice. The Kerr parameter will be conserved as the neutron star collapses with the much smaller black hole. The black hole will be rotating rapidly with respect to the accretion disc, about twice the Ω of the latter.

The rotational energy that can be extracted from the rotating black hole is

$$E_{rot} = f(\tilde{a}) M_{BH} c^2 \tag{11}$$

where

$$f(\tilde{a}) = 1 - \sqrt{\frac{1}{2}(1 + \sqrt{1 - \tilde{a}^2})}. \tag{12}$$

For the kinematics discussed aboce and a black hole mass of $2.5 M_\odot c^2$, $f(\tilde{a}) \sim 0.11$

$$E_{rot} \sim 5 \times 10^{53} \text{ ergs}. \tag{13}$$

This energy will go through the open magnetic field lines partly into powering the GRB and partly into increasing the entropy of the black hole, and through the closed field lines ("hoops") coupling the black hole to the disc. The energy budget is:

1. Open Field Lines:
   Energy following along open field lines goes about equally into a hot fireball, assumed to be in the "loading region"; i.e. in a region where the magnetic fields are perturbative, and into heating up the black hole. The equal division results when the black hole is rotating with ~ twice the angular velocity of the accretion disc, which follows from the collapse. Thus, roughly 10^{53} ergs is available for the hot fireball, the source of the GRB, and about the sams amount to increase the entropy of the black hole.

2. Closed Field Lines:
   We assume that ~ 10^{53} ergs flows through closed field lines from the black hole to the accretion disk. The more rapidly rotating black hole will provide torques, along its rotation axis, to spin up the inner disk. The magnetic field lines are frozen in the matter of the disc. With increasing centrifugal force the matter in the inner disk will move outwards, cutting down the accretion. Angular momentum is then advected outwards, so that matter can drift black inwards. More matter is then delivered outwards again. The situation is like that of the ball in a roulette wheel (R.D. Blandford, private communication). First of all it is flung outwards and then drifts slowly inwards. When it hits the hub (where matter accretes) it is again flung outwards. The viscous time scale for the fluctuation is estimated to be ~ 0.1 sec, and these are suggested to give the time structure of GRBs.

In the failed supernova explosion substantial matter is ejected, about $5M_\odot$ in the case of Nova Scorpii 1994. This ejection takes place from the black hole position, not at the center of mass of the system, so the binary is given a large space velocity, 150 km s^{-1} in the case of Nova Scorpii 1994 [10]. The F-star companion here is the most massive of the companions in the transient sources, so the space velocity is the greatest here.

From our above estimates, about 10^{53} ergs can be delivered into the accretion disc. Only $2 - 3 \times 10^{52}$ ergs of this are necessary to drive the massive supernova explosion, the hypernova. So the system is dismantled before the energy is used up. This is consistent with the fact that the black hole is left spinning with ~ 60% of the velocity of light [11].

The frequency with which transient sources are born, ~ 10^{-4} yr^{-1} in the Galaxy, provides many more progenitor systems for GRB's than the frequency of the latter. Whereas our scenario provides roughly equal energy sources of ~ 10^{53} ergs in all cases, the chief selection effect amongst them as to whether they will be seen is the magnitude of the magnetic field in the region near the black hole, we believe.

The power with our $f(\tilde{a}) = 0.11$ can be delivered at a rate

$$P = 2.5 \times 10^{50} \left(\frac{B}{10^{15}G}\right)^2 \left(\frac{M_{BH}}{M_\odot}\right)^2 \text{ erg s}^{-1}$$

$$\sim 1.6 \times 10^{51} \left(\frac{B}{10^{15}G}\right)^2 \text{ erg s}^{-1} \qquad (14)$$

for our estimated $M_{BH} = 2.5 M_\odot$ when the energy is first delivered [8]. Thus, in ~ 100 seconds the central engine will deliver 1.6×10^{53} ergs for a field of 10^{15} G. Of course, the time will be shorter for higher fields. These are not outlandishly high fields, roughly those encountered in magnetars, but we do not expect to have such high fields in many of the cases. If the magnetic fields are substantially less than 10^{15} G, the power may still be poured into the fireball and accretion disk, but the former may not be fed sufficiently fast that it gives an observable display.

ORBIT-WIDENING DUE TO MASS LOSS

Following the completion of this meeting we learned from Gijs Nelemans that the scenario we have outlined of the evolution of transient sources will not work for a $25 M_\odot$ star, due to H wind losses during core He burning. (These are included in the Schaller et al. 1992 evolutionary tracks shown in Fig. 1) The wind loss during He core burning is $\sim 31\%$ of the ZAMS mass. Since the separation a_i increases with (assumed spherically symmetrical wind loss) as

$$\frac{a'}{a_i} = \frac{M_i}{M'}, \tag{15}$$

where a' is the separation following wind loss, a low-mass main sequence star with $a_i \sim 800 R_\odot$ will be at $a' \sim 1269 R_\odot$ for a $25 M_\odot$ star. The $25 M_\odot$ star does not evolve past $1000 R_\odot$, so it will never reach the low mass main sequence companion. (The same argument works if we consider mass transfer at the Roche Lobe, rather than when the massive star meets the main sequence star) We have found that with the Schaller et al. $20 M_\odot$ star, also shown in Fig. 1, because of the lower mass loss and the greater increase in radius during He shell burning and carbon core burning, our scenario does work for a narrow range of companion masses. We thus seem to be "localizing" the formation of the transient black hole sources much more in the region of $\sim 20 M_\odot$ for the ZAMS mass of the black hole progenitor.

CONCLUSIONS

We find the great regularity in the masses of main sequence companion in the black hole transient sources. With a certain amount of detective work, we are able to show that in the progenitor binaries these companion must be between 800 and 1000 $R_\odot$. We achieve this regularity only by assuming all stars in ZAMS range $20 - 35 M_\odot$ to have an evolutionary track similar to the one we show in Fig. 1 for a $25 M_\odot$ star. We believe that with lower He wind losses and a somewhat higher value of the $^{12}C(\alpha,\gamma)^{16}O$ rate than Schaller et al. [4] used, this will at least tend to be so.

We arrive at a black hole transient source evolutionary rate of $\sim 10^{-4}$ yr^{-1} in the Galaxy. This is a substantially higher rate than that of GRBs. We discussed what we believe to be the close relation of black hole transient sources and GRBs.

ACKNOWLEDGEMENTS

I would like to thank Chang-Hwan Lee and Gijs Nelemans for much help and advice in preparation of this article. Support from the U.S. Department of Energy under grant DE-FG02-88ER40388 is gratefully acknowledged.

REFERENCES

1. Wijers, R.A.M.J., *Evolutionary Processes in Binary Star*, 327, Kluwer Acad. Publ., Eds. R.A.M.J. Wijers et al. (1996).
2. Brown, G.E., Lee, C.-H., and Bethe, H.A., *New Astrononomy* **4**, 313 (1999).
3. Wellstein, S. and Langer, N., *Astron. & Astrop.* **350**, 148 (1999).
4. Schaller, G. et al. *Astron. & Astrop. Suppl.* **96** 269 (1992).
5. Weaver, T.A. and Woosley, S.E., *Phys. Rep.* **227** 65 (1993).
6. Bethe, H.A., Brown, G.E., Applegate, J., and Lattimer, J.M., *Nuc. Phys.* **A324**, 487 (1979).
7. Bethe, H.A. and Brown, G.E., *Astrop. J.* **506**, 780 (1998).
8. Brown, G.E., Lee, C.-H., Wijers, R.A.M.J., Lee, H.K., Israelian, G., and Bethe, H.A., *New Astronomy*, accepted (2000).
9. Woosley, S.E., *Astrop. J.* **405**, 273 (1993); MacFadyen, A.I. and Woosley, S.E., *Astrophys. J.* **524**, 262 (1999).
10. Nelemans, G., Tauris, T.M., and Van den Heuvel, E.P.J., *Astron. & Astrop.* **352**, L87 (1999).
11. Gruzinov, A., *Astrop. J.* **517**, L105 (1999).
12. Kopal, Z., *Close Binary Systems: The International Astrophysics Series*, Chapman & Hall, London (1959); Pacyński, B, *Acta Astonomica* **17**, 287 (1967).

APPENDIX

We derive, following [12], a period-mass relation for the black hole transient sources. Starting from Kepler

$$P^2 = 4\pi^2 a^3/GM \tag{A1}$$

and the expression for the Roche Lobe

$$\frac{R_L}{a} = 0.46 \left(\frac{M_A}{M}\right)^{1/3} \tag{A2}$$

one finds

$$P = \left[\frac{4\pi^2}{(0.46)^3} \frac{R_L^3}{GM_A}\right]^{1/2}. \tag{A3}$$

If M_A is at its Roche Lobe then $R_A = R_L$.

Now, for $\sim 1 M_\odot$ main sequence stars $R_A \propto M^{0.88}$. Therefore,

$$R_L^3 = R_\odot^3 \left(\frac{M_A}{M_\odot}\right)^{2.64}. \tag{A4}$$

Using eq. (A4) in eq. (A3) one has

$$P_{orb} = \left(\frac{M_A}{M_\odot}\right)^{0.82} \left[\frac{4\pi^2}{(0.46)^3} \frac{R_\odot^3}{GM_\odot}\right]^{1/2}$$

$$= 9 \text{ hr} \left(\frac{M_A}{M_\odot}\right)^{0.82}. \tag{A5}$$

Iron Line Diagnostics of X-ray Binaries

Tadayasu Dotani

Institute of Space and Astronautical Science
3-1-1 Yoshinodai, Sagamihara, Kanagawa 229-8510, Japan

Abstract. ASCA archive data of low-mass X-ray binaries are analyzed to study the characteristics of iron K emission lines. In total, 20 LMXBs are analyzed, which include various categories of sources, such as Z sources, atoll sources, bursters, dippers, accretion disc coronae sources, etc. We detected significant iron K emission lines from 10 sources; the line center is 6.56 keV on average with an equivalent width of $< 10 - 170$ eV. The lines are found to be broad with a width of ~ 0.5 keV (FWHM). We discuss possible formation site of the iron emission lines. We conjecture that they may be formed through the recombination of the photo-ionized plasma located on the accretion disk. However, another mechanism, such as Compton scattering, need to be involved to explain the line width.

INTRODUCTION

X-ray binaries are close binary system consisting of a normal star and a compact star, such as a white dwarf, a neutron star and a black hole. In these binary systems, mass accretion occurs on to the compact object from the normal star through the Roche-lobe overflow or the stellar wind capture. The mass accretion on to the compact star powers its X-ray emission. X-ray binaries may be classified by the combination of the nature of the normal star and the compact star, e.g. high-mass X-ray binaries (an early type star + a neutron star or a black hole), low-mass X-ray binaries (a late-type star + a neutron star or a black hole), cataclysmic variables (a late-type star + a white dwarf). Because characteristics of the X-ray binaries show large variety depending on the category, we concentrate on the low-mass X-ray binaries (LMXBs) including a neutron star in this article.

LMXBs are relatively old system, typical age of 10^9 yr or more. They are mostly found in the galactic bulge or globular clusters. Because they are old system, neutron star in this system is believed to have only a weak magnetic field, typically $< 10^9$ G at surface. Normal star in LMXBs usually fills its Roche-lobe, and mass accretion occurs through the Roche-lobe overflow. The overflowed matter forms an accretion disk around the neutron star because of its large angular momentum. The accretion disk is optically thick and geometrically thin in most places except for the vicinity of the neutron star and the outermost regions of the disk. Outer part of the accretion disk may be inflated vertically, but the structure near the neutron star is not known. Geometrically thin accretion disk might be extended down to the neutron star surface, or it might be disrupted by the small magnetosphere of the neutron star. Other than the accretion disk, optically thin, almost completely ionized plasma is believed to exist around the neutron star. It is sometimes called accretion disk corona (ADC), but its nature and geometry is poorly known.

Iron line diagnostics of X-ray binaries are a powerful tool to investigate the geometry and physical conditions of the X-ray emitting/reprocessing regions. Iron K emission lines from LMXBs have been studied mainly using the Tenma and the EXOSAT satellites. Hirano et al. (1987) studied the iron emission lines in 10 LMXBs using the Tenma satellite, and detected an emission line feature from six LMXBs. They found that the average centroid energy of the emission lines is 6.66 ± 0.05 keV with an equivalent width (EW) of 20 – 60 eV. A significant line width of 550^{+350}_{-300} eV (FWHM) was obtained only from 4U 1608-52, though a width of ~ 1 keV could not be rejected for the other sources. The EXOSAT GSPC observations detected iron K emission lines at 6.4–6.8 keV from five out of six sources [6]. The EWs of the line were 70–170 eV, and a finite width of ~ 1 keV (FWHM) was obtained [5]. The line energy indicates that highly ionized iron ions, mainly FeXXV, are responsible for the line formation. It is generally believed that large X-ray flux from the vicinity of the neutron star makes highly ionized iron ions, and they produce line emission through recombination.

In this paper, we describe the results of the ASCA archive data analysis of 20 LMXBs. Full description of the analysis and results are found in Asai et al. (2000).

ANALYSIS AND RESULTS

Iron K lines in the energy spectra of LMXBs are weak in most cases, and careful analysis is needed to evaluate the line parameters accurately. Furthermore, because some of the LMXBs are very bright, spectral deformation due to the instrumental effects cannot be neglected. We took following steps to estimate the line parameters.

1. Calculate the energy spectra separately for SIS and GIS. Correct GIS gain shift [4] and SIS photon pile-up [2] when the source count rate is large.

2. Determine the absorption column by fitting a model function to the continuum spectra (excluding the iron energy band). We used the two component model consisting of a multicolor disk model and a blackbody. It is important to determine the absorption column accurately, because iron K edge structure due to the interstellar/circumstellar absorption sometimes affects the determination of the line parameters.

3. Determine the continuum model separately for GIS and SIS, which is appropriate to estimate the line parameters. We do not use the two component model adopted in the previous step, because the cross-over of the two components sometimes make a residual structure which can mimic the emission line.

Instead, we use a power law with an exponential cut-off model. Absorption column determined in the previous step is also included.

4. Add a gaussian line model to the continuum one by one until the reduction of χ^2 becomes not significant.

5. Compare the best-fit parameters of the line between GIS and SIS when an intrinsically narrow line is assumed. Because SIS and GIS have a different energy resolution, best-fit parameters can be inconsistent if the line is broad. When the statistics of the data is limited, it is not very useful to evaluate the line width directly. Instead, comparison of the line strength between detectors with different energy resolution may be effective.

6. Try a broad line model for the sources which gave inconsistent line strengths between SIS and GIS. In this case, we carried out a combined fitting for GIS and SIS. When the best-fit parameters are consistent between SIS and GIS even for the narrow line model, we also carried out a combined fitting for the narrow line model to improve the accuracy of the best-fit parameters.

Iron line parameters thus obtained are listed in table 1. We also include source luminosities determined from the model fitting to the continuum and the types of the sources in the table. To see how the line parameters depend on the source properties, we plot the line parameters as a function of the source luminosity in figure 1. There seems to be no clear correlation in the line parameters with the source luminosity. Similarly, source category may not be important to determine the line properties.

DISCUSSION

We analyzed 20 LMXBs in ASCA archive data, and detected significant iron K emission lines from 10 sources. Among the LMXBs we detected iron lines, X1822–371 may be an exceptional source; only X1822–371 shows 3 lines in the energy spectra. If we exclude this source, iron line parameters seem to be almost common to LMXBs. Hereafter, we discuss the common properties of the line, and do not consider the exceptional case. See White et al. (1997) for the ASCA results of X1822–371.

Iron emission lines in LMXBs are believed to be produced through the recombination of the photo-ionized plasma. Ionization degree and line emission properties are determined by the so-called ξ-parameter ($\equiv L/nR^2$; L: ionizing luminosity, n: number density of plasma, R: distance to the neutron star), when the gas is optically thin. Emissivity of iron K lines becomes maximum at $\log \xi \sim 3.65$. In this case, the center energy of the lines is about 6.68 keV. Average energy of the iron K lines we detected was found to be 6.56 keV. Although this is slightly lower than that at maximum emissivity, the difference is easily explained by the small

TABLE 1. Iron line parameters of LMXBs Determined with the ASCA data

Source	Line Parameters			Luminosity	Type[a]
	Center (keV)	Width (keV)	EW (eV)	(10^{37} erg s^{-1})	
GX 5–1	6.7 (fix)	0 (fix)	< 7	28	Z
X1820–30	6.6 ± 0.1	$0.7^{+0.2}_{-0.5}$	31^{+12}_{-11}	4.6	A,B
Cyg X-2 (93/6/18)	6.6 ± 0.1	0.5 ± 0.2	31^{+11}_{-9}	5.9	Z,B
Cyg X-2 (95/12/6)	6.67 ± 0.07	0.6 ± 0.2	57^{+13}_{-12}	5.4	
GX 9+9	$6.5^{+0.2}_{-0.1}$	< 0.7	13^{+9}_{-8}	4.6	A
GX 13+1	6.44 ± 0.05	0.3 ± 0.1	37^{+10}_{-8}	6.9	A,B
X 1636–536	6.7 (fix)	0 (fix)	< 6	0.5–2.2	A,B
X 1254–68	6.7 (fix)	0 (fix)	< 25	1.2	D,B
Aql X-1	6.5 ± 0.03[b]	< 2[b]	63^{+61b}_{-42}	0.083	T,B
EXO 0748–676	6.7 (fix)	0 (fix)	< 11	0.16	D,B
X 1916–053	$(5.9^{+0.2}_{-0.1})$	$(0.7^{+0.5}_{-0.2})$	(87^{+42}_{-35})	0.35	D,B
X 0614+091	6.7 (fix)	0 (fix)	< 51	0.089	B
X 0921–630	6.72 ± 0.05	0.5 ± 0.2	172^{+31}_{-29}	0.065	C
X 1822–371 (93/10/7)	6.38 ± 0.02	0 (fix)	58 ± 8	0.0064	D,C
	$6.63^{+0.06}_{-0.05}$	0 (fix)	28 ± 7		
	7.01 ± 0.03	0 (fix)	43^{+9}_{-10}		
X 1822–371 (96/9/26)	6.2 ± 0.1	0 (fix)	23^{+13}_{-12}	0.0064	
	$6.49^{+0.05}_{-0.04}$	0 (fix)	63^{+13}_{-17}		
	7.0 ± 0.1	0 (fix)	31^{+13}_{-12}		
X 1323–619	6.7 (fix)	0 (fix)	< 26	0.13	D,B
X 1746–371	6.7 (fix)	0 (fix)	< 16	0.45	D,B
X 1624–490	$6.4^{+0.1}_{-0.3}$	< 0.9	13^{+13}_{-9}	2.4	D,B
X 2127+119	6.7 (fix)	0 (fix)	< 17	0.42	B,C
X 1543–62	$6.8^{+0.3}_{-0.4}$	< 2	48^{+48}_{-31}	1.3	D,B
X 1850–086	6.7 (fix)	0 (fix)	< 14	0.13	B
Ser X-1	6.63 ± 0.07	0.5 ± 0.2	52^{+15}_{-13}	2.2	B

[a] Definitions of types: Z = z source; A = atoll source; C = ADC source; D = dipper; T = transient; B = X-ray burster.
[b] Only the GIS data are used.

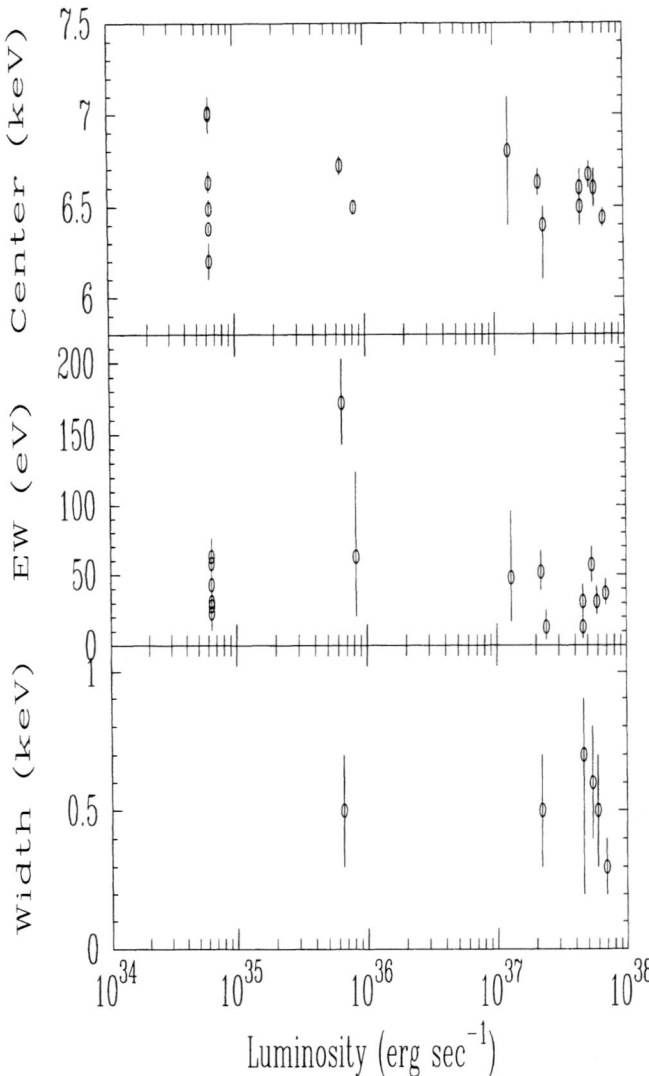

FIGURE 1. Iron line parameters of LMXBs calculated using the ASCA archive data.

changes of ξ. Thus the ASCA results is consistent to the interpretation that iron K lines are produced through the recombination of the photo-ionized plasma.

To study the nature of the line emitting region in more detail, we simulated a model spectrum using XSTAR and compared the simulated spectrum with the ASCA data. XSTAR[1] is a program for calculating the physical conditions and

---

[1] Available at http://heasarc.gsfc.nasa.gov/docs/software/xstar/xstar.html

emission spectra of photoionized gases. We found from the simulation that ξ parameter corresponding to the average line energy of 6.56 keV is roughly 10^2. If we take typical parameters of LMXBs, Roche-lobe size of the neutron star would be $\sim 10^{10} - 10^{11}$ cm. Because it is considered that the photo-ionized plasma is located within the Roche-lobe, its distance to the neutron star may be at most $R \sim 10^{10} - 10^{11}$ cm. If we adopt $L \sim 10^{37}$ erg/sec as a typical luminosity of LMXBs and substitute these parameters to the definition of ξ, we obtain $n \leq 10^{13} - 10^{15}$ cm^{-3}. This result indicates that a high density plasma is needed to keep the ionization degree of iron moderate under the irradiation of large X-ray flux from the neutron star. Such a dense plasma may be found only near the surface of the accretion disk. We consider that the surface of the accretion disk becomes highly ionized by the X-ray irradiation, and recombination of the iron ion produces the line emission. Relatively small EW of the iron line is also consistent to this interpretation, because the accretion disk subtends only a small solid angle when viewed from the neutron star.

We carried out another simulation using XSTAR to investigate the origin of the line width. It may be plausible that plasmas with different ξ contribute to the line emission, because the optically thick accretion disk extends from the vicinity of the neutron star to almost the size of the Roche-lobe. We simulated a spectrum appropriately averaged over a range of ξ. We found that, if we stick to the mean line energy of 6.56 keV, superposition of the emission lines from different ξ explains a line width of at most 0.1 keV (FWHM). This means that we need some other mechanisms to make the line width as large as 0.5 keV. Several mechanisms are conceivable; such as Doppler broadening, and Compton scattering. Doppler broadening cannot be thermal origin, because plasma temperature becomes too high (~ 1 MeV). If it is related to the Kepler motion, the plasma may be located at a distance of $\sim 10^8$ cm from the neutron star. However, it is not clear why only the plasma at this distance contribute to line formation and Doppler broadening. Thus, we consider that Doppler broadening is not plausible.

If the Compton scattering is responsible for the line width, Comptonizing plasma may have a temperature of about 1 keV, otherwise the Compton scattering shifts the line energy significantly. The optical depth of the plasma may be a few to explain the line width. The plasma may be completely ionized, in spite of its low temperature, due to the strong X-ray irradiation from the neutron star. This is consistent to the absence of low energy absorption. Presence of such plasma may be naturally explained because the X-ray irradiation enhances evaporation of plasma from the surface of the accretion disk and induce the formation of the disk wind. Such disk wind might explain the observed width of iron K lines, although the wind temperature and the optical depth need to be appropriate not to shift the line energy largely.

REFERENCES

1. Asai, K., Dotani, T., Nagase, F. & Mitsuda, K., *Astrophys. J. Suppl.* **131**, in press.
2. Ebisawa, K. et al. *Astrophys. J.* **467**, 419 (1996).
3. Hirano, T., Hatashida, S., Nagase, F., Masai, K., & Mitsuda K. *Publ. Astron. Soc. Japan* **39**, 619 (1987).
4. Makishima, K. et al. *Publ. Astron. Soc. Japan* **48**, 171 (1996).
5. White, N. E., Peacock, A., & Taylor B. G. *Astrophys. J.* **296**, 475 (1985).
6. White, N. E., Peacock, A., Hasinger, G., Mason, K. O., Manzo, G., Taylor, B. G., & Branduardi-Raymont G. *Mon. Not. R. Astron. Soc.* **218**, 129 (1986).
7. White, N. E., Kallman, T. R., & Angelini, L., in X-Ray Imaging and Spectroscopy of Cosmic Hot Plasma, ed. F. Makino, & K. Mitsuda (Tokyo: Universal Academy Press), 411 (1997)

Simulations of Radio Galaxies: Synthetic Observations

I. L. Tregillis*, Dongsu Ryu[†], and T. W. Jones*

*School of Physics and Astronomy, University of Minnesota
Minneapolis, MN 55455, USA
[†]Department of Astronomy and Space Science, Chungnam National University
Daejeon, 305-764, Korea

Abstract. Three-dimensional, magnetohydrodynamic simulations of radio galaxies, which follow the dynamical evolution of jet-driven flows as well as the acceleration, transport and cooling of cosmic-ray electrons, have been done. The simulations enable us to compute synthetic radio and X-ray observations, including spectral properties.

INTRODUCTION

The standard paradigm for radio galaxies is based on jets from active galactic nuclei, which propagate into the intergalactic medium and create giant lobes [1]. The radio emission represents synchrotron radiation from cosmic-ray electrons and magnetic fields in the regions of propagating jets and giant lobes. Numerical simulations have been used for well over a decade to shed light on the physical processes taking place within the radio galaxies. The key barrier in those studies with numerical simulations has been our lack of understanding the gap between the flows thought to drive the phenomenon and the emissions that reveal them. The flows are strongly driven systems, and so dynamically unsteady and complex. The emissions come from particles very far from thermodynamic equilibrium, so that their energy distributions depend on local microphysics and their histories. The recent progress in our ability to carry out sophisticated multi-dimensional magnetohydrodynamic simulations and model diffusive shock acceleration for cosmic-ray energization seems to provide a hope to overcome the barrier.

As the first attempt to take advantage of such progress, we have introduced a simple and economical but effective method for cosmic-ray acceleration and transport which can be combined with multi-dimensional simulations of radio galaxies [3]. With the code three-dimensional magnetohydrodynamic flow simulations have been performed where the nonthermal population of cosmic-ray electron has been followed in some detail, allowing us to compute for the first time meaningful model emissions.

The studies using simulations can be completed only after making the direct comparison between simulation results and observations. Synthetic observations can address this issue. By synthetically observing a source in simulation data whose detailed physical structure is known beforehand, we can gain insights into what real observations are actually telling us. Attempts have been made for it from purely hydrodynamic and magnetohydrodynamic simulations. The earlier versions estimated the synchrotron "pseudoemissivity" based on flow quantities and magnetic fields [2]. However, only an explicit treatment of the cosmic-ray electron population can allow us to treat the emission properties of the simulation directly.

This paper describes some initial results of our three-dimensional magnetohydrodynamic simulations along with synthetic observation. A full report is in preparation.

CODE

Flow dynamics and magnetic field evolution are treated with a second-order accurate, conservative, ideal magnetohydrodynamic code based on the total variation diminishing scheme [4]. It maintains the divergence free condition to the magnetic field to machine accuracy using an upwinded constrained transport scheme.

For cosmic-ray particle transport, we use the conventional "convection diffusion equation" [5] for the momentum distribution function, f, which follows spatial and momentum diffusion as well as spatial and momentum advection of the particles. The last of these corresponds to energy losses and gains from, for example, adiabatic expansion and synchrotron aging. However, high computational costs prohibit solving this equation through standard finite difference methods in complex flows expected in radio galaxies. To circumvent this we use a conservative finite volume approach in the momentum coordinate, taking advantage of the broad spectral character expected for $f(p)$. Particle fluxes across momentum bin boundaries are estimated by representing $f(p)$ as $f(p) \propto p^{-q(p)}$, where $q(p)$ varies in a regular way. Numerically we use the integrated number of electrons within each bin and the slope, q, within each bin. Thus, we can follow electron spectral evolution in smooth flows for all the effects mentioned above with a modest number of momentum bins. Typically we have used 8 bins to cover energies up to a few hundred GeV for electrons.

In the flows being studied, diffusive acceleration of electrons to GeV energies at shocks is effectively instantaneous within a dynamical time step. Because of that direct simulation of this physics would be prohibitively expensive. We can, however, also circumvent this difficulty if we assume the analytic, steady, test particle form for the electron distribution just behind shocks. Ignoring for this discussion some details, that spectrum will be a power law with an index, $q = 3r/r - 1$, where r is the shock compression ratio.

Together these features give us a powerful tool for numerical simulations of such complex phenomena as radio galaxies [3].

SIMULATIONS

Our simulated jet represents a light, supersonic jet, with density 10^{-2} of the ambient medium, and internal Mach number of 8. For the other parameters mentioned below this corresponds to a jet velocity of $0.1c$. The inflowing jet has a top hat velocity profile with a thin transition sheath around it. The jet inflow slowly precesses around a cone of opening angle $5°$, to break cylindrical symmetry. The incoming jet core radius is $r_j = 1$ kpc. The magnetic field is helical inside the jet and uniform in the ambient medium. On the jet axis its strength is $B_0 = 1\mu G$, with a gas pressure there 100 times greater than the magnetic pressure. This leads to a jet kinetic power of $\approx 10^{37}$ W. This numerical experiment was carried out on a $576 \times 192 \times 192$ grid. The core radius spans 15 zones.

For this simulation a relativistic electron population with a power law momentum index, $f(p) \propto p^{-4.4}$ (corresponding to a synchrotron spectral index $\alpha = 0.7$) is brought onto the grid with the jet flow. That population is subjected to acceleration at shocks, adiabatic and radiative losses. No additional nonthermal electrons were injected at shocks, however. In this simulation, the synchrotron cooling time for $\gamma = 10^4$ electrons radiating on the jet axis ($\nu \sim 300$ MHz) is 3.7×10^8 years, whereas the analogous inverse-Compton ($h\nu \sim 100$ keV) cooling time is 4.7×10^7 years. The synthetic observations presented below correspond to a time when the jet has propagated approximately 10^6 years, rendering these cooling effects negligible.

SYNTHETIC OBSERVATIONS

The combination of a vector magnetic field structure and a cosmic-ray electron energy distribution makes it possible for us to compute an approximate synchrotron emissivity, complete with spectral and polarization information, in every zone of the computational grid. Surface-brightness maps for the optically-thin emission are then easily produced by ray tracing through the computational grid, thereby projecting the source onto the plane of the sky. We have also produced X-ray surface brightness maps in the same fashion, by calculating the inverse-Compton emissivity from the interaction between the cosmic microwave background radiation and our nonthermal electrons.

The synthetic observations can be imported into any standard image analysis package and analyzed like actual observations. To make this exercise as realistic as possible we place the simulated object at an appropriate distance, currently 100 Mpc, and then convolve the computed surface brightness distribution with typically sized Gaussian beams. We can then explore a range of observable properties at multiple wavelengths, multiple angular resolutions and so on. The real power in this method comes from the fact that we can then compare synthetically "observed" source properties with the actual physical properties of the simulated objects. We can, for instance, extract the true magnetic field strength and topology in a particular region, particle and field "filling factors", or the distributions of particle spectral forms in an emitting volume.

DISCUSSION

Fig. 1 illustrates some of the important common properties in our simulations. The flow is not at all steady, since the jet terminus tends to "whip" around, sometimes forming "splatter spots" and then pinching off and redirecting itself. Here we show properties at a moment when the flow is relatively simple. The left panel reveals through volume rendering of $\nabla \cdot \mathbf{u}$ the locations of shocks in the jet and its back-flow. The strongest shocks appear blue, while relatively weaker shocks are red. The right panel shows the spatial distribution of the spectral index of ~ 10 GeV electrons at the same time. Synchrotron aging is negligible in this case, so we are seeing in the electron spectra only effects from shock acceleration and advective mixing. Shock-modified spectra flatter than the "injected spectrum" are shown in color, with the relatively flatter (steeper) spectra being blue (red). The injected spectral index, $q = 4.4$, is rendered in white. Given the absence of synchrotron aging, all spectra are at least as flat as that entering at the jet origin. Only flow regions filled with plasma originating in the jet are shown in the figure. So, for

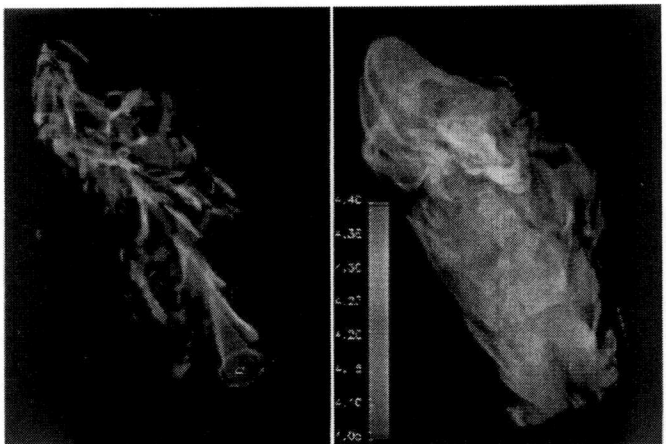

FIGURE 1. Left: Volume rendering of $\nabla \cdot \mathbf{u}$, showing the three-dimensional distribution of shocks in a precessing jet flow ~ 40 jet radii long. Stronger shocks are bluer. Only jet plasma flow is shown. Right: The distribution of 10 GeV electron momentum spectral indices, q, as defined in the text. Spectra unchanged during propagation from the jet origin are rendered in white. Shock-modified spectra are given a color related to the degree of flattening of the spectrum. Flatter spectra are bluer. The jet origin is marked by the green boxes in both images.

example, the bow shock preceding the jet is not visible in the left panel, since it occurs in the ambient medium.

These images illustrate dramatically that shock structures in jet driven flows are very complex, and also show our consequent key finding that the spectral properties of electrons emerging into the cocoon are extremely heterogeneous. Comparison of the images shows us why this finding makes sense. First, let us locate the jet flow itself. Near its base the jet can be followed through its conical internal shocks. The terminus of the jet occurs at the far upper left in a small, strong shock. It turns out that only the central core of the jet actually exits through that terminal shock. Much of the plasma emerging from the jet has passed through only weak shocks before it is redirected into the cocoon. Consequently, shocks inside and at the end of the jet have had a relatively small influence on the spectra of electrons entering the cocoon. Note next that most of the rather complex "shock web" near the jet head involves cocoon flow, in fact. Remarkably, the strongest shocks are often in the cocoon, rather than the jet. They appear to be generated by the non-axial motions of the jet head.

The distribution of electron spectra in the adjacent image does not map in an obvious way onto the shock distribution. Some insight into this complication comes from noting the apparent "streams" in the particle spectra, which are especially apparent near the jet terminus. The streams highlight flows downwind of localized, strong shocks. This feature emphasizes that the shocks are themselves very complex, and also that we are seeing a blend of many different flow histories. The relative importance of shocks inside the cocoon compared to the jet can be recognized by understanding the origin of a relatively flat spectral region visible in the figure roughly 2.5 cm to 4.5 cm from the end of the jet towards the origin (green box). There, $q \approx 4.3$, so it shows yellow in the image. Examination of flow streaklines shows that this plasma all passed through a small shock visible in the shock image (with a yellow color) about 2.5 cm from the end of the jet, but physically in the cocoon, not the jet itself. These results emphasizing the complexity of the evolution of particle spectra in jet head regions support and augment our earlier findings from two-dimensional axisymmetric simulations [3].

Figure 2 illustrates synthetically "observed" radio synchrotron and X-ray inverse-Compton images constructed from the 3-dimensional simulation. The synchrotron synthetic observation in Figure 2 corresponds to a frequency of 1.4 GHz, and the X-ray image to an energy of 10 keV.

The 1.4 GHz synchrotron luminosity νL_ν is $5.3 \times 10^{32} \times \delta_4$ W, where $\delta_4 = 10^4 \delta$, and δ is the ratio of nonthermal to thermal electron densities at the jet orifice. In what follows we have set $\delta_4 = 1$. It gives the simulated synchrotron source a relatively low luminosity compared to typical FR2 objects. It guarantees, however, that the relativistic electrons are passive. In fact, the integrated pressure in relativistic electrons ($E > mc^2$) is everywhere less than 0.1% of the thermal pressure. Given that, the synthetic observations can all be scaled simply in terms of δ, or in some cases, are independent of δ. If, for comparison, we assume from the same motivations that cosmic-ray ions have ~ 100 times more energy than electrons, those nonthermal ions could locally contribute up to ∼ 10% of the total pressure. With these assumptions our simulated source is out of equipartition between relativistic particles and magnetic fields, with the balance in favor of particles.

With the source distance assumed to be 100 Mpc the synchrotron flux over the source is about $330 \times \delta_4$ mJy. The 10 keV inverse-Compton flux is $2.4 \times 10^{-10} \times \delta_4$ Jy corresponding to a 10 keV luminosity νL_ν of $6.5 \times 10^{32} \times \delta_4$ W.

In the images one can clearly separate the jet, a terminal hot spot, a secondary hot spot and the diffuse lobe. The contrast in the radio image between the jet core and the outer lobe is higher than sometimes observed in real sources. There are two primary reasons why the jet is so prominent. First, the entire cosmic-ray electron population passes down the jet in this simulation, so the density of relativistic electrons is higher there than anyplace else. Since the magnetic field is stronger inside the jet than in much of the lobe volume, the synchrotron emissivity is relatively large there. In addition, the source in Figure 2 is quite young dynamically, so that a line of sight path length through the jet is roughly 20% of the path through the lobe. As this source aged that comparison would change significantly in favor of the lobe path length, of course.

While there is an overall correspondence between the radio and X-ray images, there are some significant differences. The jet/lobe and hot-spot/lobe contrasts are much more pronounced in the synchrotron map. Calculating the average surface brightness in the jets and in the lobes of the two maps, we find that this ratio is about three times higher in the radio map than in the X-ray map. While the radio hot spots do appear to have X-ray bright counterparts, there are also bright X-ray

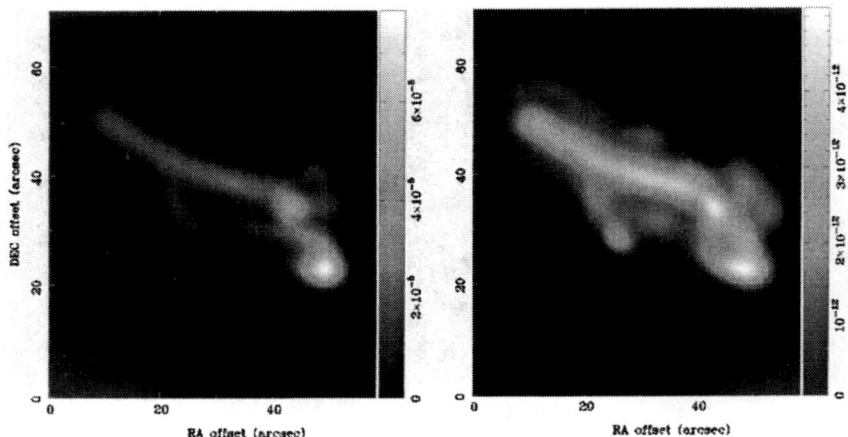

FIGURE 2. Synthetic surface brightness maps constructed as described in the text and analyzed with the MIRIAD package. Left: 1.4GHz synchrotron emission map. Right: 10keV inverse-Compton X-ray emission map. The linearly-scaled colorbar values represent Janskys per 3.5" beam.

regions without corresponding radio enhancements. The inverse-Compton emission arises from a particle population of fixed energy (~ 1 GeV at 10 keV), while the synchrotron emission at fixed frequency does not. Thus, the surface brightness of inverse-Compton emission from the cosmic microwave background simply reflects the column density of electrons in a fixed energy range, while the synchrotron is biased to regions where the fields are strongest, both because radiative power is greater there, but also because one sees emission from relatively more plentiful lower energy electrons. In cases where radiative losses are significant in the GeV range so that the spectra steepen, these differences would become even more dramatic.

Comparison of the two prominent hot spots reveals some interesting contrasts. The larger primary hot spot at the right edge of the radio lobe corresponds to an extremely complicated set of structures near the jet terminus. While it is nominally associated with the end of the jet flow, the jet terminal shock is very small and influences the spectrum of only a tiny portion of the electrons in the hot spot. In fact, much of the emergent jet flow has not passed through any strong shock before reaching the cocoon. Some of the electron population shows signs of significant shock acceleration, but the pattern is not simple. Over all, the influence of shock acceleration on the brightness and spectrum of the primary hot spot is quite small. The enhanced emission is mostly due to magnetic field amplification in the complex flows there rather than particle acceleration.

On the other hand, there is clear evidence for particle acceleration in the secondary hot spot further back in the lobe, which has a much different character. This hot spot has a distinctly flatter spectrum ($\alpha \approx 0.6$) than the jet ($\alpha = 0.7$). The surface brightness map would make it appear that the jet flows directly into this hot spot. However, this is an accident of projection. The high emissivity volume really is well outside the jet, and examination of the plasma streak lines reveals that this hot spot actually corresponds to flow downstream of a moderately strong shock formed in the cocoon "backflow", as discussed with two-dimensional simulations [3].

ACKNOWLEDGMENTS

The work by ILT and TWJ were supported in part by the NSF through grants INT95-11654 and AST96-19438, by NASA grant NAG5-5055 and by the University of Minnesota Supercomputing Institute. DR was supported in part by grant 1999-2-113-001-5 from the interdisciplinary Research Program of the KOSEF.

REFERENCES

1. Bridle, A. H. 1992, in "Testing the AGN paradigm", eds. S. Holt, S. G. Neff and C. M. Urry, (New York: American Institute of Physics), p. 386.
2. Clarke, D. A., Norman, M. L. and Burns, J. O. 1989, ApJ, 342, 700.
3. Jones, T. W., Ryu, D. and Engel, A. 1999, ApJ, 512, 105.

4. Ryu, D., Miniati, F., Jones, T. W. and Frank, A. 1998, ApJ, 509, 244.
5. Skilling, J. 1975, MNRAS, 172, 557.

The role of the outer boundary condition in accretion disk models

Feng Yuan[1,2], Qiuhe Peng[1], Jufu Lu[3] and Jianmin Wang[4]

[1] *Department of Astronomy, Nanjing University, Nanjing 210093, China*
[2] *Beijing Astrophysics Center, Beijing 100080, China*
[3] *Center for Astrophysics, University of Science & Technology of China, Hefei, 230026, China*
[4] *Laboratory of Cosmic Ray and High Energy Astrophysics,
Institute of High Energy Physics, Chinese Academy of Sciences, Beijing 100039, China*

Abstract. Taking optically thin accretion flows as an example, we investigate the effects of the outer boundary condition (OBC) on the dynamics and the emergent spectra of accretion flows. We find that OBC plays an important role. This is because the accretion equations describing the behavior of accretion flows are a set of *differential* equations, therefore, accretion is intrinsically an initial-value problem. The result means that we should seriously consider the initial physical state of the accretion flow such as its angular momentum and its temperature. An application example to Sgr A* is presented.

INTRODUCTION

It has long been assuming that the parameters describing the accretion flow include the accretion rate, the mass of the central black hole, the viscosity parameter, and the parameter describing the strength of the magnetic field in the accretion flow. Once these parameters are given, we can obtain almost all the information of the accretion flow including the dynamics and the emergent spectrum. However, from the view point of mathematics, the set of equations describing the accretion flow are nonlinear *differential* equations, therefore it is intrinsically an initial-value problem. Thus the outer boundary condition (OBC) possibly plays an important role.

On the other hand, the practical astrophysical environments are complicated therefore the physical states of the accreting gas at the outer boundary r_{out}, such as its temperature and angular momentum, are various. For example, in semi-detached binary system, where the critical Roche lobe is filled up and the accretion of matter takes place through the inner Lagrangian point, the angular momentum of the accreted gas should be high; while in detached binary system the accretion matter is stellar winds therefore their angular momenta are much lower [1]. In the nuclei of galaxies, where the supply of the accretion matter is unclear, the initial physical states of the accretion flows should be more complicated. The complexity of astrophysical environments makes it important to investigate the role of OBC in accretion disk model.

THE ROLE OF OBC IN OPTICALLY THIN ACCRETION FLOWS

Taking optically thin accretion flows as an example, we investigate the dynamics and emergent spectra, paying special attention to the effects of OBCs[2,3]. We adopt the standard procedure and assumptions widely assumed in the study of two-temperature ADAF[4]. Namely, we consider a steady axisymmetric accretion flow around a Schwarzschild black hole of mass M, the standard α-viscosity description is adopted, all the viscous dissipated energy is transferred to ions and the energy transfer from ions to electrons is provided solely by Coulomb collisions, a randomly oriented magnetic field whose strength is described by β is assumed to exist in the accretion flow. Under the above assumptions, the set of height-integrated equations

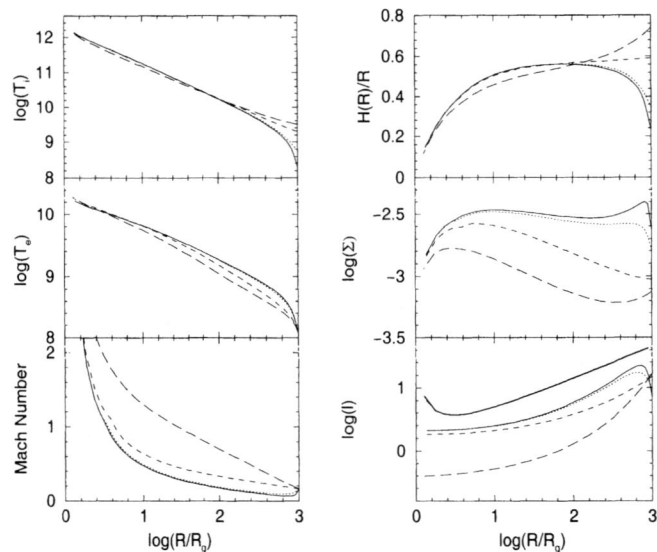

FIGURE 1. The structures of the accretion flow with different $T_{\text{out},i}$. The solid line (type I solution) is for $T_{\text{out},i} = 2 \times 10^8 K$, the dotted line (type I) for $T_{\text{out},i} = 6 \times 10^8 K$, the dashed line (type II) for $T_{\text{out},i} = 2 \times 10^9 K$ and the long-dashed line (type III) for $T_{\text{out},i} = 3.2 \times 10^9 K$. Other OBCs are $T_{\text{out},e} = 1.2 \times 10^8 K$ and $\lambda_{\text{out}} = 0.2$. The outer boundary is set at $r_{\text{out}} = 10^3 r_{\text{g}}$. Other parameters are $\alpha = 0.1, \beta = 0.9, M = 10^9 M_\odot$ and $\dot{M} = 10^{-4} \dot{M}_{\text{Edd}}$. The units of Σ and T are g cm^{-2} and K

describing the behavior of accretion flows read as follows.

$$-4\pi r H \rho v = \dot{M}, \quad \text{with} \quad H = c_s/\Omega_k \equiv \sqrt{p/\rho}/\Omega_k, \tag{1}$$

$$v\frac{dv}{dr} = -\Omega_k^2 r + \Omega^2 r - \frac{1}{\rho}\frac{dp}{dr}, \tag{2}$$

$$v(\Omega r^2 - j) = \alpha r \frac{p}{\rho}, \tag{3}$$

$$\rho v \left(\frac{d\varepsilon_i}{dr} + p_i \frac{d}{dr}\left(\frac{1}{\rho}\right) \right) = q^+ - q_{ie} = -\alpha p r \frac{d\Omega}{dr} - q_{ie}, \tag{4}$$

$$\rho v \left(\frac{d\varepsilon_e}{dr} + p_e \frac{d}{dr}\left(\frac{1}{\rho}\right) \right) = q_{ie} - q^-. \tag{5}$$

Synchrotron and bremsstrahlung emissions and their Comptonization are considered. We concentrated on the role of OBC by setting the same "general parameters" such as accretion rate, viscosity parameter and black hole mass while adopting different OBCs. We adopted the temperature T_{out} and the ratio of the radial velocity

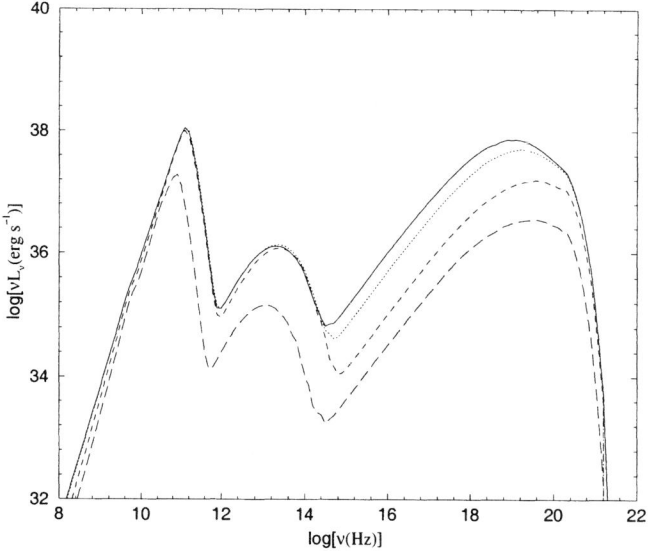

FIGURE 2. The corresponding spectra of the solutions shown in Figure 1.

to the local sound speed λ_{out} (or, equivalently, the angular velocity Ω_{out}) at a certain outer boundary r_{out} as the outer boundary conditions. We numerically solve the above set of radiation hydrodynamics equations. The global solutions must satisfy simultaneously the no-torque condition at the horizon, the sonic point condition at a sonic radius r_s and three outer boundary conditions given at a certain outer boundary r_{out}. The numerical method we adopted is the same as in [4] while the calculation of the spectrum is as in [5].

We found that both the dynamical structures and the emergent spectrum differ greatly under different OBCs. As an illustration, Figures 1 & 2 show the dynamical structures and the spectra of some solutions with different $T_{\text{out},i}$. Considering that they possess the same "general" parameters, the discrepancy among the spectra completely caused by the difference of OBC is impressive. For the effect of $T_{\text{out},e}$, see [3]. In terms of the topological structure and the profile of the angular momentum, three types of solutions are found. When T_{out} is relatively low, the solution is of type I. When T_{out} is relatively high and the angular velocity Ω_{out} is higher than a critical value Ω_{crit}, the solution is of type II. Both types I and II possess small sonic radii, but their topological structures and angular momentum profiles are different. When T_{out} is high but the angular velocity is lower than Ω_{crit}, the solution becomes of type III, characterized by a much larger sonic radius. The value of Ω_{crit} depends on the parameter of the accretion flow especially α. Figure 3 shows the transition

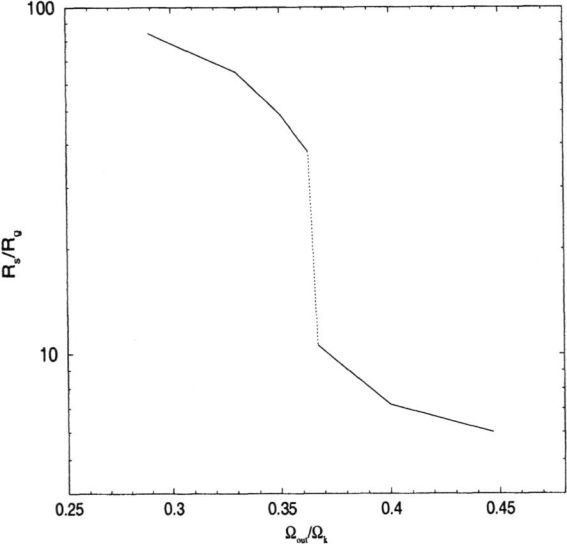

FIGURE 3. The variation of the value of the sonic radii with the angular velocity at the outer boundary. A transition is clearly shown. The OBCs are $T_{\text{out},i} = 2 \times 10^9 K$, $T_{\text{out},e} = 1.2 \times 10^9 K$ and $r_{\text{out}} = 10^3 r_g$. Other parameters are the same as Figure 1.

of the sonic radius with the angular momentum. Similar transition has been found previously in the context of adiabatic (inviscid) accretion flow[6,7]. In that case, they found that when the specific angular momentum of the flow decreased across a critical value, a transition from a disk-like accretion pattern (with small sonic radii) to a Bondi-like one (with large sonic radii) would happen. Here in this paper we find that this transition still exist when the flow becomes viscous, confirming the previous prediction[6].

AN ILLUSTRATIVE APPLICATION TO SGR A*

As an illustrative example, we apply the above results to the compact radio source Sgr A* located at the center of our Galaxy. Advection-dominated accretion flow (ADAF) model has been turned out to be of great success to explain its low luminosity and spectrum [8,9]. However, there exists a discrepancy between the mass accretion rate favored by ADAF models in the literature and that favored by the three dimensional hydrodynamical simulation, with the former ($\sim 6.8 \times 10^{-5} \dot{M}_{\rm Edd}$, in the most up to date result[10]) being 10-20 times

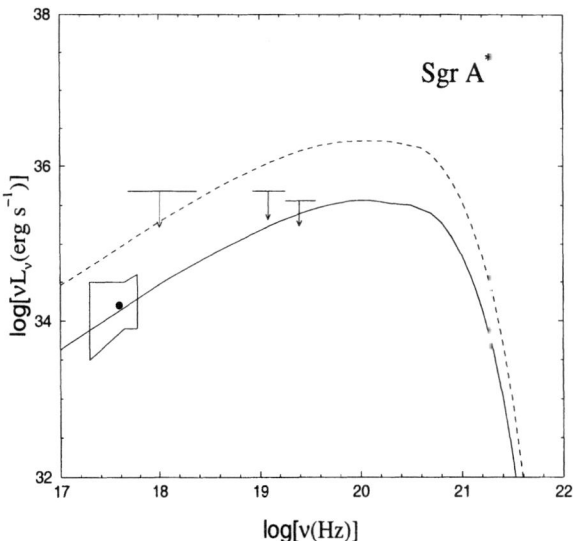

FIGURE 4. The X-ray spectrum of Sgr A*. The observational data are compiled by [9]. The spectra represented by the solid and the dashed lines are produced by the accretion flows with the same accretion rate $\dot{M} = 4 \times 10^{-4} \dot{M}_{\rm Edd}$ but different angular momentum at $r_{\rm out}$, $\Omega_{\rm out} = 0.15\Omega_{\rm K}$ for the solid line and $\Omega_{\rm out} = 0.46\Omega_{\rm K}$ for the dashed line. Due to the difference of the angular momentum of the flow at the outer boundary, the X-ray flux differs by a factor ~ 8. Adopted from [3].

smaller than the latter ($\sim 9 \times 10^{-4} \dot{M}_{\text{Edd}}$, see [11]. By seriously considering the outer boundary condition of the accretion flow, we find that due to the low specific angular momentum of the accretion gas[11], the accretion in Sgr A* should belong to type III which possesses a very large sonic radius. The surface density is very small and the bremsstrahlung emission which is responsible to the X-ray flux will correspondingly decrease compared to types I or II. Therefore, this accretion pattern can significantly reduce the discrepancy between the mass accretion rate, as Figure 4 shows (see [3] for details. for the modeling of Sgr A* from radio to hard x-ray, see [12]).

DISCUSSION

The present study is concentrated on the low-$\dot{M}$ case where the differential terms in the equation such as the energy advection play an important role therefore the effect of OBC are most obvious. How about the role of OBC when the flows become optically thick? In this case, the electron and the ion possess the identical temperature due to the strong couple between them and the local viscous dissipation and radiation loss terms in the energy balance play an important role. As a result, the temperature profile is mainly determined *locally* rather than *globally* as in the case of optically thin flows. Thus, the discrepancy of the temperature caused by OBC will lessen rapidly with the decreasing radii from the outer boundary. However, from our calculation to the one-temperature accretion flow whose temperature is principally determined locally[2], we still found OBC-dependent behavior in, e.g., the angular momentum and the Mach number profiles which are in principle determined by the *momentum* rather than the *energy* equations. We therefore predict that the optically thick accretion flow should still present such OBC-dependent behavior. When the angular momentum of the accretion flow is less than a certain critical value, the accretion pattern should become of "type III" (Bondi-like). Although these conjectures need the confirmation of detailed calculation, we note that a recent numerical simulation [13] seems to support this point.

Why the role of OBC in accretion disk models has been long neglected? In the standard thin disk model, all the differential terms in the equations are neglected and the differential equations are reduced into an algebraic one which don't entail any boundary conditions at all. In the later works on the global solutions for slim disks [14,15] and optically thin advection-dominated accretion flows[16,17], some authors did investigate the role of OBC, but failed to find its importance. The main reason is that for optically thick accretion flows (slim disk) or *one-temperature* optically thin accretion flow, the local viscous dissipation plays an important role in the energy equation, so the effect of OBC lessen rapidly away from the outer boundary. In addition, the angular momentum in their outer boundary condition was always somewhat large. This might be the reason why they didn't find the solutions with very large sonic radii.

At last, we note that the effect of OBCs depends on the location of the outer boundary r_{out}. When r_{out} is large, say $\gg 10^4 r_g$, we find the feasible range of T_{out} within which we can find a solution lessens greatly. But the effect of Ω_{out} always exists no matter how large r_{out} is. When r_{out} is small, the effect of OBCs is most obvious. This result hints us that we should pay special attention on OBC when we model black hole X-ray binaries and Seyfert 1 galaxies where a thin disk-ADAF transition scenario is usually adopted[18] since in this case r_{out} is in general very small, $\sim 100 r_g$.

REFERENCES

1. Illarionov, A.F., & Sunyaev, R.A., 1975, A&A, 39, 185
2. Yuan, F. 1999, ApJ, 521, L55
3. Yuan, F., Peng, Q. H., Lu, J.F., & Wang, J. M. 2000, ApJ, 537, 236
4. Nakamura, K. E., Kusunose, M., Matsumoto, R., & Kato, S. 1997, PASJ, 49, 503
5. Manmoto, T., Mineshige, S., Kusunose, M. 1997, ApJ, 489, 791
6. Abramowicz, M., A., Zurek, W.H., 1981, ApJ, 246, 31
7. Lu, J.F., & Abramowicz, M.A. 1988, Acta Ap. Sin., 8, 1
8. Narayan, R., Yi, I., Mahadevan, R. 1995, Nature, 374, 623
9. Narayan, R., Mahadevan, R., Grindlay, J.E., Popham, R. & Gammie, C., 1998, ApJ, 492, 554
10. Quataert, E., & Narayan, R., 1999, ApJ, 520, 298
11. Coker, R., & Melia, F., 1997, ApJ, 488, L149
12. Yuan, F. 2000, submitted to MNRAS (astro-ph/0004197)
13. Igumenshchev, I.V., Illarionov, A.F. & Abramowicz, M.A. 1999, ApJ, 517, L55
14. Abramowicz, M.A., Czerny, B., Lasota, J.P., & Szuszkiewicz, E., 1988, ApJ, 332, 646
15. Chen, X. & Taam, R. 1993, ApJ, 412, 254
16. Narayan, R., Kato, S. & Honma, F. 1997, ApJ, 476, 49
17. Chen, X., Abramowicz, M. A., & Lasota, J.-P. 1997, ApJ, 476, 61
18. Esin, A.A., McClintock, J. E., & Narayan, R. 1997, ApJ, 489, 865

Some Consequences of Magnetic Fields in High Energy Sources

Eric G. Blackman

Department of Physics and Astronomy, University of Rochester, Rochester, NY 14627

Abstract. Magnetic fields likely play a fundamental intermediary role between gravity and radiation in many astrophysical rotators. They can, among other things, 1) induce and be amplified by turbulence, 2) energize coronae, 3) launch and collimate outflows in "spring" or "fling" mechanisms The first is widely recognized to be important for angular momentum transport, but can also produce intrinsic variability and vorticity growth. The second leads to the production of high energy flares, and also facilitates a test of general relativity from AGN observations. The third can operate from rotators with a large scale fields, though the origin of the requisite large scale fields is somewhat unresolved. I discuss these three points in more detail below, emphasizing some open questions.

TURBULENCE, VARIABILITY & VORTICITY IN ACCRETION DISKS

Accretion disks are a widely accepted paradigm to explain a variety of spectral features in sources such as active galactic nuclei (AGN), X-ray binaries, cataclysmic variables (CVs), dwarf novae and protostellar [1,2,3]. As gas orbits a central massive source, internal dissipation drains the orbital energy, allowing material to move in and angular momentum out. Some fraction of the dissipated energy accounts for the observed luminosity. Micro-physical viscosities are often too inefficient so an enhanced transport mechanism, likely involving turbulence, essential. The observational evidence for an enhanced/turbulent viscosity is least direct in the case of AGN disks, though it is natural to expect that the latter would be subject to the same turbulent viscosity generating mechanisms: As astrophysical disks are likely seeded with a magnetic field, the "Balbus-Hawley" or magneto-rotational instability (MRI) [3,4] ensues. This produces self-sustaining turbulence for flows with a radially decreasing angular speed that transports angular momentum in rotationally supported disks. Turbulence driven by the MRI transports angular momentum outward, unlike convection [3,5,6]. Rossby wave vortices may also transport angular momentum outward [7].

For low enough accretion rates, the dissipated energy may be primarily advected by hot protons in advection dominated accretion flows (ADAFs) rather than radiated, possibly explaining why some sources have engines that are surprisingly quiescent [9,10,11]. For optically thin, geometrically thick ADAFs, the disks may be physically similar to the coronae above the thin disks of Seyferts [12]. The thin disk+corona is something like a hamburger sandwich with the optically thin (geometrically thick) corona being the bun and the optically thick (geometrically thin) disk is the meat. The ADAF is a sandwich with no "meat."

Alternatives to simple ADAFs suggest that quiescent luminosities might instead result from reducing the accretion rates into the inner regions [13,14,15]. Such models include advection dominated inflow outflow solutions (ADIOS) [13] and convection dominated accretion flows (CDAFs) ([14]; Narayan, these proc). These require a less than maximal viscosity [16].

Mean Field Disk Theory and Variability

In general, the ubiquity of turbulence in disks has important implications for what the accretion disk equations mean. Analytic disk equations are mean field equations. While non-linear instabilities in thin disks have been extensively simulated locally [17,18] a traditionally useful approach to disk models has been to swipe the details of the stress tensor into a turbulent viscosity of the form [19]

$$\nu_{tb} = \alpha c_s H \simeq v_{tb} l_{tb}, \tag{1}$$

where H is the disk height, c_s is the sound speed, l_{tb} is the dominant turbulent scale, v_{tb} is the eddy speed at that scale, and $\alpha_{ss} < 1$ is a constant. Use of this formalism requires a mean field theory, not a replacement of the molecular viscosity with the turbulent viscosity. Assumptions of azimuthal symmetry and steady radial inflow [1,11] require turbulent motions to be averaged over the time and/or spatial scales on which mean quantities vary.

The required mean field approach is complementary to that employed in mean field magnetic dynamo theory, where the field is split into mean and fluctuating components, $\mathbf{B} = \overline{\mathbf{B}} + \mathbf{B}'$, and the induction equation is averaged and solved. For the accretion disk case, the momentum equation must be similarly split, and the evolution equation for the mean velocity field derived. The result is that ν_{tb} represents a correlation of turbulent fluctuations, that is $\nu_{tb} = \langle \mathbf{v}'(t) \cdot \int \mathbf{v}'(t') dt' \rangle$ [20,21,22].

Mean field theory has a limited precision, and thus predicts variability. Let the mean represent a global average over azimuth and half thickness, without averaging over the radius. Since all mean velocities are scalings of the Keplerian velocity, each radius is in principle "labelled" by its Keplerian speed. However, the averages over fluctuations produce an uncertainty in this labelling. Since the luminosity at a given frequency depends on the radius of emission for accretion disk models, we can relate the uncertainty in the observed luminosity to the uncertainty in the radius. That is,

$$\Delta L_\nu/L_\nu \simeq |R\partial_R(\mathrm{Ln}L_\nu)|\Delta R/R = \Psi \Delta R/R \tag{2}$$

where Ψ can depend on temperature but ~ 1 for a range of frequencies for thin and thick disk models [23]. Let us estimate $\Delta R/R$. The "error" associated with a single turbulent fluctuation of scale l_{tb} is reduced by $N^{1/2}$ for each averaged dimension, where N is the number of eddy spatial scales averaged over in that dimension. As observational data are taken in a time averaged sense, there is also a reduction that depends on observation time, t_{obs}. The result is $\Delta R \simeq l_{tb}/[N_\phi N_z(1+t_{obs}/t_{tb})]^{1/2}$, where N_ϕ and N_z are the number of dominant eddy scales in the ϕ and z directions, and t_{tb} is the dominant correlation time.

To make further progress, note that for fully developed MHD turbulence, near equipartition between kinetic and magnetic energy is generically reached, and is $v_A \sim 2^{1/2} v_{tb}$ [3,17,18] in the case of accretion disks. The factor of $\sqrt{2}$ comes from shear that adds a bit more amplification to the field. The MRI instability onset time scale is of order the dominant eddy turnover or cascade time and is $t_{tb} \sim l_{tb}/v_{tb} \sim \Omega^{-1}$, where Ω is the rotation speed. Using these and (1) we have

$$v_A^2/2\Omega = \alpha_{ss} c_s H. \tag{3}$$

Using $\Omega H \sim c_s$ for vertical pressure support, we then have

$$\alpha_{ss} \simeq v_A^2/2c_s^2 \simeq (l_{tb}/H)^2. \tag{4}$$

The next step is to note that $N_\phi = 2\pi R/(2l_{tb}) = (\pi R/H \alpha_{ss}^{1/2})$, where the extra 2 on the bottom comes from eddy elongation in the ϕ direction from shear, and $N_z = H/2l_{tb} = \alpha_{ss}^{-1/2}$, where the 2 is from the 1/2 thickness.

Collecting all of the above into (2) gives

$$\Delta L_\nu/L_\nu \sim \Psi \alpha_{ss}(H/R)^{3/2}(1+\Omega t_{obs})^{-1/2}. \tag{5}$$

For thick disk ADAF models, $H/R \sim 1$ and $\alpha_{ss} \sim 1$ [8], large variability around the predicted luminosities can be expected unless $t_{obs} >> \Omega^{-1}$. When variability is not seen, or a systematic deviation from the theory is seen in a sample of observations at widely separated times or in different objects, the simplest ADAF type model may not be capturing the physics. Such systematic deviations seem to be evident in some of the large ellipticals [24] Other such quiescent accretor variations such as CDAFs (Narayan these proc.) or ADIOS type models [13] which involve outflows may be more appropriate there.

Vorticity

There are other effects of mean field theory in addition to the variability. If one considers the vertical disk averaging to be taken only over one hemisphere, then in addition to the scalar Shakura-Sunyaev turbulent viscosity transport term used in simple analytic accretion disk modeling, a pseudoscalar transport term also arises [22]. This term is analogous to that which appears in magnetic dynamo theory, and can lead to vorticity growth [21,25].

In the same way that the mean field magnetic dynamo characterizes an inverse cascade of magnetic helicity [26], the vorticity dynamo highlights some growth of vorticity on larger scales than the input turbulence. Enstrophy exhibits an inverse cascade in 2-D turbulence [27]. The growth of vorticity in primarily 2-D rotating fluids has been seen in nature (e.g. Jupiter [28]) as well as in simulation [29] and experiment [30]. Statistical mechanics approaches have modeled this [31]. Vorticity growth in sheared thin accretion disks has been studied less than th associated vortex evolution [32,33,34].

For an accretion disk in which mean quantities depend only on radius, Ref. [22] showed that

$$d_t\bar{\omega} = \nabla \times \alpha_0 \bar{\omega} - \nabla \times (\nu_{tb} \nabla \times \bar{\omega}) \qquad (6)$$

in the frame comoving with the cartesian velocity tangent to the mean rotation, where the coefficients are

$$\alpha_0 = (\tau_c/3)(\langle \omega'^{(0)} \cdot v'^{(0)} \rangle)$$
$$\nu_{tb} = (\tau_c/3)\langle v'^{(0)} \cdot v'^{(0)} + b'^{(0)} \cdot b'^{(0)} \rangle \qquad (7)$$

and $b'^{(0)} \equiv B'^{(0)}/4\pi\bar{\rho}$, with $\bar{\rho}$ as the mean density. (This equation presumes that 1st order cross correlation terms vanish, that is $\langle b'^{(0)} \cdot v'^{(0)} \rangle = \langle \omega'^{(0)} \cdot b'^{(0)} \rangle = \langle \omega'^{(0)} \cdot \nabla \times b'^{(0)} \rangle = 0$.) Note that ν_{tb} is not the only transport term. There is also the pseudoscalar α_0 term as in the magnetic field case. It is this pseudoscalar term which can lead to vorticity growth. Interestingly, the pseudoscalar for the vorticity, unlike for that of the mean magnetic field, has a kinetic helicity term *without* a current helicty term [22]. The α_0 can be parameterized as $\alpha_0 = q\alpha_{ss}c_s$. One necessary condition for growth turns out to be $q > H_r^* R$.

The simplest growth solutions [22] show a dominant growth scale $\sim H$, leading to intermediate scale vortices that should survive at least a vertical diffusion time, that is, $\gtrsim$ few $\times 1/\alpha_{ss}$ orbits. For $\alpha_{ss} \sim 0.01$, the resulting anti-cyclonic vortices may allow dust trapping, catalyzing planet formation when applied to star+planet system forming disks [31,33,35].

Note that this simplified model of intermediate scale vorticity growth cannot tell how many vortices grow, or where in height or azimuth these vortices are, only that there are growing solutions. This is because here the variables are averaged to depend only on radius. Note also that if the vertical averaging is taken over the full scale height, then the α_0 coefficient should vanish because the psuedoscalar reverses sign across the mid-plane. Then vorticity growth could not be identified in this over simplified formalism.

BUOYANCY, CORONAE & AGN IRON LINE PROFILES

Coronae

MHD turbulence likely involves spatial intermittency of the magnetic field [36], even in accretion disks. While the dynamics of intermittency is not fully understood, the random component of the field would be preferentially amplified at regions of strongest shear. Since small scale shear varies spatially and spectrally in a turbulent medium, the field strength would also be expected to to vary in correlation. The strongest magnetic field regions might form a kind of dynamic sponge, intermixed with "void" regions that have much weaker fields. The extreme situation, in which the magnetic field occupies a distinct volume from the thermal material, could maintain dynamical equilibrium when the ratio of the the average particle to magnetic pressure in the disk ($\equiv \beta_p$) is large. To see this, consider the field to reside in magnetically dominated flux tubes. The pressure inside balances the external particle pressure. However, if $\beta_p \gg 1$, the tubes would occupy only a small volume filling fraction ($\sim 1/\beta_p$) [37]. Since the distance between tubes would be large, intersections with other tubes would be too infrequent to significantly load particles into the tubes. In contrast, if the magnetic volume filling fraction were large, frequent reconnection events over a large fraction of the tubes' longitudinal cross section would more easily mass load the tubes and lead to one phase medium. The amount of intermittency remains to be understood. Understanding this intermittency is important because extremely evacuated tubes rise at speeds of order their internal Alfvén speeds and form coronae. Coronal magnetic dissipation is as important a problem for disks [38] as it is for the Sun [39]

While coronae have been long thought to form above disks [38,40] analytic work has not fully incorporated the turbulence and/or the origin of the initial large field. The usually invoked Parker instability favors wave modes which can be shredded by the Balbus-Hawley instability on time scales approximately equal to rise times. Simulations of turbulent disks show that coronae do form in turbulent disks [41] but the dominant mechanism, and its relationship to the process of formation in the solar corona is not fully sorted out.

If the disk were not subject to MRIs and turbulence, the induction equation tells us that the disk field would grow linearly from shear. It would saturate at a value that is higher than the saturation value of the MRI. This is because the shear can amplify the field over a much longer time in a laminar disk; the field remains coherent longer. Field strengths of order the thermal pressure could incur before buoyancy (in this case by the Parker instability) drained the field into the corona. For MRI driven disks, the shear only operates on a single magnetic field filament for a correlation time (one rotation) after which that filament loses its identity. The field energy saturates by a factor of α_{ss} lower for the turbulent case. Also for the the laminar case, the corona would be more intermittent. This is because the buoyancy and the dissipation would likely get rid of the field faster than it builds up. For a laminar disk, the field growth would be linear in time whereas for the MRI disk, the growth is exponential. A laminar disk + corona should thus exhibit longer variability periods than a turbulent disk, but with larger amplitudes. The observational evidence in the case of AGN disks/corona seems to suggest a turbulent disk. While X-ray variabilities of factors of several are observed [42], factors of orders of magnitudes are not, suggesting a steady background level of coronal dissipation.

AGN Iron Lines and Engine Geometries

Seyfert galaxy X-ray spectra, are best modeled as a combination of direct emission from a hot corona and reprocessed emission from a cold, optically thick accretion disk [43]. The direct component results from inverse Compton scattering of thermal disk UV photons. The reprocessed component incurs as photons are scattered back onto the disk.

If it weren't for the active coronae, we would lose a probe of strong gravity [44]. The reprocessed coronal emission includes the broad iron $K\alpha$ fluorescence line of rest energy 6.4 keV, which carries information about the geometry and dynamics of the reprocessing material near the black hole [44]. ASCA has observed iron lines in $\sim$ 18 Seyfert Is [45]. The best studied iron line is that of MCG-6-30-15 [46,47]. In addition to the geometry, the iron line profiles are sensitive to the disk illumination law, the disk inclination, and the inner and outer radii [44]. Most work on reprocessing in AGN has invoked thin flat disks with axially symmetric illumination laws representing a "point" X-ray source.

Some AGN line profiles like MCG-6-30-15 are consistent with flat disks [46,47] but the current data may be not precise enough to rule in or out non-axisymmetric engines in other cases. Such engines are also worth considering in order to provide robust comparisons to flat disk models. Ref [48] considered finite disk thicknesses and Ref [49] considered concave disks. Warped disks have also been considered around Schwarzchild black holes [50]. There are plausible theoretical reasons for disk warping including radiation driven warping [51] and tidal warping [52]. On the observational side, water maser emission of NGC 4258 at 0.1pc from the central engine (on larger scales than the inner accretion disk) traces a disk warp [53]. There may also be indirect evidence on these larger scales from observations of Seyfert Is which suggest that the broad line regions are not coplanar with the inner disk [54]. Dusty tori of Seyfert unification paradigms might also involve warped disks [55].

Warped disk studies reveal line features that are impossible for unobscured flat thin disks. First, shadowing of the source by the disk and shadowing of reprocessed emission by the disk blocks regions of the disk from contributing to the iron line. Sharper red than blue cutoffs or very soft blue cutoffs can also arise. The latter characteristic is seen in some profiles of Ref [45]. The sharp red cutoffs result for large inclination angles, which is consistent with some Seyfert IIs [56]. Second, non-axisymmetry of the disk means that line profiles can show time variability if the warp precesses around the disk. Third, there can be sharper peaks near the rest frequency compared to a flat disk since concavity can offer more solid angle covering fraction. Fourth, apparent misalignment of central disk plane with the obscuring torus can be accounted for.

For a concave disk, sharper peaks near the rest frequency can be accompanied by a total reprocessed emission fraction that is larger than 1/2 [49], where 1/2 is the maximum limit for a point source above a flat disk. This may play a role in ultra-soft narrow-line Seyferts (Brandt, private communication 1999), though there are other ways to achieve this enhanced reprocessing fraction.

Line profiles from a distribution of dense clouds in an optically thin, geometrically thick disk may apply to ADAFs. Dense clouds formed from thermal instability can survive long enough to produce reprocessing signatures in otherwise optically thin flows [57]. This needs more investigation.

SPRINGS, FLINGS & LARGE SCALE FIELDS

Many large scale jets and winds in astrophysics including those of young stellar objects, microquasars, gamma-ray bursts, and AGN jets may be magnetically driven. Even supernovae may also involve magnetically driven bipolar outflows (Wheeler, these proc.). How MHD jets work and where the requisite large scale magnetic fields come from are integrated questions, but are usually studied independently.

The large amount of work on MHD jet launching and collimation will not be reviewed here (see [58,59]). However, note that the launching mechanisms could be divided into "spring" [60] and "fling" [61] mechanisms. In the former class, the jet is launched initially by toroidal magnetic field pressure. Imagine for example, a dipole magnetic field anchored in a star which incurs rapid differential rotation (such as the collapse from a white dwarf to a young neutron star as invoked in [62] for gamma-ray bursts) or in a supernova core collapse (Wheeler, Meier personal comm.). As the differential rotation winds up the field, the toroidal field pressure grows quadratically in time. When the pressure reaches some critical value, the field will act something like a coiled spring and can drive a strong torsional Alfvén wave containing directed Poynting flux outward. This could in principle power a jet. Related mechanisms have been discussed for disks [60,63]. In this regard, note that in the case of AGN, it is not clear if the jet emission we see represents dissipation from instabilities at the edge of the jet, re-acceleration inside the jet along the bulk flow, or just emission from a very small number of particles carrying the currents [63] which support the magnetic fields.

In the "fling" launch mechanism [61], the initial launch is driven by centrifugal force. The rigid field lines significantly weaken the effective gravitational potential when sufficiently inclined to the normal, allowing material fling out along poloidal field lines. Subsequently, before reaching the Alfvén surface, the driving does become magnetically driven as in the "spring" mechanism. Since simulation of such launching treats the base of the jet (e.g. its initial launch point in the corona) as a boundary condition, simulators often load the field lines with a little mass flow to get the process started [64]. The initial launching is different for spring and fling, but the ultimate collimation mechanisms could be the same, e.g. hoop stresses. The extent of the collimation appears to be sensitive to the boundary conditions of the outflow however [65].

MHD jet luminosity is fueled by the rotational energy. In systems which have both central compact rotators and disks, outflows could emanate from both e.g. [66]. In black hole systems, the relative contribution to the jet power from regions within and outside of the last stable orbit of the disk has been addressed [67]. Even if a Blandford-Znajek type mechanism is operating from the hole, a jet from the disk may in fact always dominate. However, the black hole spin can still influence the jet, since it determines the inner edge of the disk, and is ultimately important for understanding the magnetospheres [68].

Where do the required magnetic fields [58] come from? The first possibility is that they are accreted. But this may not work for a turbulent accretion disk. Consider a turbulent disk threaded by a large scale vertical magnetic field. The field is subject to turbulent diffusion and may incur a net diffusion outward [69] (with some dependence on turbulent Prandtl number.) The role of reconnection may not yet be fully appreciated in this process. Without reconnection, the mean field is indeed subject to turbulent diffusion, but it cannot ultimately separate from the gas which has a systematic inward motion. In the absence of a topology change that can release field lines from the initial material they thread, the field would accrete on the diffusion time scale.

If the fields cannot be accreted then they would have to have been threading the central object before the disk formed, or be generated by a dynamo in the central object (if not a black hole) or disk. A traditional approach to the amplification of fields on scales larger than the scale of the input turbulence is the mean field dynamo [70] mentioned earlier. Only modes which have an initial seed field can be amplified. For an accretion disk dynamo, the limit of field energy density is the turbulent energy density. This limit is α_{ss} times the thermal energy density, which suggests that "spring" mechanisms are less likely from dynamo produced fields than "fling" mechanisms in turbulent disks.

It is important to distinguish between "large" scale and "mean" fields. Standard mean field theory is degenerate with respect to the topology of the field on scales smaller than the scale of the mean. Disconnected loops can have the same mean as a connected winding field line. In the formalism of mean field dynamo theory, reconnection is therefore not strictly required (though it is likely happening anyway). This is not commonly recognized. Said another way, neither the mean field nor the fluctuating component of the field are the topologically physical field. To generate jets, common wisdom holds that the mean fields actually do have to correspond to the topologically physical field. Perhaps this need not be the case if the Poynting flux driving the jet is an average quantity: $\langle \mathbf{E} \times \mathbf{B} \rangle = \langle \mathbf{e} \times \mathbf{b} \rangle + \overline{\mathbf{E}} \times \overline{\mathbf{B}}$, where the first term on the right, due to only fluctuating quantities, is usually ignored in this context (but see [71] where the collimation is non-magnetic.) If we do ignore fluctuating components, then to magnetically launch and collimate jets by the "fling" mechanism, the mean fields would need to be topologically large scale. One plausible way this could arise in a disk corona is if flux loops make their way to the surface and subsequent reconnection events inverse cascade smaller loops to larger loops [40].

The role of boundary conditions is particularly important for mean field dynamos. First, generating a net flux of the mean field in a quadrupole mode inside an object is accomplished by diffusion of the reverse flux through the boundary. Fast cycles (e.g solar cycle) of a dipole field also require boundary diffusion to change the flux inside. Incompressible simulations which employ periodic boundary conditions over the scale of the mean cannot see mean field growth because the induction equation then constrains the mean field to be time independent.

In addition, ref. [72] showed that conservation of magnetic helicity means that the growth of large scale field with one sign of magnetic helicity inside the system requires helicity of the opposite sign to diffuse out the boundary. Some results showing strong dynamo coefficient α quenching [73] may therefore highligh just an effect from the assumed boundary conditions rather than dynamical suppression. In general, to properly test large scale field formation in a turbulent disk one must really utilize a global study, with significant scale separation and diffusive boundary conditions.

Closing Comment

There is much to be learned on all of the above subjects by looking at the sun [74], as others would also advocate [58]. Note that dynamos in the Sun may operate differently than in disks. For the Sun, strong shear amplification may take place below the actual turbulent zone, whereas in disks, the turbulent zone and the shear are the same region. However the sun has a large scale wind and with an active corona [74], consistent with MHD outflows along open field lines and x-ray activity resulting from dissipation of closed field lines. We should expect the same for a wide variety of turbulent astrophysical rotators.

REFERENCES

1. Pringle J.E., ARA&A, **19**, 137 (1981)
2. Papaloizou, J.C.B. & Lin, D.N.C., ARAA, **33**, 505 (1995); Rees M.J. ARA&A, 22, 471 (1984)
3. Balbus, S.A. & Hawley, J.F., 1998, Rev. Mod. Phys, **70**, 1 (1998)
4. Balbus S.A. & Hawley J.F., ApJ, **376** 214 (1991)
5. Igumenshchev I.V.; Abramowicz, M.A. & Narayan R, ApJ **537**, L27 (2000)

6. Narayan R., Igumenshchev I.V., Abramowicz, M.A. ApJ in press, astro-ph/9912449 (2000)
7. Li H., Finn, J.M., Lovelace R.V.E., and Colgate S.A., ApJ **533** 1033 (1999); Lovelace R.V.E. et al., ApJ **513** 805 (1999)
8. Narayan R. & Mahadevan R. & Quataert E., in *Theory of Black Hole Accretion Disks* ed. M.A. Abramowicz, G. Bjornsson, J.E. Pringle Cambridge Univ Press, Cambridge, 1998, p148.
9. Ichimaru, S., Ap J, **214** 840 (1977)
10. Rees M.J., in *The Galactic Center* ed. G.R. Riegler & R.D. Blandford AIP, New York, 1980, p. 106; Rees, M.J. et al, Nature **295**, 17 (1982)
11. Narayan R., & Yi, I., **428** L13 (1994); Narayan R., & Yi, I., **452** 710 (1995)
12. Di Matteo, T., Blackman E.G. & Fabian, A.C., MNRAS **291**, L23 (1997)
13. Blandford R.D. & Begelman M.C., MNRAS **303**, L1. (1999)
14. Quatert E. & Gruzinov A., in press ApJ, astro-ph/9912440 (2000).
15. Gruzinov A., sub to ApJ, astro-ph/9809265 (1998).
16. Igumenshchev I.V. & Abramowicz M.A. MNRAS **303**, 309 (1999)
17. Brandenburg, A. et al. ApJ **446** 741. (1995)
18. Stone J.M. et al., ApJ **463** 656 (1995)
19. Shakura N.I. & Sunyaev R.A., A.&A., **24**, 337 (1973)
20. Balbus S.A., Gammie C.F., Hawley J.F., MNRAS, **271**, 197 (1994)
21. Blackman, E.G. & Chou T.C., ApJ, **489**, L95 (1997)
22. Blackman E.G., sub. to MNRAS, astro-ph/0006241 (2000)
23. Blackman E.G., MNRAS **299** L48.
24. Di Matteo T. et al., MNRAS **305** 492 (1999)
25. Moiseev S.S.et al. Sov. Phys. JETP **58** 1149 (1983); Frisch U., Zhe Z.S., Sulem P.L. Physcia D, **28** 382 (1987); Khomenko G.A., Moiseev S.S., Tur A.V., JFM, **225** 355 (1991); Kitchatinov, L. L., Rüdiger, G., & Khomenko, G., A.& A., **287**, 320 (1994); Kitchatinov, L. L., Rüdiger, G., & Kuker, M., A.& A., **292**, 125 (1994)
26. Pouquet, A., Frisch, U., & Leorat, J., JFM **77**, 321 (1976)
27. Kraichnan R.H. & Montgomery D., Rep. Prog. Phys., **43** 547 (1980)
28. Ingersoll A.P., Science 248, 308 (1990); Marcus P.S., ARAA, **31** 523 (1993)
29. Marcus P.S. JFM, 215 393 (1990); McWilliams, J.C., Phys Fl, **2** 547 (1990)
30. Sommeria J, Meyers S.D., Swinney H.L., Nat. **331** 689 (1988)
31. Chavanis P.H. & Sommeria J, JFM **356**, 259 (1998)
32. Adams F. & Watkins R., ApJ, **451** 314 (1995)
33. Godon P. & Livio M., ApJ **523** 350 (1999)
34. Chavanis P.H., A&A **356** 1089 (2000)
35. Barge P. & Sommeria J., A&A, **295** L1 (1995); Tanga P. et al. Icarus, **121** 158 (1996); Hodgson L.S. & Brandenburg A., A& A, **330** 1169 (1998); Bracco A. et al. Phys. Flu. **11**, 2280 (1999)
36. Politano, H. & Pouquet, A. Phys Rev E., **52** 636 (1995)
37. Vishniac E., ApJ 446 724 (1995); Vishniac E., ApJ **451** 816 (1995); Blackman E.G. PRL, 77 2694 (1996)
38. Galeev, A.A., Rosner, R., Vaiana, G. S., ApJ, **229** 318 (1979) Field, G.B. & Rogers, R.D., ApJ, **403** 94 (1993); Haardt F. & Maraschi, L., MNRAS, **413**, 507 (1993); Di

Matteo T., MNRAS 299 L15 (1998)
39. Priest E.R., 1998, Ap&SS, **264**, 77.
40. Romanova M.M., 1998, ApJ 500 703 (1998)
41. Miller K.A. & Stone J.M., ApJ **534** 398 (2000)
42. Brandt W.N. et al. MNRAS **303** L53, (1999).
43. Reynolds C.S., in Poutanen, J. & Svensson R., eds., *High Energy Processes in Accreting Black Holes*, ASP Conf. Series Vol. 161, p178 (1999)
44. Fabian A.C. Iwasawa K. Reynolds C.S., Young A.J. accepted to PASP, astro-ph/0004366, (2000).
45. Nandra K. et al. ApJ, **477**, 602 (1997)
46. Tanaka Y. et al., Nature, **375**, 659 (1995);
47. Iwasawa K. et al., MNRAS, **282**, 1038 (1996); Iwasawa K. et al., MNRAS, **306**, L19 (1999); Lee, J.C. et al., MNRAS, **310** 973 (1999)
48. Pariev V.I. & Bromley B.C, ApJ, **508**, 590 (1998)
49. Blackman E.G., 1999, MNRAS, **306**, L25 (1999)
50. Hartnoll S.A & Blackman E.G. MNRAS, in press, astro-ph/9908275 (2000)
51. Pringle J.E., 1996, MNRAS, **281**, 357
52. Terquem C. & Bertout C., A&A, **274**, 291 (1995)
53. Miyoshi M. et al., Nature, 373, **127** (1995); Herrnstein J.R. et al., ApJ, **468**, L17 (1996)
54. Nishiura S., Murayama T. & Taniguchi Y., PASJ, **50**, 31 (1998)
55. Phinney E.S., in F. Meyer *et al.*, eds, Theory of Accretion Disks. Kluwer, Dordrecht, p. 457 (1989); Maloney P.R., in *Highly Redshifted Radio Lines*, edited by C.L. Carilli et al., ASP Conf. Series Vol. 156, 1999, p267
56. Turner T.J. et al., ApJ, **488** 164 (1997)
57. Guilbert P.W. & Rees M.J., MNRAS, **233** 475 (1988); Celotti, A., Fabian, A. C., Rees, M.J. MNRAS **255** 419 (1992); Kuncic Z. Blackman E.G., Rees M.J. MNRAS **283**, 1322 (1996)
58. Blandford R.D., in "Proc of Discusison Meeting on Magnetic Activity in Stars, Disks and Quasars.", Ed. D. Lynden-Bell et al. Phil. Trans. Roy. Soc. A in press (2000); Ferrari A., ARAA, 36 539 (1998)
59. Konigl, A. & Pudritz, R. E., in *Protostars and Planets IV*, eds V. Mannings et al., (Tucson: University of Arizona Press); (2000)
60. Uchida Y. & Shibata K., PASJ, 37 31 (1985); Contopoulos J., ApJ 450 616 (1995); Lynden-Bell D., 1996, MNRAS, **279**, 389 (1996); Heinz S. & Begelman M.C., ApJ **535**, 104 (2000)
61. Blandford R.D. & Payne D.G. MNRAS, **199**, 883 (1982); Lovelace R.V.E., Wang J.C.L., Sulkanen M.E. ApJ, **315**, 504 (1987); Meier D.L., ApJ 522, 753 (1999)
62. Ruderman M.A., Tao L., Kluzniak W., ApJ in press, astro-ph/0003462 (2000)
63. Colgate S.A. & Li H., ApSS, **264**, 357 (1999)
64. Krasnopolsky R., Li Z-Y. Blandford R.D., ApJ **526**, 631 (2000)
65. Ustyugova G.V., et al ApJ **516**, 221 (1999)
66. Blackman E.G., Frank. A., Welch C., submitted to ApJ (2000)
67. Blandford R.D. & Znajek R.L. MNRAS, **179**, 433 (1977); Punsley B. & Coroniti F.V., ApJ **350**, 518 (1990); Livio M., Ogilvie G.I., Pringle J.E., ApJ **512**, 100 (1999);

68. Khanna R. & Camenzind M., A&A, **307**, 665 (1996); Krolik J.H., ApJ **515**, L73 (1999); Meier D.L., ApJ **522**, 753 (1999); Tomimatsu A., ApJ **578**, 972 (2000)
69. van Ballegoijen A.A., in *Accretion Disks and Magnetic Fields in Astrophysics*, Kluwer, Dodrecht 1989, p99; Lubow S. H., Papaloizou J. C. B. Pringle J.E., MNRAS, **267**, 235 (1994)
70. Parker, E. N., *Cosmical Magnetic Fields*, Clarendon Press, Oxford, 1979; Rädler, K.-H., *Generation of Cosmic Magnetic Fields*, Proc. of Mexican School On Astrophysics, Springer, Heidelberg, 1999, p1.
71. Heinz S. & Begelman M.C., ApJ **535**, 104 (2000)
72. Blackman E.G. & Field G.B., ApJ **534**, 597 (2000)
73. Cattaneo F., & Hughes, D.W. Phys. Rev. E. **54**, 4532 (1996)
74. Parnell C.E., Ap&SS, **261**, 81 (1998); Fisk L.A., *The Major Discoveries of the ULYSSES Mission* in 25th International Cosmic Ray Conference, edited by M.S. Potgieter et al., World Scientific, River Edge NJ, 1998, p.27.

Interaction of Radiation with Matter in Accretion Flow

Myeong-Gu Park

Department of Astronomy and Atmospheric Sciences
Kyungpook National University
Taegu 702-701, KOREA

Abstract. I discuss the interaction of radiation with matter in spherical and axisymmetric accretion flows onto black holes and emphasize the similarities between spherical accretion flow and disk-like flow, especially ADAF. As in spherical flow, two different solution families exist in ADAF. One is the usual ADAF where interaction of radiation with matter is minimal with negligible radiative cooling. The other is preheated ADAF in which the infalling gas is heated by Compton scattering off the photons produced at inner hot flow. The two-dimensional thermal structure of ADAF is also discussed along with the possibility of polar outflow or wind.

INTRODUCTION

Accretion flow onto black holes will assume either (quasi) spherical or axisymmetric form. Until recently most works on spherical flow have been done on strictly zero angular momentum flow while those on axisymmetric flow on geometrically thin disk. However, new accretion flow solutions that stand somewhere between pure spherical flow and flat disk-like flow have been rediscovered and extensively studied (see Narayan in this proceedings). These advection-dominated accretion flow (ADAF) solutions have a number of desirable properties that can complement the classical thin disk solutions. Most importantly, the radiation efficiency is not predetermined by the boundary conditions and can be much smaller than the usual value of 0.1. In addition, ions are in near virial temperature and electrons can reach $\sim 10^9$ K, significantly higher than the thin disk. Interestingly, most of these properties also belong to the spherical accretion flow. Hence, although ADAF is generally regarded as the disk flow with significant radial motion, it is equally helpful to consider ADAF as quasi-spherical flow with significant rotation. I will elaborate on this by reviewing the physical properties of spherical flow and axisymmetric flow.

It is rather convenient to describe the accretion flow with the dimensionless luminosity $l \equiv L/L_E$ where L_E is the Eddington luminosity, the dimensionless mass accretion rate $\dot{m} \equiv \dot{M}c^2/L_E$, and the radiation efficiency $e \equiv l/\dot{m}$.

Spherical Accretion Flow

Spherical accretion flow almost always freely falls. The thermal structure, however, changes greatly around $\dot{m} \simeq 1$ at which accretion rate the electron scattering optical depth from infinity to the horizon is exactly 1.

Lowest accretion rate flow ($\dot{m} \ll 1$) is heated by adiabatic compression, thereby maintaining $T_i \propto (r/r_s)^{-1}$ where T_i is the ion temperature and r_s the Schwarzschild radius. The flow is optically thin to both scattering and absorption. Electron temperature T_e reaches as high as $\sim 10^9$ K, heated by ions through Coulomb coupling and cooled by bremsstrahlung (and synchrotron if magnetic field exists). However, the radiation efficiency is very low due to the low density, typically $e \lesssim 10^{-6}$. This family of solutions are shown in Figures 1 and 2 as crosses [1].

Unlike $\dot{m} \ll 1$ flow, two solution families exist for $\dot{m} \gtrsim 1$. The flow is optically thick to scattering, and it can be either in low-entropy state or in high-entropy

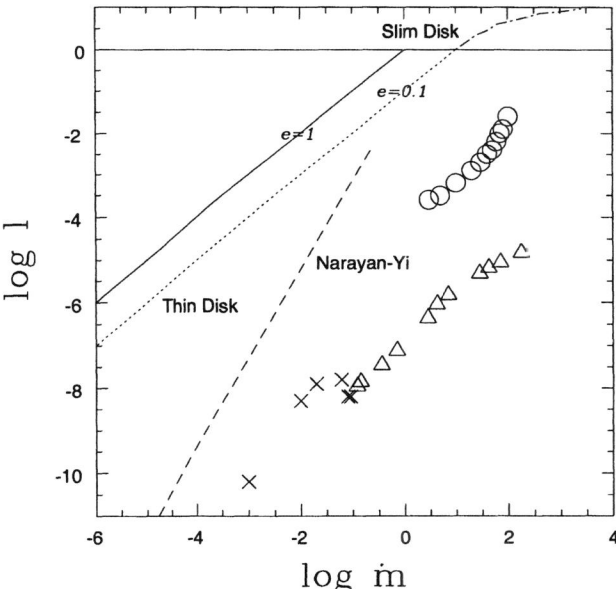

FIGURE 1. Spherical and axisymmetric accretion solutions in dimensionless mass accretion rate versus luminosity plane.

state. Low-entropy state is analogous to the stellar interior except the difference that the energy is generated by adiabatic heating (ultimately from gravitational potential) instead of nuclear burning. These solutions are triangles in Figures 1 and 2. The efficiency is still low due to the low temperature, $e \sim 10^{-7}$.

High-entropy flow is self-consistently maintained by preheating. The outer part of the flow is heated by Compton scattering off photons produced at the inner, hotter part of the flow. In the absence of preheating, the flow will collapse to the low-entropy one. Electron temperature can reach near $\sim 10^9$ K, and can produce hard photons. The luminosity correspondingly is much higher and so is the efficiency, $e \gtrsim 10^{-4}$. Although the flow is optically thick to scattering, it is optically thin to absorption, and the radiation spectrum is that of optically thin, Comptonized bremsstrahlung (and synchrotron). These solutions in the absence of magnetic field are shown in Figures 1 and 2 as open circles.

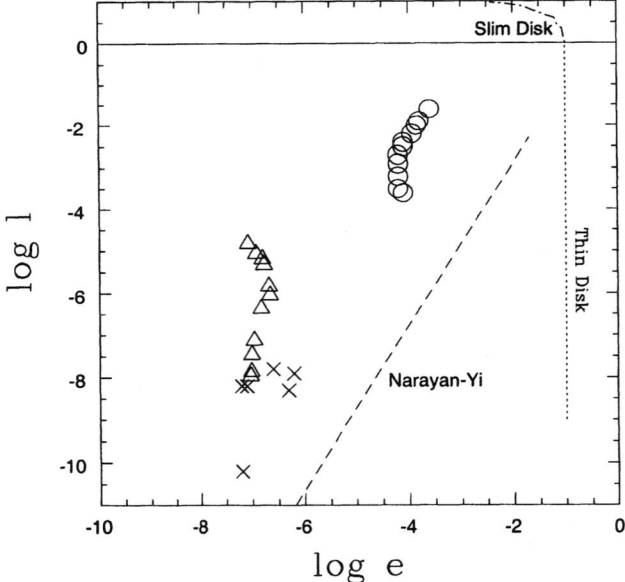

FIGURE 2. Spherical and axisymmetric accretion solutions in radiation efficiency versus dimensionless luminosity plane.

Axisymmetric Accretion Flow

Accretion flow assumes an axisymmetric shape when the flow has angular momentum. The angular momentum is transferred outward by differential rotation and viscous stress. In the thin disk solutions, the rotational speed of the flow is almost Keperian, which makes the radial velocity negligible. The flow becomes a very thin disk supported by rotation. Gravitational potential energy is transferred to the flow by viscous heating, which is balanced by radiative cooling through the surface of the disk. The total radiation efficiency of the accretion is solely determined by the position of the inner edge of the disk, at which the gas flow has a predetermined kinetic energy corresponding to Keperian rotation. Since the inner edge of the disk is generally believed to be located at the innermost stable orbit, the efficiency is $e \simeq 0.1$ regardless of the thermal state of the disk flow. Despite the high efficiency, only soft photons are produced by the thin disk because the flow is in low-temperature LTE state. Thin disk solutions appear as a straight dotted line ($e \simeq 0.1$) in Figures 1 and 2.

The accretion disk cannot maintain its thin disk shape when the luminosity approaches or exceeds the Eddington luminosity, $l \gtrsim 1$ or equally $\dot{m} \gtrsim e^{-1}$. Too much energy is generated at this high accretion rate to be readily radiated away, and trapped radiation builds up enough pressure to make the disk puffed up. Now the pressure gradient in the radial direction is important and significant radial motion ensues. The flow falls into the black hole carrying all its entropy with no time for radiative cooling. In this slim or thick disk accretion, therefore, a significant fraction of the gravitational potential energy may not be radiated, and the radiation efficiency becomes smaller than that of the thin disk accretion. The thick disk accretion with the standard α-viscosity in height-integrated formalism is called slim disk and plotted in Figures 1 and 2 as dot-dashed line that shows decreasing e as $\dot{m}$ increases. The thermal state of the thick disk resembles those of low-temperature ≥ 1 spherical accretion flow.

Quite recently, a new type of axisymmetric accretion flow has been rediscovered and extensively studied. This so called 'advection dominated accretion flow' (ADAF) is dynamically close to the thick disk accretion in that the angular momentum is sub-Keplerian and the radial velocity, which is a fraction of the free-fall velocity, can even be comparable to the rotation velocity. The energy transport due to the direct advection of entropy via radial motion of the flow dominates over the radiative cooling. Only a small fraction of the gravitational potential energy produced by the viscous dissipation is released because of inefficient cooling. The luminosity and the efficiency of ADAF are shown in Figures 1 and 2 as straight dashed line labeled as Narayan-Yi [2]. The thermal state of ADAF is quite the opposite of the thick disk. Ion temperature is almost virial and electron temperature reaches as high as $\sim 10^9$ K. The flow is optically thin to absorption, and is not in LTE state. The geometrical shape of ADAF is spheroidal or quasi-spherical due to near virial pressure. In short, ADAF is optically thin, hot spheroidal accretion flow with sub-Keplerian rotation and near virial pressure with a significant radial motion. With these properties, ADAF is very much analogous to the optically thin $\dot{m} < 1$ spherical flow, the only difference being the rotation. As with the spherical accretion flow, ADAF can exist only for $\dot{m} < 1$, otherwise the flow will become the thin disk due to increased radiative cooling. It should be stressed out that for a given accretion rate with $\dot{m} < 1$, both thin disk and ADAF exist. They are the two branches of solutions with different thermal and dynamical properties for a given $\dot{m}$.

Most works on ADAF so far have focused on the local and global properties in one-dimensional height-integrated formalism. Park and Ostriker [3] studied thermal properties of two-dimensional ADAF [4], and find that two-dimensional nature of ADAF can produce interesting properties. For example, a conical region around the polar axis cannot maintain high enough temperature due to small viscous heating while the equatorial region has no trouble to stay in high temperature state. The shaded region in Figure 3 shows the low temperature $(T \simeq 10^4 \, \text{K})$ region. This funnel extends to the equatorial plane for $\dot{m} \gtrsim 10^{-1.5}$ and almost disappears for $\dot{m} \lesssim 10^{-4}$ when the outer boundary is fixed to the virial temperature.

More detailed calculations [5] show that the usual ADAF exists only for $\dot{m} \lesssim 10^{-3.7}$ if the outer boundary is equal to the thermal equilibrium value of $T \simeq 10^4 \, \text{K}$ (star symbols for no magnetic field and diamond symbols for equipartition field in

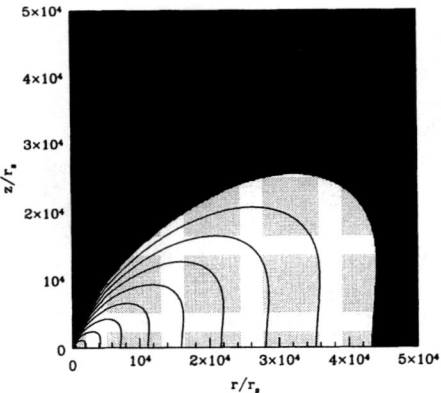

FIGURE 3. Vertical cross section of ADAF. In the shaded part of the flow, viscous heating is too small to maintain near virial temperature.

Fig. 4).

However, even in the forbidden high $\dot{m}$ region a different branch of ADAF is found that is maintained by Compton preheating of photons produced at inner hot part of the flow (crosses for no magnetic field and circles for equipartition field flow in Fig. 4). We call this preheated advection dominated flow (PADAF). PADAF is analogous to the high-temperature $\dot{m} \gtrsim 1$ spherical flow which is also sustained by Compton preheating (triangles in Figs. 1 and 2). The generic temperature profile of PADAF looks like Figure 5. Small inset on the upper right corner of the figure box shows the poloidal direction of the corresponding temperature profile: solid lines are for along the polar direction ($\theta = 0$), dotted lines for $\theta = \pi/8$ and so on. For a given type of lines, the upper one shows ion temperature and the lower one electron temperature profile. The polar part of the flow (solid lines) is heated first due to the lower density compared to the equatorial part (dashed-dot-dotted lines).

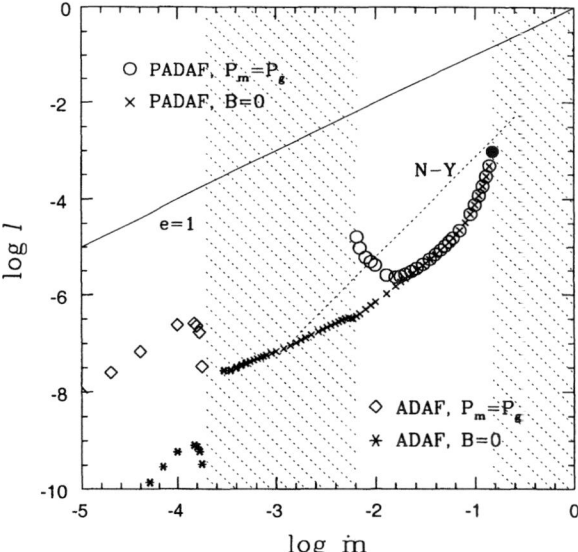

FIGURE 4. ADAF and PADAF solutions in dimensionless mass accretion rate versus luminosity plane.

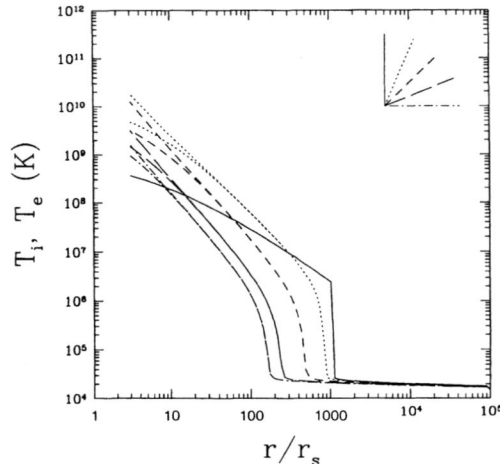

FIGURE 5. Ion and electron temperature profiles of typical PADAF. Each line correspond to different poloidal angles.

In the innermost region, equatorial region has higher temperature than the polar region due to the higher viscous heating. PADAF shows that global preheating does affect ADAF and extends the parameter space in which ADAF solution is possible. Preheating may also help the transition from the thin disk state to ADAF. It is also found that some PADAF can be overheated (above the virial temperature) near the polar region while the equatorial region accretes normally, resulting possibly in the ADAF with polar wind.

REFERENCES

1. Park M.-G., *Astrophy. J.* **354**, 83 (1990).
2. Narayan R., and Yi, I., *Astrophy. J.* **452**, 710 (1995)
3. Park M.-G., and Ostriker J., *Astrophy. J.* **527**, 247 (1999).
4. Narayan R., and Yi, I., *Astrophy. J.* **444**, 231 (1995)
5. Park M.-G., and Ostriker J., *astro-ph/0001446* (2000).

Simulations of Time-Dependent High Energy Emission from Blazars

Masaaki Kusunose[*], Fumio Takahara[†], and Hui Li[††]

[*]Department of Physics, Kwansei Gakuin University, Nishinomiya 662-8501, Japan
[†]Department of Earth and Space Science, Osaka University, Osaka 560-0043, Japan
[††]Theoretical Astrophysics (T-6, MS B288), Los Alamos National Laboratory,
Los Alamos, NM 87545, U.S.A.

Abstract. Blazars are a group of active galactic nuclei which emit a broad band continuum from radio to very high energy gamma-rays. Recent X-ray and gamma-ray observations show that the emission from blazars are variable on very short timescales. Based on the synchrotron-self-Compton model for emission mechanisms in relativistic jets, we show the results of the numerical simulations of the time evolution of electron and photon spectra in the jets of blazars, including the effect of electron acceleration.

INTRODUCTION

Blazars are known to emit broad continuum emission from radio to very high energy gamma-rays. In particular, TeV gamma-rays have been observed from Mrk 421, Mrk 501, 1ES 2344+514, and PKS 2155-304. Also short time variations on timescales such as hours or even 30 min have been observed [5,17].

It is believed that relativistically moving jets are responsible for the emission, because of short time variations and luminous emission, e.g., [2,3,5,7,9,15,17,18]. Their spectra have two peaks in ν-νF_ν representation. A peak in low energy ($\sim$keV) is probably due to synchrotron emission and the other peak in high energy is thought to be produced by inverse Compton scattering of the synchrotron photons (synchrotron-self-Compton model: SSC model), e.g., [6,13]. The effects of external soft photons, e.g., [16], and proton collisions, e.g., [12], are also proposed.

Previously the SSC model has often been used to calculate emission spectra in steady states. Since the time variation of emission is often observed, recently the SSC model is used to simulate time-varying emission spectra. However, most models assume the injection of nonthermal electrons and calculate the cooling phase of electrons [11,14].

Kirk et al. [8] introduced an acceleration term in the equation describing the time evolution of the electron energy spectrum. Our model is similar to Kirk et al., but we calculate the photon spectrum for not only synchrotron emission but also inverse

Compton scattering; the calculation by Kirk et al. [8] was limited to synchrotron emission, though they included spatial transfer of accelerated electrons, which we neglect in this paper.

In this paper, we employ the homogeneous SSC model and calculate the energy spectra of electrons and photons simultaneously. We include a simple model of electron acceleration and look for a clue to draw hints of acceleration mechanisms from observations.

MODEL

We assume a relativistically moving blob with radius R and Doppler factor $\mathcal{D} = [\Gamma(1 - \beta \cos\theta)]^{-1}$, where β is the speed of the blob normalized by the light speed, $\Gamma = (1 - \beta^2)^{-1/2}$, and θ is the angle between the jet velocity and the line of sight. The blob is divided into an acceleration region (e.g., a shock region) and a cooling region. The volume of the acceleration region is assumed to be smaller than that of the cooling region. In the acceleration region, mono-energetic electrons with Lorentz factor γ_0 are injected and subsequently accelerated on timescale $t_{\rm acc}$; we do not specify a particular acceleration mechanism here. Accelerated electrons then escape from the acceleration region with timescale $t_{e,\rm esc}$ and are injected into the cooling region. In the cooling region electrons emit photons by synchrotron radiation and Compton scattering.

Electron Spectrum

The kinetic equation describing the time evolution of electrons is given by

$$\frac{\partial N(\gamma)}{\partial t} = -\frac{\partial}{\partial \gamma}\left\{\left[\left(\frac{d\gamma}{dt}\right)_{\rm acc} - \left(\frac{d\gamma}{dt}\right)_{\rm loss}\right] N(\gamma)\right\} - \frac{N(\gamma)}{t_{e,\rm esc}} + Q(\gamma), \quad (1)$$

where γ is the Lorentz factor of electrons and $N(\gamma)d\gamma$ is the number density of electrons. We assume that monochromatic electrons with Lorentz factor γ_0 are injected in the acceleration region, i.e., $Q(\gamma) = Q_0 \delta(\gamma - \gamma_0)$; in numerical results shown below, we assume $\gamma_0 = 10$. Electrons are then accelerated and lose energy by synchrotron radiation and Compton scattering [the energy loss rate is denoted by $(d\gamma/dt)_{\rm loss}$]. The acceleration term is approximated by $(d\gamma/dt)_{\rm acc} = \gamma/t_{\rm acc}$. In the framework of diffusive shock acceleration, e.g. [1,4], $t_{\rm acc}$ can be approximated as

$$t_{\rm acc} = 20\lambda(\gamma)c/(3v_s^2) \sim 3.79 \times 10^{-6}(0.1{\rm G}/B)\xi\gamma \quad {\rm s}, \quad (2)$$

where $v_s \approx c$ is the shock speed with c being the light speed, B is the strength of magnetic fields, and $\lambda(\gamma) = \gamma m_e c^2 \xi/(eB)$ is the mean free path assumed to be proportional to the electron Larmor radius with ξ being a parameter, m_e the electron mass, and e the electron charge. For the convenience of numerical calculations, we assume $t_{\rm acc}$ does not depend on γ:

$$t_{\rm acc} = 3.79 \times 10 (0.1 {\rm G}/B)(\gamma_f/10^7)\xi \quad {\rm s}, \tag{3}$$

where γ_f is assumed to be a characteristic Lorentz factor of relativistic electrons and used as a parameter; we set $\gamma_f = 10^7$ throughout this paper. Although realistic acceleration time for the smaller values of γ should be correspondingly shorter, we make this choice because we mainly concern about the electrons with the large values of γ. We make sure that the resultant spectrum is that expected in diffusive shock acceleration by choosing $t_{e,{\rm esc}} = t_{\rm acc}$ in the acceleration region. The electron spectrum in the cooling region is calculated by Eq. (1), with $(d\gamma/dt)_{\rm acc}$ dropped. Also $Q(\gamma)$ is replaced by the escaping electrons from the acceleration region and $t_{e,{\rm esc}}$ is set to be $3R/c$.

Photon Spectrum

The photon spectrum is calculated by

$$\frac{\partial n_{\rm ph}(\epsilon)}{\partial t} = \dot{n}_{\rm C}(\epsilon) + \dot{n}_{\rm em}(\epsilon) - \dot{n}_{\rm abs}(\epsilon) - \frac{n_{\rm ph}(\epsilon)}{t_{\gamma,{\rm esc}}}, \tag{4}$$

where $n_{\rm ph}(\epsilon)d\epsilon$ is the photon number density. The term $\dot{n}_{\rm C}(\epsilon)$ denotes Compton scattering. Photon production and self-absorption by synchrotron radiation are included in $\dot{n}_{\rm em}(\epsilon)$ and $\dot{n}_{\rm abs}(\epsilon)$, respectively. External photon sources are not included. The rate of photon escape is estimated as $n_{\rm ph}(\epsilon)/t_{\gamma,{\rm esc}}$. We set $t_{\gamma,{\rm esc}} = R_{\rm acc}/c$ and R/c in the acceleration and cooling regions, respectively. Here $R_{\rm acc}$ is the thickness of the acceleration region; we assume that the acceleration region is a slab with radius R and thickness $R_{\rm acc}$, and in numerical calculations below we set $R_{\rm acc} = c\,t_{\rm acc}/2$. We solve Eqs. (1) and (4), simultaneously.

RESULTS

We assume $\mathcal{D} = 10$, $B = 0.2$ G, and $R = 1.5 \times 10^{16}$ cm. Electrons with $\gamma = 10$ are injected steadily in the acceleration region at a rate of 0.008 electrons/cm^3/s. For the acceleration timescale, $\xi = 2 \times 10^3$ and $\gamma_f = 10^7$ are used. The initial condition is that there are no electrons and photons in the blob.

Figure 1 shows the time evolution of the electron distribution for $t = 0 \sim R/c$ in the blob's frame. With increasing time, the value of the maximum Lorentz factor increases and the spectrum becomes flat, i.e., $N(\gamma) \propto \gamma^{-2}$. The value of the maximum Lorentz factor is determined by the balance among acceleration, cooling, and escape. In Figure 1, the steady state is also shown by a dashed curve. This curve shows a break at $\gamma \sim 2 \times 10^3$, and the break is caused by radiative cooling. Below $\gamma \sim 2 \times 10^3$, the curve is flat, which is expected by the assumption

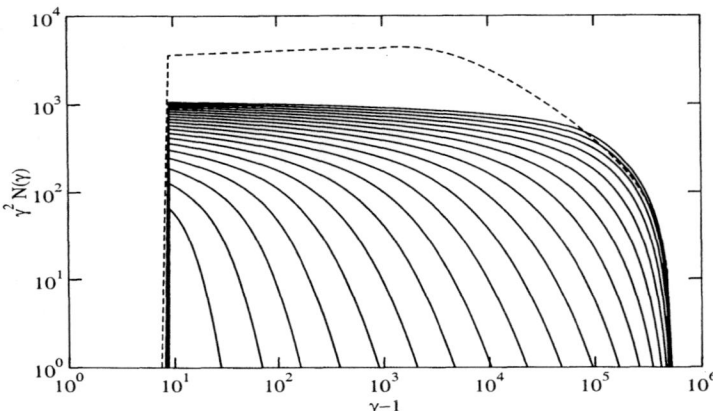

FIGURE 1. Time evolution of the electron spectrum in the cooling region for $t = 0 - R/c$ with equally spaced time span $0.05R/c$. Electrons with $\gamma = 10$ are steadily injected in the acceleration region. The spectrum evolves from lower curves to upper curves. The spectrum at $t = 10R/c$ (steady state) is also shown by a dashed curve.

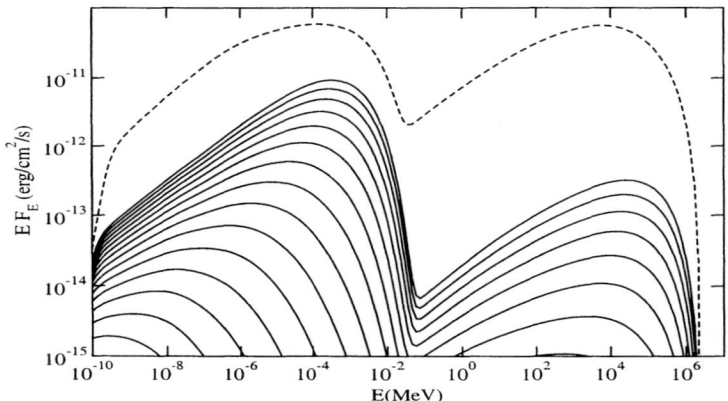

FIGURE 2. Time evolution of the photon spectrum for $t = 0 - R/c$ with equally spaced time span $0.05R/c$. The spectrum evolves from lower curves to upper curves with time. The spectrum at $t = 10R/c$ (steady state) is also shown by a dashed curve.

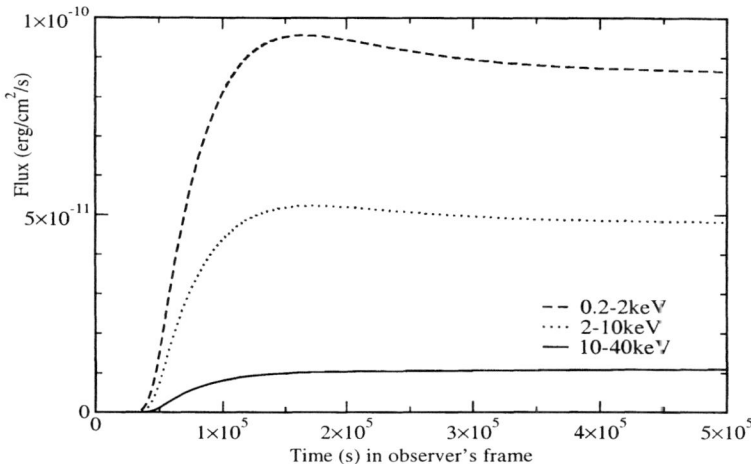

FIGURE 3. Light curves for $t = 0 - 10R/c$.

that $t_{e,\mathrm{esc}} = t_{\mathrm{esc}}$ in the acceleration region (the curve is slightly increasing with increasing γ and this is due to a small numerical error).

The photon spectrum emitted from the cooling region, corresponding to Figure 1, is shown in Figure 2. The steady state spectrum is also shown by a dashed curve. This photon spectrum is calculated in the observer's frame, and red shift $z = 0.03$ and Hubble constant $H_0 = 75$ (km/s/Mpc) were assumed to calculate the flux.

Light curves are shown in Figure 3 for three energy bands; the fluxes are from the cooling region. During electron acceleration, the energy flux increases and attain a steady state after $t \sim 3R/c$ in the jet frame, on which timescale photons escape from the cooling region because $t_{\gamma,\mathrm{esc}} = 3R/c$ is assumed.

SUMMARY

We calculated the emission spectrum from a relativistically moving blob, including electron acceleration. Then the time variations of electron and photon spectra are obtained when mono energetic electrons are injected. More detailed results have been published in [10,11]. Our code is capable of including various acceleration models by changing the dependence of t_{acc} on electron energy and magnetic field, etc. Comparison with observations and our numerical results will set some constraints on the acceleration mechanisms working in blazars.

REFERENCES

1. Blandford, R. D., and Eichler, D., *Phys. Rep.*, **154**, 1 (1987).
2. Blandford, R. D., and Rees, M. J., *Pittsburgh Conf. on BL Lac Objects*, ed. A. M. Wolfe (Pittsburgh: Univ. Pittsburgh Press), 328 (1978).
3. Catanese, M. et al., *Astrophys. J. Letters*, **487**, 143 (1997).
4. Druly, L. O.' C., *Rep. Prog. Phys.*, **46**, 973 (1983).
5. Gaidos, J. A. et al., *Nature*, **383**, 319 (1996).
6. Inoue, S., and Takahara, F., *Astrophys. J.*, **463**, 555 (1996).
7. Kataoka, J. et al., *Astrophys. J.*, **514**, 138 (1999).
8. Kirk, J. G., Rieger, F. M., and Mastichiadis, A., *Astron. & Astrophys.*, **333**, 452 (1998).
9. Krennrich, F. et al., *Astrophys. J.*, **511**, 149 (1999).
10. Kusunose, M., Takahara, F., and Li, H. 2000 *Astrophys. J.*, **536**, 299.
11. Li, H., Kusunose, M. 2000, *Astrophysical. J.*, **536**, 729.
12. Mannheim, K., *Astron. & Astrophys.*, **269**, 67 (1993).
13. Maraschi, L., Ghisellini, G., and Celotti, A., *Astrophys. J. Letters*, **397**, L5 (1992).
14. Mastichiadis, A., and Kirk, J. G. 1997, *Astron. & Astrophys.*, **320**, 19 (1997).
15. Mukherjee, R. et al., *Astrophys. J.*, **490**, 116 (1997).
16. Sikora, M., Begelman, M. C., and Rees, M. J., *Astrophys. J.*, **421**, 153 (1994).
17. Takahashi, T., Tashiro, M., Madejski, G., Kubo, H., Kamae, T., Kataoka, J., Kii, T., Makino, F., Makishima, K., and Yamasaki, N., *Astrophys. J. Letters*, **470**, L89 (1996).
18. Ulrich, M.-H., Maraschi, L., and Urry, C. M., *Ann. Rev. Astron. & Astrophys.*, **35**, 445 (1997).

Galactic Center ADAF Ruled out By Polarization

Eric Agol

Physics and Astronomy, Johns Hopkins University, Baltimore, MD, 21218
Currently at: Theoretical Astrophysics, Caltech MS 130-33, Pasadena, CA 91125

Abstract. Recent observations of linear polarization at high frequency towards the galactic center radio source, Sgr A*, place strong constraints on the nature of the radio emission region. We discuss how these constraints rule out the larger accretion rate need for low efficiency accretion flows, such as advection-dominated accretion flows, and we present a toy model which can explain the radio to sub-mm spectrum and polarization of Sgr A*.

INTRODUCTION

Recent proper motion measurements of stars near the Galactic center have shown there to be a central mass of $3 \times 10^6 M_\odot$ located within 0.03 arcseconds of the central radio source, Sgr A*, with an inferred mass density of $\sim 10^{13} M_\odot/pc^3$ (Genzel et al. 1998, Ghez et al. 1998, Ghez et al. 2000). The radio source is quite faint, with a bolometric luminosity of $\sim 10^{37} erg/s$, so that the mass to light ratio of the central object is $\sim 10^{-3} L_\odot/M_\odot$. The high mass density and low luminosity are strong evidence that a black hole lurks at the dynamical center. The accretion rate of matter by the black hole should be as large as $10^{-(4-5)} M_\odot/yr$ (Quataert, Narayan, & Reid 1999; Coker & Melia 1999). This is rather surprising given the rather low bolometric luminosity of the point source associated with Sgr A*; the inferred accretion efficiency is as low as 2×10^{-6}. Low efficiency might be achieved if viscous heating pumps energy into protons, which cannot cool as efficiently as electrons and proceed to carry their thermal energy into the black hole, a so-called "Advection-Dominated Accretion Flow" (or ADAF, Ichimaru 1977, Narayan & Yi 1994; Narayan, Yi, & Mahadevan 1995), or if the gas has low angular momentum, and accretes spherically (Melia 1992). Alternatively, the accretion rate may be overestimated, or most infalling gas may be driven mechanically outwards by energy released from the gas which does accrete, either in a jet, wind, or convectively or magnetically driven outflow (Falcke, Mannheim, & Biermann 1993, Igumenshchev & Abramowicz 1999, 2000; Stone, Pringle, & Begelman 1999; Quataert & Gruzinov 2000; Narayan, Igumenshchev, & Abramowicz 2000; Igumenshchev, Abramowicz,

& Narayan 2000, Stone & Pringle 2000). The actual accretion rate onto the black hole may be as low as $10^{-9} M_\odot/yr$.

One possible way to distinguish low and high accretion rates of the matter near the black hole is to measure the polarization and self-absorption frequency of the radio emission which is thought to be due to synchrotron-emitting plasma (Bower et al. 1999b, Agol 2000, Quataert & Gruzinov 2000). Higher accretion rates imply higher gas densities and magnetic field strengths, leading to more Faraday rotation which will depolarize the synchrotron radiation and cause a higher self-absorption frequency. There is no definitive detection of linear polarization or self-absorption frequency of Sgr A*, consistent with the idea of a high accretion rate. However, recent tantalizing polarization observations by Aitken et al. (2000, hereafter A00) with the SCUBA array reveal that near 1 mm, the radio source may be linearly polarized, and that the polarization angle changes with wavelength by about 90°, possibly indicating that the self-absorption frequency occurs around 0.5mm. If these observations are confirmed with higher angular resolution to avoid confusion from polarized dust emission, then the polarization will be strong evidence against a high accretion rate. We discuss synchrotron polarization in section 2, the limits on the accretion rate in section 3, a toy model which can explain the current observations (if taken at face value) in section 4, and summarize in section 5.

SYNCHROTRON THEORY

In the synchrotron limit ($\gamma \gg 1$) for an isotropic electron velocity distribution, some analytic results have been derived, which we now summarize (Ginzburg & Syrovatskii 1965 & 1969: GS). For a uniform slab of electrons with a power-law distribution, $dn_e/d\gamma \propto \gamma^{-\xi}$ (with $\gamma_{min} \leq \gamma \ll \gamma_{max}$ such that electrons with γ_{min} and γ_{max} do not contribute to the frequency of interest), we can relate the magnetic field strength and electron density in the slab to the fluid-frame brightness temperature and the spectral turnover due to self-absorption. For $\xi = 2$ and a uniform field $B_\perp$ (projected into the sky plane) we find $B_\perp \sim 2T_{11}^{-2}\nu_{12}$ G and $\tau_C \sim 3 \times 10^{-2} T_{11}^4 \nu_{12} \gamma_{min}^{-1}$, where T_{11} is the brightness temperature in units of 10^{11} K at the self-absorption frequency $\nu_t = 10^{12}\nu_{12}$ Hz and τ_C is the Compton scattering optical depth of the emission region. For $\nu < \nu_t$, the emission is self-absorbed so $F_\nu \propto \nu^{5/2}$, while above this frequency the emission is optically-thin and $F_\nu \propto \nu^{(1-\xi)/2} \exp(-\nu/\nu_{max})$ where $\nu_{max} = 3B_\perp e \gamma_{max}^2/(4\pi m_e c)$.

In the optically-thin regime, the polarization plane is perpendicular to the magnetic field with polarization $\Pi = (\xi + 1)/(\xi + 7/3)$, up to 100% for $\xi \gg 1$. In the optically-thick regime, $\Pi = -3/(6\xi + 13)$ (for $\xi > 1/3$); the radiation polarized perpendicular to the magnetic field is absorbed more strongly than the opposite polarization, causing the radiation polarized along the magnetic field to dominate, switching the polarization angle by 90°, which changes the sign of Π. Numerical calculations show that the optically-thick polarization peaks at $|\Pi| = 20\%$ for $\xi = 1/3$, but remains large for $0 < \xi < 2$.

To compute the polarization near the self-absorption frequency requires a knowledge of the polarized opacity and emissivity, $\mu_{\perp,\|}, \epsilon_{\perp,\|}$ For $\xi = 2$, these can be approximated as (GS): $\mu_{\perp,\|} = r_s^{-1}(\nu/\nu_t)^{-3}(1 \pm \xi/4)$ and $(\epsilon_{\perp,\|}/\mu_{\perp,\|}) = 2S_t/9(\nu/\nu_t)^{5/2}(13 \pm 9)/(4 \pm 1)$ where r_s is the size of the emission region, ν_t is the frequency for which the total source has an optical depth of unity (i.e. $\tau = \mu r_s = 1/2(\mu_\perp + \mu_\|)r_s = 1$), S_t is the source function near the frequency ν_t, and the $+$ or $-$ signs go with the radiation emitted $\perp$ or $\|$ to the magnetic field, respectively. GS then express the polarization and emission for a slab with uniform magnetic field strength and direction, constant density, and size r_s: $I_\perp = (\epsilon_\perp/\mu_\perp)(1 - \exp(-\mu_\perp r_s))$, $I_\| = (\epsilon_\|/\mu_\|)(1 - \exp(-\mu_\| r_s))$, and $\Pi = (I_\perp - I_\|)/(I_\perp + I_\|)$, where $I_\perp, I_\|$ are the intensities (erg/cm^2/s/Hz/sr) with polarization perpendicular and parallel to the projected direction of the magnetic field on the sky.

For electron distributions which are highly peaked at a single energy (such as mono-energetic or relativistic Maxwellian) the polarization for $\nu \lesssim \nu_t$ is zero.

The Faraday effect rotates the polarization vector of photons emerging from different optical depths by different amounts, causing a cancellation in polarization (Agol & Blaes 1996). The differential Faraday rotation angle within the source scales as $\Delta\theta = 3.6 \times 10^{28} \tau_{phot} B \nu^{-2} \gamma_{min}^{-2}$ (Jones & O'Dell 1977), where τ_{phot} is the Compton optical depth of the photosphere. When optically thin, $\tau_{phot} \sim \tau_C$ is constant, so rotation is largest at the self-absorbed wavelength. When self-absorbed, τ_{phot} of the photosphere scales as $\nu^{\xi/2+2}$, so the differential Faraday rotation angle $\propto \nu^{\xi/2}$ (for $\xi > 1/3$), again largest at the self-absorption wavelength. The differential rotation at ν_t is $\Delta\theta \sim 2\pi g(\xi)(\theta_b/\gamma_{min})^\xi/\gamma_{min}$, where γ_{min} is the minimum electron Doppler factor, $g(\xi)$ is a dimensionless factor of order unity, and θ_b is the brightness temperature in units of $m_e c^2/k_B$.

LIMITS ON THE ACCRETION RATE

The observations of polarization in Sgr A* provide the following constraints on emission models:

1) The differential Faraday rotation angle near ν_t must be $\ll \pi$.

2) The electron distribution must be non-thermal since the polarization due to a thermal electron distribution is suppressed when self-absorbed by a factor of $\exp(-\tau)$. If the beam correction by A00 is correct, then $\Pi \sim 12\%$ at self-absorbed wavelengths, requiring $\xi \lesssim 2$.

3) The self-absorption frequency must lie near the change in polarization angle.

4) The component at lower frequencies must have no linear polarization.

5) The magnetic field must be ordered to prevent cancellation of polarization.

These constraints rule out several models proposed in the literature, as will be discussed in turn.

The low efficiency of an ADAF implies a higher accretion rate and thus higher density than for a high efficiency flow of the same luminosity and geometrical thickness. For Sgr A*, an accretion rate of $\sim 10^{-(4-5)} M_\odot/\text{yr}$ is inferred due to capture of gas in the vicinity of the black hole (Quataert, Narayan, & Reid 1999; Coker & Melia 1999), which is the value assumed in ADAF models. Assuming that the gas falls in at near the free-fall speed, one infers an electron density $n_e = 10^{10}$ cm$^{-3}\dot{m}_{-5}x^{-3/2}$ and a magnetic field strength of $B = 10^3 \text{G} \dot{m}_{-5}^{1/2} x^{-5/4}(v_A/0.1v_{ff})$, where x is the radius of the emission region in units of $r_g = GM/c^2$, $\dot{m}_{-5}$ is the accretion rate in units of $10^{-5} M_\odot/\text{yr}$, and v_A/v_{ff} is the ratio of the Alfvén speed to the free-fall speed. These values imply a total Faraday rotation angle at the self-absorption frequency ν_t of $\Delta\theta \sim 10^4 \dot{m}_{-5}^{3/2} \nu_{12}^{-2}(v_A/0.1v_{ff})$. This value is so large that rotation of the emitted radiation leads to zero net polarization, so ADAFs are in direct conflict with the observed polarization. Only significant modifications of the model, such as a reduction in the accretion rate by a factor of 10^{-3}, can reduce the Faraday rotation angle $\ll \pi$. An accretion rate of $10^{-8} M_\odot/\text{yr}$ is consistent with the observed luminosity if the accretion flow has a higher efficiency $\sim 2\%$, no longer "advection-dominated." In addition, ADAF models assume a Maxwellian electron distribution, which cannot produce the observed switch in polarization angle. Finally, ADAFs predict a higher self-absorption frequency: Özel et al. (2000) find that $\nu_t \sim 5 \times 10^{12} \dot{m}_{-5}^{5/9}$ Hz, which implies $\dot{M} \sim 4 \times 10^{-7} M_\odot/\text{yr}$ to be consistent with the observed $\nu_t \sim 5 \times 10^{11}$ Hz. The accretion rate might be reduced if there is significant gas lost by a wind or jet (Begelman & Blandford 1999; Quataert & Narayan 1999) or if the Bondi rate is reduced by heating the infalling gas with heat carried outwards by a convection-dominated accretion flow, or "CDAF" (Igumenshchev & Abramowicz 1999, 2000; Stone, Pringle, & Begelman 1999; Quataert & Gruzinov 2000; Narayan, Igumenshchev, & Abramowicz 2000; Igumenshchev, Abramowicz, & Narayan 2000).

The model of Melia (1992) is rather similar to the ADAF model, and thus suffers the same problems: the high accretion rate implies high density which is inconsistent with the observed polarization.

Beckert & Duschl (1997) considered several 1-zone, quasi-monoenergetic and thermal emission models for the synchrotron emission. These electron distributions do not produce a swing in polarization angle by 90 degrees since the polarization is suppressed when self-absorbed. Their model does produce a self-absorption frequency near the correct frequency, however. Falcke, Mannheim, & Biermann (1993) present a disk-plus-jet model which assumes a tangled magnetic field topology which would erase any polarization. However, an ordered magnetic field would be a small change to their model which might bring it into line with the polarization observations.

TOY MODEL

Now, we attempt to construct a model consistent with all of the data, using uniform emission regions for simplicity. Typical optically-thin AGN spectra show $\xi \sim 2-3$; since $\xi = 2$ is consistent with the polarization from A00, we fix $\xi = 2$ in our model fits. The model parameters for the polarized component are $S_t = 6$ Jy, $\nu_t = 550$ GHz (corresponding to $\lambda = 0.55$ mm), and $\nu_{max} \sim 5000$ GHz (Figure 1).

To explain the lack of polarization and spectral slope flatter than 5/2, we require an additional component which is unpolarized and has a cutoff near 1 mm so that it doesn't dilute the polarization at shorter wavelengths. Since Sgr A* has a spectral slope of 1/3 at mm wavelengths and appears to have a spectral turnover at 1 GHz, we model the spectrum as a monoenergetic electron distribution with energy γ and zero polarization (due to Faraday depolarization or tangled magnetic field) which becomes self-absorbed at low frequency (Beckert & Duschl 1997). For the unpolarized component, we find $F_\nu = 1.3(\nu/\nu_{max})^{1/3} \exp(-\nu/\nu_{max})$ Jy with $\nu_{max} \sim 50$ GHz, and $\nu_t \sim 1$ GHz (Figure 1).

Figure 1 compares the model to the data. To compare the polarization, we have plotted the Stokes' parameter that lies at 83°. Remarkably, the polarization should rise to $\sim 100\%$ at even shorter wavelengths.

DISCUSSION

Krichbaum et al. (1998) report a source radius of 55μas at 1.4 mm from VLBI observations; this corresponds to $19r_g$. The source size may be smaller at higher frequencies, but we expect the radius of the emission region to be greater than the size of the event horizon of the black hole, which has an apparent size of $\sim 5r_g \sim 15\mu$as projected on the sky (including gravitational bending, Bardeen 1973), so we use an intermediate size in further estimates. The flux of the fitted model at the self-absorption frequency, $\nu_t = 550$ GHz, is ~ 9 Jy. This implies a brightness temperature in the emission frame $T_b \sim 1.6 \times 10^{10}(r_s/10r_g)^{-2}\Gamma^{-1}$ K, where r_s is the size of the source (we have assumed the area of the source is πr_s^2) and Γ is the bulk Doppler boost parameter. For a steeply falling electron number distribution, $kT_b \sim 4\gamma m_e c^2$ (for $\xi = 2$), where $\gamma m_e c^2$ is the energy of the emitting electrons, implying $\gamma \sim 10(r_s/10r_g)^{-2}\Gamma^{-1}$ for the electrons at the self-absorption frequency. Using the formulae from §2, we find: $B_\perp = 350(r_s/10r_g)^4\Gamma$ G, $\tau_C = 10^{-5}(r_s/10r_g)^{-8}\Gamma^{-5}$, and $\gamma_{max} = 50(r_s/10r_g)^{-2}\Gamma^{-1}$, implying $n_e \sim 6 \times 10^6(r_s/10r_g)^{-9}\Gamma^{-5}$cm^{-3}. The ratio of magnetic to rest-mass energy density is $B^2/(8\pi n_e m_p c^2) \sim 1(r_s/10r_g)^{17}\Gamma^{9/2}$ for an electron-proton plasma, indicating a relativistic Alfvén speed. The Faraday rotation angle at $\nu_t = 5.5 \times 10^{11}$ Hz is $\Delta\theta \sim 350(r_s/10r_g)^{-4}\Gamma^{-2}\gamma_{min}^{-3}$, assuming $B_\parallel \sim B_\perp$. For $r_s \sim 10r_g$, γ_{min} can be as large as 4, reducing $\Delta\theta$ to 5; for $r_s \sim 5r_g$, γ_{min} can be as large as 20 reducing $\Delta\theta$ to ~ 0.6. Alternatively, if the synchrotron emission is due to a pair plasma, Faraday rotation will be reduced by the ratio of the proton number density to the pair number density. The rotation angle is further reduced at the observed wavelengths by a factor $\sim \nu/\nu_t$. The high energy cutoff for the electron distribution may be due to synchrotron cooling since $t_{cool} = 8 \times 10^8 \gamma_{max}^{-1} B^{-2} \sim 6(r_s/10r_g)^{-6}\Gamma^{-3}$ sec, similar to the dynamical time, $t_D \sim 13 x^{-3/2}$ sec. Given the strong scaling of quantities with the unknown r_s and Γ, the above estimates can only be improved with future observations.

The unpolarized emission component dominates at ∼ 7 mm, where Lo et al. (1998) measure a source size of ∼ 5×10^{13} cm. The self-absorption frequency then requires $\gamma \sim 400$, $B \sim 0.1$ G, and $n_e \sim 4 \times 10^5$ cm^{-3}. Though somewhat ad-

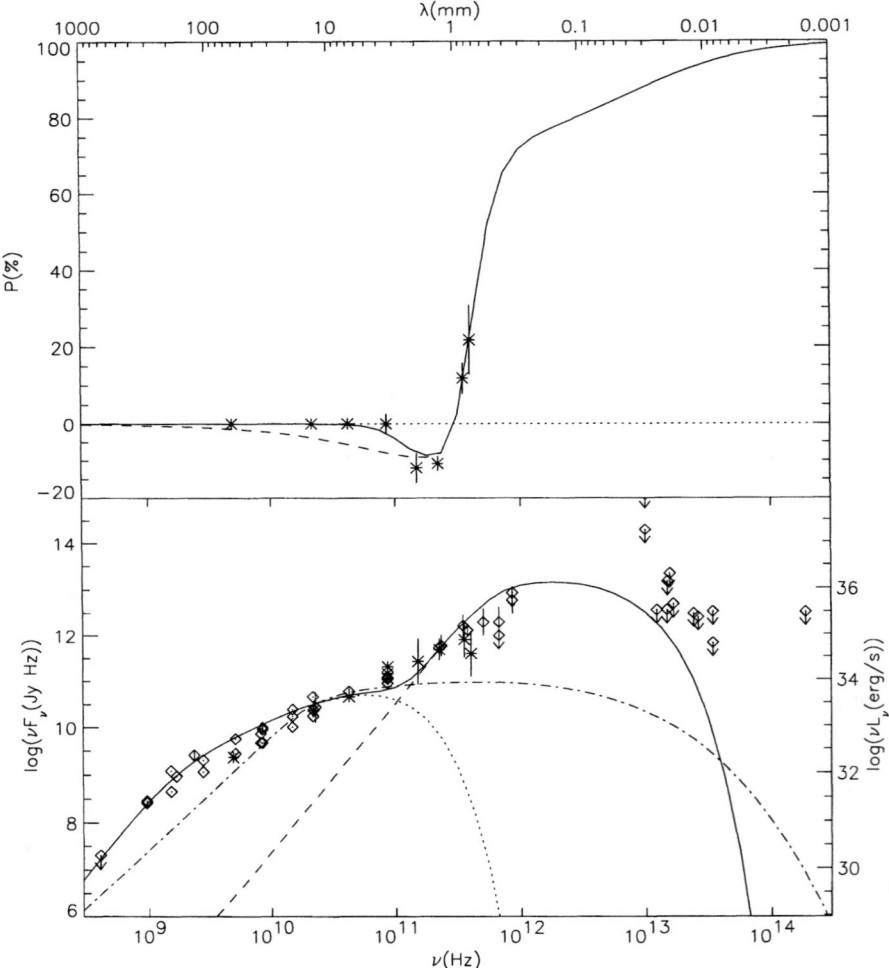

FIGURE 1. Fig. 1: Polarization and spectral energy distribution of Sgr A* compared to model. The dashed line shows the polarized component, the dotted line the unpolarized, mono-energetic component, and the solid line the sum of the two. The dot-dash line shows the maximum CDAF model (assumed to be unpolarized; the total polarization is similar if the CDAF replaces the monoenergetic component). The diamonds are the data compiled by Narayan et al. (1995), while the asterisks are the data from Bower et al. (1999a,b) and A00.

hoc, this model reproduces the spectrum well. The Faraday rotation parameter is rather small, so depolarization requires field which is tangled on a scale ~ 100 times smaller than the size of the emission region.

Since the self-absorption frequency occurs at $\sim 500\mu m$, it will be possible to image shadow of a black hole from the ground using VLBI, providing a direct confirmation of the existence of an event horizon (Falcke, Melia, & Agol 2000). Future sub-mm polarimetric VLBI observations might show rotation of the polarization angle near the black hole, a general relativistic effect which becomes stronger for a spinning black hole (Connors, Stark, & Piran 1980).

This contribution is an abridged version of Agol (2000).

REFERENCES

1. Agol, E., 2000, ApJL, 538, 121
2. Aitken, D. K., et al., 2000, ApJ, in-press, astro-ph/0003379, A00
3. Begelman, M. C. & Blandford, R. D., 1999, MNRAS, 303, L1
4. Bower, G. C., Backer, D. C., Zhao, J. H., Goss, M., & Falcke, H. 1999a, ApJ, 521, 582
5. Bower, G. C., Wright, M. C. H., Backer, D. C., & Falcke, H. 1999b, ApJ, 527, 851
6. Beckert, T. & Duschl, W. J., 1997, A&A, 328, 95
7. Coker, R. & Melia, F., 1999, ApJ, 511, 750
8. Falcke, H., Mannheim, K., & Biermann, P. L., 1993, A&A, 278, L1
9. Falcke, H., Melia, F., & Agol, E., 2000, ApJ, 528, L13
10. Genzel, R., Eckart, A., Ott, T., & Eisenhauer, F., 1997, MNRAS, 291, 219
11. Ghez, A. M., Klein, B. L., Morris, M., & Becklin, E. E. 1998, ApJ, 509, 678
12. Ghez, A. M., Morris, M., Becklin, E. E., Tanner, A., & Kremenek, T., 2000, Nature, Sept. 21 issue, astro-ph/0009339
13. Ginzburg, V. L. & Syrovatskii, S. I., 1965, ARA&A, 3, 297, GS
14. Ginzburg, V. L. & Syrovatskii, S. I., 1969, ARA&A, 7, 375, GS
15. Ichimaru, S., 1977, ApJ, 214, 840
16. Igumenshchev, I. V. & Abramowicz, M. A., 1999, MNRAS, 303, 309
17. Igumenshchev, I. V. & Abramowicz, M. A., 2000, astro-ph/003397
18. Krichbaum, T. P. et al. 1998, A&A, 335, L106
19. Lo, K. Y., Shen, Z. Q., Zhao, J. H., & Ho, P. T. P. 1998, ApJ, 508, L61
20. Melia, F., 1992, ApJ, 387, L25
21. Narayan, R. & Yi, I., 1994, ApJ, 428, L13
22. Narayan, R., Yi, I., & Mahadevan, R., 1995, Nature, 374, 623
23. Narayan, R., Igumenshchev, I. V. & Abramowicz, M. A., 2000, astro-ph/9912449
24. Özel, F., Psaltis, D., & Narayan, R., 2000, ApJ, in press, astro-ph/0004195
25. Quataert, E., Narayan, R., & Reid, M. J., 1999, ApJ, 517, L101
26. Quataert, E. & Gruzinov, A., 2000, ApJ, in press, astro-ph/9912440
27. Stone J. M., Pringle J. E., & Begelman M. C., 1999 MNRAS, 310, 100
28. Stone J. M. & Pringle J. E., 2000, astro-ph/0009233

Models for Type Ia Supernovae and Evolutionary Effects with Redshift

P. Höflich

Department of Astronomy, University of Texas, Austin, TX 78681, USA

Abstract. Based on detailed models for the explosions, light curves and NLTE-spectra, evolutionary effects of Type Ia Supernovae (SNe Ia) with redshift have been studied to evaluate their size on cosmological time scales, how the effects can be recognized and how one may be able to correct for them.

In the first part, we summarize the current status of scenarios for Type Ia Supernovae, including the explosion of a Chandrasekhar mass white dwarf (M_{Ch}-WD), the merging scenario and the helium-triggered explosions of low-mass WDs. We show that delayed detonation models can account for the majority of observations of spectra and light curves. IR observations are a new and powerful tools to constrain explosion models and details of the flame propagation in the WD. A strong Mg II line at 1.05 μm shows that nuclear burning takes place at the outer, low density layers. This requires a transition from the deflagration to the detonation regime of the nuclear burning front, or a very fast deflagration close to the speed of sound. We put the models into context with the empirical brightness decline relation which is widely applied to use SNe Ia as yardsticks on cosmological distance scales. This relation can be well understood in the framework of M_{Ch}-WDs as a consequence of the opacity effects in combination with the amount of ^{56}Ni which determines the brightness. However, the amount of ^{56}Ni actually produced depends on a combination of free parameters such as central density and the chemical composition of the WD, and the propagation of the burning front. We get a spread of $\approx 0.4^m$ around the mean relation which is larger than currently favored by observations ($\approx 0.12^m$, [40]) which may hint of an underlying coupling of the progenitor, the accretion rates and the propagation of the burning front.

In a second part, we investigate the possible evolutionary effects in SNe Ia both with respect to changes in the sample of SNe Ia and individual variations, and how they can be identified. We find that evolution may produce an offset in the brightness decline relation is restricted to a few tenth of a magnitude. The effects reveal themself by changes in the U and UV fluxes, and in a change in the maximum brightness/decline relation by $\Delta M \approx 0.1 \times \Delta t$ where Δt is the difference between local and distant SN-samples. According to new data [1], $\Delta t \leq 1d$ and, thus, evolution is unlikely to eliminate a need for Λ.

INTRODUCTION

Two of the important developments in observational SN-research in the last few years were to establish the long-suspected correlation between the peak brightness of SNe Ia and their rate of decline by means of modern CCD photometry [38], and the exact distance calibrations provided by an HST Key Project (e.g. [41]). Both allowed a empirical determinations of H_o with unprecedented accuracy [11]. Independent from these calibrations and empirical relations, H_o has been determined by comparisons of detailed theoretical models for light curves and spectra with observations [27,16,35]. All determinations of the Hubble constant are in good agreement with one another. Recently, the routine successful detection of SNe at large redshifts, z [39,40], has provided an exciting new tool to probe cosmology. This work has provided results that are consistent with a low matter density in the Universe and, most intriguing of all, yielded hints for a positive cosmological constant. The cosmological results rely on empirical brightness-decline relations which are calibrated locally. One of the main concerns are systematic, evolutionary effects in the properties of SNe Ia. Both from theory and observations, there are some hints for the presents of evolution.

Observational evidence include the finding of Branch et al. [5] who have shown that the mean peak brightness is dimmer in ellipticals than in spiral galaxies. Wang, Höflich & Wheeler [47] found that the peak brightness in the outer region of spirals is similar to those found in ellipticals, but that in the central region both intrinsically brighter and dimmer SNe Ia occur. This implies that the progenitor populations are more inhomogeneous in the inner parts of spirals which contain both young and old progenitors.

From theory, time evolution is expected to produce the following main effects: (a) a lower metallicity will decrease the time scale for stellar evolution of individual stars by about 20 % from Pop I to Pop II stars [42] and, consequently, the progenitor population which contributes to the SNe Ia rate at any given time. The stellar radius also shrinks. This will influence the statistics of systems with mass overflow; (b) early on, we expect that systems with shorter life time will dominate the sample whereas, today, old system are contributing which may have not occurred early on. In addition, some scenarios with a life time comparable to the age of the univers such as two merging WDs may be absent a few Gyrs ago, but they may contribute today. (c) The initial metallicity will determine the electron to nucleon fraction of the outer layers and hence affects the products of nuclear burning; (d) Systems with a shorter life time may dominate early on and, consequently, the typical C/O ratio of the central region of the WD will be reduced; (e) The properties of the interstellar medium may change; (f) The influence of Z on the structure of WDs may change, but this effect has remains very small (e.g. 2% when comparing solar with 0.01 solar, [19]) (g) The distribution of C and O will depend on Z as it influences the 'normal' stellar evolution and the properties of the C/O core [42] (h) The metallicity will effect nuclear burning during the accretion phase of the progenitor [34].

Based on theoretical models described in §III, we want to study the possible effects of the change of the progenitor and its metallicity on the light curves and spectra of SNe Ia. Note that most of the results have been obtained in serveral collaborations over the years (see acknowledgments).

NUMERICAL TOOLS

Most of the results discussed in the following sections are based on our calculations which are consistent with the explosion, light curves and spectra. In some cases, the stellar evolution and the accretion on the WD is treated in detail.

Stellar Evolution: The stellar models have been calculated using the evolutionary code Franec (e.g. [43,6,8]) or are provided by Nomoto's group [46]. Subsequently, the evolution of the C/O core is followed up by accreting H/He rich material at a given accretion rate on the core by solving the standard equations for stellar evolution using a Henyey scheme. Nomoto's equation of state is used [32]. Crystallization is neglected. For the energy transport, conduction [22], convection in the mixing length theory, and radiation is taken into account. Radiative opacities for free-free and bound free transitions are approximated in Kramer's approximation and by free electrons. A nuclear network of 35 species up to ^{24}Mg is used.

Hydrodynamics: The explosions are calculated using a one-dimensional radiation-hydro code, including nuclear networks ([16] and references therein). This code solves the hydrodynamical equations explicitly by the piecewise parabolic method [7] and includes the solution of the frequency averaged radiation transport implicitly via moment equations, expansion opacities (see below), and a detailed equation of state. Nuclear burning is taken into account using a network which has been tested in many explosive environments (see [44] and references therein).

Light Curves: Based on the explosion models, the subsequent expansion and bolometric as well as monochromatic light curves are calculated using a scheme recently developed, tested and widely applied to SN Ia (e.g. [13,19]). The code used in this phase is similar to that described above, but nuclear burning is neglected and γ ray transport is included via a Monte Carlo scheme. In order to allow for a more consistent treatment of scattering, we solve both the (two lowest) time-dependent, frequency averaged radiation moment equations for the radiation energy and the radiation flux, and a total energy equation. At each time step, we then use $T(r)$ to determine the Eddington factors and mean opacities by solving the frequency-dependent radiation transport equation in the comoving frame and integrate to obtain the frequency-averaged quantities. About 1000 frequencies (in one 100 frequency groups) and about 500 to 1000 depth points are used. The averaged opacities have been calculated under the assumption of local thermodynamic equilibrium. Both the monochromatic and mean opacities are calculated using the Sobolev approximation. The scattering, photon redistribution and thermalization terms used in the light curve opacity calculation are calibrated with NLTE calculations using the formalism of the equivalent-two-level approach [14].

Spectral Calculations: Our non-LTE code (e.g. [14,19]) solves the relativistic radiation transport equations in comoving frame. The energetics of the SN are calculated. The evolution of the spectrum is not subject to any tuning or free parameters. The spectra are computed for various epochs using the chemical, density and luminosity structure and γ-ray deposition resulting from the light curve code providing a tight coupling between the explosion model and the radiative transfer. The effects of instantaneous energy deposition by γ-rays, the stored energy (in the thermal bath and in ionization) and the energy loss due to the adiabatic expansion are taken into account. The radiation transport equations are solved consistently with the statistical equations and ionization due to γ radiation for the most important elements (C, O, Ne, Na, Mg, Si, S, Ca, Fe, Co, Ni). Besides 40,000 lines treated in full non-LTE, $\approx 10^6$ additional lines are included assuming LTE-level populations and an equivalent-two-level approach for the source functions.

MODELS FOR THERMONUCLEAR SUPERNOVAE

There is general agreement that SNIa result from the thermonuclear explosion of a degenerate white dwarf (HF60). Within this general picture, three classes of models have been considered: (1) An explosion of a CO-WD, with mass close to the Chandrasekhar mass, which accretes mass through Roche-lobe overflow from an evolved companion star [30]. The explosion is then triggered by compressional heating near the WD center. (2) An explosion of a rotating configuration formed from the merging of two low-mass WDs, caused by the loss of angular momentum due to gravitational radiation [48,21,37]. The explosion may also be triggered near the center by the compression. (3) Explosion of a low mass CO-WD triggered by the detonation of a helium layer [31,51]. Only the first two models appear to be viable. The third, the sub-Chandrasechar WD model, has been ruled out on the basis of predicted light curves and spectra [18,36].

From the theoretical standpoint, one of the key questions is how the flame propagates through the WD. Several burning models have been proposed in the past, including detonation [2,12], deflagration [23,29], and the delayed detonation model (DD) [26,50,52]. The DD-model assumes that burning starts as a subsonic deflagration with a certain speed $S_{\text{def}} < a_s$, where a_s is the sound speed, and then undergoes a transition to a supersonic detonation. The detonation speed follows directly from the standard Hugoniot relations. However, due to the one-dimensional nature of current model, the speed of the subsonic deflagration and the moment of the transition to a detonation are free parameters, or calibrated by 3-D calculations. The moment of deflagration-to-detonation transition (DDT) is conveniently parameterized by introducing the transition density, ρ_{tr}, at which DDT happens. Recently, significant progress has been made toward a better understanding of the propagation of nuclear burning fronts during the deflagration phase. The models provide good results during the phase when the Ryleigh-Taylor instabilites dominate but the results depend sensitively on the assumed sub-grid models at later phase of the flame propagation [24,25,28], and the DDT-transition is not understood quanitatively. Therefore, further constraints must come from the observations.

What we observe as a SN is not the explosion itself, but the light emitted from a rapidly expanding ejecta produced by the explosion. As the photosphere recedes, deeper layers of the ejecta become visible. A detailed analysis of multi-band light curves and spectra gives us the opportunity to determine the density, velocity and composition structure of the ejecta, and to constrain the physical model of the explosion. The major results of one-dimensional modeling and comparison with observations are summarized below.

The best current explosion model, the delayed detonation and pulsating delayed detonation explosion models of Chandrasekhar mass carbon-oxygen WDs can account for the spectral and light curve evolution of both "normally bright" and subluminous SNIa in the optical and IR (see Fig. 1, [14–16,49]. Within this framework, we can understand for often used brightness decline relation. For a plausible range in the transition density, $\rho_{tr} \simeq (1.5 - 2.5) \times 10^7$ g cm^{-3} [16], sufficient thermonuclear energy is generated by burning nearly the entire WD to provide the observed expansion velocities, a small spread in the explosion energy [14]. The amount of ^{56}Ni varies between $\simeq 0.1 - 0.7$ M$_\odot$. The variation of M_{Ni} gives a range in maximum brightness that matches the observations (Fig. 2). The models with less nickel are not only dimmer, but are cooler and have lower opacity, giving

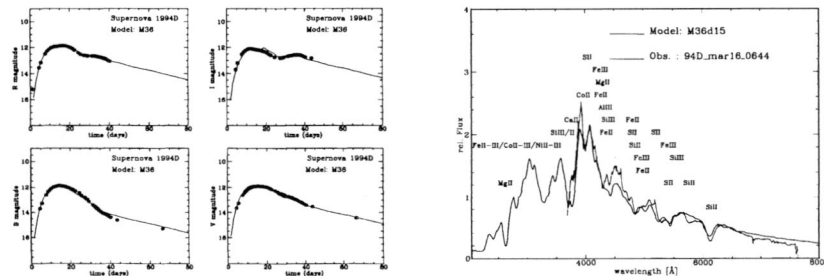

FIGURE 1. Observed LCs of SN 1994D in comparison with the theoretical LCs of a typical delayed detonation model, and the corresponding synthetic spectrum around bolometric maximum in comparison to the observations of SN1994D at Mar. 16th (from Höflich 1995).

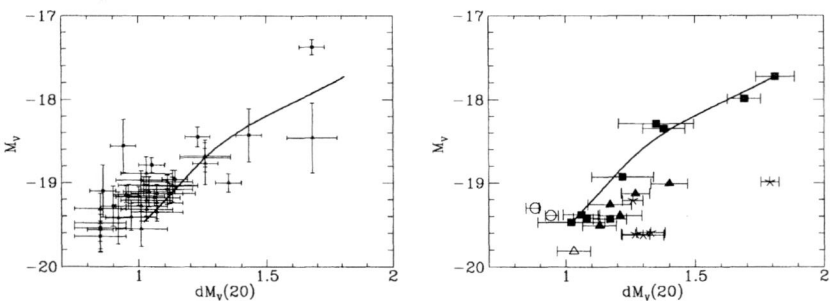

FIGURE 2. (Left panel) Observed light curve maximum brightness - decline rate relation. M_V is presented as a function of the decline from maximum at 20 days. (Right panel) The predicted relation for an array of models of SNe Ia representing delayed detonations (open triangles), pulsating delayed detonations (filled circles), merging models (open circles) and helium detonations (asterisks). For both the delayed detonation and merger scenarios models are only considered if they allow for a representation of some of the observed SNe Ia (Höflich et al. 1996).

them redder, more steeply declining light curves, in agreement with the observations [13,17]. The amount of ^{56}Ni depends primarily on ρ_{tr}, and to a much lesser extent on the assumed value of deflagration speed, initial central density of the exploding star, and the initial chemical composition (ratio of carbon to oxygen). This is the basis of why, to first approximation, SNIa appear to be a one-parameter family. Nonetheless, variations of the other parameters also lead to some variations of the predicted properties of SNIa, which indicate that the assumption of a one-parameter family is not strictly valid. We get a spread around the mean relation of $\approx 0.4^m$ which is consistent with the spread based on the CTIO data published by Hamuy et al. [11] but larger than suggested by recent observations ($\approx 0.12^m$ [40]). This narrow a spread cannot be understood in the context of current models but it may hint of an underlying coupling of the progenitor, the accretion rates and the propagation of the burning front.

During the last 2 to 3 years, near IR spectra became available and provided a new insights. Firstly, the comparison with observed IR spectra with model predictions was an important test for the models since there was no a priori guarantee that a model that matched optical light curves would produce satisfactory IR spectra (Fig. 3, [49]). The broad feature between 1.5 and 1.8 μm is produced by a large number of Fe, Co and Ni lines and indicates the size of the ^{56}Ni rich regions. A very strong Mg II line at about 1.05 μm at early times can be understood as a natural consequence of burning in DD-models. Because Mg is produced in the region of explosive carbon burning, the Doppler shift provides a unique tool to determine the transition zone from explosive carbon to oxygen burning. Its high, minimum velocity clearly demonstrates the need for a DDT transition or a deflagration speed well in excess of those of the deflagration model W7 [33], and it rules out strong

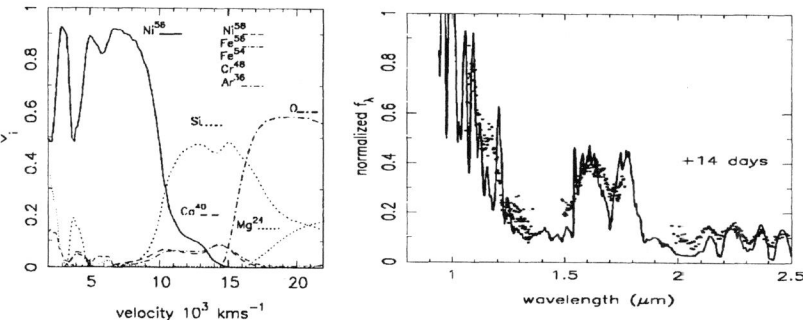

FIGURE 3. Final chemical structure of a delayed detonation model as calculated in 1-D in comparison and the theoretical IR-spectrum in the IR at about 2 weeks after maximum light in comparison with the observed spectrum of 1976g at about +10 days by Bowers et al. (1997). The emission feature between 1.5 and 1.8 μm is produced by a large number of Fe, Co and Ni lines. The ^{56}Ni layers are optically thick at these wavelengths. It appears as soon as the 'photosphere' at adjoining wavelengths recedes to velocities smaller than the outer edge of the ^{56}Ni layers (from Wheeler, Höflich, Harkness and Spyromillo 1998).

mixing as a possible explanation for the presence of high velocity Si frequently seen in 'normal' bright SNe Ia [3,9].

Recently, IR for the very subluminous SN1998by by Gerardy et al. [10] shows no Ni at layers with expansion velocities larger than $\approx 5000 km/sec$. Progress in combustion in SNeIa predicted (e.g. [24]) that, during the deflagration phase, some Ni bubbles may rise to about half of the WD-radius corresponding to 0.4 $M_\odot$ which may be inconsistent with results of the 1-D spectra (Fig.4) because, in subluminous SNe Ia, most of the ^{56}Ni is produced during the deflagration phase. The solution to this problem may prove to be the Rosetta stone for understanding the influence of the preconditioning of the WD, i.e. the progenitor star, its rotation etc., and of propagation of burning fronts. Eventually, it may help to answer the question whether a SNIe Ia is subluminous because its special initial conditions influence the burning front, or whether current models for the deflagration need to be modified.

EVOLUTIONARY EFFECTS

A change of the main sequence mass of the typical progenitor will change the size the region of central He burning during the stellar evolution and, consequently, the size and C/O ratio in the WD which, eventually, may explode as a SNe Ia. The initial metallicity Z of the WD is inherented from the molecular cloud from which the star has been formed. The influence of Z and C/O ratio on light curves and spectra has been studied for the example of a set of DD-models with the basic properties as follows: central density of the WD, $\rho_c = 2.6 \times 10^9 g/cm^{-3}$, $v_{burn} = \alpha * v_{sound}$ with $\alpha = 0.03$ during the deflagration phase and a transition to detonation at $\rho_{tr} = 2.7 \times 10^7$. The quantities of Table 1 in columns 2 to 5 and 6

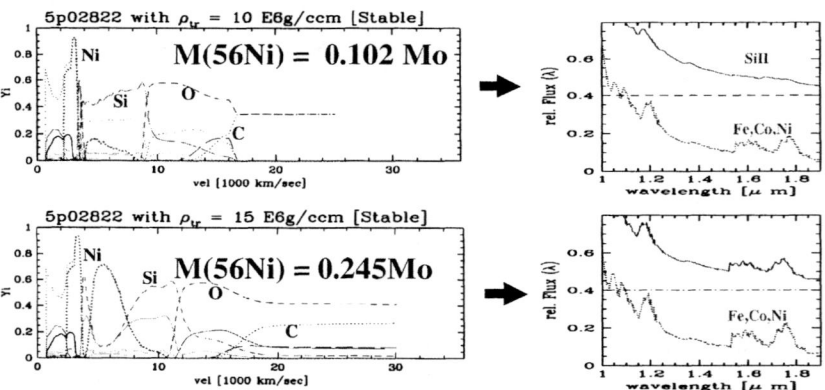

FIGURE 4. Final chemical structure of two strongly subluminous delayed detonation model and the corresponding, theoretical IR-spectrum at 1 and 2 weeks after maximum light. In the strongly subluminous SN1998by, the 1.5 to 1.8 μm feature was not observed until about 2 weeks after maximum light. To be consistent with this observation, the abundance ^{56}Ni must be less than $\approx 10\%$ at layers with expansion velocities in excess of $\approx 5000 km/sec$ Gerardy et al. (2000).

TABLE 1. Basic parameters for the delayed detonation models.

Model	C/O	R_Z	E_{kin}	M_{Ni}	Model	C/O	R_Z	E_{kin}	M_{Ni}
DD21c	1/1	1/1	1.32	0.69	DD25c	1/1	3/1	1.32	0.69
DD23c	2/3	1/1	1.18	0.59	DD26c	1/1	1/10	1.32	0.73
DD24c	1/1	1/3	1.32	0.70	DD27c	1/1	10/1	1.32	0.69

to 9 are: C/O ratio; R_Z the Z relative to solar by mass; E_{kin} kinetic energy (in $10^{51} erg$); M_{Ni} mass of ^{56}Ni (in solar units). The parameters are close to those which reproduce both the spectra and light curves reasonably well [33,14].

Direct influence of C/O: As the C/O ratio of the WD is decreased from 1 to 2/3, the explosion energy and the ^{56}Ni production are reduced and the Si-rich layers are more confined in velocity space (Fig. 5). A reduction of C/O by about 60 % gives slower rise times by about 3 days and an increased luminosity at maximum light, a somewhat faster post-maximum decline and a larger ratio between maximum light and the ^{56}Co tail ([19], see also Fig.8). The slight increase in luminosity at maximum light is caused by the smaller expansion rate. Consequently, less energy stored early on is wasted in expansion work, but contributes to the luminosity. The smaller ^{56}Ni production causes the reduction of the luminosity later on.

A reduction of the C/O ratio has a similar effect on the colors, light curve shapes and element distribution as a reduction in the deflagration to detonation transition density but, for the same light curve shape, the absolute brightness is larger for smaller C/O. Moreover, the kinetic energy is reduced by about 10 % (Table 1) and, consequently, the expansion velocity derived by the Doppler shift in the spectra becomes smaller by about 5% . An independent determination of the initial C/O ratio and the transition density is possible for local SN if detailed analyses of both the spectra and light curves are performed simultaneously.

Direct influence of the metallicity: To test the influence of the metallicity for nuclei beyond Ca, we have constructed models with parameters identical to DD21c but with initial metallicities between 0.1 and 10 times solar (Table 1). The energy release, the density and velocity structure are virtually identical to that of DD21c

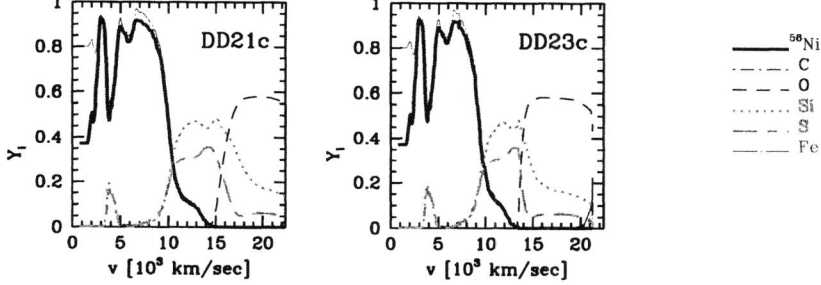

FIGURE 5. Abundances as a function of the final expansion velocity for the delayed detonation models DD21c and DD23c. Both the initial ^{56}Ni and the final Fe profiles are shown.

and, consequently, the bolometric (and also monochromatic) optical light curves are rather insensitive. The main influence of Z is a slight increase of the ^{56}Ni mass with decreasing Z due to a higher Y_e. The reason is that Z mainly effects the initial CNO abundances of a star. These are converted during the pre-explosion stellar evolution to ^{14}N in H-burning and via $^{14}N(\alpha,\gamma)^{18}F(\beta^+)^{18}O(\alpha,\gamma)^{22}Ne$ to nuclei with N=Z+2 in He-burning. The result is that increasing Z yields a smaller proton to nucleon ratio Y_e throughout the pre-explosive WD. Higher Z and smaller Y_e lead to the production of more neutron-rich Fe group nuclei and less ^{56}Ni (Fig. 6). For lower Z and, thus, higher Y_e, some additional ^{56}Ni is produced at the expense of ^{54}Fe and ^{58}Ni [45]. The temperature in the inner layers is sufficiently high during the explosion that electron capture determines Y_e. In those layers, Z has no influence on the final burning product. The main differences due to changes in Z are in regions with expansion velocities in excess of ≈ 12000 km/sec. Most remarkable is the change in the ^{54}Fe production which is the dominant contributor to the abundance of iron group elements at these velocities since little cobalt has yet decayed near maximum light (Fig. 6).

The initial WD composition has been found to have rather small effects on the overall LCs. The ^{56}Ni production and hence the bolometric and monochromatic optical and IR light curves differ only by a few hundredths of a magnitude. This change is almost entirely due to the small change in the ^{56}Ni production and hardly

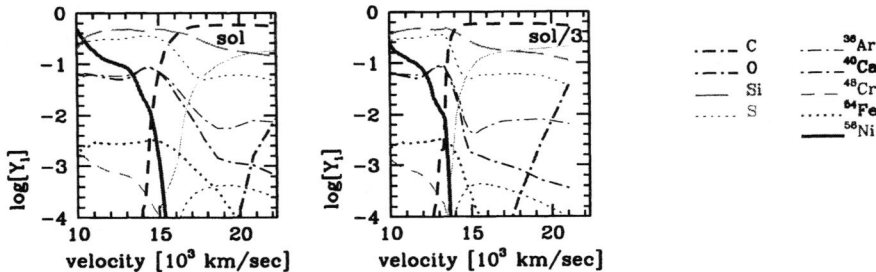

FIGURE 6. Abundances of different isotopes as a function of the expansion velocity for DD21c with initial abundances of solar and solar/3.

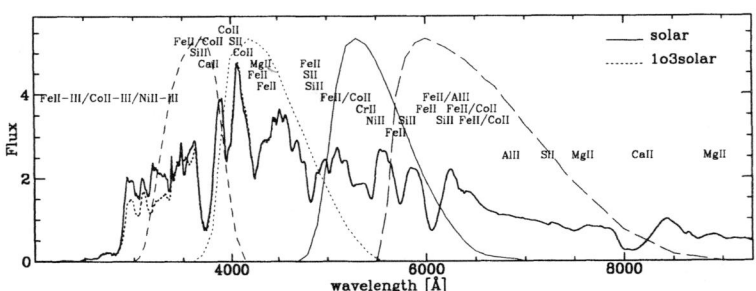

FIGURE 7. Comparison of synthetic NLTE spectra at maximum light for Z solar (DD21c) and 1/3 solar (DD24c). The standard Johnson filter functions for UBV, and R are also shown.

due to a change in the opacities because the diffusion time scales are governed by the deeper layers where burning is complete. However, the short wavelength part of the spectrum ($\lambda \leq 4000\text{Å}$) at maximum light is affected by a change in Z (Fig.7). This provides a direct test for Z of local SNe and, thus, may give a powerful tool to unravel the nature (and lifetime) of SNe Ia progenitors.

By 2 to 3 weeks after maximum, the spectra are completely insensitive to the initial Z because the spectrum is formed in even deeper layers where none of the important abundances are effected by Z. Thus, for two similar bright SNe with similar expansion velocities, a comparison between the spectral evolution can provide a method to determine the difference in Z, or it may be used to detect evolutionary effects for distant SNe Ia if high quality spectra are available.

Influence of the Stellar Evolution on the WD Structure: Up to now, we have neglected the influence of the metallicity and the mass of the progenitor on the structure of the initial WD for a given mass on the main sequence. Such dependencies may become of important if SNe are observed at large distance. On cosmological distance scales, Z is expected to be correlated with redshift. At the time of the explosion, the WD masses are close to the Chandrasekhar limit. The WD has grown by accretion of H/He and subsequent burning from the mass of the central core of a star with less than $\approx 7 M_\odot$ (Fig. 6). In the accreted layers, the C/O ratio is close to 1; however, the initial mass of the C/O WD is determined by the stellar evolution. The core mass depends on M_{MS} of the progenitor and on Z.

Here, we want to discuss the size of the metallicity effect using the example of a 7 $M_\odot$ model with Z=0.02 and 0.004 (Fig.8). Z mainly effects the convection during the stellar Helium burning and, consequently, the size of the C/O core and the central C/O ratio. We note that the exact size of the effect and its sign depends on mass of the progenitor at the zero age main sequence, and Z. Even for a given mass, the changes are not monotonic, but may change sign from Pop I to Pop II to Pop III [8,46]. In addition, the tendency depends sensitively on the assumed physics such as the $^{12}C(\alpha,\gamma)^{16}O$-rate (e.g. [43]). Therefore our example can serve as a guide to estimate the size of this effect.

In agreement with above, the total C/O mass determines the explosion energy and, consequently, has the main effect on the light curves. After accretion on the initial core M_C, the total mass fraction of M_C/M_{Ch} is 0.75 and 0.61 for the models with Z=0.04 and 0.02 respectively. At the time of the explosion, $\rho_c = 2.4 \times 10^9 g/cm^{-3}$. For the burning, $\alpha = 0.02$ and $\rho_{tr} = 2.4\ 10^7 g\ cm^{-3}$).

Monochromatic LCs are shown in Fig. 8. As expected from the last section, the main effect on the light curves is caused by the different expansion ratio determined by the integrated C/O ratio. The change in the maximum brightness remains small ($M_V(Z = 0.02) - M_V(Z = 0.04) = -0.02^m$) and the rise times are different by about 1 day ($t_V((Z = 0.02) = 17.4d$ vs. $t_V(Z = 0.04) = 16.6d$. The most significant effect is a steeper decline ratio and a reduced ^{56}Ni production for the model with solar Z. This is mainly due to the slower expansion ratio. This translates into a systematic offset of $\approx 0.1^m$ in the maximum brightness decline ratio [11]. Using either the streching method or the LCS-method gives similar offsets.

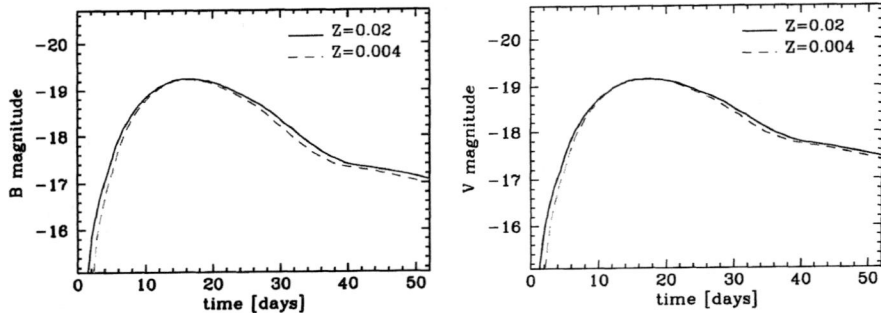

FIGURE 8. Comparison of the final chemical profile of the C/O core of a star with $M_{MS} = 7 M_\odot$ (right) with Z=0.02 and 0.004, and the corresponding V light curves (left) of delayed detonation models with identical central density and descriptions for the burning front (see text).

Worth noting is the following trend: For realistic cores, the mean M_C/M_O tends to be smaller than the canonical value of 1 used in all calculations prior to 1998 (e.g. [33,50,16]). Consequently, as a general trend, the rise times are about 1-3 days slower in all models based on the detailed WD structure compared to the models published prior to 1996 ([16], and references therein).

ACKNOWLEDGMENTS

Most of the results reviewed here have been obtained in collaborations with collegues too numerous to be listed all as coauthors. Therefore, I would like to thank here, in particular, A. Chieffi, I. Dominguez, R. Fesen, C. Gerardy, A. Khokhlov, M. Limongi, E. Müller, K. Nomoto, J. Spyromillo, Y.Stein, O. Straniero, F.K. Thielemann, H.Umeda, L. Wang & J.C. Wheeler. This work is supported in part by NASA Grant LSTA-98-022.

REFERENCES

1. Aldering G., Knop R., Nugent P. 2000, AJ, accepted & astro-ph/0001049
2. Arnett W. D. 1969, Ap. Space Sci., 5, 280
3. Benetti B. 1989, PhD-thesis, University of Padua/Italy
4. Bowers, E.J.C. et al., 1997, MNRAS, 290, 663
5. Branch D., et al. 1998, Phys. Rep., in press
6. Chieffi, A., Limongi, M., and Straniero, O. 1998, ApJ 502, 737
7. Collela P., Woodward P.R. 1984,J.Comp.Phys. **54**, 174
8. Dominguez I. Höflich P., Straniero O., Wheeler J.C. 1998, in the Cosmos V, Eds. N. Prantzos & S. Harissopulos, Editions Frontieres, Paris, p. 248
9. Fisher E. et al. 1998, MNRAS 304, 67
10. Gerardy C.L., Höflich P., Fesen R., Wheeler J.C. 2000, ApJ submitted
11. Hamuy M. et al. 1996,AJ **112**, 2438

12. Hansen C.J., Wheeler J.C. 1969, Ap. Space Sci., 3, 464Hansen C.J., Wheeler J.C. 1969, Ap. Space Sci., 3, 464
13. Höflich, P.A, Khokhlov, A.M., Müller, E., 1993, AA, 259, 549
14. Höflich P. 1995, ApJ 443, 89
15. Höflich P., Khokhlov A., Wheeler J.C. 1995, ApJ 444, 831
16. Höflich P., Khokhlov A. 1996, ApJ 457, 500
17. Höflich P., Khokhlov A., Wheeler J.C., Phillips M., Suntzeff N., Hamuy M., 1996, ApJ 472, 81
18. Höflich P., Dominik C., Khokhlov A., Nomoto K., Thielemann K., Wheeler J.C. 1996b, in: Type Ia Supernovae, eds. R. Canal et al., Kluver, p. 659
19. Höflich P., Wheeler J.C., Thielemann F.K 1998, ApJ 495, 617
20. Hoyle F., Fowler W. A. 1960, ApJ 132, 565
21. Iben I.Jr., Tutukov A.V. 1984, ApJS, 54, 335
22. Itoh N., Mitake S., Iyetomi H., Ichimaru S. 1983, ApJ 273, 774
23. Ivanova I.N., Imshennik V.S., Chechetkin V.M. 1974, ApSS, 31, 497
24. Khokhlov A. 1995, ApJ, 449, 695
25. Khokhlov A., Oran E.S., Wheeler J.C. 1997a, ApJ, 478, 673
26. Khokhlov A. 1991ab , A&A 245, 114 & L25
27. Müller E., Höflich P. 1994, A&A 281, 51
28. Niemeyer J.C., Hillebrandt W. 1995, ApJ, 452, 779
29. Nomoto K., Sugimoto S., & Neo S. 1976, ApSS, 39, L37
30. Nomoto K., Sugimoto D. 1977, PASJ, 29, 765
31. Nomoto K. 1980, IAU-Symp., 93, D. Reidel, p. 295
32. Nomoto K. 1982, ApJ, 253, 798
33. Nomoto K., Thielemann F.-K., Yokoi K. 1984, ApJ 286, 644
34. Nomoto K., Umeda H., Hachisu I., Kato M., Kobayashi C., Tsujimoto T. 2000, in "SNeIa: Theory and Cosmology", eds. J. Truran & J. Niemeyer, Cambridge University Press
35. Nugent P., et al. 1996, Phys. Rev. Let. 75, 394 & 1974E
36. Nugent P., Baron E., Branch D., Fisher A., Hausschild P. 1997, ApJ 485, 812
37. Paczyński B.,1985, in: Cataclysmic Variables and Low-Mass X-Ray Binaries, eds. D.Q. Lamb & J. Patterson, Reidel,Dordrecht, p. 1
38. Phillips M.M. et al. 1987, PASP 90, 592
39. Perlmutter S. et al. 1998, ApJ, in press and astro-ph/9812473
40. Riess A. et al. 1998, AJ 116, 1009
41. Saha A. et al. 1997, ApJ 486, 1
42. Schaller G., Schaerer D., Meynet G., Maeder A. 1992, A&A Suppl 96, 269
43. Straniero, O., Chieffi, A., and Limongi, M. 1997, ApJ 490, 425
44. Thielemann F.K., Nomoto K., Hashimoto M. 1996, ApJ 460, 408
45. Thielemann F.K., Nomoto K., Yokoi K. 1986, A&A 158, 17
46. Umeda H., Nomoto K., Yamaoka H., Wanajo S. 1999, ApJ 513, in press
47. Wang L., Höflich P., Wheeler, J.C. 1997, ApJ 487, L29
48. Webbink R.F. 1984, ApJ, 277,355
49. Wheeler, J.C., Höflich, P.A., Harkness R., Spyromillo J. 1998, ApJ, 496, 908
50. Woosley S. E. & Weaver T. A. 1994, Proc. of Les Houches Session LIV, eds. Bludman

et al., North-Holland, 63
51. Woosley S.E., Weaver T.A., 1986,ARAA,24
52. Yamaoka H., Nomoto K., Shigeyama T., Thielemann F., 1992, ApJ, 393, 55

ASTRO-HADRON PHYSICS
AND
PARTICLE ASTROPHYSICS

Little Bang at Big Accelerators: Heavy Ion Physics from AGS to LHC

J. Schukraft

CERN Div. PPE, CH-1211 Geneva 23

Abstract. The field of ultra-relativistic heavy ion physics, which started some 10 years ago at the Brookhaven AGS and the CERN SPS with fixed target experiments, has entering today a new era with the recent (July 2000) start-up of the Relativistic Heavy Ion Collider RHIC and preparations well under way for a new large heavy ion experiment at the Large Hadron Collider LHC. This overview, which is the combined write-up of talks given at this conference [1] and in [2], will sketch a rough picture of the heavy ion program at current and future machines and concentrate on a few important topics, in particular the question if current results show any of the signs predicted for the phase transition between normal hadronic matter and the Quark-Gluon Plasma.

INTRODUCTION

The aim of high-energy heavy-ion physics is the study of strongly interacting matter at extreme energy densities. Statistical QCD predicts that, at sufficiently high density, there will be a transition from hadronic matter to a plasma of deconfined quarks and gluons — a transition which in the early universe took place in the inverse direction some 10^{-5} s after the Big Bang and which might play a role still today in the core of collapsing neutron stars. The study of the phase diagram of nuclear matter (see Fig. 1), utilising methods and concepts from both nuclear and high-energy physics, constitutes a new and interdisciplinary approach in investigating matter and its interactions. In high-energy physics, interactions are derived from first principles (gauge theories), and the matter concerned consists mostly of single particles (hadrons/quarks). In contrast, on nuclear physics scales the strong interaction is shielded and can, therefore, to date only be described in effective theories, whereas matter consists of extended systems with collective features. Combining the *elementary-interaction* aspect of high-energy physics with the *macroscopic-matter* aspect of nuclear physics, the subject of heavy-ion collisions is QCD thermodynamics, i.e. the study of bulk matter consisting of strongly interacting particles (hadrons/partons). The formalism to be used would ideally be the one of thermodynamics, where complex multi-particle states are described in terms of a few macroscopic variables.

The study of the QGP is of interest to explore and test QCD on its natural scale (Λ_{QCD}) and addresses the fundamental questions of confinement and chiral-symmetry breaking, which are connected to the existence and properties of the quark-gluon plasma. Moreover, it is of general relevance in understanding the dynamical nature of phase transitions involving elementary quantum fields, as the QCD phase transition is the only one accessible to laboratory experiments.

CURRENT STATUS AND RESULTS

Initial conditions and global features

The predictions of lattice QCD are rather firm in that a transition to the QGP should exist in the vicinity of a critical temperature T_c of $\approx 150 - 200$ MeV (whether the transition is of first order, second order, or only 'rapid' is still a matter of debate). However, whether the QGP is actually created in heavy-ion collisions at current energies is a different question and will depend on the dynamics of the reactions and in particular on the initial conditions of the system shortly after the collision. In order to reach the QGP, or even only to use macroscopic concepts (such as 'phase transition') and the language and variables of thermodynamics (such as 'temperature' or 'density'), the system has to be *extended* — i.e. its dimensions ought to be much larger than the typical scale of strong interactions — it has to

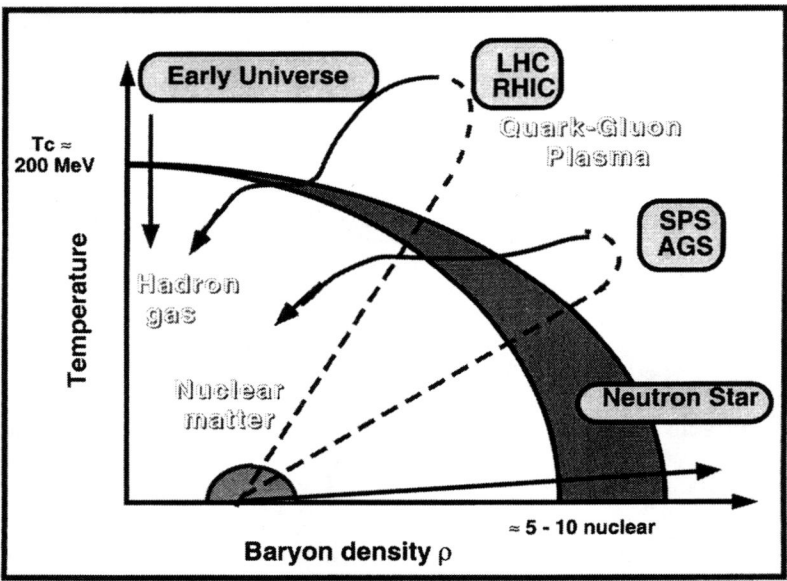

FIGURE 1. The phase diagram of hadronic matter and the hadron gas - quark-gluon plasma phase transition.

be in (or near) *equilibrium* — i.e. its lifetime has to be larger than the relevant relaxation times — and the *energy density* ϵ has to exceed the critical threshold for QGP formation. This threshold is predicted by lattice QCD to be of the order of 1 – 3 GeV/fm^3, equivalent to a temperature T_c of 150 – 200 MeV or a baryon density ρ_c of 5 to 10 times normal nuclear matter density (see Fig. 1).

Present results from the ongoing fixed-target program indicate that the initial conditions realized in these reactions could indeed be favourable for QGP formation. In head-on central collisions, hundreds of particles are produced per unit of rapidity, the system expands to a size of the order of 1000 fm^3 (as measured by particle interferometry), and initial energy densities are estimated to exceed 2 GeV/fm^3. However, the expansion is also extremely fast, with an estimated total lifetime of only a few fm/c from the first instance of the collision until the final freeze-out of hadrons.

While these results show that we are certainly *close* to the requirements listed above for QGP formation, they are by no means *sufficient*. In particular the energy density estimates are inversely proportional to the assumed 'formation time', i.e. the time needed to reach thermal equilibrium, and might well be smaller (or bigger?) by a factor of the order of two. Also, the lifetime of the system seems marginal, and even if a QGP is formed it might simply not live long enough for its signals to clearly stand out from the background created in later, hadronic phases of the evolution. The existence of a QGP phase can only be settled experimentally by searching for direct and specific signals.

Recent experimental highlights

The following sections will concentrate on three main topics which are at the heart of the quest for the QGP, and in which significant progress has been achieved over the last years, i.e. are there experimental indications for *equilibrated hadronic matter, chiral symmetry restoration*, and *deconfinement?*

Equilibrium hadronic matter? While in principle the study of non-equilibrium hadronic matter might be of considerable interest, in practice the huge number of largely unknown dynamical parameters governing the evolution of heavy ion reactions would make the analysis of such a complex system very difficult. The powerful laws of thermodynamics can reduce this complexity and make definite and testable predictions, largely independent of the microscopic dynamics, for those degrees of freedom which evolve in equilibrium. The price to pay is a loss of information concerning events preceding the equilibrium, as the memory of earlier (and possibly more interesting) stages of the evolution is largely lost.

In reality, we will have to deal with a hierarchy of processes and scales, some of which have large cross-sections and correspondingly small relaxation times and therefore might evolve close to equilibrium, and others which decouple early from a thermal evolution and are sensitive to the hot initial phase of the reaction. Prime candidates for the former are hadronic observables, like momentum spectra and particle ratios, and for the latter hard probes and electromagnetic signals.

In a purely thermal system of hadrons, the momentum distributions, when expressed as a function of the transverse mass m_T ($m_T = \sqrt{m^2 + p_T^2}$), will be independent of the particle mass with a slope inversely proportional to the temperature T. In an expanding system, an additional collective flow component can develop which blue-shifts the momentum spectra with a common transverse velocity β_T leading

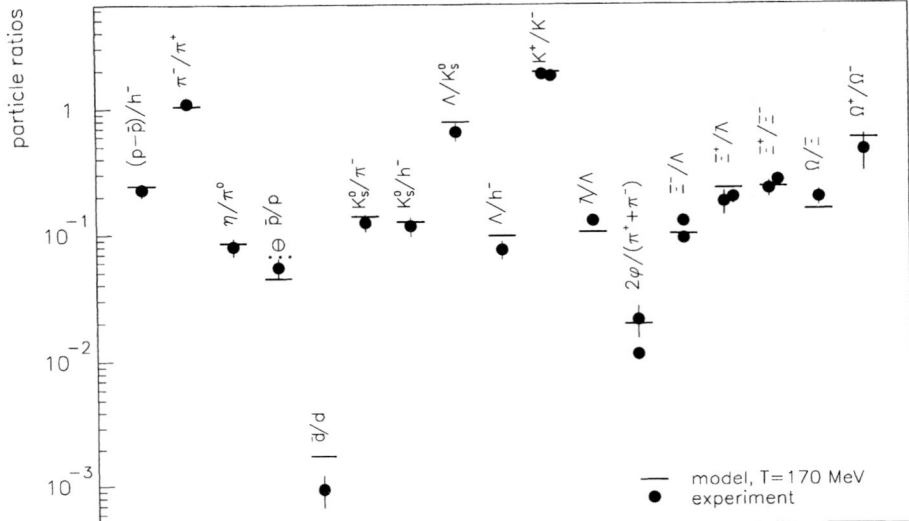

FIGURE 2. Hadrochemical equilibrium model calculation of hadron yields (full lines, calculated for a temperature T of 168 MeV and a baryochemical potential μ_B of 266 MeV) compared to data from CERN SPS.

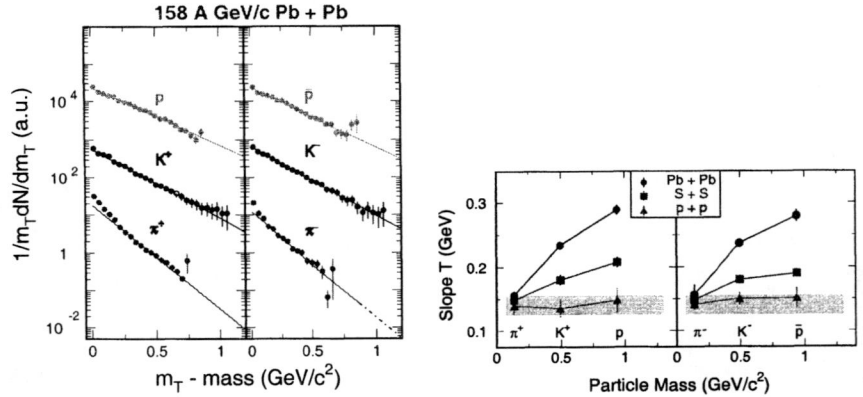

FIGURE 3. Transverse mass spectra (left) and inverse slope parameters (right) of pions, kaons and protons near midrapidity from NA44.

to a mass dependent component. Likewise, the abundance of particle species in equilibrium hadronic matter is given by two independent parameters, i.e. the temperature T and a baryochemical potential μ_B (which reflects the baryon asymmetry in the initial state). A hadronic system in both 'thermal' (momentum) and 'chemical' (particle abundance) equilibrium is therefore fully determined by only three independent parameters: T, β_T and μ_B.

Such a simple prescription seems to be indeed borne out by the data. This is illustrated in Fig. 2, which shows a comparison of measured particle ratios with predictions based on chemical equilibrium [3]. These ratios are in rather good agreement with the equilibrium calculations for a temperature of about 170 MeV and a baryon chemical potential of $\approx$ 250 MeV, corresponding to 1/3 nuclear matter density (at the AGS the corresponding values are T $\approx$ 140 MeV and $\mu_b \approx$ 500 MeV).

Momentum spectra of different particles in Pb-Pb reactions [4] are also well described by a thermal distribution, if, in addition, a common (to all particle types) flow velocity of $\beta_T \approx 0.4c - 0.6c$ is introduced (see Fig.3). On the right part of Fig.3,

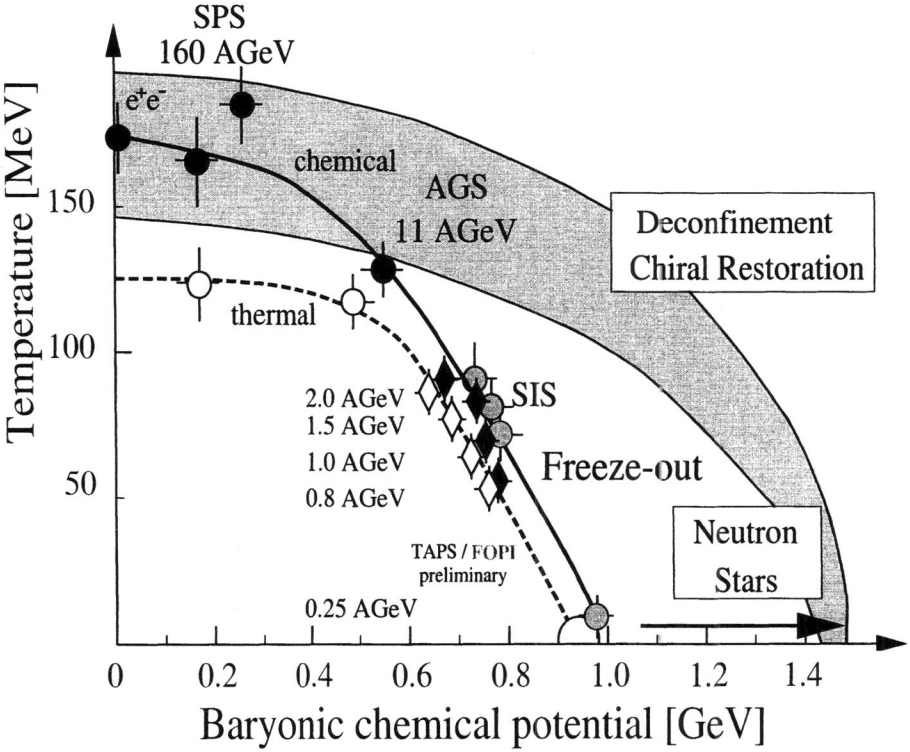

FIGURE 4. Compilation of chemical (particle ratios) and thermal (momentum spectra) freeze-out points from SIS to SPS energies

the inverse m_T-slopes are shown for pp, S+S and Pb+Pb reactions at comparable energies ($\sqrt{s} \approx 20$ GeV). While the slopes in pp reactions are independent of particle type, i.e. exhibit 'm_T-scaling', the slope parameter increases proportional to the particle mass for heavier reaction systems as expected for collective flow. The temperature extracted from momentum spectra at SPS is of the order of 120 MeV, i.e. significantly lower than the one extracted from the particle ratios mentioned above.

A large set of independent hadronic observables, i.e. momentum spectra, particle ratios and HBT correlation results (which are also sensitive to T and β_T [5]), seems to be consistent with a surprisingly simple picture of the late stages of heavy ion reactions: Different particle species are created in relative abundance consistent with chemical equilibrium ratios at a 'temperature' of T $\approx$ 170 MeV; this dense hadronic system then expands and cools to a temperature of about 120 MeV, converting random 'thermal' motion into ordered collective flow until the final freeze-out, when the system is so dilute that all interactions cease. This experimental phase diagram, with both chemical and thermal freeze-out locations as determined from data ranging from very low energies (SIS) up to SPS, is shown in Fig. 4 (taken from ref. [6]). The location of the particle ratio freeze-out point in the temperature-density plane is located very close to the expected phase boundary between hadronic matter and the QGP, and the distance between chemical and thermal freeze-out increases with the beam energy, indicating an increasing dynamical path in the hadronic phase for larger systems (more final state particles).

However, before this intuitively appealing scenario can be taken as fully established, a number of experimental and conceptual questions will have to be clarified. On the experimental side, resolving some inconsistencies between different experiments and better statistics over a large range of impact parameters (in particular for Hyperons) will be needed to come to a more quantitative test of predictions. On the conceptual side, the most puzzling observation is that already very elementary reactions look practically as 'thermal' as heavy ion reactions. While it has been known, but never 'understood', that momentum spectra in hadronic reactions look 'thermal' (obey m_T-scaling, see Fig.3), a recent re-analysis of pp and e^+e^- reactions has shown that also the particle ratios can be extremely well described with thermodynamics [7]. How can this be possible in systems containing only a few hadrons where a dynamical path from arbitrary initial conditions to thermal distributions via interactions (rescattering) is very unlikely? The success of thermal models to describe particle ratios in reactions ranging from e^+e^- at LEP to Pb-Pb at the SPS could be a hint for a universal feature of the parton-to-hadron (phase?) transition, which might be governed by statistics and phase space at the time of particle creation rather than by dynamical features [6,8].

Assuming that some satisfactory answers to these questions can eventually be found, we could then go on to analyse the hadronic data in more detail to look for information on the dynamics preceding the freeze-out. Relaxation times in a partonic and a hadronic medium are likely to be different, and therefore the questions *how* and *how fast* did the system reach equilibrium in different channels are of interest, particularly in the Hyperon sector, where hadronic relaxation times are estimated to be extremely long. Flow patterns should be sensitive to the equation of state of matter and therefore contain indirect evidence for a QGP phase transition preceding freeze-out.

Chiral symmetry restoration? Weakly interacting electromagnetic probes (photons or leptons) are a direct means of gaining information on the early dense and hot stages of the collision, as they leave the interaction volume without being altered by final state effects. While, so far, only upper limits exist for direct (thermal) photon production, recent data on lepton pairs show an unexpectedly large yield at low masses, below the ϱ meson.

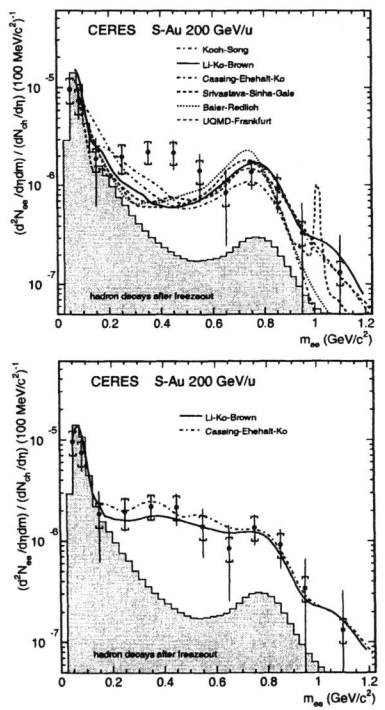

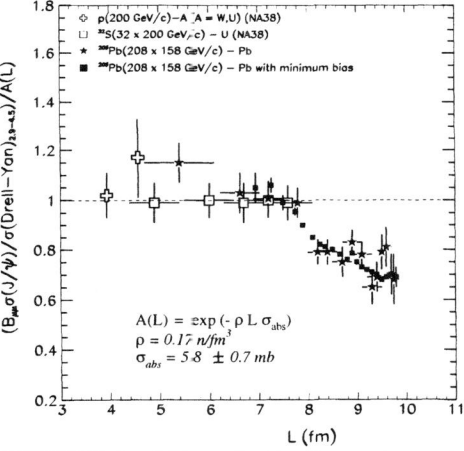

FIGURE 5. Di-electron invariant mass distribution measured by the NA45 experiment in central S+Au collisions, compared to calculations including hadronic decays and effects expected for high pion densities (top) and calculations incorporating in addition a density dependent mass shift of the ϱ meson (bottom).

FIGURE 6. J/Ψ production for proton, sulphur and lead induced collisions relative to the Drell-Yan yield as a function of the thickness L of matter traversed on average. The data are divided by the normal nuclear absorption (exponential in L) which is consistent with the results up to peripheral Pb+Pb reactions. A sudden onset of an additional 'anomalous' suppression is observed for central Pb+Pb.

Figure 5 shows the electron pair mass spectrum observed in central S+Au collisions by NA45 [9]. The upper part summarizes model calculations which include contributions from hadronic decays (shaded area) and from in-medium pion annihilation and bremsstrahlung. An excess at $0.2 < m(e^+e^-) < 0.6$ remains unexplained. The lower panel exhibits perfect agreement with the data obtained in models which include in addition an in-medium modification of the ϱ and ω masses. A similar excess, consistent with the same model calculations, has been found in the $\mu^+\mu^-$ mass spectrum by NA34/3 [9].

In-medium modification of vector mesons, if experimentally confirmed by more conclusive data, could be a direct consequence of the chiral symmetry transition at the phase boundary between hadronic matter and the QGP. The rapidly varying quark condensate should lead to changes in the properties of hadrons (masses, width) in the vicinity of the phase transition, which will be observable in the lepton mass spectrum for mesons decaying in the dense transition regime. This would indeed be a spectacular verification of the concept underlying the generation of light hadron masses in QCD.

An excess in the intermediate mass range (1.5 - 2.5 GeV) observed in muon pairs by NA50 has so far not found any convincing interpretation [10]. Speculations concerning its origin range from enhanced open charm production and final state rescattering of produced charm quarks to thermal radiation of virtual photons. An experiment is currently proposed at the SPS to address this question.

Deconfinement? Signals originating from hard-scattering processes at the very beginning of the reaction are an ideal tool to probe the state of the surrounding QCD matter. The original idea [11] that J/Ψ production should be suppressed in a QGP relies on a Debye screening mechanism which renders colour interactions short ranged in a dense medium ('deconfinement') and therefore prevents the formation of bound resonances.

J/Ψ suppression with similar characteristics as predicted for a QGP was indeed one of the first results reported from heavy ion experiments in 1987. Its subsequent interpretation, alternating repeatedly between 'trivial' and 'exciting', is probably the best example on how our understanding of nuclear collisions has progressed in a constant interplay between theory and experiment, new explanations and new data. An up-to-date compilation of J/Ψ production relative to the Drell-Yan continuum in pA and AB reactions is shown in Fig. 6 versus the average path length L traversed by the $c\bar{c}$ pair after its creation inside the target and projectile nuclei [12]. Up to and including central S-U collisions, this ratio decreases exponentially with L, consistent with a nuclear absorption cross-section of 6 mb. It took the better part of the last ten years, a variety of data for J/Ψ, Ψ' and Υ — from low energy pp and pA reactions to photoproduction and high p_T production at the Tevatron — and a good measure of other ingredients (nuclear structure functions, initial and final state scattering, formation time) to come to a consistent and theoretically substantiated interpretation [13,14]. The exponential attenuation is today seen as resulting from the interaction between the nuclear medium and a pre-resonance state, a coloured $c\bar{c}$-gluon configuration which evolves only later (and outside the nucleus) after some finite formation time into the physical colour neutral J/Ψ or Ψ' hadron. So J/Ψ suppression has provided a lot of insight into the dynamics of charmonium production, hadron formation and using the nucleus as a tool to measure short time scales, but leaving no room for QGP effects.

The extrapolation of this model to central Pb-Pb reactions was straightforward, essentially parameter free, and completely wrong (see Fig. 6)! The Pb-Pb data, whether plotted as a function of L or any other variable, shows significantly less J/Ψ's than hadronic absorption models would predict by extrapolating from light ion and pA results. While some debate still persists if the additional suppression is really 'anomalous' or not, new precision data which is currently being analysed should settle this question in the near future. Most likely, some additional physics will have to be included in order to describe the Pb data. However, whether this 'new physics' will require deconfinement, dynamical pre-cursor phenomena of the QGP transition, or just some overlooked hadronic effect, remains to be seen.

Future fixed target program

Given the recent exciting developments, the future directions are perfectly clear. The current SPS fixed target program is unique in the world, it addresses a well focused set of fundamental questions, it has entered an extremely productive phase and it is now being brought to its full potential. With the exception of the Hyperon and the low mass lepton pair measurements, statistics is, in general, not a problem. A run at the lowest possible SPS energy, around 40 GeV/nucleon and carried out end of 1999, should increase the baryon density, possibly close to its maximum value. Signals related to chiral symmetry restoration, in particular the low mass lepton pairs, will in general be rather sensitive to baryon density. The low energy run can also make contact with the AGS regime and will allow the CERN experiments, which are quite distinct in their capabilities from the AGS detectors, to compare with and complement the program at lower energies. The SPS program is currently foreseen to terminate in the year 2000, however, extensions to address in particular charm production and the intermediate mass lepton pair excess have been conditionally approved. Later, around the turn of this century, the new generation of heavy ion colliders will come into operation.

HEAVY ION PHYSICS OF THE NEXT CENTURY

The study of ultra-relativistic heavy-ion collisions is a rather new, but rapidly evolving field. After the pioneering experiments at the BEVALAC and in DUBNA with relativistic heavy ions ($E/m \approx 1$), the first experiments started in 1986 with light ions almost simultaneously in Brookhaven (AGS) and at CERN (SPS). Really heavy ions ($A \approx 200$) have been available in the AGS since the end of 1992 and at the SPS since the end of 1994. With the colliders RHIC ($\sqrt{s} = 200$ GeV/n) and LHC ($\sqrt{s} = 5.5$ TeV/n) starting operation in 2000 and 2005, respectively, the available energy in the centre-of-mass will have increased by almost five orders of magnitude within 20 years. This unprecedented pace was made possible only by (re)using accelerators, and to some extent even detectors, built over a much longer time scale for use in high-energy physics. The following sections will summarise the physics and experiments to come in these latest (and possible last) heavy-ion machines.

Heavy ion collisions at the colliders are expected to provide a qualitatively different environment from existing accelerators by creating a very hot, and therefore more clearly detectable QGP, via hard initial parton scatterings that can be calculated rather precisely. Extrapolating from present results to LHC, all parameters relevant to the formation of the QGP will be more favourable: the energy density, the size and lifetime of the system and relaxation times should improve by a large factor, typically by an order of magnitude compared to Pb+Pb collisions at the SPS. We expect particle densities of several thousand per unit of rapidity, a freeze-out volume approaching 100,000 fm^3 and an initial energy density orders of magnitude larger than the one of normal nuclear matter. The initial temperature might be close to 1000 MeV, as compared to a value of 400 – 500 MeV at RHIC and about 200 MeV at the SPS. The energy densities and temperatures at LHC should be far above the deconfinement threshold, allowing us to probe the QGP in its asymptotically free 'ideal gas' form.

The analysis of extended strongly interacting matter at both colliders will move from the fixed target regime dominated by soft phenomena into a domain where hard interaction between the primary partons will lead to a rapid production of further partons, and the interactions within this dense partonic medium, with the resulting strong increase in entropy, are expected to produce the QGP. The abundant formation of 'minijets' with transverse momenta of a few GeV plays an essential role in this process. Perturbative QCD calculations can be used to construct and evaluate such parton interaction cascades; they indeed show the expected rapid rise towards thermalisation, on time scales considerably below 1 fm/c, to the extremely high initial temperatures mentioned.

In order to verify that a QGP was produced and to study its properties, we need probes sensitive to the earliest and hottest stages of the medium. Three such probes are currently known: Bound heavy quark resonances (quarkonia), hard jets, and thermal dileptons/photons. Only charmonia as deconfinement probes have been studied successfully at the SPS as discussed above; for all others, higher incident energies appear necessary to find a positive signal.

The deconfinement analysis of hot and dense matter must be extended to bottonium states. The Υ, with its very small radius, can be dissociated only at the highest energy density attainable at LHC (of order 30 GeV/fm^3), while the excited states Υ' and Υ'' are comparable to the charmonium resonances and will serve as important consistency checks.

Hard jets probe the produced medium through the energy loss of partons passing through dense matter. The theoretical aspects of this problem were recently studied in considerable detail. In particular, the rate of energy loss was found to depend quite sensitively on the size of the medium and there are now indications that the energy loss in cold nuclear matter is much smaller than that in a hot QGP. The production and subsequent attenuation of fast partons will add a crucial new penetrating probe to diagnose the nature of the strongly interacting matter produced in heavy ion collisions.

The temperature of the primordial medium could be best determined through measurement the of spectra of real or virtual photons. The thermal photon spectrum will be an integral over the temperature history of the system. Superimposed are the soft photons from the late hadronic stage as well as the primary Drell-Yan or hard QCD photons. Whether there is a window in transverse momentum (around one to a few GeV) to actually measure such thermal dileptons or photons depends crucially on the density of the produced system; fortunately, conditions could be quite favourable at the high energy densities predicted for RHIC and LHC.

Another unique feature of heavy-ion collisions at the colliders is the possibility to measure a large number of observables with very good accuracy on an event-by-event basis: impact parameter, multiplicity, particle ratios and spectra and, of particular importance, size and lifetime from interferometry. Single event analysis, currently pioneered by NA49 at the SPS, will become a precision tool at very high multiplicity. One of the important design considerations for both the STAR

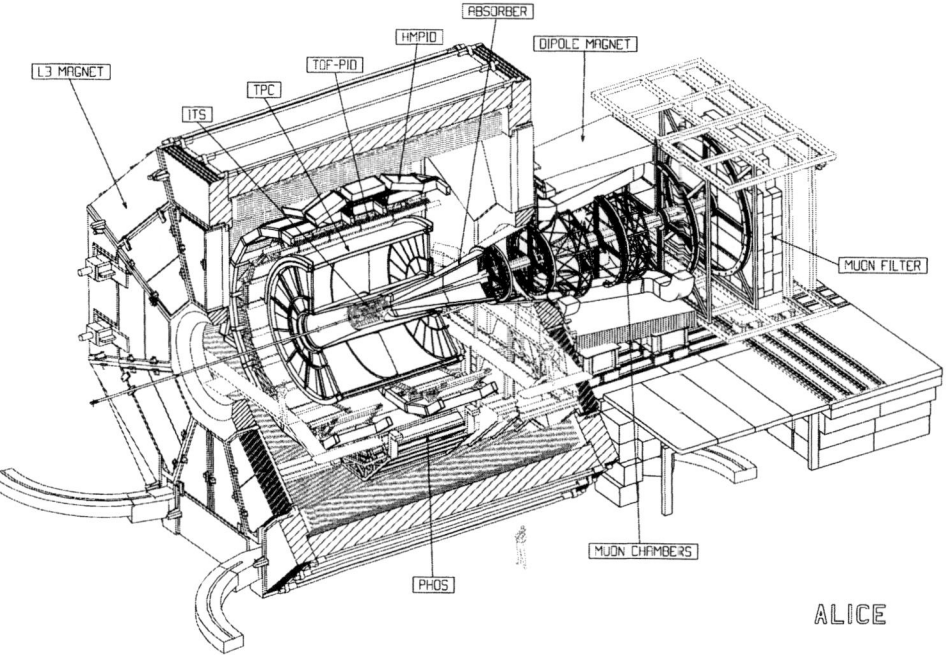

FIGURE 7. Schematic view of the ALICE detector which combines a central barrel for electron pair, photon, and multihadron studies, with a forward dimuon spectrometer.

detector at RHIC and the ALICE detector at LHC (Fig. 7) is to make full use of this opportunity. It will allow the study of correlations and non-statistical fluctuations which would otherwise be washed out when averaging over many events. Such fluctuations are, in general, associated with critical phenomena in the vicinity of a phase transition.

SUMMARY

The still very young field of ultra-relativistic heavy ion physics has proceeded since its inception in 1986 through three essential phases:

The initial round of 'exploratory' experiments has shown that appropriate detectors and analysis procedures can cope with the extreme particle densities produced in heavy ion collisions. They have qualitatively shown that an extended, interacting and very dense system has been formed that differs in many observables from the more elementary hadron-hadron reactions investigated in the past. Falling short of striking discoveries, this phase has nevertheless provided a *'principle proof of feasibility'* and has substantiated the expectation that heavy ion collisions are an appropriate tool to create equilibrium hadron matter and eventually the quark-gluon plasma.

The next phase was characterized by efforts to get a comprehensive and precise set of data and to come to a quantitative understanding of the experimental results. A close and very effective interaction between theory and experiment, models and data, has led to significant progress in identifying relevant ingredients and important microscopic processes.

The field is currently in its third, and most dramatic phase. Results from both AGS and SPS with really 'heavy' ion beams have produced puzzling results which strongly hint at a picture of high energy nuclear reactions almost too good to be true: i) a premordial phase of deconfined partons – the QGP ? – responsible for quarkonium suppression, followed by ii) a transition regime with gradual onset of chiral symmetry breaking, leading to changes in the properties of light hadrons, concluded by iii) a gas of hadronic matter governed by the simple laws of thermodynamics. On the short term, the ongoing experiments should provide us with additional and more complete data in order to substantiate (or refute) this scenario.

In the longer term, making use of RHIC and LHC for heavy ion collisions provides a unique opportunity for exploring the physics of QCD matter in a qualitatively very different region of extremely high energy density. RHIC and its four major experiments (STAR, PHENIX, BRAHMS and PHOBOS) have made the first step this year with Au+Au collisions at an energy an order of magnitude above what is currently available at the SPS. The LHC, with a centre-of-mass energy almost thirty times above RHIC, will lead by 2005 into a region comparable only to the highest energy cosmic ray events. Its single dedicated detector, the ALICE experiment, represents *the* long term future of the ultra-relativistic heavy ion program. Building a detector of the size and complexity required for LHC will be an unprecedented challenge, and its successful completion will need the continued, strong and emphatic support and participation of the heavy ion community world wide.

REFERENCES

1. Explosive Phenomena in Astrophysical Compact Objects, Seoul, South Korea 2000.
2. Particles and Fields: Seventh Mexican Workshop, Merida, Mexico, 1999 (Editors: Alejandro Ayala, Guillermo Contreras, and Gerardo Herrera), AIP Conference Proceedings 531, 2000, pp. 3-15.
3. P. Braun-Munzinger, J. Stachel, H. Wessels, N. Xu, Phys Lett. B 366 (1996) 1, J. Stachel, Nucl. Phys A610 (1996) 509c and P. Braun-Munzinger, I. Heppe, J. Stachel Phys. Lett. B 465 (1999) 15.
4. N. Xu et al, Nucl. Phys A610 (1996) 175c.
5. H. Appelshauser et al, Eur. Phys.J. C2 (1998) 661.
6. U. Heinz, J. Phys.G 25 (1999) 263.
7. F. Becattini, U. Heinz, Z. Phys. C 76 (1997) 269.
8. R. Stock, Phys. Lett. B 456 (1999) 277.
9. A. Drees, Nucl. Phys A610 (1996) 536c.
10. E. Scomparin et al, Nucl. Phys A610 (1996) 331c.
11. T. Matsui, H. Satz, Phys. Lett. B 178 (1986) 416.
12. M.C. Abreu et al, Phys. Lett. B 450 (1999) 456.
13. C. Lourenco, Nucl. Phys A610 (1996) 552c.
14. D. Kharzeev, Nucl. Phys A610 (1996) 418c.

Physics of Dense and Superdense Matter

Mannque Rho[a,b]

[a] Service de Physique Théorique, CE Saclay, 91191 Gif-sur-Yvette, France
[b] School of Physics, Korea Institute for Advanced Study, Seoul 133-791, Korea

Abstract. I discuss a few aspects of dense hadronic matter and superdense QCD matter that are considered to be relevant to the physics of compact astrophysical systems. The connection between a "bottom-up approach" and a "top-down approach" is made with the help of an effective field theory strategy. Topics treated are meson condensation going up from low density and color superconductivity going down from asymptotic density with the approach to the chiral phase transition made in terms of Brown-Rho scaling.

INTRODUCTION

Super-dense hadronic and/or quark matter is considered to be relevant for understanding compact stars in the context of supernova explosions, hypernovae, merging of compact (neutron and black-hole) stars, gamma-ray bursts etc. See [1] for a recent review. It is not known at present precisely which densities are relevant to such systems but the region of the density involved is presumably somewhere in-between very low and very high density regimes, the two extreme regimes being in principle fairly well controlled. To study the physics of the density regime involved which cannot be systematically accessed as I will describe below, we have at our disposal basically two possible theoretical approaches: One, "bottom-up" one going up from very low density and the other, "top-down" going down from very high (asymptotic) density. The former relies on laboratory (experimental) data available at low density in nuclear and hadronic physics and the latter on the theoretical machinery available at super-high density at which QCD becomes weak-coupling. We would eventually like to bring the two approaches together and have them match. But both going up beyond nuclear matter density and going down from the asymptotic density to the relevant regime are difficult because of lack of reliable theoretical tools and of experimental data. In this talk I will discuss some aspects of both and indicate how the two "extremes" could be brought together at a density regime supposedly relevant to the astrophysical processes of interest. This talk will be qualitative, intended to give a personal overview of the present situation. Certain aspects of the topic somewhat lying outside of the main theme of this meeting are discussed in a separate note [2].

PHASE STRUCTURE

In many of the current meetings in nuclear and hadronic physics, one is shown a canonical diagram depicting the phase structure of hadsronic matter in terms of temperature (T) vs. chemical potential μ or baryon number density ρ. In heavy-ion meetings, the principal focus is put on the phase structure at high temperature. There, the phase below $T_c \sim 150$ Mev is in hadronic state and the one above in quark-gluon plasma. There is strong evidence from lattice gauge calculations that the phase transition does indeed take place at a given T_c although above the T_c, things are not as simple as the naive form of quark-gluon plasma that people have thought of before. In any event, it seems likely that one will be able to follow temperature properties of the systems with small non-zero density both theoretically and empirically, the former on lattice and the latter in the heavy-ion machines that are operating and/or are to come in the near future [3,4]. The most recent phase diagram taken from Braun-Munzinger's article [5] – which was also seen in Schukraft's talk – is shown in Fig.1 [5]. It gives a perspective in this matter. There are many laboratories involved in this endeavor, such as "early universe," LHC, RHIC, ..., SIS as indicated in the figure.

It is a different story with matter at high density. Fcr technical reasons, it turns out to be extremely difficult to put dense matter on the lattice except for the academic case of $N_c = 2$ systems, accounting for the paucity of theoretical guidance as to what happens to hadronic systems at non-zero density. The knowledge we have up to date is based on various models. Models indicate that as density increases toward the point at which QCD (chiral) phase transition is supposed to take place, there can be a multitude of phase transitions that may or may not be in concordance with chiral phase transition. As I shall discuss below, there are some experimental data around nuclear matter density but there are no terrestrial data much beyond. Thus there is need to resort to astrophysical observations. Neutron stars are the only laboratory available for this purpose as indicated in Fig.1. The aim of this note is to discuss how to proceed along the density axis of the figure.

PROBING DENSE MATTER

Matter at high temperature which may be relevant for the early Universe can be – and is being – probed in heavy-ion collisions at relativistic energy. Furthermore QCD at high temperature can be simulated on lattice which will eventually give a solid theoretical information. Now what about probing matter at high density? Up to nuclear matter density, ample information is available from nuclear systems. Thus combined with reasonable models, one could attempt to go beyond nuclear matter by extrapolating what we learn at nuclear matter density. This constitutes the "bottom-up" approach. At low density, hadrons are the proper variables in QCD and the "bottom-up" approach consists of doing systematic calculations in the strong-coupling regime aided by experimental data. However one cannot push this strategy too far: this scheme must break down at some density.

At asymptotic density, the theory is in principle well-defined since QCD is weak-coupling there and perturbation theory is well-defined. One may therefore start with the weak-coupling QCD and adopt the "top-down" approach. As we go down in density, however, the coupling increases, so at some point the perturbative approach must also break down. The ultimate goal would then be to have the two approaches come together at some intermediate density. The difficulty is that the gap between the two break-down points is not narrow and it is in this gap that interesting astrophysical phenomena seem to take place. Furthermore, the region in question cannot be probed in terrestrial laboratory experiments. Thus no reliable predictions can be made. We believe that the compact stars such as neutron stars are the only systems from which we can gain empirical information for the regime concerned and since no lattice QCD is available at finite density at least in the near future, the compact stars seem to be the only direct sources for dense matter physics.

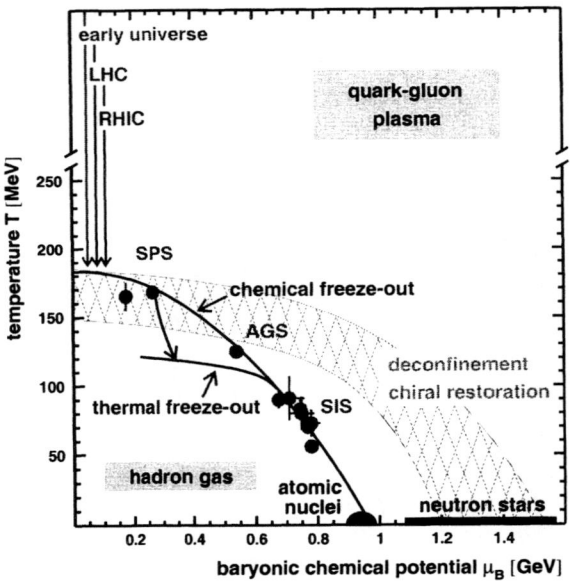

FIGURE 1. The conjectured QCD phase diagram. It shows the recent development including the hadro-chemical freeze-out points determined from thermal model analyses of heavy-ion collision data at SIS, AGS and SPS energy. The hatched region represents the plausible phase boundary delineating two different phases of QCD indicated by lattice QCD calculations at $\mu_b = 0$.

BOTTOM-UP APPROACH

Pions Do Not Condense

Beyond the normal matter density, the first phase change that is expected to occur is meson condensation. The lightest meson is the pion so one would think that pions would condense first [6]. What happens can be understood in a simple and caricatured way as follows.

Pions may condense in S or P waves. The S-wave condensation can occur due to a change in the pion mass in medium. This is a simple reflection of the fact that the mass term in scalar field theory is a "relevant" term. Since the pion mass in nature is ~ 140 MeV which is small compared with the chiral symmetry scale, $\Lambda_\chi \sim 1$ GeV, a small attraction in the scalar direction could push the mass down to zero, so one would think that it would be easy to trigger the phase transition with the order parameter $\langle \pi^a \rangle \neq 0$. Baryon matter density $\rho \sim \langle B^\dagger B \rangle$ where B is the baryon field could induce such a change which will be an S wave. In nature this phase transition does not seem to occur mainly because there is chiral symmetry protection. Although the pion is massive, its mass is small because chiral symmetry is broken explicitly by a small amount, i.e., the tiny quark mass which arises from the symmetry breaking at the electro-weak scale. Thus in some sense chiral symmetry is still operative and hence protects the pions from undergoing a phase change in the S-wave channel.

On the other hand, the P-wave condensation can occur due to the P-wave attraction that pions feel in the presence of nucleons. The interaction dictated by chiral symmetry is of the derivative type $\sim \frac{g_A}{f_\pi} \overline{N}^\dagger(x) \tau^a \boldsymbol{\sigma} N(x) \cdot \boldsymbol{\nabla} \pi^a(x)$ where g_A is the axial coupling constant and f_π is the pion decay constant, both of which are associated with the weak interaction. This P-wave interaction with the nucleon is strong as evidenced by the huge resonance $\Delta(1230)$ associated with the strong πN interaction. The attraction is greater the greater the coupling g_A. The P-wave attraction which increases with density in nuclear medium can generate instability against P-wave pion condensation signaled by a space-dependent order parameter $\langle \pi^a(x) \rangle$. What happens as density increases can be analyzed with a Lagrangian that preserves chiral symmetry structure of dense matter [7]. The crucial element in this formulation of pion condensation is that the phase transition is extremely sensitive to how g_A – which is associated with the space component of the axial current – behaves in nuclear medium. It turns out that the critical density goes like $\rho_c^\pi \propto (g_A^\star - 1)^{-1}$. One can therefore get a qualitative idea as to how the P-wave condensation occurs by looking at how the effective coupling $g_A^\star$ behaves in nuclear medium.

There has been much controversy on the effective g_A in nuclei and nuclear matter and its relevance to chiral phase structure. Experiments in nuclei exciting giant Gamow-Teller resonances have been interpreted in terms of this effective Gamow-Teller coupling. Whatever may be the valid interpretation, the outcome is that the *effective* g_A (that is, $g_A^\star$ which is density dependent)[1] tends to the limit

$$g_A^\star \to 1 \quad \text{as} \quad \rho \to \rho_0. \tag{1}$$

This argument does not pinpoint the critical density but one can say that ρ_c^π is pushed beyond the regime where the theoretical framework makes sense.

Kaon Condensation

As we will hear in the next talk by Chang-Hwan Lee, kaon condensation can occur in, and make an important impact on, neutron-star formation. I will give a very simplified argument for how and why kaons will condense in dense neutron-star matter. More rigorous discussion is given by Chang-Hwan Lee in [8].

In a nut-shell, a kaon, particularly the negative charged K^- which contains an anti-strange quark $\bar{s}$, can condense in neutron-star matter precisely because the strange quark is massive at the scale of low-energy physics we are concerned with. Since the strange quark has a current mass of $\sim$ 150 MeV [2], chiral symmetry is explicitly broken in the strange sector and the low-energy theorems associated with Goldstone bosons sometimes fail badly. What the matter density does is to "rotate" the explicitly broken symmetry in such as way that the kaon mass gets lighter to partially restore the explicitly broken symmetry. Suppose that all Goldstone boson masses were zero. Then the same chiral symmetry protection argument that applies to pions will apply to kaons as well so that baryon number density cannot do anything to the energy density of the system. If on the other hand all Godlstone bosons were massive and degenerate, the rotation around the chiral circle (see Fig.2) would not affect the energy density and so nothing would happen. Indeed, things happen precisely because the potential is tilted due to the symmetry breaking and the baryon density does just the un-tilting to restore the symmetry triggering kaon condensation [10]. The transition is essentially triggered by the "relevant" mass term in the Lagrangian, so is S-wave. There is a beautiful support for this dropping-mass scenario from laboratory experiments [9].

---

[1] There has been much debate as to whether the quenching of g_A in nuclei and nuclear matter as evidenced in Gamow-Teller transitions is *fundamental* or some garden-variety nuclear phenomenon. I believe this is a non-issue. There is no question that taken as an *effective* parameter in a precisely defined sense, the g_A is quenched in nuclei. Whether it can be described by many-body effects (e.g., "core polarizations" due to tensor forces) or Δ-hole effects or even partial chiral restoration is mostly semantic – they are of the same physics – and is not relevant for the issue. What is relevant is that the *effective* Gamow-Teller matrix element can be described in dense medium by a quenched $g_A^\star$ in the chiral Lagrangian and that it goes to the value of 1 as density increases.

[2] The quark mass is not renormalization-group invariant and so has to be defined with a given scale. Even at a given scale, it is not well-determined. Various methods of analysis give the range 100 $\sim$ 200 MeV for the strange current-quark mass. This uncertainty makes the estimate of the critical density somewhat uncertain.

In strong interactions, strangeness has to be conserved and if kaons were to condense, the pairs $K^\pm$ would have to do so. If it were to happen at a low enough density, say, $\rho/\rho_0 \sim 3$, then condensation could occur only if the repulsive term that counteracts the scalar attraction associated with the symmetry breaking term, namely, the so-called vector coupling, gets decoupled. In this case, both $K^\pm$ mesons become lighter and eventually may condense. But if this were to occur at much higher density than say $\sim 3\rho_0$, then kaons would be prevented from condensing in heavy-ion process for the reason to be mentioned below.

In the neutron-star matter, the kaon mass does not have to go all the way to zero. In fact when the electron chemical potential that increases with density – for non-relativistic nucleons – crosses the dropping kaon mass, the electron will be converted by weak interaction to the kaon

$$e^- \to K^- + \nu_e \qquad (2)$$

and the condensing of kaons is favored for energetic reasons. Here K^+ will play no role. For K^- the vector interaction is attractive, so it speeds up the drop of the kaon mass.

The theoretical arguments that go into this treatment can be reliable only if the condensation occurs at low enough density. Were it to occur at much higher density, then there would be no reason to believe that the approximation makes any sense. Indeed the physics at higher density coming up in the bottom-up approach will be obstructed by the inability to handle the short-distance physics referred in nuclear physics to as short-range correlations. For instance a straightforward embedding of two-nucleon interactions at short distance into many-body systems would predict that the phase transition will take place at a much higher density than that relevant to neutron stars [11]. However this consideration is based on the Hartree approximation in a potential model with a KN potential that is not realistic at high density [12]. Nonetheless the issue of short-range correlations is open.

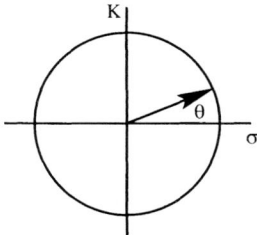

FIGURE 2. Projection onto the σ, K plane. The angular variable θ on the "chiral circle" represents fluctuation toward kaon mean field. Kaon condensation corresponds to "rotating" into the K direction (slightly) away from the chiral circle.

BR Scaling

Once kaons condense at some density ρ_{Kc}, going up further in density up to (QCD) chiral restoration is like opening a Pandora's box. As mentioned, we have no firm proof for *any* transition with density since lattice calculations are not feasible at present. There are a large number of different and conflicting ideas published in the literature. I neither have time nor wish to go into details here. It is fair to say that nothing *additional* that is illuminating or startling is expected to emerge in the physics of compact stars in this regime, once the kaon condensation and related processes have done what Chang-Hwan describes in his talk. In this murky circumstance, I will take an option that allows me to go ahead all the same. That is to heed what is said in Chinese fortune cookies in North America: *"There are only two ways of doing things: One is the wrong way and the other is our way."* I shall therefore take "our way" which is to invoke "Brown-Rho (BR) scaling" [13,14].

Consider symmetric nuclear matter. (Asymmetric matter can be treated similarly.) At a density found in nuclear matter, one finds that a description in which all degrees of freedom are *quasi-particles* treated in the mean field works well provided that those quasi-particles have masses and coupling constants that respect chiral symmetry and run with density [2,15]. In particular if one considers an effective field theory with chiral symmetry with the nucleons, pions, scalar meson and vector mesons as the appropriate degrees of freedom, then with density, the masses run with density as [3]

$$\Phi(\rho) \approx \frac{f_\pi^\star(\rho)}{f_\pi} \approx \frac{m_\sigma^\star(\rho)}{m_\sigma} \approx \frac{m_V^\star(\rho)}{m_V} \approx \frac{M^\star(\rho)}{M}. \qquad (3)$$

Here M is the nucleon mass, and the subscripts σ and V stand respectively for the scalar σ and the vector mesons ρ, ω. Since the pion decay constant can be taken as an order parameter for chiral symmetry phases, the masses are all seen to follow the order parameter in density. Thus when $f_\pi^\star \to 0$ in the chiral limit as $\rho \to \rho_\chi$ (where ρ_χ is the as yet undetermined critical density for chiral restoration), the masses approach zero in the same way. In nature, the chiral limit is not reachable, so the masses will not strictly go to zero but they will drop smoothly (in the scenario of BR scaling) as density increases with interesting physical consequences.

Following this BR scaling behavior all the way to the chiral transition clearly sidesteps the Pandora's box alluded above. What happens in this scenario is then that as one approaches the chiral phase transition density, the relevant degrees of freedom will be quasi-quarks (rather than quasi-nucleons), pions, scalar σ (i.e., a chiral partner of the pion) etc. These will go *smoothly* over to the other side of the phase boundary. This smooth transition could also be witnessed when one comes down in the top-down way from asymptotic density as I will discuss below.

---

[3] These are not *pole* masses. They can however be related to physical masses given the appropriate corrections.

The BR scaling as postulated in [13] has support both from theory [2,15] and from experiments, e.g., in inelastic electron scattering from nuclei and in dilepton production in heavy-ion collisions [12]. It seems to also account for the J/ψ suppression observed at CERN that is heralded to be the signal for the formation of quark-gluon plasma [16]. This may be one more evidence for "quark-hadron" continuity and/or "Cheshire Cat phenomenon" mentioned below.

TOP-DOWN APPROACH

Color superconductivity

When the density is asymptotically high, QCD in weak coupling is applicable so one can compute things reliably. Because of asymptotic freedom, the gauge coupling becomes weak and the dynamics of the quarks is governed by one gluon exchange. In this situation, diquarks are mostly likely to condense in the scalar channel giving rise to color superconductivity. There have been many papers written recently on this subject [17]. I would like to discuss a few aspects of the phenomenon that may be relevant to the physics of compact stars.

When the quark chemical potential is so high that one can ignore the u-, d- and s-quark masses, the situation becomes simple and elegant. But as one goes down to the regime where the chiral phase transition takes place, the situation gets very muddy and we have no idea what's really happening. Even color superconductivity scenarios differ depending upon how many flavors come into play. At non-asymptotic chemical potential, other excitations than diquark, such as quark-hole modes (e.g., Overhauser modes), can be equally or even more important. I shall therefore focus on the ideal case of infinite chemical potential with three flavors of quarks and hope for the best as we go down to the density relevant to compact stars.

A left-handed quark is characterized by three colors (i.e., group $SU(3)_c$) and three flavors (i.e., group $SU(3)_L$), so can be denoted q_{La}^α with α standing for color and a for flavor and similarly for the right-handed quark. It turns out that it is most favorable to condense the diquark in the color-flavor-locked (CFL) state as

$$\langle q_{Li\alpha}^a(\vec{p}) q_{Lj\beta}^b(-\vec{p}) \rangle = - \langle q_{Ri\alpha}^a(\vec{p}) q_{Rj\beta}^b(-\vec{p}) \rangle = \kappa(p_F) \epsilon_{\alpha\beta} \epsilon^{abI} \epsilon_{ijI} \qquad (4)$$

where i,j are the spin indices. This means that both color and chiral symmetries are spontaneously broken by the condensate as [4]

$$SU(3)_c \times SU(3)_L \times SU(3)_R \to SU(3)_{c+L+R}. \qquad (5)$$

As a consequence, seventeen Goldstone bosons are excited i.e., nine scalars and eight pseudoscalars. Eight scalars get eaten up by the gluons which become massive

---

[4] The $U(1)_V$ corresponding to the fermion number is also broken but the $U(1)_A$ symmetry is restored at high density. There is also a Z_3 invariance that we need not worry about here.

and the remaining scalar gets condensed to give rise to superfluidity. The eight pseudoscalars constitute the lowest modes remaining in the system that pick up a mass proportional to the current quark mass that goes to zero as $\mu \to \infty$.

The most remarkable conjecture made by Schäfer and Wilczek [18] posits that there is one-to-one correspondence between the spectrum of excitations in the zero-density regime and that of excitations in the CFL phase. Apart from the 8 Goldstone pseudoscalars that are present in both regimes, the eight massive gluons possess the same quantum numbers as the octet vector mesons in free space and the 8 color-flavor-locked quarks (which may be simply understood as quark solitons or qualitons [19]) can be mapped to the octet baryons of zero density. The conjecture is that the infinite-density spectrum is *continuously connected* to the zero-density spectrum, implying "quark-hadron continuity" or in the more pictorial term, "Cheshire Cat phenomenon" [20]. This aspect shows that going toward high density beyond ρ_χ may reveal a richer and quite different variety than going to high temperature. There is however an obstacle to this intriguing possibility: there may be a series of phase changes as one goes down in density from infinite to the regime relevant to compact stars. First of all, even in color superconductivity (CSC), scenarios differ depending upon the scale of the strange quark mass relative to the chemical potential [17]. Not less importantly, there can be other phases than color superconductivity such as Overhauser effect [21], ISB [22] and others [23] that may intervene before going into the Goldstone phase of lower density. Despite the plethora of other possibilities, it is remarkable – and perhaps meaningful – that both pion condensation [24] and kaon condensation [25] could occur in the high density regime. If confirmed, this could be another support for the notion of quark-hadron continuity.

Astrophysical implications

The mere existence of pions and kaons in the CFL phase is a highly intriguing possibility one would like to explore. Although quarks/baryons are gapped in the CFL phase with a gap $\Delta \sim 10 - 100$ MeV, there can still be light fermions in some situations. These degrees of freedom could play an important role in astrophysical compact objects. Since it does not look feasible to make firm predictions for both terrestrial and astrophysical observables, one would like to turn the process around and *learn* from astrophysical observations about the state of matter in which such exotic phenomena take place.

There have been a series of papers addressing the role of color superconductivity (CSC) in the cooling of neutron stars [26]. The situation is quite complex and a complete analysis is probably not yet possible. The complexity comes from the fact that we do not really know what should be taken into account in the cooling since there are meson condensations and other processes that could intervene at the same time. Even within the framework of CSC, there can be several competing factors to consider. First there is the effect on the electro-weak interaction proper. A simple calculation shows that inelastic neutrino processes (i.e., neutrino production) can be enhanced in the CFL phase [27]. Next, the produced neutrinos may have longer mean-free path in the CFL phase than in normal phase [28], thereby increasing the emissivity. On the other hand, the gap Δ can be large so that there will be a strong suppression from the Boltzmann factor $exp(-\Delta/T)$. All these competing factors would have to be carefully accounted for before one can come to any estimate of the cooling rate. A lot more work and help from observations will be needed for a clearer picture. We heard in this conference about a very interesting data coming from a point source in Cassiopeia A which could be a valuable clue to the issue [29].

Another interesting effect is on magnetic fields in compact stars [30,31]. Since vector symmetry is spontaneously broken (actually in all scenarios of color superconductivity (CSC)), both color and ordinary electromagnetism are spontaneously broken. However although electrically charged diquarks condense as do electron pairs in ordinary superconductivity, they do not produce the familiar Meissner phenomenon of ordinary superconductivity. The reason is that there is an unbroken $U(1)$ symmetry with a massless gauge boson. What happens is that as long as the color gauge coupling g_c is much bigger than the electromagnetic coupling e, magnetic fields can penetrate into the system largely un-expelled from the CSC phase. As a consequence, if rotating compact stars are in the CSC phase, the magnetic field will be stable against decay on time scales longer than, say, 10^7 years set by observation [32]. In [30], it is even suggested that if an isolated pulsar were observed to lose its magnetic field as it spins down, it would rule out the presence of the CSC (or even quark) phase in the pulsar. Although quite interesting, this negative evidence is not too exciting. It would be a lot more exciting if the CFL or other CSC phases were associated with a positive consequence. Again, given the various other possibilities in the density regime involved, the situation may be different from this simplified consideration. It remains an open issue.

SUMMARY

The approach to the density regime relevant to compact stars is made in both ways: "bottom-up" and "top-down." The two extremes, namely, bottom and top, are fairly well described, the former based on effective field theory of QCD combined with experimental data and the latter based on weak-coupling theory of QCD. The two approaches leave a rather wide gap in between where they have no overlap and it is in this range that interesting phenomena seem to take place in dense matter in compact stars. A bridge is suggested in terms of BR scaling coming from bottom-up. Combined with what we learn from the top-down approach and from astrophysical observations, it would be possible to establish the intriguing "quark-hadron continuity" alias "Cheshire Cat." The main aim of this talk was to stress that the compact stars are the only source for physics at high density, the domain that cannot be accessed in terrestrial laboratories.

ACKNOWLEDGMENTS

It is a pleasure to acknowledge the generous help and hospitality of Korea Institute for Advanced Study for my participation in co-organizing, and for attending, this first KIAS workshop on astrophysics and astro-hadron physics. This paper was prepared while I was visiting Korea Institute for Advanced Study.

REFERENCES

1. Balberg, S., and Shapiro, S.L., astro-phys/0004317; Heiselberg, H., and Pandharipande, V., astro-phys/0003276.
2. Rho, M., nucl-th/0007060
3. See for a recent review with references, Satz, H., "Phase transitions in QCD," hep-ph/0007209.
4. Schukraft, J., talk in this conference.
5. Braun-Munzinger, P., nucl-ex/0007021.
6. Migdal, A.B., Saperstein, E.E., Troitsky, M.A., and Voskresensky, D.N., Phys. Rept. **192**, 179 (1990).
7. See, e.g., Baym, G., and Campbell, D.K., "Chiral symmetry and pion condensation," in *Pions in nclei* ed. Rho, M., and Wilkinson, D.H., North-Holland, Amsterdam, 1978, pp. 1031-1094.
8. Lee, C.-H., Phys. Rep. **275**, 256 (1996)
9. Li, G.Q., Lee, C.-H., and Brown, G.E., Phys. Rev. Lett. **79**, 5214 (1997).
10. Brown, G.E., Kubodera, K., and Rho, M.. Phys. Lett. **B175**, 57 (1987).
11. Carlson, J., Heiselberg, H., and Pandharipande, V., nucl-th/9912043.
12. Brown, G.E., and Rho, M., to appear in Phys. Rep.
13. Brown, G.E., and Rho, M., Phys. Rev. Lett. **66** (1991) 2720.
14. Brown, G.E., and Rho, M., Phys. Rep. **269**, 333 (1996).
15. Song, C., nucl-th/0006030, Phys. Rep., in press.
16. Tsushima, K., Sibirtsev, A., Saito, K., Thomas, A.W., and Lu, D.H., nucl-th/0005065; Phys. Lett. **B484**, 23 (2000).
17. For reviews, see Wilczek, F., " QCD in Extreme Conditions," hep-ph/0003183; Schäefer, T., "Color Superconductivity," nucl-th/9911017; Rajagopal, K., " Mapping the QCD Phase Diagram," hep-ph/9908360, Nucl. Phys. **A661**, 150c (1999).
18. Schäfer, T., and Wilczek, F., Phys. Rev. Lett. **82**, 3956 (1999).
19. Hong, D.-K., Rho, M., and Zahed, I., Phys. Lett. **B468**, 261 (1999).
20. Nowak, M.A., Rho, M., and Zahed, I., *Chiral Nuclear Dynamics* (World Scientific, Singapore, 1996).
21. Park, B.-Y., Rho, M., Wirzba, A., and Zahed, I., Phys. Rev. **D62**, 034015 (2000).
22. Langfeld, K., and Rho, M., Nucl. Phys. **A660**, 475 (1999).
23. Kim, Y., and Rho, M., nucl-th/0005069.
24. Son, D.T., and Stephanov, M.A., hep-ph/0005225.
25. Schäfer, T., nucl-th/0007021.

26. Page, D., Prakash, M., Lattimer, J.M., and Steiner, A., hep-ph/0005094; Blaschke, D., Klähn, T., and Voskresensky, D.N., astro-ph/9908334.
27. Hong, D.-K., Lee, H.K., and Rho, M., to appear.
28. Carter, G.W. and Reddy, S., hep-ph/9995228.
29. Umeda, H., talk in this conference; Umeda, H., Nomoto, K., Tsuruta, S., and Nineshige, S., astro-ph/9910113
30. Alford, M., Berges, J., and Rajagopal, K., Nucl. Phys. **B571**, 269 (2000).
31. Blaschke, D., Sedrakian, D.M., and Shahabasyan, K.M., astro-ph/9904395.
32. Makashima, K., "Magnetic fields of binary X-ray pulsars," in *The structure and eveolution of neutron stars* eds. Pines, D. et al., Addison-Wesley, New York, p.86.

Chirally Protected Pion Mass

Tae-Sun Park[a], Hong Jung[b] and Dong-Pil Min[c]

[a] *Department of Physics and Astronomy, University of South Carolina, Columbia SC 29208, USA*
[b] *Department of Physics, Sookmyung Women's University, Seoul 140-742, Korea*
[c] *Department of Physics, Seoul National University, Seoul 151-742, Korea*

Abstract. The pion mass in the nuclear matter is protected by the chiral symmetry, which makes the existence of the S-wave pion condensation in the neutron star most unlikely. Some subtleties related to the calculation by the heavy baryon chiral perturbation theory of the pion mass in the dense matter are discussed.

Inside of the neutron star, the density of the matter can reach to the value almost 4 times bigger than the normal nuclear density, $\rho_0 = 0.17 fm^{-3}$ [1]. Whether there exist the exotic states such as the meson condensation and quark-gluon plasma at the inner core of the neutron star and how they do show up their existence form the central issues for our understanding of the life of stars as well as the evolution of the universe. The pion and kaon condensations have been considered upon our present understanding of numerous data in nuclear and particle physics. But the P-wave pion condensation is practically impossible to take place even in such a dense system. [2] Now I am going to give the reasons why the S-wave pion condensation follows also the same fate. In fact, the P-wave pions can be created inside the nucleus from the vertex of $np\pi^-$ while the S-wave pion can be created directly by the weak vertex $\pi\nu e$. In order for the S-wave pion condensation to occur, the effective mass of the pion inside the dense matter must be smaller than the chemical potential of the electron, so that the creation of the pion is energetically favored in the balance of $\pi^- \leftrightarrow e^- + \nu$. We are interested in evaluating the variation of the effective mass of pion in the dense matter as a function of the density. Many efforts along this line, of course, have been given theoretically and experimentally. [3–5] Before going into the discussion on the computation of the effective mass, I wish to mention about the particularity of the S-wave condensation first.

Unlike the P-wave condensation, pions in the S-wave do not carry the momentum. As pion is the pseudoscalar particle, it can by no means couple with the magnetic field, so that the S-wave pion condensation cannot influence the magnetic field of the star contrarily to the claims. [6] Important is the mass pole of the propagator. The pole position of the in-medium pion propagator with vanishing three-momentum defines the effective,

$$w^2 - m_\pi^2 - \Pi(w^2) = 0 \tag{1}$$

or

$$m_\pi^{*2} = m_\pi^2 + \Pi(m_\pi^{*2}). \tag{2}$$

Recent measurement of the pionic state [5], $1S$ and $2P$ of ^{207}Pb through the reaction ^{208}Pb$(d,^3$He$)$ provides the information on the pionic *deeply* bound state, which are,

$$B_{1s} = 7.1 \pm 0.2 \text{ MeV} \tag{3}$$
$$B_{2p} = 5.31 \pm 0.09 \text{ MeV} \tag{4}$$

Using the results of these measurements, Friedman and Gal [3] verified the existence of the S-wave repulsion due to the two-body correlation in the nucleus. The presence of the S-wave repulsion in the nucleus makes the pionic state to have sharper width, and to be less bound. Their fit is made using the S-wave part of the empherical optical potential, which corresponds to the following pion self energy,

$$\Pi_s = -4\pi(1 + \frac{\mu}{M})[a_0(\rho_n(r) + \rho_p(r)) + a_1(\rho_n(r) - \rho_p(r)) + 4B_0\rho_n(r)\rho_p(r)] \tag{5}$$

where μ is the reduced mass of the pion(m_π) and nucleon(M), and $\rho_{n(p)}$ is the density of neutron(proton), and the isospin even(odd) scattering length are denoted by $a_0(a_1)$ with the correlation strength B_0,

$$a_0 = -0.0077 \pm 0.0072 \ m_\pi^{-1} \tag{6}$$
$$a_1 = -0.0962 \pm 0.0071 \ m_\pi^{-1} \tag{7}$$
$$B_0 = -0.062 + i0.056 \ m_\pi^{-4}. \tag{8}$$

Fitting the binding energy of 1S state, they obtained the strength of the optical potential at the center of the nucleus as about 27 MeV. This equation contains the density dependence of the self energy on the linear and quadratic power. The relevant diagram for the linear dependence is depicted in Fig. 1a, whose contribution reads

$$\Pi_s = -4\pi(1 + \frac{\mu}{M})[a_{\pi N}(\rho_n(r) + \rho_p(r))] \tag{9}$$

for the symmetric nuclear matter with the πN scattering length $a_{\pi N}$. We remark in eq.(5) that the self energy has the quadratic dependence of the density. On the other hand, Waas, Brockman and Weise [4] suggested the $\rho^{\frac{4}{3}}$ dependence of the self energy which stems from the Ericson-Ericson re-scattering in addition to the quadratic dependence. Encouraged by the successful examples [2] of the heavy baryon chiral perturbation theory(HBChPT) to explain/predict the nuclear reactions, Tae-Sun Park, Hong Jung and D.-P. Min try to obtain the dependence of the self energy with respect to the density of the nuclear matter. [7] Main target is to know what dependence can be predicted from HBChPT, and how those terms show their dominance as the nuclear density increases. Calculating up to the next-to-next-to the leading order of the HBChPT as shown in Fig.1, which corresponds to the two loop order, they arrive at the conclusion that the actual dependence is quite complicated but confirms the expected dependence as discussed in previous works mentioned above. In order to evaluate the pion self-energy systematically, they employ Weinberg's formalism of HBChPT. In this scheme, irreducible diagrams are characterized to be order of $\mathcal{O}(Q^\nu)$, where Q is the typical size of the momenta involved and/or pion mass, which is regarded as small compared to the chiral scale $\Lambda \sim 1$ GeV. Fermi momentum k_F is also counted as of order of Q. Figure 1 shows the nuclear diagrams in computing the pion self-energy. There nucleon loops in *free* space are forbidden, therefore all the nucleon lines (in solid line) drawn in Fig.1 run only up to k_F. The leading order (LO) contribution to the S-wave pion self-energy is from the one-loop graph with the Weinberg-Tomozawa term, which vanishes however for symmetric nuclear matter. If such a term survives, it would be $\mathcal{O}(Q^4)$. The next-to-leading order (NLO) contribution comes from the one-loop graph with the $\pi\pi NN$ vertices from the higher chiral order Lagrangian. This is the consequence that the chiral symmetry protects the change of pion mass in the nuclear medium. At NNLO, we have also genuine two-body contributions (Fig. 1($b-d$)) as well as loop correction to the one-body contribution. Including all of them, the pion self energy up to NNLO or up to $\mathcal{O}(Q^6)$ reads

$$\Pi(\omega^2) = -\left[4\pi(1+\mu/M)a_{\pi N} + (w^2 - m_\pi^2)\left(\frac{4c_{23}}{f_\pi^2} + \zeta\frac{3g_A^2 m_\pi}{64\pi f_\pi^4}\right)\right]\frac{2k_F^3}{3\pi^2}$$
$$+ \frac{1}{16\pi^4 f_\pi^4}\left[2\omega^2 I_F(\sqrt{m_\pi^2 - \omega^2}) + \frac{g_A^2}{2}(m_\pi^2 + \zeta(w^2 - m_\pi^2))J_F(m_\pi)\right], \qquad (10)$$

where

$$I_F(m) = k_F^4 - \frac{k_F^2 m^2}{6} + \left(\frac{k_F^2 m^2}{2} + \frac{m^4}{24}\right)\log\frac{4k_F^2 + m^2}{m^2} - \frac{4}{3}mk_F^3\tan^{-1}\frac{2k_F}{m},$$

$$J_F(m) = \frac{\partial}{\partial m^2}m^2 I_F(m), \qquad (11)$$

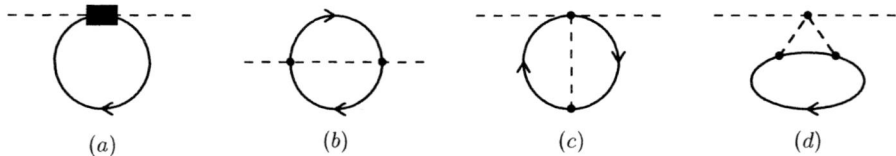

FIGURE 1. Diagrams that contribute to the pion's self energy. Solid lines are nucleons, and dashed lines are pions. The rectangle in (a) denotes whole the diagrams (including loop corrections) that contribute to the πN scattering amplitude.

where $c_{23} \equiv \frac{1}{2}\left(c_2 + c_3 - \frac{g_A^2}{8m_N}\right)$ with the low energy constants c_i [7] and ζ is an off-shell parameter for the definition of the pion field which is not fixed by the chiral symmetry alone. When we employ the back-ground-field method which is quite convenient to remove the artificial contributions from the loop diagrams, it corresponds to taking $\zeta = 4$. [8] Here, one should note an important character that, due to the chiral protection,

$$m_\pi^{*2} - m_\pi^2 = \mathcal{O}\left(Q^5\right) \tag{12}$$

instead of order of Q^2, because the non-vanishing-leading-order (or NLO) of the pion self-energy is $\mathcal{O}\left(Q^5\right)$. Therefore, the NNLO part of eq.(10) is not of $\mathcal{O}\left(Q^6\right)$, but of orders from $\mathcal{O}\left(Q^6\right)$ to $\mathcal{O}\left(Q^9\right)$. This is a peculiarity of pion mass in medium due to the chiral protection for the Goldstone boson. Therefore, one should reconsider the ordering of the self energy of pion of eq.(10) based on the protection.

Inserting this into eq.(10) with $\omega^2 = m_\pi^{*2}$, we see that all the ζ-dependent terms in $\Pi(m_\pi^{*2})$ are of order of Q^9. There is no off-shell ambiguity in $\Pi(m_\pi^{*2})$ up to $\mathcal{O}(Q^8)$. Expanding $\Pi(m_\pi^{*2})$ following to the *chirally protected* counting rule, the two-loop order calculation determines the first two leading order contributions unambiguously,

$$\Pi^{(5)}(m_\pi^{*2}) = -4\pi(1+\mu)a_{\pi N}\frac{2k_F^3}{3\pi^2}, \tag{13}$$

$$\Pi^{(6)}(m_\pi^{*2}) = \frac{m_\pi^2}{16\pi^4 f_\pi^4}\left[2k_F^4 + \frac{g_A^2}{2}J_F(m_\pi)\right]. \tag{14}$$

These expressions show how the pion mass depends on the density of matter, somewhat complicated than the previous works, but quite transparent in figuring out the importance of each term as a function of the density.

Furthermore, we can show that there is no contribution at $\mathcal{O}(Q^7)$ order,

$$\Pi^{(7)}(m_\pi^{*2}) = 0, \tag{15}$$

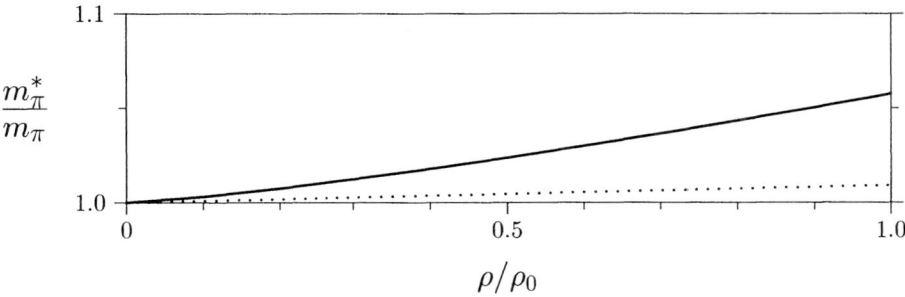

FIGURE 2. The cumulative graphs for the m_π^*/m_π with respect to ρ/ρ_0. The dotted and solid lines stand for the results up to Q^5 and Q^6.

by following arguments. All the one-body contributions are already absorbed into $\Pi^{(5)}$ at on-shell, and the difference $\Pi(m_\pi^{*2}) - \Pi(m_\pi^2)$ is $\mathcal{O}(Q^8)$, which can be read from eq.(10). And since three-body contributions begin only from $\mathcal{O}(Q^8)$, the only possibile source for $\mathcal{O}(Q^7)$ contribution is from the two-body diagrams; two-loops with one inclusion of the so called "$1/m_N$" corrections. While there are some such diagrams, all of them are isovector and cannot contribute to the pion's self energy in symmetric matter. Therefore, indeed, up to the next-to-next to the leading order after the reordering, one can obtain the effective pion mass which is free from the off-shell ambiguity. At $\mathcal{O}(Q^8)$, which is still ζ-independent, there appear one-, two- and three-body contributions, which have not been studied yet.

By expanding $\Pi(m_\pi^{*2})$ with detailed analysis on the counting on the gap $(m_\pi^{*2} - m_\pi^2)$, we thus observe that the in-medium pion mass increases only very mildly with respect to the density; and at normal nuclear density, m_π^* is bigger than m_π by about 6 %. This result has little dependence on the parameter set used, and completely independent of the off-shell parameter ζ.

While higher order than considered, the dependence on the off-shell parameter invites some more discussion. The Brown-Rho (BR) scaling [9] may shed some insight in this problem. If we follow the BR scaling, the whole parameters of the pion self-energy should be replaced by the in-medium effective values. Let $\Pi^*(w^2)$ be the BR scaled pion self-energy, and the equation for the m_π^* reads

$$m_\pi^{*2} = m_\pi^2 + \Pi^*(m_\pi^{*2}). \tag{16}$$

with

$$\Pi^*(m_\pi^{*2}) = -4\pi(1+\mu)a_{\pi N}^* \frac{2k_F^3}{3\pi^2} + \frac{m_\pi^{*2}}{16\pi^4 f_\pi^{*4}} \left[2k_F^4 + \frac{g_A^{*2}}{2} J_F(m_\pi^*) \right]. \tag{17}$$

Now we see that this is essentially the same with eqs.(13-14). Indeed, the effective on-shell condition is $\omega = m_\pi^*$ and the $\Pi^*(m_\pi^{*2})$ is effectively an on-shell quantity. It is quite clear that the pion mass increases in the nuclear matter, though small, so that the expectation of the S-wave pion condensation is very unlikely.

We thank to Mannque Rho for his comments and helpful discussions and also to other organizers. This work is supported in part by KOSEF grant 985-0200-001-2 and by KRF grant 1999-015-DI0023.

REFERENCES

1. R.B. Wiringa, V. Fiks and A. Fabrocini, Phys. Rev. **C38**, 1010 (1988).
2. Mannque Rho, talk given in this workshop.
3. E. Friedman and A. Gal, "On the determination of the pion effective mass in nuclei from pionic atoms", nucl-th/9805004, Phys. Lett.**B432**, 235 (1998).
4. T. Waas, R. Brockmann, and W. Weise, Phys. Lett. **B405**, 215 (1998).
5. T. Yamazaki *et al.*, Phys. Lett. **B418**, 246 (1998).

6. In-Saeng Suh and G. J. Mathews, arXiv:astro-ph/9912358.
7. Tae-Sun Park, Hong Jung and Dong-Pil Min, in preparation.
8. G. Ecker, Phys. Lett. **B 336**, 508 (1994).
9. G.E. Brown and M. Rho, Phys. Rev. Lett. **66**, 2720 (1991).

Kaon Production in Heavy Ion Collisions and Kaon Condensation in Neutron Stars

Chang-Hwan Lee

Dept. of Physics & Astronomy, State Univ. of New York, Stony Brook, NY 11794, USA
Korea Institute for Advanced Study, Seoul 130-012, Korea

Abstract. We discuss the possibility of dropping masses of kaons in dense matter. Kaon production in heavy ion collisions provide a unique environment where we can test the kaonic properties in dense system. We found that the current experiment is consistent with reduced kaon masses. This supports the idea that reduced kaon mass can soften the neutron star equation of states, reducing the maximum mass of neutron star.

INTRODUCTION

The importance of interplay among various fields in physics is growing. The dense environment is one of the most interesting place where all the fields can work together to understand the nature.

Here we give some example of the hot topics, which are connected to the dense matter, in various fields of physics. Since the discovery of quarks and gluons, the theoretical and observational evidences of quark deconfinment, quark-gluon plasma, color superconductivity, etc., were desperately searched for in particle physics. The dense environment created by the collision of relativistic heavy ions is essential for the observation of these exotic phenomena. For the densities smaller than the critical density of the quark-gluon plasma formation, the strong interactions among hadrons play the dominant role. Strong density dependences in the hadronic properties, e.g., masses and sizes of mesons and baryons, are the main issues in current nuclear physics. Since the dense system is consist of many particles, the current developments in condensed matter physics, e.g. renormalization group equations, are widely used in order to understand the physics of dense matter.

Because of the extreme conditions, the dense environment which we are interested is not a part of our common experience. They have to be created in the lab, even though they can survive only in a very short time scale, or they have to be searched for from the universe. On the ground based experiments, the (relativistic) heavy ion collisions provide the unique dense environment where we can test our theories.

In the universe, neutron stars and black holes provide a unique enviroment with densities greater than the normal nuclear matter density.

The density-dependence in the properties of kaons, especially negatively charged kaons, is one of key issues in all these fields. Because K^- consists of anti-up quark and strange quark, it has strong attraction in dense matter.[1] This strong attraction can be also understood from the work Schäfer [1] that the masses of kaons become extremely small compared to that of free space in the limit where color-flavor locked phase does thte dominant role. Recent experiments from GSI support the idea of reduced kaon masses in dense matter [3]. The astrophysical implication of this dropping kaon mass appears in the neutron star equation of state. The reduced kaon mass makes the equation of state softer, which provides us smaller maximum neutron star mass. The current observation on the radio pulsars are consistent with our reduced neutron star mass limit.

In this work, we summarize the basics of kaon condensation and its implication in astrophysics in connection with the kaon production in heavy ion collisions.

KAON PRODUCTION IN HEAVY ION COLLISIONS

Since we cannot directly test the kaon properties in neutron stars, GSI (Germany) has been performing the heavy ion experiments to produce kaons in $Ni + Ni$ collisions. The large enhancement in K^- production compared to K^+ was observed, and standard scenario could not explain the enhancement. By introducing the dropping kaon mass scenario in the simulation of transport calculation, we could successfully reproduce the empirical results. Our work strongly supports that the kaon effective masses indeed drop in dense system [3].

In Fig. 1, we summarize the $K^\pm$ effective masses and the $K^\pm$ production ratio in GSI experiments. Because of the cancellation between the scalar (attractive for both quark-quark and quark-qnti-quark interactions) and vector interaction (repulsive for quark-quark interaction), the mass of K^+ ($u\bar{s}$) slightly increases with density. Since the vector interaction is also attractive bewteen quark and anti-quarks, the scalar and vector interaction add up to give very strong attraction for K^- as shown in the left panel of Fig. 1. In the right panel of Fig. 1, one can confirm that the reduced kaon mass can explain the experimental data. In fact, upto now, our work provides the unique successful solution for the large enhancement of K^- production. We believe that the K^- effective mass really drops in dense system.

KAON CONDENSATION IN NEUTRON STARS

The properties of hot and dense matter have been among the most interesting issues in theoretical astro-hadron physics. Especially, the equation of state of neu-

---

[1] The attraction between quarks (nucleon) and anti-quarks(anti-up quarks in kaon) does the most important role. The strange quark doesn't interaction much with up and down quarks.

tron star was one of the main topics in the study of dense system. As the density increases, the electron chemical potential increases due to the β-equilibrium to balance the positive charge induced by protons, and above a critical density, the production of exotic mesons like kaon is energetically favorable than production of electrons. (In addition to the up and down quarks, we have one more degree of freedom, strange quark, in the center of neutron stars.) The density dependence of electron chemical potential and the kaon effective masses are summarized in the left panel of Fig. 2. Because of the bosonic property of kaons, the ground state energy can be substantially reduced by creating condensed kaons in dense system. The released energy can be carried out by neutrinos almost without rescattering with hadrons in dense system. This reduces the outward pressure, making the neutron star softer. Finally it will enhance the formation of small mass black holes.

In the standard equation of state, the maximum neutron star mass was about $2-3$ times that of the sun $(M_\odot)$. By introducing kaon condensation, however, we can reduce the maximum masses down to 1.5 $M_\odot$ as in the right panel of Fig. 2. In Fig. 3, the well observed binary pulsar masses indicate that the neutron star masses are all consistent with this mass boundary. Furthermore, supernova 1987A is believed to have dense core with mass around 1.5 $M_\odot$, which was theoretically confirmed by the neutrino detection in Kamiokande and IMB. However, the existence of a pulsar in the center of the supernova remnants have not been confirmed yet, this indicates that there is a possibility that the core went into black hole, supporting our scenario. Whether there exist neutron star in the core SN1987A is still an open question.

In our work [4], we clarified the issue of kaon condensation in chiral perturbation theory, a QCD effective theory. Before our work, chiral approach including kaons

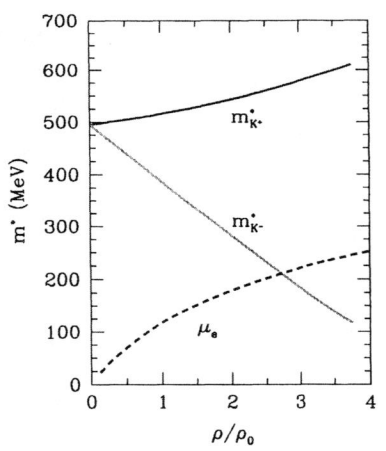
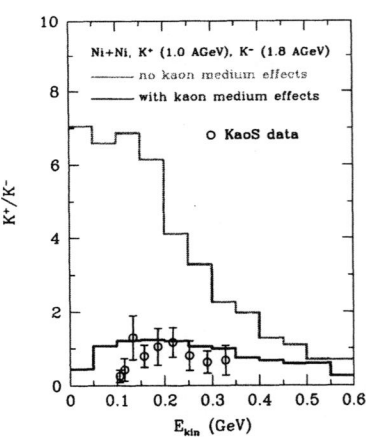

FIGURE 1. $K^\pm$ effective masses and the $K^\pm$ production radio.

has discrepancies with the experiments. They cannot reproduce the threshold KN scattering data. In our approach, however, by introducing $\Lambda(1405)$ as a elementary particle field, we could describe both the KN scattering data and the kaon condensation. In our work, we showed that the physical results of the reduced kaon masses and the kaon condensation are solid against the uncertainties in the introduction of $\Lambda(1405)$.

Recently, Schäfer [1] worked on the kaon condensation from top-down approach, where he started from very high density limit in which color-flavor locking phase plays dominant role. He support the possibility of kaon condensation at high densities.

MAXIMUM NEUTRON STAR MASS

There is by no means agreement about maximum and minimum neutron star masses in the literature. The mass determination of Vela X-1 [6] have been consistently higher than the Brown & Bethe 1.5 $M_\odot$ [5] which is consistent with well measured neutron star masses in Fig. 3. Taking the maximum mass to be 1.87 $M_\odot$ [6], however, one cannot explain why only a few circular neutron-star, carbon/oxygen-white-dwarf $((ns, co)_c)$ which survived the common envelope evolution are seen, since this mass is high enough to give those of the white dwarf companions [7].

Van Kerkwijk [6] lists the many difficulties encountered in the neutron star mass measurement in Vela X-1. Certainly the fact that the center of mass of the system lies well within the B-star companion and the fact that the O-star is "floppy" complicates the measurement. The deviations of the center of luminosity from the center of mass also complicates the data analysis.

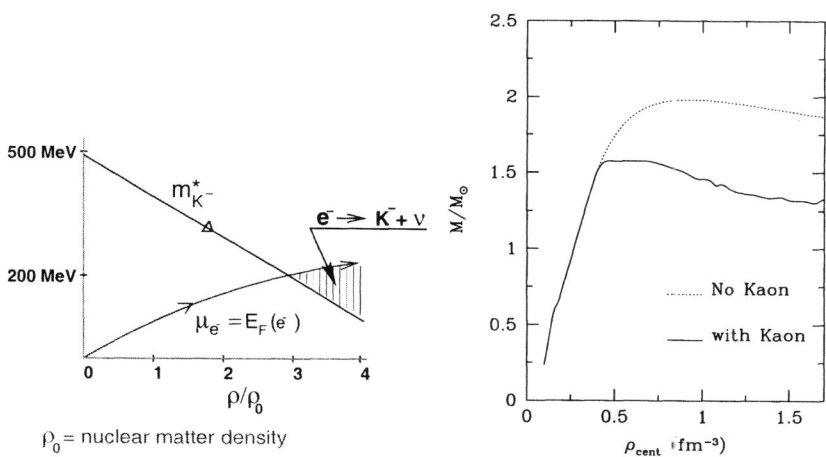

FIGURE 2. Kaon Condensation and Neutron Star Equation of State.

TABLE 1. Comparisons between Neutron Stars and Heavy Icon Collisions.

	Neutron Star	Heavy Ion Collision
Temperature	~ 0 MeV	~ 100 MeV (10^{12} K)
Time Scale	$\gg 1$ sec	$\lesssim 20$ fm/c (6×10^{-23} sec)
Central Density	$> 3\rho_0$	$\lesssim 3\rho_0$
Size	~ 10 km	$\sim 5 fm$ (5×10^{-13} cm)
Strangeness	violated	conserved
Weak Interaction (neutrinos)	Yes	No
Isospin asymmetry	neutron dominated	$\sim$ symmetric nuclear matter

Upper limits of higher neutron star masses are also suggested from QPO (Quasi Periodic Oscillations) signals, but the origin of QPO is still controversial.

DISCUSSION

Even though some part of Kaon condensation and kaon production is understood, there still remain many open issues in addition to the maximum neutron star mass discussed before. First of all, the kaon production in heavy ion condensation is an in-direct evidence of kaon condensation. In Table 1, the basic physical parameters of

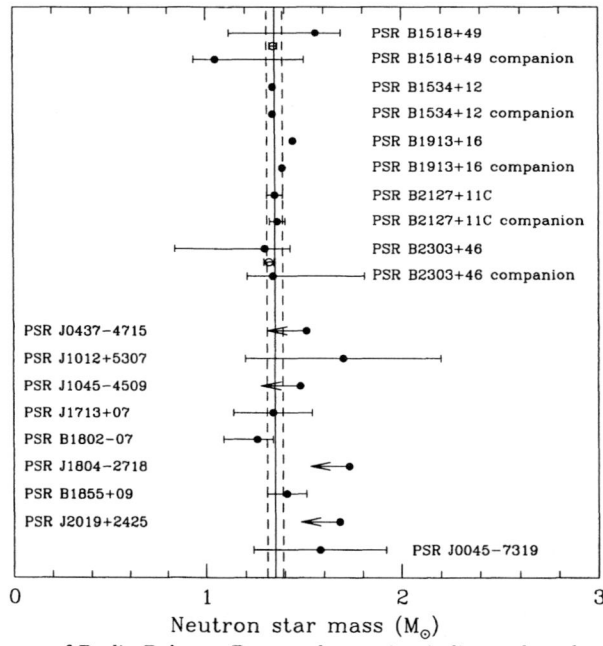

FIGURE 3. Masses of Radio Pulsars. Recent observation indicate that the neutron star mass of PSR J2019+2425 is less than 1.49 $M_\odot$ (Tauris & Savonije 1999).

these two systems are summarized. The uncertainties in the neutron star equation of state is still an open issue, which are discussed in detail by J. Lattimer in this same volume.

There are many prospects in the future observations. Among them, the idenfication of the radius of existing neutron stars is one of the top priority issues. Upto now, there is no clear identification of neutron star radius. A clever theoretical idea which can determine the radius of neutron star is necessary.

The existence of compact core in Cas A was recently identified. Because of the short age (~ 500 years), the identification of its core as a neutron star or a black hole will give very important hints for the cooling processes of neutron star. This can provide some constraints on the possibility of kaon condensation.

ACKNOWLEDGMENTS

I acknowledge the support from the U.S. Department of Energy under grant DE-FG02-88ER40388.

REFERENCES

1. Schäfer, T., arXiv:nucl-th/0007021 (2000).
2. Brown, G.E., Lee, C.-H., Rho, M., and Thorsson, V., *Nucl. Phys.* **A567**, 937 (1994).
3. Li, G.Q., Lee, C.-H., Brown, G.E., *Phys. Rev. Lett.* **79**, 5214 (1997).
4. Lee, C.-H., *Phys. Rep.* **275**, 255 (1996).
5. Bethe, H. A., & Brown, G. E. 1995, ApJ, 445, L129
6. Van Kerkwijk 2000, Proceedings of ESO Workshop on "Black Holes in Binaries and Galactic Nuclei", Garching, Sep. 1999, eds. L. Kaper, E.P.J. van den Heuvel, and P.A. Woudt, Springer-Verlag; astro-ph/0001077
7. Brown, G.E., Lee, C.-H., Portegies Zwart, S., and Bethe, H.A., 2001, ApJ, to appear.
8. Tauris, T.M. & Savonije, G.J. 1999, A&A, 350, 928

Neutrino Interactions In Color-Flavor-Locked Dense Matter

Deog Ki Hong[a,b1], Hyun Kyu Lee[c2], Maciej A. Nowak[a,d3] and Mannque Rho[a,e4]

[a] *School of Physics, Korea Institute for Advanced Study, Seoul 130-012, Korea*
[b] *Department of Physics, Pusan National University, Pusan, Korea*
[c] *Department of Physics, Hanyang University, Seoul, Korea*
[d] *Marian Smoluchowski Institute of Physics, Jagellonian University, 30-059 Krakow, Poland*
[e] *Service de Physique Théorique, CE Saclay, 91191 Gif-sur-Yvette, France*

Abstract. At high density, diquarks could condense in the vacuum with the QCD color spontaneously broken. Based on the observation that the symmetry breaking pattern involved in this phenomenon is essentially the same as that of the Pati-Salam model with broken electroweak–color $SU(3)$ group, we determine the relevant electroweak interactions in the color-flavor locked (CFL) phase in high density QCD. We briefly comment on the possible implications on the cooling of neutron stars.

INTRODUCTION

Recent developments on high-density QCD [1] suggest that diquarks condense in superdense hadronic matter giving rise to a color-flavor locked state [2]. Among the excitations on this broken phase vacuum are massive color gluons metamorphosed to vector mesons and integrally charged quarks behaving as baryons [3,4]. Although such a state may not be formed in heavy-ion collisions, it may be relevant in the physics of compact stars such as the cooling of neutron stars.

The ultimate goal of studying the state at high density and at low temperature is to explore the astrophysical implication of this broken phase: What is particularly interesting is the role of neutrinos in the cooling of neutron stars. In preparation for such a study, we consider how matter in the color-flavor locked state responds to electroweak (EW) probes. In this talk, we exploit the observation that the symmetry breaking pattern in the color-flavor locked (CFL) state is basically the same as that of the broken color gauge theory, originally proposed by Pati and

---

[1]) E-mail: dkhong@pnu.edu
[2]) E-mail: hklee@hepth.hanyang.ac.kr
[3]) E-mail: nowak@kiwi.if.uj.edu.pl
[4]) E-mail: rho@spht.saclay.cea.fr

Salam [5] for a grand unification scheme, to study how neutrinos (or generally weak current) interact in the CFL matter.

GAUGE THEORETICAL MODEL FOR THE CFL PHASE

In the minimal Pati-Salam model, $SU(3)_{color}$ gauge group as well as the flavor group, $SU(2)_L \times U(1)_R$, are spontaneously broken. One of the features of this model is the integrally charged quarks (i.e., Han-Nambu quark model [6]) and the charged massive gluons, both of which are analogous to the excitations in the color-flavor locked QCD phase. The gauge bosons get mixed to form mass eigenstates: for example the neutral gauge bosons, photon A, weak boson Z_0, and a combination of gluon $\tilde{V}_0$ are a mixture of the original gauge bosons, B of $U(1)_R$, W_0 of $SU(2)_L$ and a combination, $V_3 + V_8/\sqrt{3}$, of $SU(3)_{color}$. Consequently it is very natural to expect weak interactions from the colored objects. An interesting feature of this model is that not only photon [7] but also neutrinos (in general, weak current) can probe the color. Despite its intrinsic elegance, this model has not been considered as a relevant model for Nature since no experiments performed at low density have shown any evidence for broken $SU(3)_{color}$ gauge symmetry. However recent theoretical developments indicate the possibility of spontaneous symmetry breaking of $SU(3)_{color}$ in the hadronic matter at very high density and more intriguingly, the symmetry breaking pattern – when suitably reinterpreted – is identical to that of the minimal Pati-Salam model. This suggests that in the color-flavor locked phase, the Pati-Salam model could be exploited to infer the *structure* of electroweak interactions involving the colored objects. This does not mean that this is the only way of deriving the interaction forms. As discussed in [8], one can do a systematic weak-coupling calculations valid at high density taking into account the color-flavor locking and consequent non-perturbative effects. The point of this paper is that the symmetry breaking pattern shares its generic characteristic with other schemes already available in the literature for different reasons that allows a simple understanding of the electroweak couplings.

For our purpose we shall choose the gauge group to be $SU(2)_L \times U(1)_R \times SU(3)_{color}$ as in the minimal version of Pati-Salam model. One can easily extend it to $SU(3)_L$ or $SU(4)_L$ [5]. The basic structure remains the same. [5] In addition to the standard Higgs doublet ϕ, which is responsible for the electro-weak symmetry breaking transforming as $(\{2\}, \{1\})$ for $(SU(2)_L, SU(3)_c)$, there are color-flavored

---

[5] Operationally the symmetry scheme can be extended to incorporate the left-right global symmetry of nature:

$$(SU(2)_L \times U(1)_R \times SU(3)_c)_{local} \times (SU(3)_L \times SU(3)_R)_{global} \qquad (1)$$
$$\to (U(1)_{em})_{local} \times (SU(3)_{c+L+R})_{global}. \qquad (2)$$

Higgs, σ, which transform as $(\{2\}+\{2\},\{3^*\})$ with nonzero vacuum expectation values for the color flavor diagonal elements:

$$\langle \sigma_{aj} \rangle = \sigma \text{ for } a = j \ (a = 1, 2, 3, 4, \ j = 1, 2, 3) \tag{3}$$
$$= 0 \text{ otherwise} \tag{4}$$

where $a = 1, 2$ and $3, 4$ are the doublets of $SU(2)_L$ and j is the color index. In fact, a are identical to the indices for global flavor symmetry of the strong interaction when Cabbibo mixing is understood. Hence the color-flavored Higgs transforms as a fundamental representation under the global flavor transformation. The charm quark with the flavor index $a = 4$ is too heavy for consideration at high density, so we will be primarily concerned with $SU(3)_f$. The flavor index $a = 4$ will not figure in our discussion. We note that it is analogous to the color-flavor locked diqaurk condensate :

$$\langle q_{Li\alpha}^a(\vec{p}) q_{Lj\beta}^b(-\vec{p}) \rangle = -\langle q_{Ri\alpha}^a(\vec{p}) q_{Rj\beta}^b(-\vec{p}) \rangle = \kappa(p_F)\epsilon_{\alpha\beta}\epsilon^{abI}\epsilon_{ijI}, \tag{5}$$

where $\alpha, \beta = 1, 2$ are Weyl indices, $a, b = 1, 2, 3$ flavor indices, and $i, j = 1, 2, 3$ color indices. (We will specify the CFL phenomenon more precisely later.) For now we focus on the essential point. The diquark transforms under color as $\{3^*\}$ or $\{6\}$. One can see that $\{3^*\}$ [6] is equivalent to the color-flavored Higgs σ. This observation leads us to propose that the symmetry breaking pattern in the CFL phase can be directly mapped to that in the Pati-Salam model. Of course we are not implying that the Pati-Salam model is effective in the zero-density regime.

Let us imagine that both electroweak and color-flavor locking symmetry breakings have taken place, with v representing the VEV for the former and σ the VEV for the latter. Denote the electroweak mixing angle (or Weinberg angle) by θ_W and the QCD mixing angles by β and δ. In the standard procedure, the gauge covariant couplings of the gauge bosons to the Higgs scalars give rise to the mass terms for the gauge bosons. After diagonalizing the mass terms, we have a massless photon given by [7]

$$\tilde{A} = \cos\beta(\sin\theta_W W_3 + \cos\theta_W B_0) + \sin\beta V_0 \tag{6}$$

where

$$\tan\beta = \frac{2}{\sqrt{3}} \frac{g}{g_s} \sin\theta_W \tag{7}$$

$$\tan\theta_W = \frac{g'}{g} \tag{8}$$

and

---

[6] It has been argued that the $\{3^*\}$ scalar channel dominates in the diqaurk condensate [9–11].
[7] For simplicity, we omit the Lorentz index μ in the vectors. The charge operator and gauge fields effective in the CFL phase are denoted by tilde.

$$\tilde{Z} = Z - \frac{4}{\sqrt{3}} \frac{g_s g}{g^2 + g'^2} \frac{\sigma^2}{v^2} \cos\theta_W V_0 + \mathcal{O}(\sigma^2/v^2), \tag{9}$$

$$\tilde{V}_0 = -\sin\beta(\sin\theta_W W_3 + \cos\theta_W B_0) + \cos\beta V_0 + \mathcal{C}(\sigma^2/v^2), \tag{10}$$

where

$$Z = \cos\theta_W W_3 - \sin\theta_W B_0, \tag{11}$$

$$V_0 = \sqrt{\frac{3}{4}}(V_3 + \frac{1}{\sqrt{3}} V_8). \tag{12}$$

Here g_s, g and g' are respectively the gauge coupling constants for $SU(3)_c$, $SU(2)_L$ and $U(1)$. Now $\tilde{W}^\pm$ and $\tilde{V}^\pm$ are mixed states of $W^\pm$ and $V_{\rho\pm}$:

$$\tilde{W}^\pm = \cos\delta\, W^\pm - \sin\delta\, V_{\rho\pm}, \tag{13}$$

$$\tilde{V}^\pm = \sin\delta\, W^\pm + \cos\delta\, V_{\rho\pm} \tag{14}$$

where

$$\tan\delta = \frac{g}{g_s}(\frac{M_V}{M_W})^2. \tag{15}$$

The other four colored gauge bosons remain unmixed:

$$V_{K^*}^+, V_{K^*}^-, \tag{16}$$

$$V_{K^*}^0, V_{\bar{K}^*}^0. \tag{17}$$

The subscripts ρ, K^* etc. represent the Higgsed gluons with the corresponding quantum numbers [3]. The masses of the gauge bosons are given by

$$M_V \sim g_s \sigma/\sqrt{2}, \tag{18}$$

$$M_W \sim gv/\sqrt{2}. \tag{19}$$

Taking $M_V \sim$ few 100 MeV, we expect that

$$\frac{M_V}{M_W} \sim 10^{-3}. \tag{20}$$

We also expect that

$$\frac{g_s}{g} \sim \frac{g_s}{g'} \sim 10.$$

The unit of electromagnetic charge is defined as [8]

$$\tilde{e}^2 = \frac{g_s^2 g^2 \sin\theta_W}{g_s^2 + 4g^2 \sin^2\theta_W/3}. \tag{21}$$

---

[8] We use $\tilde{e}$ to distinguish it from the zero-density charge e.

The left-handed quarks and leptons are classified as $(\{2\},\{3^*\})$ and $(\{2\},\{1\})$ respectively. The right-handed quarks are classified as $(\{1\},\{3^*\})$. Using eqs.(6) and (21), the electromagnetic charge can be obtained by

$$\tilde{Q}_{em} = Q_{f(lavor)} + Q_{c(olor)} \tag{22}$$
$$Q_f = (I_3 + Y/2)_f, \tag{23}$$
$$Q_c = (I_3 + Y/2)_c \tag{24}$$

which gives integer charges of the Han-Nambu type to the quarks:

$$\begin{pmatrix} \tilde{Q}_u \\ \tilde{Q}_d \\ \tilde{Q}_s \end{pmatrix} = \begin{pmatrix} 0 & 1 & 1 \\ -1 & 0 & 0 \\ -1 & 0 & 0 \end{pmatrix}$$

where the (three) columns represent the (three) colors.

In order to identify the charge states of the quarks, which due to the locking of color with flavor make up a nonet (an octet plus a singlet), with the baryons in the CFL phase, we have to take the tensor product of the color and the flavor [9]. One can identify the two $\tilde{Q} = 1$ states with $\tilde{p}$ and $\tilde{\Sigma}^+$, the two $\tilde{Q} = -1$ states with $\tilde{\Sigma}^-$ and $\tilde{\Xi}^-$ and three $\tilde{Q} = 0$ with $\tilde{n}$, $\tilde{\Sigma}^0$ and $\tilde{\Xi}^0$. The remaining $\tilde{Q} = 0$ state is the singlet baryon which is assumed to be very massive and hence to decouple.

NEUTRINO INTERACTIONS WITH CFL EXCITATIONS

Given the gauge bosons (6), (9) and (10), it is straightforward to read off their weak interaction vertices. We are interested in processes that probe the dense CFL phase. The electromagnetic interaction may not be useful for probing the dense phase because the coupling with the matter must remain strong and most likely lose its memory of the broken phase when observed by the outside detector. However neutrino interactions are very weak and can be a good probe.

Weak Interaction Mediated by Colored Gauge Bosons

Consider the color-gluon annihilation into the lepton pair, $l\bar{l}$, as an example of weak neutral current interactions:

$$\tilde{V}^+ + \tilde{V}^- \to \tilde{Z} \to l\bar{l}, \tag{25}$$
$$\tilde{V}^+ + \tilde{V}^- \to \tilde{V}_0 \to l\bar{l}. \tag{26}$$

---

[9] The simplest mnemonic for this operation is to arbitrarily assign colors so that the first column corresponds to a flavor equivalent of $\bar{u}$, the second column to $\bar{d}$ and the third column to $\bar{s}$. One would obtain the nonet equivalent to the octet and singlet of mesons.

The coupling at the $\tilde{V}\tilde{V}\tilde{Z}$ vertex in the process mediated by $\tilde{Z}$, eq.(25), is given by

$$f\cos^2\delta \frac{4}{\sqrt{3}} \frac{g_s g}{g^2+g'^2} \frac{\sigma^2}{v^2} \cos\theta_W \sim \frac{4}{\sqrt{3}} g \cos^3\theta_W (\frac{M_V}{M_W})^2 \tag{27}$$

which gives a suppression factor

$$\sim (\frac{M_V}{M_W})^2 \tag{28}$$

compared to the conventional $\nu\bar{\nu}$ production. However if there are substantial amounts of gluon excitations confined in dense hadronic matter at nonzero T before it cools down completely, it may overcome the suppression factor and affect the cooling process appreciably.

The suppression factor in the process mediated by $\tilde{V}_0$, eq (26), due to the vertex $\tilde{V}_0 l\bar{l}$ is given by

$$\sim \sin\beta \sim g/g_s. \tag{29}$$

The propagator in the low energy limit $Q^2 \ll M_V^2$ is greater than in eq.(25), i.e.,

$$\frac{1}{Q^2 - M_V^2} \sim \frac{1}{M_V^2}. \tag{30}$$

However the amplitude for fusion is enhanced at the strong interaction vertex, $\tilde{V}\tilde{V}\tilde{V}_0$, by a factor of f, and we get the factor for the amplitude

$$\sim Q_f g \frac{g}{g_s} \frac{1}{M_V^2} g_s = Q_f \frac{g^2}{M_V^2} \tag{31}$$

with Q_f given by eq.(23). One can now see that the gluon fusion into the charged flavor $l\bar{l}$ pair is greater than the weak neutral current by a factor of $\sim (\frac{M_V}{M_W})^{-2} \sim 10^6$ and hence comparable to photon mediated processes [7]. However this enhancement does not apply to gluon-mediated $\nu\bar{\nu}$ processes because Q_f is zero for neutrino. In general, for the neutral current with neutrinos, the contribution from color-gluon mediated processes in the broken phase vanishes since the amplitude is proportional to Q_f(neutrino) which is $= 0$. We arrive at the same conclusion for $q\bar{q} \to \nu\bar{\nu}$.

The charged current weak interaction in the process mediated by $\tilde{V}_0$ is also comparable to the ordinary weak interaction strength for the neutrino-quark interaction in the low-energy limit. Consider the following processes in matter,

$$q + l \to q' + \nu(\bar{\nu}), \tag{32}$$
$$q \to q' + l + \nu(\bar{\nu}). \tag{33}$$

As in the gluon annihilation processes, there are two amplitudes that can be decomposed into three parts: quark gauge boson vertex, propagator, gauge boson-lepton-neutrino vertex,

$$qq'\tilde{W}^{\pm} \to \tilde{W}^{\pm} \to l\nu\tilde{W}^{\pm}, \tag{34}$$
$$qq'\tilde{V}^{\pm} \to \tilde{V}^{\pm} \to l\nu\tilde{V}^{\pm}. \tag{35}$$

In the low energy limit, eq.(34) gives the ordinary weak amplitude

$$\sim \frac{g^2}{M_W^2}. \tag{36}$$

It is easy to see that the contribution of the color gauge-boson-mediated process, eq.(35), also gives an amplitude comparable to that of the $W^{\pm}$ mediated process,

$$\sim g\frac{g}{g_s}(\frac{M_V}{M_W})^2 \frac{1}{M_V^2} g_s \sim \frac{g^2}{M_W^2}. \tag{37}$$

It should be noted however that the quark decay mediated by $\tilde{V}_0$ in eq.(33) cannot take place because of the energy conservation: the quarks with different colors but with same flavor have the same mass. Therefore the neutrino production mediated by the color-changing weak current is limited to the process in eq.(32)

$$q_r + e^- \to q_b + \nu \tag{38}$$
$$q_b + e^+ \to q_r + \bar{\nu}. \tag{39}$$

To keep the system in a color-singlet state in the cooling process, these processes should occur equally to compensate the color change in each process. It implies that these processes depend on the abundance of positrons in the system. At finite temperature in the cooling period, it is expected that there will be a substantial amount of positrons as well as electrons as long as the temperature is not far below $\sim MeV$. Of course the additional enhancement of the neutrino production due to the CFL phase depends on the abundance of positrons in the system which depends mainly on the temperature. If confined colored gluons are present in the matter in the CFL phase, the same amplitude can be obtained in eq.(35) when qq' is replaced with VV'.

The result obtained above can be summarized as predicting an enhancement of the effective four-point coupling constant for the neutrino production process in the low energy limit. The enhancement due to the neutrino-color interaction is suppressed by factors of $e^{-\Delta/T}$ or $e^{-M_V/T}$, since it depends on the unpaired excitations above gap which can participate into neutrino-color interaction. Hence for the cooling process at low temperatures as $\sim 10^9$K it is not so effective. However during the early stage of proto-neutron star the temperature is expected to be high enough $\sim 20 - 50$MeV [12]to see the effect of the enhancement due to color excitations.

In the next section, the evolution of the effective coupling constant is used with the help of a renormalization group analysis to show that there is an additional enhancement of the coupling constant down to the low temperature.

Effective Four-point Fermi Coupling Constant

In this subsection, we calculate how the weak coupling constant runs in dense matter.

In dense matter the gluons are screened since soft gluons can decay into particles and holes near the Fermi surface. The one-loop screening effect at zero temperature has been calculated in the literature [13]. In the high density limit, the one-loop vacuum polarization density is given, with $M^2 = N_f g_s^2 \mu^2/(2\pi^2)$ and $V^\mu = (1, \vec{v}_F)$, $\bar{V}^\mu = (1, -\vec{v}_F)$, by

$$\Pi_{ij}^{\mu\nu}(p) = -\frac{iM^2}{2}\delta_{ij} \int \frac{d\Omega_{\vec{v}_F}}{4\pi}\left(\frac{-2\vec{p}\cdot\vec{v}_F V^\mu V^\nu}{p\cdot v + i\epsilon \vec{p}\cdot \vec{v}_F} + g^{\mu\nu} - \frac{V^\mu \bar{V}^\nu + \bar{V}^\mu V^\nu}{2}\right) + \mathcal{O}(1/\mu). \tag{40}$$

The terms involving quark-anti-quark pair creation and gluon loops are suppressed by $1/\mu$ and are ignored. At low energy, $p_0 \ll |\vec{p}| \sim p_F$ for the gluon and hence the gluon propagators take the following form in the Landau gauge [14]:

$$i\mathcal{D}_{\mu\nu}^{ij}(p_0, \vec{p}) \simeq \delta^{ij}\frac{|\vec{p}|}{|\vec{p}|^3 + i\pi M^2 p_0}O_{\mu\nu}^{(1)} + \delta^{ij}\frac{1}{p_0^2 - |\vec{p}|^2 - M^2}O_{\mu\nu}^{(2)}. \tag{41}$$

The projectors for the magnetic and electric modes are respectively

$$O^{(E)} = P^\perp + \frac{(u\cdot p)^2}{(u\cdot p)^2 - p^2}P^u, \quad O^{(M)} = -\frac{(u\cdot p)^2}{(u\cdot p)^2 - p^2}P^u, \tag{42}$$

with

$$P_{\mu\nu}^\perp = g_{\mu\nu} - \frac{p_\mu p_\nu}{p^2}, \quad \text{and} \quad P_{\mu\nu}^u = \frac{p_\mu p_\nu}{p^2} - \frac{p_\mu u_\nu + u_\mu p_\nu}{(u\cdot p)} + \frac{u_\mu u_\nu}{(u\cdot p)^2}p^2, \tag{43}$$

where $u_\mu = (1, 0, 0, 0)$. At low energy, the electric gluons are Debye screened with a screening mass M and the magnetic modes are dynamically screened (or Landau damped) at $p_0 \neq 0$ [15].

As argued by Alford et al [2], at high density with three light flavors, the $SU(3)_c$ gauge symmetry is spontaneously broken by forming a color-flavor-locked diquark condensate, eq. (5). Then, by Higgs mechanism all eight gluons get mass of order of $g_s\mu$, which can be easily seen by calculating the one-loop vacuum polarization tensor with quarks with Majorana mass (or a gap) Δ generated by the diquark condensate [16–19]:

$$\Pi_{ij}^{\mu\nu}(p) = -g_s^2 \int \frac{d^4 q}{(2\pi)^4}\text{tr}\left[T^A\gamma^\mu \frac{q_\parallel \cdot \gamma + \Delta}{q_\parallel^2 + \Delta^2}T^A\gamma^\nu \frac{(q+p)_\parallel \cdot \gamma + \Delta}{(p+q)_\parallel^2 + \Delta^2}\right] \tag{44}$$

$$\simeq 0.86\frac{iN_f}{3}\frac{g_s^2\mu^2}{2\pi^2}\delta_{ij}\left(g^{\mu\nu} - \frac{p^\mu p^\nu}{p^2}\right) + \cdots \tag{45}$$

where $q_\parallel^\mu = (q_0, \vec{v}_F \vec{q} \cdot \vec{v}_F)$ and the ellipses denote terms containing more powers of momentum. At low momentum all gluons get a dynamical mass, $M_V \simeq 0.2 g_s \mu$ for $N_f = 3$, independent of the gap, Δ, though the relevant scale for the dynamical mass generation is of order of Δ. Let us consider the weak decay of light quasi-quarks, described by the four-Fermi interaction:

$$\mathcal{L}_{Fermi} = \frac{G_F}{\sqrt{2}} \sum_{\vec{v}_F} \bar{\psi}_L(\vec{v}_F, x) \gamma^\mu \psi_L(\vec{v}_F, x) \bar{\nu}_L(x) \gamma_\mu \nu_L(x) \tag{46}$$

$$= \frac{G_F}{\sqrt{2}} \sum_{\vec{v}_F} \psi_L^\dagger(\vec{v}_F, x) \psi_L(\vec{v}_F, x) \bar{\nu}_L(x) \slashed{v} \nu_L(x) \tag{47}$$

where $G_F = 1.166 \times 10^{-5}$ GeV^{-2} is the Fermi constant and ψ denotes the quasi-quark near the Fermi surface, projected from the quark field Ψ as in [20],

$$\psi(\vec{v}_F, x) = \frac{1 + \vec{\alpha} \cdot \vec{v}_F}{2} e^{-i\vec{v}_F \cdot \vec{x}} \Psi(x). \tag{48}$$

Since the four-Fermi interaction of quarks with opposite momenta are marginally relevant and gets substantially enhanced at low energy, it may have significant corrections to the couplings to quarks of those weakly interacting particles [20]:

$$\delta \mathcal{L}_{vq} = \frac{G_F}{\sqrt{2}} \psi_L^\dagger(\vec{v}_F, x) \psi_L(\vec{v}_F, x) \bar{\nu}_L(x) \slashed{v} \nu_L(x)$$
$$\times \frac{i g_{\bar{3}}}{2 M_V^2} \delta^A_{tv;us} \int_y \left[\bar{\psi}_t(\vec{v}_F, y) \gamma^0 \psi_s(\vec{v}_F, y) \bar{\psi}_v(-\vec{v}_F, y) \gamma^0 \psi_u(-\vec{v}_F, y) \right] \tag{49}$$
$$= \frac{4}{3} \frac{g_{\bar{3}}}{2\pi} \frac{G_F}{\sqrt{2}} \psi_L^\dagger(\vec{v}_F, x) \psi_L(\vec{v}_F, x) \bar{\nu}_L(x) \bar{v} \cdot \gamma \nu_L(x),$$

where $\vec{v}_F$ and $\vec{v}_F'$ are summed over and $g_{\bar{3}}$ is the value of the marginal four-quark coupling at the screening mass scale M. In terms of the renormalization group (RG) equation at a scale E

$$\frac{dG_F(t)}{dt} = \frac{4}{3} \frac{g_{\bar{3}}(t)}{2\pi} G_F(t), \tag{50}$$

where $t = \ln E$. The scale dependence of the marginal four-quark coupling in the color anti-triplet channel is calculated in [11,20]. At $E \ll \mu$

$$\bar{g}_{\bar{3}}(t) \simeq \frac{4\pi}{11} \alpha_s(t). \tag{51}$$

Since $\alpha_s(t) = 2\pi/(11t)$, we get

$$G_F(E) \simeq G_F(\mu) \left(\frac{\mu}{E}\right)^{\frac{16\pi}{363}}. \tag{52}$$

Since the RG evolution stops at scales lower than the gap, the low energy effective Fermi coupling in dense matter is therefore

$$G_F^{\text{eff}} = G_F \cdot \left(\frac{\mu}{\Delta}\right)^{\frac{16\pi}{363}}. \tag{53}$$

We emphasize that this enhancement applies equally to the β decay of quarks and other neutrino production processes described in the previous section.

NEUTRON-STAR COOLING PROCESSES

At asymptotic density and low temperature ($T \neq 0$), the relevant excitations are quasi-quarks that are not Cooper-paired, and 17 Nambu-Goldstone bosons. All other massive particles, Higgsed gluons and other massive excitations [21] are expected to be out of thermal equilibrium and decoupled.[10] Thus the main cooling processes must be the emission of weakly interacting light particles like neutrinos or other (weakly interacting) exotic light particles (e.g. axions) from the quasi-quarks and Nambu-Goldstone bosons in the thermal bath.

We can think of two processes. The first one is the weak decay of quasi-quarks considered in the previous subsection where the running weak coupling indicates a modest enhancement of the process. Since neutrinos interact weakly, it can effectively carry away the energy of quark matter. For the neutrino emissivity from quasi quarks, the so-called Urca process is relevant. The neutrino emissivity by the direct Urca process in quark matter, which is possible for most cases in quark matter, was calculated by Iwamoto [22]. For the CFL superconductor, we expect the calculation goes in parallel and the neutrino emissivity is

$$\epsilon_{\text{direct}} \propto \alpha_s \rho Y_e^{1/3} T^6, \tag{54}$$

where ρ is the density, T is the temperature of the quark matter, and Y_e is the ratio between the electron and baryon density. On the other hand, the neutrino emissivity by the modified Urca process, which is the dominant process in the standard cooling of neutron stars [23], is suppressed by $(\Delta/\mu)^4$, since the pion coupling to quarks is given by $g_{qq\pi} \sim \Delta/\mu$ [24]. Thus, the neutrino emmisivity by the modified Urca process in the CFL quark matter is greatly suppressed in the CFL quark matter, compared to normal quark matter. Futhermore, since the pion-pion interaction in the CFL quark matter are also suppressed by Δ/μ [16,25], we note that all the low energy excitations in the CFL quark matter are extremely weakly coupled, implying that the CFL quark matter has large heat capacitance and cools down very slowly at temperatures lower than the gap.

---

[10] If density is not too high, that is, in the regime relevant for such compact stars as neutron stars, there may take place Goldstone boson condensation from the top-down point of view as from the bottom-up. There may also be excitations of generalized (bound-state) mesons discussed in [21] that become low-lying and hence participate in the cooling process. We cannot address these issues in this paper.

The second process that appears to be important is the bremsstrahlung emission of neutrino pairs from massless colored Nambu-Goldstone bosons: $\phi\phi \to \phi\phi\nu\bar{\nu}$.

The relevant terms *before* the gauge boson mass matrix is diagonalized can be written as

$$\mathcal{L}^{int} = \cdots + \frac{g_s}{2}\phi^{\dagger i}_{a'} \overleftrightarrow{\partial}_\mu \phi^j_{a'}(\lambda^A)_{ij} V^A_\mu + \frac{g}{\cos\theta_W} Z_\nu \left(T_3 - \sin^2\theta_W Q_f\right)_{ab} \bar{l}_{La}\gamma^\nu l_{Lb} + \cdots \quad (55)$$

where ϕ's are the scalar Nambu-Goldstone bosons and $l_L = (\nu, e)^T_L$ is the left-handed lepton isospin doublet. Then the coupling for $\phi\phi\nu\bar{\nu}$ is induced by the $\tilde{Z}$ exchange

$$(\phi\phi\tilde{Z})\, G_{\tilde{Z}}\, (\nu\bar{\nu}\tilde{Z}) \quad (56)$$

where $G_{\tilde{Z}}$ is the $\tilde{Z}$ propagator. This can be written as an effective vertex given by

$$\mathcal{L}^{eff} = \cdots + -\frac{4}{\sqrt{3}} \frac{g_s g}{g^2 + g'^2} \cos\theta_W \frac{\sigma^2}{v^2} \frac{g_s}{2} \frac{1}{M_W^2} \frac{g}{\cos\theta_W} \phi^{\dagger i}_{a'} \overleftrightarrow{\partial}_\mu (\lambda^8)_{ij} \phi^j_{a'} \bar{\nu}_L \gamma^\mu \nu_L + \cdots \quad (57)$$

A similar result was given in eq.(27) for $V^+V^-\nu\bar{\nu}$. The result (57) can be easily understood by noting that the Golstone bosons in eq.(57) are nothing but the longitudinal components of the massive gluons. This process is again suppressed by a factor of $\sim (M_V/M_W)^2$ compared to the conventional $\nu\bar{\nu}$ production.

DISCUSSION AND SUMMARY

In this talk, we argue that the symmetry pattern of the color-flavor-locked phase of QCD at high density in the presence of electroweak interactions is mapped to the Pati-Salam model of grand unification. Then, we have shown that this is a simple way of deducing the electroweak coupling of the CFL degrees of freedom. We find that the neutrino interaction with matter in the color-flavor locked phase can be enhanced by additional gluon-mediated processes. It remains to be verified that one can arrive at the same result in the weak-coupling QCD calculation of the type performed in [8,21].

It is perhaps useful to further comment on the idea of mapping the EW responses of the CFL phase to the Pati-Salam model. The symmetry (breaking) pattern is presumably encoded in the effective potential of the scalars ϕ and σ, $\mathcal{V}^{eff}(\phi, \sigma)$. At low density, the physical vacuum has minimum at $\langle\phi\rangle \neq 0$, $\langle\sigma\rangle = 0$ and defines the electro-weak gauge theory, i.e., the established Standard Model. At some high density, however, the effective potential could develop a VEV of σ, as suggested by model calculations [1]. One of the possibilities is the color-flavor-locked phase considered here. (Other possibilities discussed in the literature can also be addressed similarly.) There are two issues to be resolved in the symmetry breaking schemes with scalar particles. As in the electro-weak theory at zero density, there are massive Higgs particles for which the effective potential is yet to be derived or explained. Presumably the effect of density might be marginal for these excitations. As for the color-flavor symmetry breaking, while there is rather compelling renormalization-group-flow argument to suggest that the vacuum expectation value of σ is non-zero at some (asymptotic) density, the explicit form of the effective potential (with density dependence) is yet to be derived.

The essential feature of the neutrino production (except for the fusion processes) is that the color changes as neutrinos(or anti-neutrinos) are produced by the processes mediated by the color gluon exchange. One might therefore think that such processes are forbidden due to the color singlet requirement of the system. However at finite temperature the presence of positron excitations in the system induces processes which preserve the color singlet status as explained in the text. It is noted that the neutrino-color interaction is suppressed at low temperature cooling stage of neutron star but is expected to be effective at the early stage of proto neutron star.

Together with the general enhancement of the effective four-point coupling constant in RG analysis, the enhancement of the neutrino production implies that the cooling process speeds up as the CFL phase sets in dense hadronic matter near the critical temperature. But, at temperature much below the critical temperature, the interaction of quasi-quarks and pions and kaons is extremely weak, suppressed by Δ/μ, and the CFL quark matter cools down extremely slowly.

For a realistic calculation of the cooling rate of compact stars, we need to also consider the neutrino propagation in the CFL matter before the neutrinos come out of the system. A recent study [26] suggests that the presence of the CFL phase can accelerate the cooling process because neutrino interactions with matter are reduced in the presence of a superconducting gap Δ. However this result is subject to modification by the effect of additional interactions – not taken into account in this work – mediated by the colored gluons on the quark polarization. It would be interesting to see how the enhancement of the neutrino production correlates with the neutrino-medium interaction. This is one of the physically relevant questions on how the enhanced neutrino interaction could affect neutron-star(proto neutron star) cooling following supernova explosion. This issue is currently under investigation.

ACKNOWLEDGMENTS

We would like to acknowledge the hospitality of KIAS where this work was initiated as a part of KIAS program for astro-hadron physics. The work by D.K.H. is supported by KOSEF grant number 1999-2-111-005-5 and HKL is supported in part by the interdisciplinary research program of the KOSEF, grant no. 1999-2-003-5.

REFERENCES

1. See for recent reviews, F. Wilczek, " QCD in Extreme Conditions," hep-ph/0003183; T. Schäfer, "Color Superconductivity," nucl-th/9911017; K. Rajagopal, " Mapping

the QCD Phase Diagram," hep-ph/9908360.
2. M. Alford, K. Rajagopal, and F. Wilczek, Nucl. Phys. B **538**, 443 (1999), hep-ph/98044233.
3. T. Schäfer and F. Wilczek, Phys. Rev. Lett. **82**, 3956 (1999), hep-ph/9811473
4. D.K. Hong, M. Rho and I. Zahed, Phys. Lett. **B468**, 261 (1999), hep-ph/9906551
5. J.C. Pati and A. Salam, Phys. Rev. **D8**, 1240(1973); D10, 275(1974)
6. M.Y. Han and Y. Nambu, Phys. Rev. **B139**, 1006 (1965)
7. H.K. Lee and J.K. Kim, Phys. Rev. Lett. **40**, 485 (1978)
8. M. Rho, A. Wirzba and I. Zahed, Phys. Lett. **B473**, 126 (2000); M.A. Nowak, M. Rho, A. Wirzba and I. Zahed, hep-ph/0007034
9. M. Alford, J. Berges and K. Rajagopal, Nucl. Phys. **B558**, 219 (1999) [hep-ph/9903502].
10. R. D. Pisarski and D. H. Rischke, nucl-th/9907094.
11. D.K. Hong, Phys. Lett. **B473**, 118 (2000), hep-ph/9812510.
12. J.A. Pons, S. Reddy, M. Prakash, J.M. Lattimer and J.A. Miralles, Astrophyaical J. **513**, 780(1999)
13. J. Kapusta and T. Toimela, Phys. Rev. **D37**, 3731 (1988); *ibid.* **D39**, 3197 (1989); C. Manuel, Phys. Rev. **D53**, 5866 (1996); G. Alexanian and V.P. Nair, Phys. Lett. **B390**, 370 (1997).
14. D. K. Hong, V. Miransky, I. Shovkovy, and L.C.R. Wijewardhana, Phys. Rev. **D61**, 056001 (2000), hep-ph/9906478.
15. D.T. Son, Phys. Rev. **D59**, 094019 (1999), hep-ph/9812287.
16. D. T. Son and M. A. Stephanov, Phys. Rev. **D61**, 074012 (2000) [hep-ph/9910491].
17. S. R. Beane, P. F. Bedaque and M. J. Savage, Phys. Lett. **B483**, 131 (2000) [hep-ph/0002209].
18. K. Zarembo, Phys. Rev. **D62**, 054003 (2000) [hep-ph/0002123].
19. D. H. Rischke, Phys. Rev. **D62**, 054017 (2000) [nucl-th/0003063].
20. D. K. Hong, Nucl. Phys. **B582**, 451 (2000) [hep-ph/9905523].
21. M. Rho, E. Shuryak, A. Wirzba and I. Zahed, hep-ph/0001104
22. N. Iwamoto, Phys. Rev. Lett. **44**, 1637 (1980).
23. B. L. Friman and O. V. Maxwell, Astrophys. J. **232**. 541 (1979).
24. D. K. Hong, T. Lee and D. Min, Phys. Lett. **B477**, 137 (2000) [hep-ph/9912531]; D. K. Hong, Phys. Rev. **D62**, 091501 (2000) [hep-ph/0006105].
25. M. Rho, A. Wirzba and I. Zahed, Phys. Lett. **B473**, 126 (2000) [hep-ph/9910550].
26. G.W. Carter and S. Reddy, hep-ph/0005228

Magnetoelastic Pulsations of Neutron Stars

Sergey Bastrukov[†‡], Jongmann Yang[†], Dmitry Podgainy[‡], and Fridolin Weber[*]

[†]*Center for High Energy Astrophysics and Isotope Studies,*
Ewha Womans University, Seoul 120-750, Korea
[‡]*Computer Physics Division, Joint Institute for Nuclear Research, 141980 Dubna, Russia*
[*]*Nuclear Science Division, Lawrence Berkeley National Laboratory, Berkeley, CA 94720, USA*

Abstract. We discuss non-radial pulsations of a non-rotating neutron star brought to equilibrium in the state of superparamagnetic magnetization of stellar material. Highlighted are equations of magneto-elastodynamics underlying macroscopic description of large-scale motions of magnetically anisotropic neutron matter possessing properties of elastic Fermi-solid. It is shown that incompressible permanently magnetized nuclear matter can transmit perturbations by transverse magnetoelastic waves. The unique feature of oscillatory behavior of superparamagnetic neutron star is its capability of supporting torsional, differentially-rotational magnetoelastic pulsations. Based on the energy variational principle, analytic form is derived for period of non-radial magnetotorsion pulsations triggering Alfvénic hydromagnetic waves in the neutron star magnetosphere. Our order of magnitude estimates for period of this axial, odd parity magnetotorsion mode, referred to as m/t-mode, suggest that magnetoelastic pulsations may affect curvature radiation from pulsars by shaking magnetic field lines frozen in the neutron star surface, and, thus, producing periodic fluctuations in the beam direction. It is argued that developed model may be worthy of consideration in physics of soft gamma repeaters whose variability may be powered by magnetotorsion pulsations exhibiting superparamagnetism of nuclear matter constituting interior of magnetars.

INTRODUCTION

One of the puzzles of pulsar astrophysics is that neutron star matter possesses capability of sustaining highly intense magnetic field [1]. The present belief is that fossil magnetic field of collapsed massive star resides in the core of an isolated pulsar whose post-formation magnetic field is supported by entirely convective crustal currents [2]. Together with this evolutionary calculations of neutron star structure indicate that material content of neutron star mass is dominated by neutron matter which could undergo, as was argued in recent paper [3], second order phase transition in magnetically ordered ferromagnetic-like state. Critical discussion of early works regarding ferromagnetism of neutron star matter can be found in [4]. In this communication we discuss possibility of paramagnetic magnetization of neutron star matter induced by implosive contraction in the presence of seed magnetic field and its impact on oscillatory behavior of neutron star.

The existing scenario of the neutron star birth [5] in supernova event implies that catastrophic contraction of weakly magnetized massive progenitor star is accompanied by intensive neutronization of stellar matter primarily due to backward β-process responsible for the fast cooling of pulsar by neutrino emission [6]. Since the neutronization of matter is governed by parity violating process, the number of parallell and antiparallel spin magnetic moments of neutrons directed along seed magnetic field may be highly different. So that major part of neutron star mass may be brought to gravitational equilibrium with macroscopically anisotropic distribution of permanent magnetization $\mathbf{M}$ (magnetic moment per unit volume of neutron star matter) typical of paramagnetic materials. The constitutive equation of paramagnetics has the form $\mathbf{M} = \chi \mathbf{B}$, with χ being the magnetic susceptibility, substantially positive dimensionless parameter expressing the fact that paramagnetic materials cause a strengthening of magnetic field when placed in external field. The spin paramagnetic susceptibility of neutron Fermi-continuum (under conditions typical of neutron star interior: the average density and temperature $\rho \approx 10^{13}$ g/cm^3 and $T \approx 10^8$ K, respectively) is estimated to be $\chi = \rho \mu_N^2/(3 m_n k_B T) \approx 0.1$, where m_n is the mass of neutron, μ_N is the nuclear magneton and k_B stands for the Boltzmann constant. Computed with this χ the total dipole magnetic moment of neutron star at the magnetic pole $\mu_{\rm NS} = (4\pi/3) M R^3$ (with $M = \chi B$) fairly good agrees with the standard estimate $\mu_{\rm NS} = 1/2 B R^3$. Notice that so high value of χ is typical of terrestrial superparamagnetics. Therefore, in our search for dynamical manifestation of permanent magnetization of pulsars, the magnetically polarized neutron star matter is referred to as para and/or superparamagnetic material on equal footing. One of such manifestations might be non-radial pulsations associated with large-scale fluctuations of magnetization. Circumstantial evidence for neutron star pulsations is the coherence of micropulses in millisecond region of pulsar spectrum. The discovered in [8] quasi-periodicity of the micropulses is very close to theoretical estimates for periods of non-radial gravitational-elastic pulsations [9–11]. The present work focuses on eigenmode attributed to superparamagnetic magnetization of neutron star matter and, in accord with generally accepted ideology of the theory of neutron star pulsations [12], we confine our discussion to solely oscillations generic to permanently magnetized neutron star.

MAGNETO-ELASTODYNAMICS OF NEUTRON STAR MATTER

We start by introducing equations of magneto-elastodynamics describing macroscopic motions of magnetically saturated neutron star matter, treated as a highly robust elastic material [10,11]. The basic suggestion underlying our analysis is that continuum mechanics of superparamagnetic nuclear matter can be described in terms of the bulk density $\rho(\mathbf{r}, t)$, the velocity of elastic displacements $\mathbf{u}(\mathbf{r}, t)$, and the field of magnetization $\mathbf{m}(\mathbf{r}, t)$ which are considered as independent dynamical variables of dissipative-free motions governed by coupled *equations of magneto-elastodynamics*

$$\frac{d\rho(\mathbf{r}, t)}{dt} + \rho(\mathbf{r}, t) \nabla \cdot \mathbf{u}(\mathbf{r}, t) = 0, \tag{1}$$

$$\rho \frac{d\mathbf{u}(\mathbf{r}, t)}{dt} = \frac{1}{2} \nabla \times [\mathbf{m}(\mathbf{r}, t) \times \mathbf{B}], \qquad \mathbf{M} = \chi \mathbf{B}, \tag{2}$$

$$\frac{d\mathbf{m}(\mathbf{r}, t)}{dt} = [\boldsymbol{\omega}(\mathbf{r}, t) \times \mathbf{m}(\mathbf{r}, t)], \qquad \boldsymbol{\omega}(\mathbf{r}, t) = \frac{1}{2}[\nabla \times \mathbf{u}(\mathbf{r}, t)], \tag{3}$$

which have been derived in [13] by systematic application of the conservation laws of the electrodynamics and mechanics of elastic continuum media. In above equations $d/dt = \partial/\partial t + \mathbf{u} \cdot \nabla$ is the convective derivative. The most important point to be stressed here is that driving force in equation (2) results from body-toque density $\mathbf{m}(\mathbf{r}, t) \times \mathbf{B}$ characterizing magnetoelastic interaction between magnetic field and magnetization mediated by rotational deformations of elastic medium under which the direction of $\mathbf{m}$ changes but the magnitude does not. In the model under consideration the equations of magneto-elastodynamics are regarded to be fundamental equations governing continuum mechanics of magnetically polarized elastic material (which may be non-conducting like neutron matter). They have the same physical significance for the motions of matter possessing highly pronounced properties of paramagnetic polarizability as equations of magneto-hydrodynamics governing motions of perfectly conducting medium in the presence of uniform magnetic field. In [14] these equations have been utilized to describe wave motions in poorly conducting magnetically polarized interstellar medium of dark molecular clouds.

For the former we consider propagation of linear perturbations in unbounded uniformly magnetized nuclear matter. Making use of standard procedure of linearization: $\mathbf{u} \to \mathbf{u}_0 + \delta \mathbf{u}(\mathbf{r}, t)$, $\mathbf{m} \to \mathbf{m}_0 + \delta \mathbf{m}(\mathbf{r}, t)$, where $\mathbf{u}_0 = 0$ and $\mathbf{m}_0 = \mathbf{M}$, and considering constitutive equation $\mathbf{B} = \mathbf{M}/\chi$, we obtain

$$\nabla \cdot \delta \mathbf{u}(\mathbf{r}, t) = 0, \qquad \nabla \cdot \delta \mathbf{m}(\mathbf{r}, t) = 0, \tag{4}$$

$$\rho \frac{\partial \delta \mathbf{u}(\mathbf{r}, t)}{\partial t} = \frac{1}{2\chi} \nabla \times [\delta \mathbf{m}(\mathbf{r}, t) \times \mathbf{M}], \tag{5}$$

$$\frac{\partial \delta \mathbf{m}(\mathbf{r}, t)}{\partial t} = \frac{1}{2}[[\nabla \times \delta \mathbf{u}(\mathbf{r}, t)] \times \mathbf{M}]. \tag{6}$$

The left of equations (4) expresses incompressibility and the right one is the requirement of the absence of induced density of magnetic poles. Let us consider plane-wave form of perturbation

$$\delta \mathbf{u} = \mathbf{u}' \exp(i\omega t - i\mathbf{k} \cdot \mathbf{r}), \qquad \delta \mathbf{m} = \mathbf{m}' \exp(i\omega t - i\mathbf{k} \cdot \mathbf{r}), \tag{7}$$

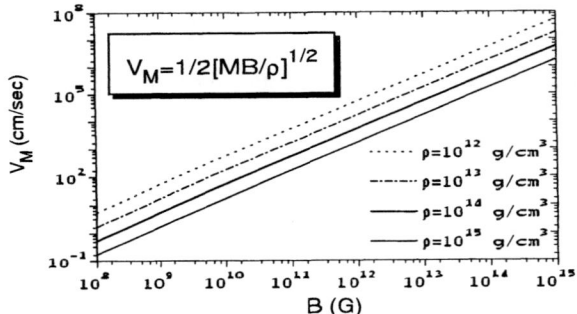

FIGURE 1. Velocity of transverse magnetoelastic wave v_M in superparamagnetic neutron star matter vs intensity of magnetic field B.

where amplitudes $\mathbf{u}'$ and $\mathbf{m}'$ are some small constant vectors, $\mathbf{k}$ stands for the wave vector and ω is the frequency of magnetoelastic oscillations. Inserting (7) into (4) one gets

$$\mathbf{k} \cdot \delta \mathbf{u} = 0, \quad \mathbf{k} \cdot \delta \mathbf{m} = 0. \tag{8}$$

Substitution of (7) into (5) yields $\omega \rho \delta \mathbf{u} = -(1/2\chi)(\mathbf{k} \cdot \mathbf{M})\delta \mathbf{m}$. After substitution of (7) into (6) one has $\omega \delta \mathbf{m} = -(1/2)[(\mathbf{k} \cdot \mathbf{M}) \delta \mathbf{u} - \mathbf{k}(\delta \mathbf{u} \cdot \mathbf{M})]$. However, taking scalar product of last equation with $\mathbf{k} \neq 0$ and considering (8), one finds $(\delta \mathbf{u} \cdot \mathbf{M}) = 0$. Given this, the result of above plan-wave transformation of equations (4)-(6) is expressed by equations

$$\omega \rho \delta \mathbf{u} + \frac{1}{2\chi}(\mathbf{k} \cdot \mathbf{M})\delta \mathbf{m} = 0, \tag{9}$$

$$\omega \delta \mathbf{m} + \frac{1}{2}(\mathbf{k} \cdot \mathbf{M})\delta \mathbf{u} = 0, \tag{10}$$

exhibiting strong magnetization-flow coupling. In spirit of magneto-hydrodynamical treatment of "frozen-in" lines of magnetic force in hydrodynamic flow, the above magnetoelastic coupling between magnetization and rate of elastic displacement can be thought of as vibrations of filaments of magnetization frozen in elastodynamical flow traveling in the direction of equilibrium magnetic anisotropy, so that $\mathbf{k} \parallel \mathbf{M}$. From (9) and (10) it follows

$$\omega^2 = v_M^2 k^2, \quad v_M^2 = \frac{MB}{4\rho} = \frac{M^2}{4\chi\rho} = \frac{\chi B^2}{4\rho}. \tag{11}$$

The obtained dispersion equation describes transverse magnetoelastic wave in which the vectors of magnetization and velocity of elastic displacements undergo coupled oscillations in the plane perpendicular to the direction of propagation. Quadratic form of resultant dispersion relationship exhibits the fact that for shear magnetoelastic waves both $\mathbf{M}$ and $-\mathbf{M}$ directions are energetically equivalent. It worth emphasizing that in diamagnetic matter, $\chi < 0$, the above magnetoelastic fluctuations are unstable. The kinematic character of the magnetoelastic wave has many features in common with transverse hydromagnetic wave in magnetoactive plasma. However, the very existence of Alfvénic wave is the feature generic to perfect conductivity of plasma, whereas the magnetoelastic waves owe their existence to the magnetic polarizability, and, therefore, they can be transmitted by non-conducting neutron matter.

To get a feeling about magnitude of velocity of magnetoelastic wave in paramagnetic nuclear matter, in Fig.1 we plot v_M as a function of intensity of magnetic field B for different values of density ρ. These figure shows that transverse magnetoelastic wave is slowly propagating agitation as compared to sound wave whose speed in neutron star matter $c_s \sim 10^8 - 10^9$ cm/sec.

NON-RADIAL MAGNETOELASTIC PULSATIONS OF NEUTRON STAR

It is the major goal of this communication to describe eigenmode of non-radial pulsations of neutron star unique to state of paramagnetic saturation of incompressible neutron star matter possessing properties of highly robust elastic Fermi-solid [11]. To evaluate frequency we take advantage of the energy variational principle which has been efficiently utilized in our previous studied of non-radial spheroidal and torsional gravitational-elastic [11] and hydromagnetic [15] pulsations of neutrons stars. The starting point of the method is equation of energy balance

$$\frac{\partial}{\partial t} \int \frac{\rho \delta \mathbf{u}^2}{2} dV = \int [\delta \mathbf{m} \times \mathbf{B}] \cdot \delta \boldsymbol{\omega} \, dV, \tag{12}$$

which is obtained after scalar multiplication of (5) with $\delta \mathbf{u}$ and integrating by part and ignoring surface integrals having in mind that the material content of crustal region is dominated by electron-proton-nuclear plasma (Ae-phase) whose magnetic properties are suppressed. The next step is to apply the Rayleigh factorization procedure for the velocity field of elastic displacement

$$\delta \mathbf{u}(\mathbf{r}, t) = \mathbf{a}(\mathbf{r}) \dot{\alpha}(t), \quad \text{and hence,} \quad \delta \boldsymbol{\omega}(\mathbf{r}, t) = \frac{1}{2} [\nabla \times \mathbf{a}(\mathbf{r})] \dot{\alpha}(t), \tag{13}$$

where $\mathbf{a}(\mathbf{r})$ is the field of instantaneous elastic displacements and $\alpha(t)$ is the harmonic in time temporal amplitude of elastic fluctuations. Inserting (13) into (6) and eliminating time derivative one has

$$\delta \mathbf{m}(\mathbf{r}, t) = \frac{1}{2} [[\nabla \times \mathbf{a}(\mathbf{r})] \times \mathbf{M}] \, \alpha(t), \tag{14}$$

Substitution of (13) and (14) into (12) leads to

$$\frac{dH}{dt} = 0 \qquad H = \frac{J\dot{\alpha}^2}{2} + \frac{K\alpha^2}{2}, \qquad (15)$$

where the inertia J and the stiffness K of magnetoelastic vibrations are given by

$$J = \int \rho \mathbf{a}^2 \, dV \qquad K = \frac{1}{4\chi} \int [[\nabla \times \mathbf{a}] \times \mathbf{M}]^2 \, dV. \qquad (16)$$

Thus, to compute the frequency of the magnetoelstic mode $\omega^2 = K/J$, it is necessary to specify the field of instantaneous displacements $\mathbf{a}$. The unique feature of magnetic pulsations under consideration is that they are accompanied by elastic displacements in the neutron star bulk which are of substantially rotational character. With this in mind, we consider case of torsional long-wavelength vibrations around the polar axis z excited in paramagnetic neutron star with the constant magnetization inside pointing to the same direction: $\mathbf{M} = [M_x = 0, M_y = 0, M_z = M]$. The kinematics of elastic torsional deformations of neutron star is described in details in [11,16], and we take advantage of the explicit form for the velocity given in this latter paper

$$\delta \mathbf{u}(\mathbf{r},t) = \frac{1}{2}[\delta\boldsymbol{\omega}(\mathbf{r},t) \times \mathbf{r}], \quad \delta\boldsymbol{\omega}(\mathbf{r},t) = N_t \nabla r^L \, P_L(\mu)\dot{\alpha}(t), \quad N_t = \frac{1}{R^{L-1}}, \quad \mu = \cos\theta. \quad (17)$$

Hereafter $P_L(\mu)$ stands for the Legendre polynomial of the multipole degree L. The corresponding field of instantaneous torsional displacements has the form of the toroidal vector field

$$\mathbf{a}(\mathbf{r}) = N_t \nabla \times [\mathbf{r} \, r^L \, P_L(\mu)]. \qquad (18)$$

Inserting (18) in (16) we obtain

$$J = 4\pi\rho R^5 \frac{L(L+1)}{(2L+1)(2L+3)}, \qquad K = \frac{\pi}{\chi} M^2 R^3 \frac{L(L^2-1)(L+1)}{(4L^2-1)}. \qquad (19)$$

At $L = 1$, the magnetotorsion moment of inertia equals to moment of inertia of rigid sphere: $J(L=1) = (2/5)\mathcal{M}R^2$, where $\mathcal{M} = (4\pi/3)\rho R^3$ is the star mass. The case of $L = 1$ corresponds to rigid body rotations of the star as a whole without changing its intrinsic state, so that the multipolarity of lowest vibration L=2. The full spectrum of magnetotorsion non-radial pulsations, below referred to as m/t-mode, is given by

$$\omega_M(L) = \omega_M \left[\frac{(2L+3)(L^2-1)}{(2L-1)}\right]^{1/2}, \qquad \omega_M = \frac{v_M}{R} \qquad (20)$$

where ω_M is the natural unit of frequency of magnetotorsion pulsations, and v_M is given by (11).

Fig.2 displays dependence of period $P = 2\pi/\omega$ of m/t-mode upon multipole degree L of vibration. For a neutron star with magnetic field of Crab-pulsar, period of magnetotorsion pulsations is of the order of $P \sim 30$ seconds. It is remarkable that for the magnetic field intensity typical of magnetars, the period of $m/t-mode$ falls into interval 5-15 second which is the realm of observed variability of Soft Gamma Repeaters (SGR) [17,18].

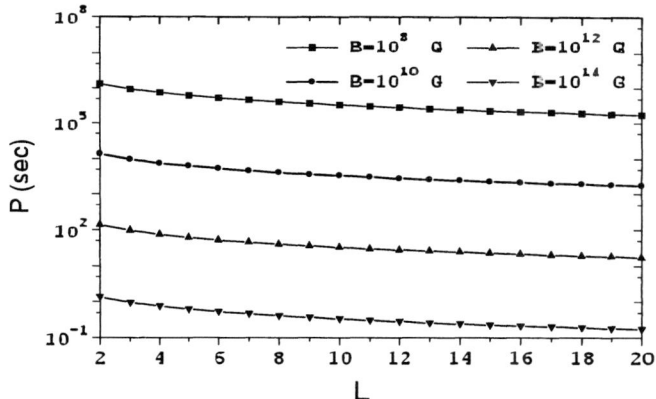

FIGURE 2. Period P of magnetoelastic torsion pulsations vs multipole degree of vibration L.

SUMMARY

This communication presents arguments that considerable contribution to the pulsar magnetic field may originate from superparamagnetic magnetization of neutron star matter that might be induced by seed magnetic field inherited by neutron star from massive weakly magnetized progenitor and highly intensified by catastrophic implosion. We have found that superparamagnetic neutron star can undergo non-radial magnetoelastic torsional pulsations presumably localized in the neutron star core, beneath the perfectly conducting crust. Torsion oscillatory deformations progressed toward peripheral perfectly conducting region may be coupled with toroidal fluctuations of magnetic field lines resulting in hydromagnetic oscillations of crustal material which, in turn, may serve as an efficient trigger of Alfvénic waves in the neutron star magnetosphere. Sticking to this line of arguments one might expect that magnetotorsion pulsations in question should affect curvature radiation from pulsars by shaking magnetic field lines frozen in the neutron star surface and, thus, producing periodic fluctuations in the beam direction. Our numerical estimates for period of magnetotorsion pulsations indicate that the hypothesis of superparamagnetic magnetization of neutron star matter is not inconsistent with present knowledge of pulsar magnetic field. In normal pulsars the magntotorsion mode might be observed as long-periodic (30-50 second) modulation of the main pulse train. It should be noted that above suggestive discussion regarding possible conversion of magnetoelastic pulsations of superparamagnetic neutron star into magnetospheric hydromagnetic waves has some features in common with arguments of Duncan and Thompson [19,20] underlying their treatment of SGR activity, and we suspect that the developed model may be worthy consideration in physics of magnetars. In particular, our order of magnitude estimates for period of $m/t - mode$ suggest that 7.5-second variability of SGR1806-20 might be powered by non-radial magnetoelastic pulsations of superparamagnetic magnetar.

REFERENCES

1. Chanmugam G., *Ann. Rev. Astron. Astrophys.* **30**, 143 (1992).
2. Bhattacharya D., in *Pulsar Timing, General Relativity and the Internal Structure of Neutron Stars*. Eds. Z. Arzoumanian, F. Van der Hooft, and E.P.J van der Heuvel, Royal Netherlands Academy of Art and Sicence, Amsterdam, 1999, p. 235.
3. Akhiezer A.I., Laskin N.V., and Peletminskii S.V., *JETP* **82**, 1066 (1996).
4. Michel F. C., *Theory of Neutron Star Magnetospheres*, Chicago Univ. Press, Chicago, 1991.
5. Burrows A., and Lattimer J.M., *ApJ* **307**, 178 (1986).
6. Weber F., *Pulsars as Astrophysical Laboratories for Nuclear and Particle Physics*, IOP, Bristol, 1999.
7. Kittel C., *Introduction to Solid State Physics*, 7th edn., Wiley, New York, 1996.
8. Strohmayer T. E., Cordes J. M., and Van Horn H.M., *ApJ* **389**, 685 (1992).
9. Van Horn H.M., *ApJ* **236**, 899 (1980).
10. McDermott P.N., Van Horn H.M., and Hansen C. J., *ApJ* **325**, 725 (1988)
11. Bastrukov S. I., Weber F., and Podgainy D. V., *J. Phys.* **G 25**, 107 (1999).
12. Andersson N., Kojima Y., and Kokkotas K.D., *ApJ* **402**, 855 (1996).
13. Tiersten H.F., *J. Math. Phys.* **5**, 1298 (1964).
14. Yang J., and Bastrukov S., *JETP Letters* **71**, 395 (2000).
15. Bastrukov S.I., and Podgainy D. V., *Phys. Rev.* **E 54**, 4465 (1996).
16. Bastrukov S., Molodtsova I., Podgainy D., Weber F., Papoyan V., *Phys. Part. Nucl.* **30**, 436 (1999).
17. Kouveliotou C., et. al., *Nature* **239**, 235 (1998).
18. Kulkarni S.R., Thompson C., *Nature* **239**, 215 (1998).
19. Duncan R. C., and Thompson C., *MNRAS* **275**, 225 (1995).
20. Thompson C., and Duncan R. C., *ApJ* **473**, 322 (1996).

Neutron Star Equation of State

James M. Lattimer*

*Department of Physics & Astronomy, State University of New York at Stony Brook[1]
Stony Brook, NY 11794-3800

Abstract. Experimental information concerning the equation of state in neutron stars is lacking, because of the necessary extrapolations in both density and neutron excess from the nearly symmetric nuclear matter observed in nuclei. However, the combination of new developments in the theory of neutron star structure and in astronomical observations provides important constraints. From a theoretical perspective, it is argued that the extrapolation in neutron excess is more crucial for neutron star structure than is the density extrapolation. For example, the radius of neutron stars is primarily a function of the pressure of matter in the vicinity of nuclear matter density, which is essentially determined by the isospin properties of dense matter. In the absence of extreme softening in the dense matter equation of state, a measurement of the radius of a neutron star more accurate than about 1 km will usefully constrain the equation of state. In addition, the moment of inertia and the binding energy of neutron stars are nearly universal functions of the star's compactness. The potential constraints that can be deduced from observations of thermal emission from young neutron stars, neutrinos from newly born neutron stars, Quasi-Periodic Oscillations from X-ray emitting neutron stars in binaries, and glitches from pulsars are discussed.

INTRODUCTION

This talk focuses upon which aspects of neutron stars should be observed in order to definitively discriminate among the multitude of equations of state (EOSs) that have been proposed for dense matter. Among the most important observables are the neutron star radius and the cooling history of a neutron star. It is demonstrated that, unless the EOS of dense matter contains a phase transition accompanied by extreme softening, the neutron star radius primarily provides information about the pressure of neutron-dominated matter around the nuclear saturation density. In turn, this pressure is essentially determined by the symmetry properties of the nuclear force. Measurements of a radius with approximately a 1 km precision are necessary to provide useful discrimination among dense matter models. The state of matter at higher densities might be more fruitfully revealed by the determination of neutron star cooling histories, which are sensitive to the star's internal composition.

---

[1] Research supported in part by USDOE Grant # DOE/DE-FG02-87ER-40317.

MASS AND RADIUS LIMITS

A theoretical upper limit to the mass of a neutron star is set by the condition of causality: assuming that the sound velocity is limited to be no larger than c above some fiducial density ρ_f, the maximum neutron star mass is [24] $M_{max} < 4.2\sqrt{\rho_s/\rho_f}$ $M_\odot$ where $\rho_s = 2.7 \cdot 10^{14}$ g cm^{-3} is the density of nuclear equilibrium matter. On the other hand, there is a practical, albeit theoretical, lower mass limit for neutron stars, about $1.1 - 1.2$ $M_\odot$, which follows from the minimum mass of a protoneutron star. This is estimated by examining a lepton-rich configuration with a low-entropy inner core of ~ 0.6 $M_\odot$ and a high-entropy envelope [9]. This is also in agreement with expectations from supernova calculations, in which the inner homologous collapsing core material comprises at least 1 $M_\odot$.

To date, several accurate mass determinations of neutron stars are available from radio binary pulsars [27], and they all lie in a narrow range ($1.25 - 1.44$ $M_\odot$). One neutron star in an X-ray binary, Cyg X-2, has an estimated mass of 1.8 ± 0.2 $M_\odot$ [17], but this determination is not as clean as for a radio binary. Another X-ray binary, Vela X-1, has a reported mass around 1.9 $M_\odot$ [28], although a later analysis suggests it is about 1.4 $M_\odot$ [26]. A third object, the eclipsing X-ray binary 4U 1700-37, apparently contains an object with a mass of 1.8 ± 0.4 $M_\odot$ [10], but since this source does not pulse and has a relatively hard X-ray spectrum, it may contain a low-mass black hole instead [3]. It would not be surprising if neutron stars in X-ray binaries had larger masses than those in radio binaries, since the latter have presumably accreted relatively little mass since their formation. Alternatively, Cyg X-2 could be the first of a new and rarer population of neutron stars formed with high masses which could originate from more massive, rarer, supernovae.

Another promising technique for estimating masses has come from the discovery of Quasi-Periodic Oscillations (QPOs) from X-ray emitting neutron stars in binaries provides a possible way of limiting neutron star masses and radii. These oscillations are manifested as quasi-periodic X-ray emissions, with frequencies ranging from tens to over 1200 Hz. Some QPOs show multiple frequencies, in particular, two frequencies ν_1 and ν_2 at several hundred Hz. These frequencies are not constant, but tend to both increase with time until the signal ultimately weakens and disappears. In the beat frequency model (Alpar & Shaham 1985, Psaltis et al. 1998), the highest frequency ν_2 is associated with the Keplerian frequency ν_K of inhomogeneities or blobs in an accretion disc. The lower frequency ν_1 is associated with a beat frequency between ν_2 and the spin frequency of the star. The largest frequency measured to date is $\nu_{max} = 1230$ Hz. However, general relativity predicts the existence of a maximum orbital frequency, since the inner edge of an accretion disc must remain outside of the innermost stable circular orbit, located at a radius of $r_s = 6GM/c^2$ in the absence of rotation. This corresponds to a Keplerian orbital frequency of $\nu_s = \sqrt{GM/r_s^3}/2\pi$ if the star is non-rotating. Equating ν_{max} with ν_s, and since $R < r_s$, one deduces

$$M \lesssim 1.78 \ M_\odot; \qquad R \lesssim 8.86(M/M_\odot) \text{ km}. \tag{1}$$

Corrections due to stellar rotation, $\nu_2 - \nu_1 \approx 250 - 350$ Hz, alter the metric from a Schwarzschild geometry, and increases the theoretical mass limit in equation (1) to about 2.1 M$_\odot$ [23]. This value remains strictly an upper limit, however, unless further observations support the interpretation that ν_{max} is associated with orbits at precisely the innermost stable orbit.

The association of the frequency $\nu_2 - \nu_1$ with the spin of the neutron star implies it should remain constant in time. However, recent observations reveal $\nu_2 - \nu_1$ changes with time in a given source. Osherovich & Titarchuk [18] proposed a model in which ν_1 is the Keplerian frequency and ν_2 is a hybrid frequency of the Keplerian oscillator under the influence of a magnetospheric Coriolis force. In this model, the frequencies are related to the neutron star spin frequency ν by

$$\nu_2 = \sqrt{\nu_1^2 + \nu^2}. \qquad (2)$$

This relation fits the observed variations of ν_2 and ν_1 in several QPOs, but the Keplerian frequency (ν_1) is at most 850 Hz. This leads to an upper mass limit that is nearly 3 M$_\odot$ and is therefore of little practical use to further limit either the star's mass or radius.

An alternative model [25] associates ν_2 with the Keplerian frequency of the inner edge of the disc, ν_K, and $\nu_2 - \nu_1$ with the precession frequency of the periastron of slightly eccentric orbiting blobs at radius r in the accretion disc. In a Schwarzschild geometry, $\nu_1 = \nu_K\sqrt{1 - 6GM/rc^2}$. Note that $(\nu_K - \nu_1)^{-1}$ is the timescale that an orbiting blob recovers its original orientation relative to the neutron star and the Earth, so that variations in flux are expected to be observed at both frequencies ν_K and $\nu_K - \nu_1$. Presumably, even eccentricities of order 10^{-4} lead to observable effects. This model predicts that

$$\nu_1/\nu_2 = 1 - \sqrt{1 - 6(2\pi GM\nu_2)^{2/3}/c^2}, \qquad (3)$$

a relation that depends only upon M and which fits observations of QPOs, but only if $1.9 \lesssim M/M_\odot \lesssim 2.1$. This result is not very sensitive to complicating effects due to stellar rotation: the Lense-Thirring effect and oblateness.

A relatively firm restriction to neutron star radii is also set by causality [11,7]:

$$R > 3.04GM/c^2. \qquad (4)$$

A somewhat more stringent constraint for the radius of the Vela pulsar can be obtained from observations of neutron star glitches, assuming that glitches originate in the neutron star crust [14]:

$$R > 3.9 + 3.5M/M_\odot - 0.08(M/M_\odot)^2 \text{ km}. \qquad (5)$$

The radius limits are compared in Fig. 1 with theoretical mass-radius relations for several recent EOSs that are tabulated in Table 1. Clearly, if high masses, or extremely small radii, for any neutron stars are confirmed, significant constraints on the EOS would be realized.

EQUATIONS OF STATE

The composition of a neutron star chiefly depends on the nature of strong interactions, which are not well understood in dense matter. Most models that have been investigated can be conveniently grouped into three broad categories: non-relativistic potential models, relativistic field theoretical models, and relativistic Dirac-Brueckner-Hartree-Fock models. In each of these approaches, the presence of additional softening components such as hyperons, Bose condensates or quark matter, can be incorporated. A representative sample of EOSs, including references and typical compositions, are listed in Table 1. Details of these approaches are summarized in Ref. [12].

The rationale for considering a wide variety of EOSs, even some that are relatively outdated or which have been superceded, is to provide contrasts among widely different theoretical paradigms and to illuminate general relationships between the $P - \rho$ relation and the macroscopic properties of the star, such as the radius. In all cases, except for PS [20] which has no charged components, the pressure is evaluated assuming zero temperature and beta equilibrium without trapped neutrinos. We

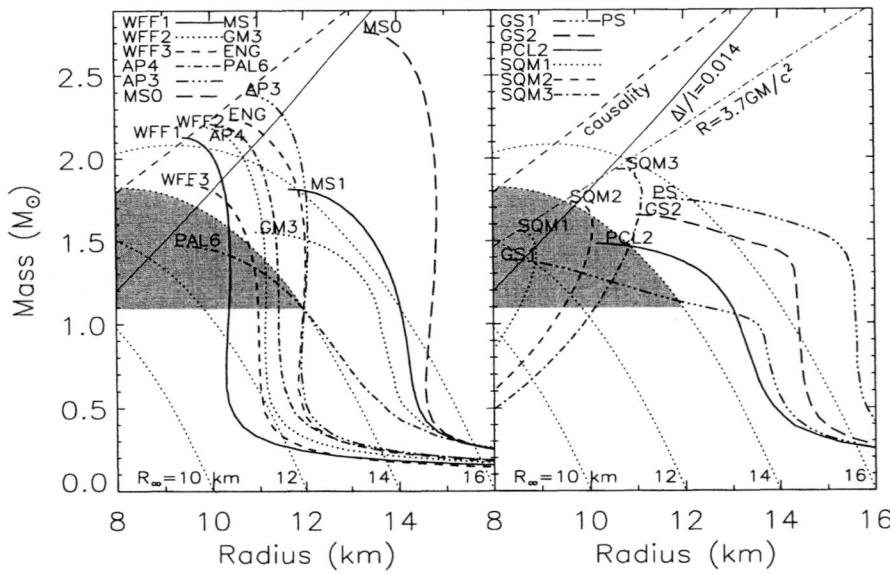

FIGURE 1. Mass-radius relations for recent EOSs (see Table 1). The right panel exhibits models with extreme softening. The dashed line indicates causality (Eq. 4), the solid line is the glitch limit (Eq. 5), the dashed-dot line is the SQM limt (Eq. 6), and dotted lines are contours of $R_\infty = R/\sqrt{1 - 2GM/Rc^2}$. Inferred allowed regions for RX J185635-3754 are shaded.

TABLE 1. Equations of State. Approach refers to the underlying theoretical technique; numbers refer to distinct parameter sets. Composition refers to strongly interacting components (n=neutron, p=proton, H=hyperon, K=kaon, Q=quark).

Symbol	Reference	Approach	Composition
FP	[5]	Variational	np
PS	[20]	Potential	$n\pi^0$
WFF(1-3)	[32]	Variational	np
AP(1-4)	[1]	Variational	np
MS(1-3)	[15]	Field Theoretical	np
MPA(1-2)	[16]	Dirac-Brueckner HF	np
ENG	[4]	Dirac-Brueckner HF	np
PAL(1-6)	[21]	Schematic Potential	np
GM(1-3)	[6]	Field Theoretical	npH
GS(1-2)	[8]	Field Theoretical	npK
PCL(1-2)	[22]	Field Theoretical	npHQ
SQM(1-3)	[22]	Quark Matter	$Q\ (u,d,s)$

chose to include PS, despite its supercedence by more sophisticated calculations [32,1], because it produces extremly large radius neutron stars.

The pressure-density relations for some of the selected EOSs are shown in Figure 2. There are obviously two general classes of EOSs. First, *normal* EOSs have a pressure which vanishes as the density tends to zero. Second, *self-bound* EOSs have a pressure which vanishes at a significant finite density. Below a transition density of 1/3 to 1/2 ρ_s, the ground state of matter consists of heavy nuclei in equilibrium with a neutron-rich, low-density gas of nucleons. As far as the mass-radius relation is concerned, the EOS in this region is not relevant. However, the value of the transition density, and the pressure there, are important ingredients for the determination of the size of the superfluid crust of a neutron star that is believed to be involved in the phenomenon of pulsar glitches [14].

The best-known example of self-bound stars results from Witten's [33] conjecture that strange quark matter is the ultimate ground state of matter. Here, the self-bound EOSs are represented by strange-quark matter models SQM1–3, using perturbative QCD and an MIT-type bag model. The most compressed strange-quark matter star has $B = 94.92$ MeV fm^{-3} together with zero strange quark mass and no interactions (SQM1). It was shown in Ref. [11] that the loci of maximum mass strange-quark matter configurations satisfy

$$R \approx 3.7 GM/c^2. \tag{6}$$

There is a fairly wide range of predicted pressures for beta-stable matter in the density domain $n_s/2 < n < 2n_s$. For the EOSs displayed, the range of pressures covers about a factor of five, but this survey is by no means exhaustive. The higer pressure group is primarily composed of relativistic field-theoretical models, while the lower pressure group is primarily composed of non-relativistic potential models. That such a wide range in pressures is found is somewhat surprising, given that each of the EOSs provides acceptable fits to experimentally-determined nuclear matter properties. It is straightforward to show that the pressure of neutron star matter near ρ_s is approximately

$$P \simeq \rho_s^2 (\partial S_v / \partial \rho)_{\rho_s}, \tag{7}$$

where $S_v(\rho)$ is the (density-dependent) bulk nuclear symmetry energy parameter. Clearly, the extrapolation of the pressure from symmetric matter to nearly pure neutron matter depends upon the density dependance of S_v and is poorly constrained. Nuclear experiments that attempt to constrain S_v include binding energies, fission, and giant resonances. It is not suprising that relativistic field-theoretical models generally have symmetry energies that increase more rapidly with density than do non-relativistic potential models.

A few of the plotted normal EOSs have considerable softening at high densities, especially PAL6, GS1, GS2, GM3, PS and PCL2. PAL6 has an abnormally small value of incompressibility ($K_s = 120$ MeV). GS1 and GS2 have phase transitions

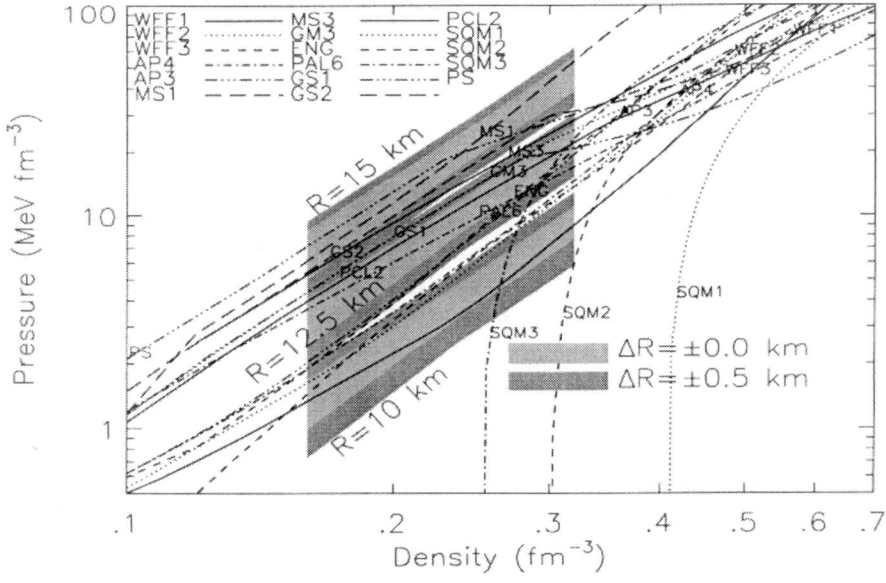

FIGURE 2. Pressure-density relations for recent EOSs (see Table 1). The shaded regions indicate error ranges in predicted EOSs for three assumed radius measurements. Light (dark) bands assume 0.0 (0.5) km observational error in the radius measurement.

to matter containing a kaon condensate, GM3 has a large population of hyperons appearing at high density, PS has a phase transition to a neutral pion condensate and a neutron solid, and PCL2 has a phase transition to a mixed phase containing strange quark matter. These EOSs can be regarded as representative of the many suggestions of the kinds of softening that could occur at high densities.

Although the $M - R$ trajectories for normal stars can be strikingly different, in the mass range from 1 to 1.5 $M_\odot$ or more it is usually the case that the radius has relatively little dependence upon the stellar mass. The major exceptions illustrated are the model GS1, in which a mixed phase containing a kaon condensate appears at a relatively low density and the model PAL6 which has an extremely small nuclear incompressibility (120 MeV). Both of these have considerable softening and a large increase in central density for $M > 1$ $M_\odot$. Pronounced softening, while not as dramatic, also occurs in models GS2 and PCL2, which contain mixed phases containing a kaon condensate and strange quark matter, respectively. All other normal EOSs in this figure, except PS, contain only baryons among the hadrons.

While it is generally assumed that a stiff EOS implies both a large maximum mass and a large radius, many counter examples exist. For example, GM3, MS1 and PS have relatively small maximum masses but have large radii compared to most other EOSs with larger maximum masses. Also, not all EOSs with extreme

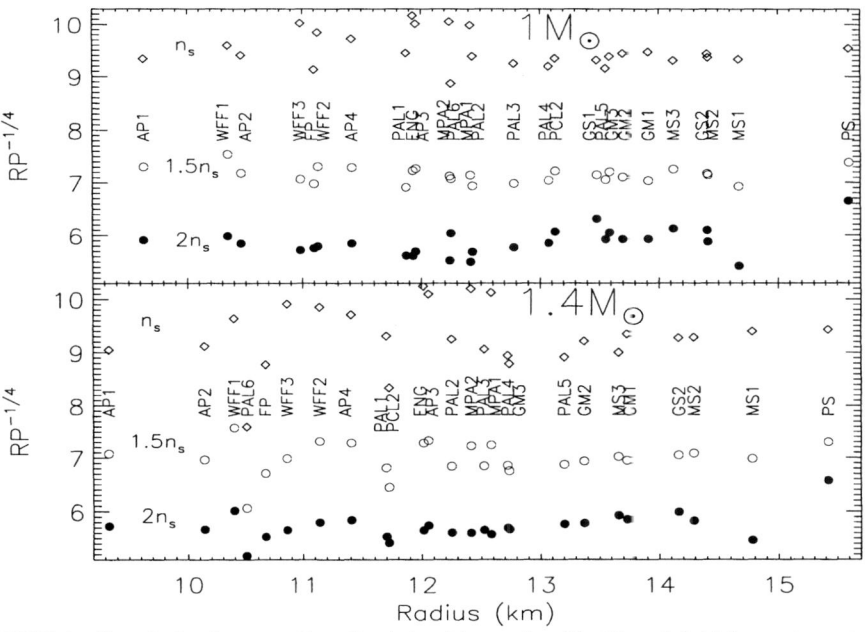

FIGURE 3. Correlation between P at fixed densities and radii of 1 and 1.4 $M_\odot$ neutron stars.

softening have small radii for $M > 1\,M_\odot$ (e.g., GS2, PS). Nonetheless, for stars with masses greater than $1\,M_\odot$, only models with a large degree of softening (including strange quark matter configurations) can have radiation radii $R_\infty < 12$ km.

The relative insensitivity of the radius to the mass for normal neutron stars can be traced to properties of Newtonian polytropes:

$$R \propto K^{n/(3-n)} M^{(1-n)/(3-n)}, \tag{8}$$

where the EOS is $P = K\rho^{1+1/n}$ and K is a constant. Fig. (2) indicates that $n \approx 1$, which implies that R is independent of M but proportional to $\sqrt{K}$. In other words, there should be a quantitative relationship between the radius of say, a $1.4\,M_\odot$ star, and the pressure at a given density (which determines K). Lattimer and Prakash [12] verified that this conjecture is true. Fig. (3) shows the remarkable empirical correlation which exists between the radii of 1 and $1.4\,M_\odot$ normal stars and the matter's pressure evaluated at fiducial densities of 1, 1.5 and $2\,n_s$. This correlation is valid only for cold, catalyzed neutron stars, i.e., not for protoneutron stars which have finite entropies and might contain trapped neutrinos.

Numerically, the correlation has the form of a power law:

$$R_M \simeq C(n, M)\, [P(n)]^{0.25}, \tag{9}$$

where $P(n)$ is the total pressure inclusive of leptonic contributions evaluated at the density n, and $C(n, M)$ is a number that depends on the density n at which the pressure was evaluated and the stellar mass M. The fact that the exponent in Eq. 9 is ≈ 0.25 and not $1/2$ as predicted by Eq. 8 is due to General Relativity [12]. *The immediate consequence of this correlation is that an inversion yields the approximate equation of state in the vicinity of ρ_s should the radius of a neutron star ever be measured:* $P(n) \simeq [R_M/C(n, M)]^4$.

Any radius measurement will have observational uncertainty. Combining this with the standard deviation in the empirical relation between P and R determines how accurately the EOS might be established from an eventual radius measurement. The inferred ranges of pressures, as a function of density and for three possible values of $R_{1.4}$, are shown as the shaded regions in Figure 1. The results are only slightly sensitive to the actual mass, as can be gleaned from Fig. 3. These results suggest that a useful restriction to the equation of state is possible if the radius of a neutron star can be measured to an accuracy better than about 1 km.

Note that useful constraints might be obtained from just a single measurement of a neutron star radius, rather than requiring a series of simultaneous mass-radius measurements as Lindblom [13] proposed.

RADIUS MEASUREMENT OF RX J185635-3754

The best prospect for measuring a neutron star's radius may be the nearby object RX J185635-3754, a relatively bright non-pulsing neutron star which is not associated with a supernova remnant. Discovered [29] with the ROSAT X-ray satellite, it has since been observed with the EUVE satellite and the Hubble Space Telescope [30]. Its X-ray spectrum is well-fit with a blackbody with an effective temperature $T_{eff} = 57$ eV and $R_\infty/d = .061$ km/pc, as shown in Fig. 4. The abrupt drop in flux for energies between 0.01 and 0.1 keV is due to interstellar H absorption. This star's projection against the ≈ 130 pc-distant R CrA molecular cloud and inferred low H column absorption density ($N_H \sim 10^{20}$ cm^{-2}) thus implies $R_\infty < 8$ km, a value so small that $M < 1.04$ M$_\odot$. However, the optical flux is about a factor 4 greater than the X-ray blackbody predictions (see curve BB in Fig. 4). This could be explained either by assuming a non-uniform temperature distribution, or by incorporating the effects of an atmosphere, either of which can effectively increase R_∞/d by factors of 2 or more.

This can be seen by recognizing that the bulk of the emitted energy is in X-rays, near the spectrum's peak; hence the X-ray flux and total flux vary as T^4. However, the optical flux is on the Wien tail and thus varies as T. Therefore we can write

$$L = 4\pi\sigma R^2 T^4 \approx 4\pi\sigma R_x^2 T_x^4, \quad L_{opt} = AR^2 T = AfR_x^2 T_x \qquad (10)$$

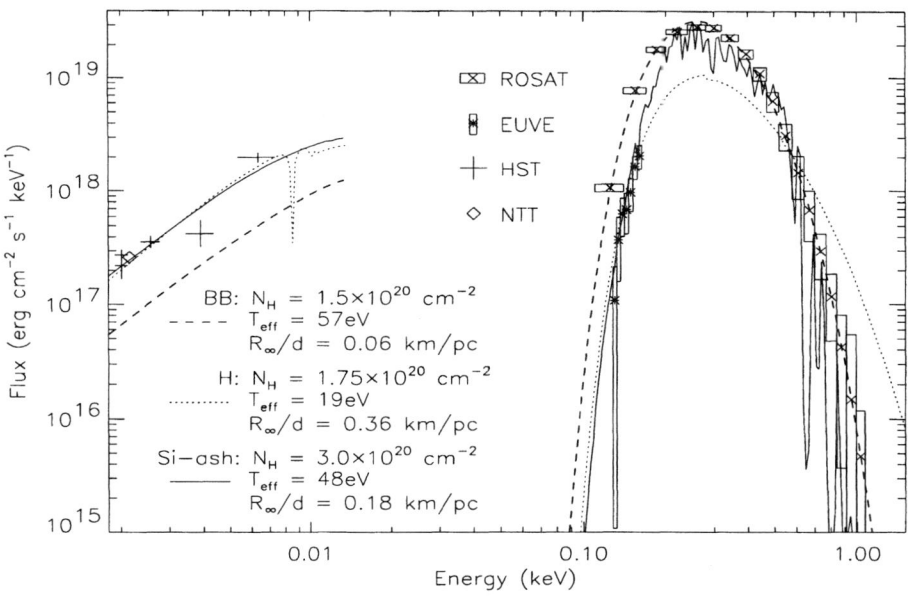

FIGURE 4. Observed and predicted spectra for RX J185635-3754. Symbols and error boxes show the X-ray (ROSAT), EUVE and optical (HST, NTT) data. BB is the best-fit blackbody to the X-ray data; H and Si-ash are best-fit H and Si-burning product atmospheres fitted to all data. The absorption column density N_H, T_{eff} and R_∞/d for each model are shown.

where (R_x, T_x) refer to the blackbody X-ray fit, (R, T) refer to the atmosphere fit, and A is a constant. Solving for R, one finds

$$R/R_x = f^{2/3}, \qquad T/T_x = f^{-1/3}. \tag{11}$$

For a value $f \approx 4$, the increase in inferred radius compared to the blackbody X-ray fit is $f^{2/3} \approx 2.5$, suggesting that $R_\infty/d \approx 0.153$ km/pc or $R_\infty < 20$ km.

An et al. [2] have fit several atmosphere models to this object, but found that only heavy-element atmospheres can fit the multiwavelength data. In particular, while a simple blackbody underestimates the optical flux, a light-element atmosphere, either pure H or He, generally underestimates the X-ray flux if the optical data is fitted (see Fig. 4). This is the first time that the composition of a neutron star atmosphere has been deduced from observations. This result is not surprising: in order for H to dominate in the atmosphere requires a total mass of about 10^{-15} $M_\odot$, and to accumulate this much mass in a million years requires an accretion rate in excess of 10^{11} g s^{-1}. From the Bondi-Hoyle accretion rate,

$$\dot{M} = 1.4 \cdot 10^5 (M/M_\odot)^2 \rho v^3 \text{ g s}^{-1}, \tag{12}$$

where $\rho \sim 10^{-24}$ g cm^{-3} is the interstellar medium density and v is the relative velocity, the star would have to be moving less quickly than 10 km s^{-1} to accrete enough H. Since the average pulsar velocity is more than 100 km s^{-1}, it would appear rather unlikely that an isolated neutron star would have a H-dominated atmosphere. An et al. find that $T_{eff} \approx 48 \pm 6$ eV and $R_\infty/d \approx 0.18 \pm 0.04$ km pc^{-1}, for heavy-element atmospheres. Note that these values are consistent with the decrease in T_{eff} and the increase in R_∞ deduced from simple arguments above.

The first parallax information for this object is now available [31], and the inferred distance is $d \approx 62^{+10}_{-7}$ pc. Combined with the atmosphere models, this suggests that $R_\infty \approx 10.6 \pm 3.4$ pc. This constraint, shown as the shaded region in Fig. 1, is interesting. It implies that $M < 1.82$ M$_\odot$ and that $R < 11$ km if $M > 1.4$ M$_\odot$. It, however, is not at present very restrictive when applied to the pressure-density relation (Fig. 3). Further spectral observations with HST and with the Chandra and XMM X-ray facilities should reduce the errors of the atmospheric modeling and refine the estimate of R_∞/d. In addition, spectral lines might be identified that would not only yield the gravitational redshift, but would also further confirm the atmospheric composition. This includes the interesting possibly of a detection of the Hα feature shown in Fig. 4 if a sufficiently dense, yet optically thin, layer of H is present. Because the redshift depends on a different combination of M and R than R_∞, this would permit *both* M and R to thereby be estimated. It is remotely possible that an estimate of the surface gravity of the star can be determined from atmospheric modeling, and this would provide a further check on M and R.

Together with the parallax, Walter [31] also determined the proper motion for RX J185635-3754, which is about 0.3 seconds of arc a year with a position angle of 100 degrees (just south of east). This motion is carrying it away from the Upper Sco association of young stars, and the star would have been passing through this association about 1-1.3 million years ago. Moreover, a coincidence exists with the inferred position at that time of the runaway star ζ Oph, which is currently at a distance of about 140 pc with a proper motion of about 0.028 seconds of arc a year that *also* carries it away from Upper Sco with a similar age. It is believed that runaway stars like ζ Oph originated in close binaries that were disrupted when their companions became supernovae. If the neutron star and ζ Oph have a common origin, the neutron star must have been given a kick of approximately 200 km s^{-1} when it was born [31].

Our estimates for T_∞ and the age allow this star to be placed on the traditional neutron star cooling diagram. Fig. (5) shows the cooling of 1.4 M$_\odot$ stars under various assumptions [19] about superfluid gaps for nucleonic stars and hybrid quark-nucleon stars. The uppermost curve corresponds to the so-called standard cooling scenario; other curves include various rapid cooling processes, modulated by superfluidity. RX J185635-3754 falls above the standard cooling curve by about a factor of 2 in temperature. The space velocity of this star is too large for significant accretional heating, but a number of other heating processes have been suggested

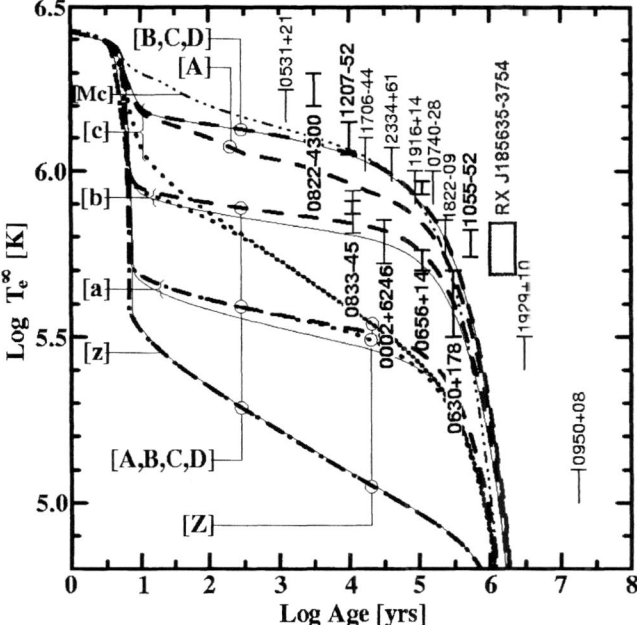

FIGURE 5. Cooling of 1.4 M$_\odot$ stars with np matter (continuous curves) and npQ matter (dashed and dotted curves). The curves labelled [a], [b], and [c] correspond to n $^3P_2$ gaps ranging from small to large; [z] corresponds to zero n gap. Models labelled [A], [B], [C] and [D] correspond to quark gaps ranging from small to large; [Z] corresponds to zero quark gap. M is the standard cooling curve. Presumed thermal emission from pulsars and RX J185635-3754 are shown.

that could explain its high temperature. Another possibility is that the age of the star could be in error if there is no connection with ζ Oph or Upper Sco.

ACKNOWLEDGMENTS

I thank KIAS for the invitation to Korea, the gracious hospitality we found there, and the opportunity to try Yak butter tea. I acknowledge M. Prakash and D. Page for their important contributions to the research summarized herein.

REFERENCES

1. Akmal A., and Pandharipande V.R., *Phys. Rev.* **C56**, 2261 (1997).
2. An, P., Lattimer J.M., Pons J., Prakash M., and Walter F., in preparation (2000).
3. Brown G.E., Weingartner J.C., and Wijers R.A.M.J. *Astrophys. J.* **463**, 297 (1996).
4. Engvik L., Hjorth-Jensen, M., Osnes E., Bao G., and Østgaard E. *Astrophys. J.* **469**, 794 (1996).
5. Friedman B., and Pandharipande V.R., *Nucl. Phys.* **A361**, 502 (1981).
6. Glendenning N.K., and Moszkowski S.A. *Phys. Rev. Lett.* **67**, 2414 (1991).
7. Glendenning N.K. *Phys. Rev.* **D46**, 1274 (1992).
8. Glendenning N.K., and Schaffner-Bielich J. *Phys. Rev.* **C60**, 025803 (1999).
9. Goussard J.-O., Haensel P., and Zdunik J.L. *Astron. Astrophys.* **217**, 137 (1986).
10. Heap S.R., and Corcoran M.F. *Astrophys. J.* **387**, 340 (1992).
11. Lattimer J.M., Prakash M., Masak D., and Yahil A. *Astrophys. J.* **355**, 241 (1990).
12. Lattimer J.M., and Prakash M. *Astrophys. J.*, in press (2000).
13. Lindblom L. *Astrophys. J.* **398**, 569 (1992).
14. Link B., Epstein R.I., and Lattimer J.M. *Phys. Rev. Lett.* **83**, 3362 (1999).
15. Müller H., and Serot B.D. *Nucl. Phys.* **A606**, 508 (1996).
16. Müther H., Prakash M., and Ainsworth T.L. *Phys. Lett.* **B199**, 469 (1987).
17. Orosz J.A., and Kuulkers E. *Mon. Not. Roy. Ast. Soc.* **305**, 132 (1999).
18. Osherovich V., and Titarchuk L. *Astrophys. J.* **522**, L113 (1999).
19. Page D., Prakash M., Lattimer J.M., and Steiner A.W. *Phys. Rev. Lett.*, submitted (2000).
20. Pandharipande V.R., and Smith R.A. *Nucl. Phys.* **A237**, 507 (1975).
21. Prakash M., Ainsworth T.L., and Lattimer J.M. *Phys. Rev. Lett.* **61**, 2518 (1988).
22. Prakash M., Cooke J.R., and Lattimer J.M. *Phys. Rev.* **D52**, 661 (1995).
23. Psaltis D., et al. *Astrophys. J.* **501**, L95 (1998).
24. Rhoades C.E., and Ruffini R. *Phys. Rev. Lett.* **32**, 324 (1974).
25. Stella L., and Vietri M. *Phys. Rev. Lett.* **82**, 17 (1999).
26. Stickland D., Loyd C., and Radzuin-Woodham A. *Mon. Not. Roy. Ast. Soc.* **286**, L21 (1997).
27. Thorsett S.E., and Chakrabarty D. **Astrophys. J. 512**, 288 (1999).
28. van Kerkwijk J.H., van Paradijs J., and Zuiderwijk E.J. *Astron. Astrophys.* **303**, 497 (1995).
29. Walter F., Wolk S.J., and Neuhäuser R. *Nature* **389**, 358 (1996).

30. Wakter F., and Matthews, L.D. *Nature* **389** 358 (1997).
31. Walter F., preprint (2000).
32. Wiringa R.B., Fiks V., and Fabrocine A. *Phys. Rev.* **C38**, 1010 (1988).
33. Witten E. *Phys. Rev.* **D30**, 272 (1984).

GAMMA-RAY BURSTS

Very High Energy Phenomena in GRBs

Tomonori Totani

National Astronomical Observatory, Theory Division
Mitaka, Tokyo 181-8588, Japan

Abstract. Here I review production of very high energy particles and expected radiations from them in the standard relativistic blastwave model of GRBs. Electrons and protons can be accelerated in the internal shock of GRBs up to $\sim 10^{14}$ and 10^{20} eV, respectively, and GRBs are one of the possible sites of the production of ultra high energy cosmic rays observed on the earth. Three radiation processes can be considered for very high energy gamma-ray production: inverse-Compton, photo-pion production, and proton synchrotron, all of which could be an efficient energy loss process of GRBs. Photo-pion production is also interesting as the high energy neutrino emission mechanism. Internal absorption of hadron-origin gamma-rays may be crucial for the energetics of GRBs in the soft gamma-ray bands. The recent evidence of TeV emission from GRB 970417a can best be interpreted by the synchrotron radiation of 10^{20} eV protons.

INTRODUCTION

Gamma-ray bursts (GRBs) are the most energetic explosions in the universe. The central engine driving these energetic phenomena is still a mystery, but from the opacity argument, there is a widely accepted picture that gamma-rays are emitted from the shock wave generated by the relativistic outflow from the central engine with a Lorentz factor of $\Gamma \gtrsim 10^2$–10^3. Two shocks are considered as the site of gamma-ray production: internal shock generated by relative velocity difference of ejected matter and external shock generated by the interaction between the ejected matter and intestellar/circumstellar medium. Afterglows are generally believed to be external shocks, but internal shocks are favored for the origin of prompt bursts because of the complicated time structure of GRBs. See, e.g., Piran (1999) [1] for a recent review. Here we discuss very high energy phenomena in GRBs expected from the above standard picture of GRBs.

PARTICLE ACCELERATION AND HIGH ENERGY RADIATION FROM GRBS

Given the above picture of GRBs, it is possible to discuss particle acceleration in GRBs and emission from them in the internal shock or external shock. Here we concentrate to the prompt emission from internal shocks. Basic parameters are the total (isotropic) energy of GRBs ($E_{\rm iso}$), the Lorentz factor Γ, and time scale of GRBs. The time scale may be either that of the shortest variability ($\sim msec$) or width of pulse ($\sim$ 1–10 sec). In addition to these parameters, there should be two important parameters for the energy fraction of accelerated electrons and magnetic fields, and they are denoted as ξ_e and ξ_B. (Here we have assumed that the energy fraction of accelerated protons is of order unity for the simplicity.) The electron parameter ξ_e is often assumed to be of order unity, but it is not guaranteed. Since GRBs are generated by kinetic shocks, this parameter could be as small as $\xi_e \sim m_e/m_p$ if the energy transfer from baryons to electrons is inefficient. The magnetic field strength is also highly uncertain, but again it is typically assumed to be of order unity.

If the particle acceleration time is given by $\sim r_L$, i.e., the Lamor radius, it is easy to show that electrons and protons can be accelerated to $\sim 10^{14}$ and 10^{20} eV, respectively, with the typical parameters [2,3]. The maximum energy of electrons is constrained by the synchrotron cooling while the cooling constraint and acceleration time constraint are comparable for the maximum proton energy.

The most conservative (less exotic) radiation process for very high energy radiation is probably the Inverse-Compton scattering of the synchrotron photons by electrons (so-called synchrotron-self-Compton (SSC) process). This process is known as the best explanation of the high energy emission from active galactic nuclei of the blazar class (e.g., [4,5]). As is well known, the luminosity ratio of the synchrotron to IC radiation is given by the ratio of the magnetic field energy density to target photon energy density. (It should be noted, however, the Klein-Nishina effect reduces the IC luminosity beyond $\sim$ 50 GeV.) Therefore, if $\xi_e \sim \xi_B \sim 1$, roughly the same luminosity is expected from synchrotron and IC, as observed for blazars. For the SSC model of GRBs, see, e.g., [6].

Photo-pion production is the most efficient for protons of about $\sim 10^{15}$ eV, because the protons of this energy efficiently interact by the Δ-resonance with the soft gamma-rays around MeV. Therefore, the decay products of pions, i.e., gamma-rays and neutrinos are most efficiently radiated at around $\sim 10^{14}$ eV. Depending on the model parameters, the energy loss fraction of 10^{15} eV protons within the GRB time scale can be of order unity. (See [7,8] for neutrinos by this process from GRBs).

The synchrotron cooling time scale of protons becomes comparable with the GRB time scale at the maximum energy of 10^{20} eV, and the synchrotron photon energy is around TeV [9,13,14]. Therefore this process can be an efficient emission process for TeV gamma-rays from GRBs. The characteristic of this process that is different from the above two is that it does not depend on the electron or target photon energy density. In fact, even if the energy transfer from baryons into electrons is negligible, i.e., $\xi_e \sim 10^{-3}$, this process still can be an efficient process provided that $\xi_B \sim 1$. In this case TeV luminosity can be much stronger than the ordinary sub-MeV emission from GRBs [13,14]. We suggest that this process gives a natural explanation for the recent evidence of TeV emission from GRB 970417a by the Milagrito project.

In the following of this paper we discuss the energetics of GRBs in the sub-MeV band [15] and the interpretation of the Milagrito result by the proton-synchrotron model [16].

ENERGETICS OF GRBS

It has now been confirmed that there is a wide dispersion in the observed total energies of GRBs. Relatively weak GRBs such as GRB 970508 at $z = 0.835$ typically have a total energy of $E_{\gamma,\text{iso}} \lesssim 10^{52}$ erg [10], where $E_{\gamma,\text{iso}}$ is the isotropic total energy of observed gamma-rays assuming isotropic radiation. On the other hand, there is a population of GRBs with quite a large amount of energy with $E_{\gamma,\text{iso}} \sim 3 \times 10^{54}$ erg such as GRB 990123 at $z = 1.6$ [11]. Despite this large dispersion in total energy, there is no significant change in other overall properties of GRBs such as spectra or light curves. If all GRBs are triggered by a similar event, it is somewhat strange that GRB total energies have such a wide dispersion. The difference of beaming may explain this dispersion; however, in this case it is expected that the GRB luminosity and afterglow luminosity is correlated while observations suggest that there is almost no correlation between the two.

Currently the most popular explanation for the GRB phenomenon is dissipation of the kinetic energy of ultra-relativistic bulk motion with a Lorentz factor of $\Gamma \gtrsim 10^{2-3}$, in internal shocks which are generated by relative velocity difference of relativistic shells ejected from a central engine. All the total energy ejected as relativistic bulk motion cannot be dissipated in internal shocks, and hence the total energy truly emitted as kinetic motion (E_{iso}) should be larger than the observed $E_{\gamma,\text{iso}}$, at least by a factor of several. Therefore, some of GRBs must emit quite a large amount of energy, $E_{\text{iso}} \gtrsim 10^{55}$ erg. If the efficiency of the internal shock is not so high, we may have to consider an isotropic energy reaching $\sim 10^{56}$ erg. Therefore, if GRBs are produced by stellar death events, GRBs must be strongly beamed at least by a factor of 100. In this paper we refer to the isotropic energy E_{iso} for convenience, but a strong beaming is implicitly assumed to reduce the actual energy emission.

Here we propose a theoretical model to explain the wide dispersion of total GRB energies, in which all GRBs emit roughly the same amount of energy, $E_{\text{iso}} \sim 10^{55-56}$ erg with the same beaming factor. The observed difference of total GRB energies is attributed to difference of efficiency of gamma-ray production. This model predicts strong emission above TeV energies from some fraction of GRBs, in which the total energy emitted as very high energy (VHE) gamma-rays is comparable with or even larger than that in the ordinary sub-MeV range. In this paper we briefly describe this model, and then try to interpret the interesting report by the Milagro group in this conference, which suggests strong emission of TeV gamma-rays from GRB 970417a with a TeV fluence at least 10 times greater than the sub-MeV fluence [12].

THE PROTON-SYNCHROTRON MODEL FOR THE ENERGETICS OF GRBS

Full description of this model has already been given in Ref. [15], and here I summarize the qualitative feature of the model. Since the origin of the GRB energy is relativistic bulk motion, protons should carry a much larger amount of energy than electrons by a factor of $m_p/m_e \sim 2,000$, at least in the initial stage of the internal shock generation. It is uncertain what fraction of the proton energy is converted into electrons, but the simplest Coulomb interaction cannot transfer the proton energy into electrons within the time scale of GRBs. The soft γ-rays are generally considered to be generated by electrons, because of the short time variability of GRBs. Therefore it is not unreasonable that, in some GRBs, only $(m_e/m_p) \sim 10^{-3}$ of the total kinetic energy is carried by electrons and then emitted as soft γ-rays. On the other hand, it may also be possible that a physical process works as an energy conveyor from the hidden energy reservoir (i.e., protons) into electrons (or positrons). If the energy transfer is almost complete in a GRB, a significant fraction of E_{iso} can be radiated as soft γ-rays.

We suggest that this new energy transfer channel from protons into electrons is $e^{\pm}$-pair creation by very high energy photons of proton-synchrotron. When $E_{\text{iso}} \gtrsim 10^{55}$ erg, synchrotron radiation of protons accelerated to 10^{20} eV can be a very efficient emission process because the cooling time of such protons is comparable with the typical GRB duration (~ 10 sec). The energy of these synchrotron photons for an observer is about 1–10 TeV, and strong TeV emission from GRBs is possible. On the other hand, such proton-synchrotron photons may interact with low energy electron-synchrotron photons and create $e^{\pm}$ pairs. The opacity of this reaction is of order unity, and strongly depends on the Lorentz factor of GRBs, as $\tau \propto \Gamma^{-5}$ in a simple model. The GRB luminosity is determined by the efficiency of energy transfer from protons into $e^{\pm}$ pairs, i.e., the opacity of pair-creation reaction. If the energy transfer is sufficient, we observed a strong GRB such as GRB 990123, while we observe a weak GRB such as GRB 970508 in case of almost no energy transfer. A modest dispersion in Γ from one GRB to another results in drastic change in the energetics of GRBs in sub-MeV range. It can also be shown that the photon energy band of the synchrotron radiation of the created pairs becomes about MeV, giving an explanation for the sub-MeV GRB phenomenon. This mechanism may be similar to a see-saw between sub-MeV and TeV energies, in which the total kinetic energy of GRBs is roughly the same for all GRBs and difference of GRB energetics is whether dominant emission in TeV or MeV energies.

AN INTERPRETATION OF THE MILAGRITO RESULT

The Milagro group reported that they found an excess of gamma-rays above several hundreds GeV from one GRB 970417a out of the 54 BATSE GRBs in the field of view of the Milagrito detector (proto-type of Milagro), and estimated the chance probability of this excess after examining 54 GRBs as 1.5×10^{-3} [12]. If this signal is truly from the GRB, the TeV fluence must be at least 10 times greater than the sub-MeV fluence of this GRB. The impact on the GRB energetics would be quite strong. Our model has predicted such an extreme phenomenon, if the energy transfer from protons into electrons is inefficient [13,14]. Here we discuss whether the Milagrito result can be explained by our model.

In order to observe such TeV emission much stronger than sub-MeV emission, there are two necessary conditions in our scenario: 1) GRB 970417a must be faint in sub-MeV range like the GRB 970508 rather than the most energetic GRBs such as GRB 990123, and 2) the redshift of GRB 970417a must be relatively low. If GRB 970417a is a strong GRB in the sub-MeV range like the GRB 990123, the even larger total energy emitted in the TeV range cannot be explained by stellar death models even with a strong beaming. If the redshift of GRB 970417a is too high, TeV gamma-rays would be strongly absorbed by the interaction with the cosmic infrared background radiation. The fluence of GRB 970417a in all the four energy channels of the BATSE (> 20 keV) is 3.9×10^{-7} erg cm^{-2} which is not specially bright, and hence it seems possible that this GRB has relatively low redshift and small total energy.

Quantitative analysis requires the absorption optical depth of VHE gamma-rays due to the cosmic infrared background. We have calculated this optical depth as a function of the source redshift and observed photon energy, by using a model of luminosity density evolution of stellar lights in the universe [17]. The emission component of dust is calculated assuming that the dust emission spectrum is the same with that of the solar neighborhood and fraction of stellar lights absorbed by dust is determined to reproduce the observed far infrared background radiation measured by the COBE satellite [18]. This model of optical depth is quantitatively consistent with other publications within the model uncertainties (e.g., [19,20]).

Figure 1 shows the total energies emitted from GRB 970417a in MeV and TeV bands as a function of redshift assumed for this GRB (solid and dashed lines). The energy in the MeV band is simply calculated by the observed BATSE fluence. We have calculated the energy in the TeV band taking into account the absorption of TeV gamma-rays in the intergalactic field, as mentioned above. We estimated the optical depth of absorption at the observed photon energy of 200 GeV, considering that the threshold energy of the Milagrito is a few hundreds GeV [12]. The estimate of the fluence of TeV gamma-rays observed by the Milagrito depends on the higher cut-off energy and power-law index of the spectrum (see Fig. 5 of ref. [12]). We assume the power-law index of -1.5 (standard synchrotron index corresponding to the particle power-law index of -2), and estimated the cut-off energy as a function of redshift by the energy at which the optical depth of the intergalactic absorption becomes the unity.

The fraction (1/54) of GRBs possibly detected in the TeV band in all the BATSE GRBs in the field-of-view of the Milagrito is also important information. We should consider that the GRB 970417a was the nearest one among the 54 GRBs, because the opacity of intergalactic absorption of TeV gamma-rays rapidly increases with redshift. Then we can estimate the likely redshift of this GRB from the event rate of $\sim 1/54$. In our model the total energy in the BATSE range is expected to widely distribute in $E_{\rm iso} \sim 10^{51}$–10^{54} erg, and hence we assume the total energy distribution in the $\log E_{\rm iso}$ space, $dN/d(\log E_{\rm iso})$, is constant in a range of $E_{\rm iso} = 10^{51}$–3×10^{54} erg. We also assume that the GRB rate is proportional to the star formation rate in the universe. Then we can calculate the probability that a GRB is nearer than a given redshift z, in the whole sample of the BATSE GRBs. (We have estimated the threshold fluence of the BATSE GRBs by the published BATSE catalog.) This is shown in Fig. 1 by the dot-dashed line (see the right-hand-axis scale of the figure). We can infer from this result that the likely range of the redshift of GRB 970417a is $0.6 \lesssim z \lesssim 1$. (The vertical lines show the redshift corresponding to the detection rate 1/54 with 1σ statistical uncertainties.)

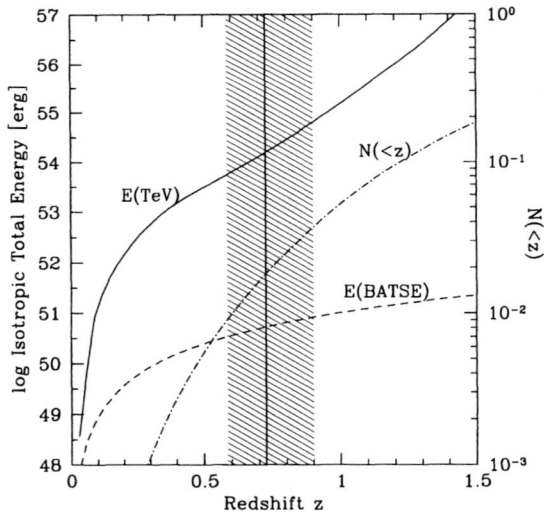

FIGURE 1. TeV/MeV Energies of GRB 970417a and Detection Probability. See the right-hand-axis scale for $N(< z)$ (fraction of BATSE GRBs within a given z). (See text for detail.)

It is interesting to note that the energy emitted in the TeV range is $E_{\rm iso} \sim 10^{54-55}$ erg for this likely redshift range. This energy scale is comparable with that of the most energetic GRBs such as GRB 990123, and hence this energy is not too large as the total energy budget of GRBs. In fact, our model assumes that all GRBs emit $E_{\rm iso} \gtrsim 10^{55}$ erg as the total kinetic energy, and considerable fraction of this energy is converted first into TeV gamma-rays by proton-synchrotron radiation. If the opacity of the pair-creation reaction is not high, these TeV gamma-rays are emitted without conversion into the MeV band. In this case the energy emitted in the MeV band by the synchrotron radiation of electrons originally loaded in the internal shock could be as low as $\sim 10^{-3}$ of the TeV band, because they carry only very small fraction of kinetic energy by the electron-proton mass ratio. We suggest that GRB 970417a with $E_{\gamma,\rm iso} \sim 10^{51}$ erg in the sub-MeV band was such a class of GRBs. Therefore, our model gives an explanation for the TeV and MeV fluence of GRB 970417a and the detection rate of BATSE GRBs by the Milagrito.

REFERENCES

1. Piran, T. *Phys. Rep.* **314**, 575 (1999)
2. Waxman, E. *Phys. Rev. Lett.* **75**, 386 (1995)
3. Vietri, M. *ApJ*, **453**, 883 (1995)
4. Inoue, S. & Takahara, F. *ApJ*, **463**, 555 (1996)
5. Kataoka, J. et al. *ApJ*, **514**, 138 (1999)
6. Dermer, C.D., Böttcher, M., & Chiang, J. *ApJ*, **537**, 255 (2000)
7. Waxman, E. & Bahcall, J.N. *Phys. Rev. Lett.* **78**, 2292 (1998)
8. Vietri, M. *Phys. Rev. Lett.* **80**, 3690 (1998)
9. Vietri, M. *Phys. Rev. Lett.* **78**, 4328 (1997)
10. Metzger, M.R. et al., *Nature* **387**, 878 (1997)
11. Kulkarni, S.R. et al., *Nature* **393**, 35 (1999)
12. Atkins, R. et al., *ApJ*, **533**, L119 (2000)
13. Totani, T., *ApJ* **502**, L13 (1999)
14. Totani, T., *ApJ* **509**, L81 (1999)
15. Totani, T., *MNRAS* **307**, L41 (1999)
16. Totani, T., *ApJ* **536**, L23 (2000)
17. Totani, T., Yoshii, Y., & Sato, K., *ApJ* **483**, L75 (1999)
18. Hauser, M.G. et al., *ApJ*, **508**, 25 (1998)
19. Salamon, M.H & Stecker, F.W., *ApJ*, **493**, 547 (1998)
20. Primack, J.R. et al., *Astroparticle Phys.*, **11**, 93 (1999)

Crustal Shear Oscillations, Magnetar Spindown, and the 1998 August 27 Flare

Robert C. Duncan

Dept. of Astronomy and McDonald Observatory
University of Texas at Austin

Abstract. Pure shear oscillations in neutron star crusts are strongly excited by magnetically-induced starquakes. These *toroidal modes* could play a key role in the physics of Soft Gamma Repeaters (SGRs) and Anomalous X-ray Pulsars (AXPs). Toroidal modes drive relativistic Alfvén waves into the magnetosphere. Energy is lost along field lines that are blown open by this outflow pressure. This helps damp the crust vibrations while driving a relativistic wind of waves and particles away from the star. When channeled by a strong magnetic field, this outflow carries away angular momentum. Frequent small-scale fractures maintain a quasi-steady shear mode excitation in a SGR's crust, accelerating SGR spindown. During energetic bursts or flares, large-scale fractures excite the vibrations to much higher amplitudes. Shear oscillations with energy $\sim 10^{44} E_{44}$ erg were probably excited during the 1998 August 27 event, comparable to the observed energy in hard photons. This inevitably drove transient accelerated spindown, with a net period shift $\Delta P/P = 10^{-4} E_{44}^{1/3}$. Such a period shift was found in SGR 1900+14 following the August 27 flare. RXTE observations made one day after the event show that the enhancement of the spindown rate had nearly decayed away by that time; this constrains the dipole field of the star to $B_\star \gtrsim 10^{14}$ Gauss. Anomalous X-ray Pulsars may be inactive magnetars, without strongly vibrating crusts. There is evidence that AXP 2259+586 experienced two active episodes during the past 20 years with crust excitations $\sim 10^{39}$–10^{40} ergs during each episode, driving accelerated spindown by an amount $\Delta P/P \sim 3 \times 10^{-6}$. We describe how $P(t)$ varies following a sudden, strong excitation. Because the excess $\dot P$ decays on a time scale $\tau \sim B_\star^{-2}$, SGR or AXP timing data following a strong excitation could be used to measure the stellar dipole field $B_\star$.

INTRODUCTION: NEUTRON STAR CRUST DEFORMATIONS

The crust of a neutron star has several components: (1) a Fermi sea of relativistic electrons, which provides most of the pressure in the outer layers; (2) another Fermi sea of free neutrons, present only at depths below the neutron drip level, where $\rho > \rho_{drip} \approx 4.6 \times 10^{11}$ gm cm^{-3}; (3) an array of nuclei or ions, which forms a body-centered cubic (bcc) Coulomb lattice in much of the crust, and which condenses into rod-like and slab-like structures near the base of the crust: "nuclear spaghetti and lasagna" or "nuclear pasta" (e.g., ref. [1] and references therein).

In a magnetar, the crust is subject to strong, evolving magnetic stresses [2,3]. Magnetic evolution in the crust is mostly due to Hall drift; while ambipolar diffusion of magnetic flux through the liquid interior can strain the crust from below [4–6]. The crust and field together evolve through a sequence of equilibrium states in which magnetic stresses are balanced by material restoring forces, punctuated by starquakes whenever the crust is strained past its breaking point.

Restoring forces in the crust are provided by different components depending on the kind of deformation. Pure shear deformations along equipotential surfaces experience restoring forces only from the Coulomb forces of the lattice (or pasta), as quantified by the shear modulus. This is much weaker than the degeneracy/gravity forces which resist either bulk compressions or vertical displacements in hydrostatic equilibrium. Because the magnetic field can drive pure shear deformations through a larger range of motion than any other deformations, it does more work against shear restoring forces, engendering greater energy transfer between the the field and crust with more interesting physical consequences. In particular, crustal *twists*, with spiral patterns of shear strain, may be a common type of magnetically-driven deformation. Such deformations excite toroidal modes in the crust when released by fractures. This suggests that toroidal oscillations play an important role in magnetar physics [7]. Giant flares from SGRs, like the 1979 March 5 and 1998 August 27 events, might involve large-scale *circular* fractures, wherein a cap of the crust twists along a circular fault line to whatever position minimizes the total (interior and exterior) magnetic energy [8,3,7].

In this paper, we begin by reviewing the physics of toroidal oscillations, following refs. [7,9,10]. We then discuss some ways in which these modes damp, and how this process could affect the spindown histories of magnetars. We outline some potentially observable consequences, and make comparisons with SGR and AXP observations.

First, a few caveats to keep in mind. Studies of neutron star internal modes have (thus far) adopted the all-lattice idealization. However, nuclear pasta in the deep crust responds very anisotropically to shear: it has the elastic properties of a liquid crystal [11]. A realistic treatment of shear oscillations depends upon the coherence lengths of the pasta domains, about which little is known. We will introduce scaling parameters in the equations which may help adjust for the pasta. A second caveat concerns defects: dislocations and impurities in the lattice. These are sites for fracturing: their presence significantly diminishes the strains that the lattice can bear [12]. In extreme cases, a rapidly-cooled neutron star crust could contain zones of amorphous (glassy) solid [13], which would flow plastically under stress more readily than would any crystal. However, this super-cooled amorphous phase is metastable, and studies of cooling, newborn neutron stars favor widespread crystallization [14]. Plastic flow could still occur in some stars, where the crust is hot, within a factor ~ 0.1 of the melt temperature [12], or in places where the magnetic field is strong enough to overwhelm lattice stresses ($B > B_\mu$ in eq. [2]

below). The very young, high-field AXP 1841-045 [15,16] is a good candidate *plastic magnetar* which may need to experience field decay and cooling before seismic activity and its observable manifestations—SGR bursts and accelerated spindown—can begin.

CRUSTAL SHEAR OSCILLATIONS

The toroidal modes are the only neutron star internal modes which involve pure shear deformations of the crust, thus they may be strongly excited in magnetars [7]. Toroidal modes are indicated by $_\ell t_n$, where $_2t_0$ is the fundamental since $\ell = 1$ deformations do not conserve angular momentum [9,10]. The overtone number n gives the number of radial nodes in the eigenfunction; all $n \gtrsim 1$ modes have frequencies $\gtrsim 1$ kHz. Fractures which extend through the depth of the crust tend to excite $n = 0$ modes. The frequencies of $n = 0$ t-modes are

$$\nu_\ell \approx (V_\mu/2\pi R_\star)\,[\ell(\ell+1)]^{1/2} \tag{1}$$

where $V_\mu = (\mu/\rho)^{1/2}$ is the shear wave velocity and $R_\star$ is the stellar radius.

At $\rho > \rho_{drip}$, $V_\mu = 1.1 \times 10^8 \rho_{14}^{-0.1}$ cm s^{-1} based on the lattice shear modulus of ref. [17] and a power-law fit to the crust equation of state in ref. [18]. This varies only weakly with density, a circumstance which explains why eq. (1) works well.[1]

Each t-mode has $(2\ell + 1)$ values of azimuthal mode number: $-\ell < m < \ell$. These are degenerate in frequency when lattice restoring forces dominate (or, for nuclear pasta, when the mode wavelength $\sim R/\ell$ greatly exceeds the pasta coherence length). A crustal magnetic field causes frequency shifts of magnitude $|\Delta P/P| \sim (B/B_\mu)^2$, which can resolve the degeneracy in m [7]. Here B_μ is related to the shear modulus μ by

$$B_\mu = (4\pi\mu)^{1/2} \approx 4 \times 10^{15} \rho_{14}^{0.4}\text{ Gauss.} \tag{2}$$

In this paper we will assume that $B \ll B_\mu$ in the crust so that magnetic frequency shifts can be neglected.

For a modal displacement $\xi = \xi(r,\theta,\phi)\,e^{i2\pi\nu t}$, the mode energy is approximately

$$E = \mathcal{E}\,\nu^2\,\langle\xi^2\rangle, \tag{3}$$

where $\mathcal{E} = 6.4 \times 10^{32}\,\eta\,R_{10}^4\,M_{1.4}^{-1}$ gm is the mass of the crust, multiplied by some numerical factors, and the stellar mass is $M_\star = 1.4\,M_{1.4}M_\odot$. For the fundamental mode, $\xi[_2t_o] = \xi_o \sin 2\theta$, so $\langle\xi^2\rangle = (8/15)\xi_o^2$, and eq. (3) is consistent with eq. (4) in ref. [7]. Note that the mass of the crust scales as $\sim R_\star^4 M_\star^{-1}$ since the area goes as $R_\star^2$ and the thickness scales inversely with $g_s = GM_\star/R_\star^2$ the acceleration of gravity

---

[1] Indeed, for $\ell = 2$, eq. (1) implies $\nu_2 = 39\,R_{10}^{-1}$ Hz, where $R_{10} \equiv (R_\star/10$ km$)$. This is close to the real $_2t_0$ frequency for a star with mass $M_\star = 1.4\,M_\odot$, as quoted in eq. (2) of ref. [7]. Note that the *observable* frequency in ref. [7] is redshifted by $(1 - 2GM/Rc^2)^{1/2} \approx 0.765$.

at the surface [7]. The factor η is introduced to scale for theoretical uncertainties. Nuclear pasta can significantly reduce the amount of energy stored in shear strains [11], suggesting $\eta < 1$, perhaps $\eta \sim 0.5$. On the other hand, magnetic restoring forces in the crust, and restoring forces due to field lines penetrating the crust from below, tend to drive η to larger values.

The kinetic energy density and lattice strain potential energy densities have equal amplitudes in the crust: $\frac{1}{2}\rho(2\pi\xi_o\nu)^2 = \frac{1}{2}\mu\psi_o^2$. Using eq. (1), this implies a strain amplitude $\psi_o \approx (\xi_o/R_\star)[\ell(\ell+1)]^{1/2}$. For $\langle \xi_o^2 \rangle \approx \frac{1}{2}\xi_o^2$ and $\ell \gg 1$, this is $\psi_o = (2\pi/V_\mu)(2E/\mathcal{E})^{1/2}$ or $\psi_o \approx 0.03\, E_{44}^{1/2}\,\eta^{-1/2}\, R_{10}^{-2}\, M_{1.4}^{1/2}$, assuming a modal energy of $10^{44}\, E_{44}$ erg. Such large strains may dynamically *trigger* fractures [7], although the dynamical yield strain can significantly exceed the static yield strain.

SHEAR MODE DAMPING

Crustal shear oscillations could damp via several mechanisms. Here we focus on damping via an Alfvén-driven wind.[2]

The power in escaping Alfvén waves, and in the relativistic particles which support associated currents, is $L_A \simeq cR_\star^2 \int_{open} d\Omega [(\delta B)^2/8\pi]$, where the integration runs over the the footpoints of all open field lines. The amplitude of transverse magnetic vibrations at these footpoints is $(\delta B/B) = (2\pi\nu\xi/c)$, thus $L_A = 4\pi^2 c^{-1} R_\star^2 B_{open}^2 \Delta\Omega_{open}\, \nu^2 \langle \xi^2 \rangle$. Here the angle-brackets $\langle \rangle$ indicate an average over footpoints which span a solid angle $\Delta\Omega_{open}$, where the field strength at these footpoints is B_{open}. Now, field lines which extend beyond the Alfvén radius R_A are blown open by the pressure of the outflow, where $(L_A/4\pi R_A^2 c) = (B_\star^2/8\pi)(R_A/R_\star)^{-6}$, or $R_A = R_\star (2L_A/B_\star^2 R_\star^2 c)^{-1/4}$. Here $B_\star$ is the *dipole* field strength at the star's magnetic pole; higher-order magnetic multipoles can be neglected when estimating R_A because they diminish more rapidly with distance. If we idealize that the field geometry is dipolar all the way down to the surface[3] then there exist two caps of open field lines at the magnetic poles:

---

[2] Several alternative damping mechanisms are worth noting. Microscopic viscosity in the crust is dominated by electron-phonon interactions; it is comparatively negligible [19]. At high-ν and high wave amplitudes, nonlinear Alfvén wave interactions can support an energy cascade to small scales, engendering dissipation on *closed* field lines [20]. This plays an important role in the X-ray emissions of bright SGR bursts and flares [3], but when the seismic energy spreads out over the star, this cascade luminosity is comparatively small: $(L_{cas}/L_A) \lesssim 10^{-2}\, E_{44}^{2/3}\, (\nu/\text{kHz})$ (cf. eqs. 7.24 and 7.25 in [20]). Fractures triggered by t-modes at sites of pre-existing strain may tend to *boost* excitation rather than damp it [7]. Finally, damping by coupling to the liquid interior depends upon the geometry of field lines penetrating the crust from below, and is relatively negligible in a plausible range of conditions.

[3] The magnetic field of a SGR probably has strong high-order multipole moments, as evinced by the four-peaked emission during the 1998 August 27 event [21]. The footpoints of the far-reaching field lines which are blown open by the outflow will generally cover some small fraction of the star's surface. This area is plausibly comparable to the area of the two polar caps, as assumed here. The field strength B_{open} need not be strictly equal to $B_\star$ either, although it is likely to be

$\Delta \Omega_{open} = 2\pi \theta_{cap}^2$, with $\theta_{cap} = (R_\star/R_A)^{1/2}$. Together these equations imply [20]

$$L_A = (2\pi^8)^{1/3} B_\star^2 R_\star^2 c^{-5/3} [\nu^2 \langle \xi^2 \rangle]^{4/3}. \tag{4}$$

For crustal shear modes, eq. (3) and (4) imply

$$L_A = \kappa\, E^{4/3}, \quad \text{where} \quad \kappa \equiv B_\star^2 R_\star^2 (2\pi^8/\mathcal{E}^4 c^5)^{1/3}. \tag{5}$$

If this energy loss dominates toroidal mode damping, then $\dot{E} = -\kappa E^{4/3}$. Since this is independent of mode frequency, we will assume that this damping law applies to any pattern of toroidal crust excitation.[4] Following a sudden excitation, $E(0) = E_o$,

$$E(t) = E_o\,(1 + t/\tau_o)^{-3}, \quad \text{and} \quad L(t) = L_o\,(1 + t/\tau_o)^{-4}, \tag{6}$$

where $\tau_o = 3E_o/L_o$. Equation (5) implies

$$L_o = 1.4 \times 10^{40}\, E_{44}^{4/3}\, \eta^{-4/3}\, R_{10}^{-10/3}\, M_{1.4}^{4/3} \left(\frac{B_\star}{10\,B_Q}\right)^2 \text{ erg s}^{-1}, \tag{7}$$

where $B_Q \equiv m_e^2 c^3/\hbar e = 4.4 \times 10^{13}$ Gauss. The decay time is

$$\tau_o = \frac{3}{B_\star^2 R_\star^2}\left(\frac{\mathcal{E}^4 c^5}{2\pi^8 E_o}\right)^{1/3} = 5.6\, E_{44}^{-1/3}\, \eta^{4/3}\, R_{10}^{10/3}\, M_{1.4}^{-4/3}\left(\frac{B_\star}{10\,B_Q}\right)^{-2} \text{ hrs}. \tag{8}$$

SPINDOWN DRIVEN BY CRUST VIBRATIONS

The Alfvén-driven outflow of waves and relativistic particles carries off angular momentum [20,22,23]: $I\dot{\Omega} = -\Lambda\,(L_A/c^2)\,\Omega\,R_A^2$, where $\Omega = 2\pi/P$ and $\Lambda \sim 1$ is a numerical constant.[5] This implies

$$\frac{\dot{P}}{P} = \frac{\Lambda\, B_\star\, R_\star^3}{\sqrt{2}\, I\, c^{3/2}}\, L_A^{1/2}. \tag{9}$$

---

comparable.
[4] This is a good approximation. If multiple modes are excited incoherently, $(\delta B/B)^2 = \sum_i (2\pi \nu_i \xi_i/c)^2$, and the excitation energy is $E = \mathcal{E} \sum_i \nu_i \langle \xi_i^2 \rangle$ generally because of mode orthogonality. But note that $\langle\,\rangle$ here represents an average over the whole crust, whereas in eq. (4) it is an average over the footpoints of open field lines. This will lead to corrections of order unity in eq. (5), and can contribute to $\dot{P}$ variability in SGRs with low-order excitations; see Concluding Discussion below.
[5] Pure Alfvén waves ascending along a bundle of field lines with constant magnetic flux have amplitudes $(\delta B/B) \propto B^{-1/2}$, and would go nonlinear at $R \gtrsim R_A$. But a relativistic particle wind is blown out along with the waves, since electrodynamic forces on surface charges tremendously exceed gravity: $F_s \sim e\,\delta B \sim 20\, E_{44}^{1/2}\,(B/10\,B_Q)$ dynes. These charged particles carry Alfvénic currents, and wave energy is transferred to the particles in the outflow.

Using eq. (5) we can derive how much mode energy must be present to drive spindown with a given $(\dot{P}/P)$:

$$E = \frac{\sqrt{2}}{\pi^2} \frac{\mathcal{E} \, c^{7/2}}{B_\star^3 \, R_\star^6} \left(\frac{I \, \dot{P}}{\Lambda \, P}\right)^{3/2} \tag{10}$$

SGR 1806-20 and SGR 1900+14 have been found to spin down at a rate $(\dot{P}/P) \approx 1 \times 10^{-11}$ during quiescent (non-bursting) episodes [24–29]. If driven by crust vibrations, this implies a quiescent, quasi-steady-state excitation level of

$$E_{qs} = 0.9 \times 10^{40} \left(\frac{\dot{P}/P}{10^{-11}}\right)^{3/2} \left(\frac{B_\star}{10 \, B_Q}\right)^{-3} \eta \, \Lambda_{2/3}^{-3/2} \, I_{45}^{3/2} \, R_{10}^{-2} \, M_{1.4}^{-1} \text{ erg}, \tag{11}$$

where we scale to a value $\Lambda = \Lambda_{2/3} \times (2/3)$ (cf. ref. [23]) and a stellar moment of inertia $I_\star = 10^{45} \, I_{45}$ gm cm^2.

This mechanism of spinning down requires that crustal vibrations be continually re-excited by small-scale fractures. Fractures can be triggered by the shear oscillations at sites where the magnetic strains are already near critical. Such triggered events tend to bolster modes that are already excited: the "self-exciting" property of t-modes [7]. The mean power of excitation in quiescence is

$$\dot{E}_{qs} = 6 \times 10^{34} \left(\frac{\dot{P}/P}{10^{-11}}\right)^2 \left(\frac{B_\star}{10 \, B_Q}\right)^{-2} \Lambda_{2/3}^{-2} \, I_{45}^2 \, R_{10}^{-6} \text{ erg s}^{-1}, \tag{12}$$

which is comparable to the observed X-ray power from these stars. Indeed, non-thermal X-ray emissions from SGRs and AXPs may come from the Alfvén-driven outflow, and from Alfvén-energized electrons trapped in the magnetosphere. The characteristic decay time is (eqs. [8] and [11])

$$\tau_{qs} = 5.2 \left(\frac{\dot{P}/P}{10^{-11}}\right)^{-1/2} \left(\frac{B_\star}{10 \, B_Q}\right)^{-1} \eta \, \Lambda_{2/3}^{1/2} \, I_{45}^{-1/2} \, R_{10}^4 \, M_{1.4}^{-1} \text{ days.} \tag{13}$$

Fractures must occur more often than this to avoid $\dot{P}$ fluctuations on this timescale. Thus the energy per fracture satisfies $E_{fr} \ll \dot{E}_{qs}\tau_{qs}$, or $E_{fr} \ll 3 \times 10^{40} \, (\dot{P}/10^{-11} \, P)^{3/2} \, (B\star/10 \, B_Q)^{-3}$ erg. This condition may be easily satisfied for most crust fractures, which plausibly occur on the Hall length-scale $\lambda_H \sim 0.1$ km, releasing energy $E_{fr} \sim 10^{36}$ erg; see eq. (54) and (67) of ref. [5]. Such fractures would not produce observable SGR bursts; they tend to impart almost all their energy to crustal shear modes. The average time between fractures is $\Delta t = E_{fr}/\dot{E}_{qs} \approx 17(E_{fr}/10^{36} \text{ erg})$ s. There are $\sim 3 \times 10^4 \, E_{fr,36}^{-1}$ such events in a decay time, τ_{qs}.

This mechanism of angular momentum loss predicts a correlation of $\dot{P}$ with burst activity. However, only rare events which inject $\gtrsim 10^{40}$ erg into crustal modes will strongly affect $\dot{P}$ (eq. [11]), and then only transiently (eq. [13]). *Decaying* modes don't tend to trigger fractures because all near-critical strains were previously released. This suggests that enhancements of $\dot{P}$ due to bursts may tend to die out quickly. Nevertheless, they could sometimes be observable, especially in very energetic events. We now turn to this issue.

THE 1998 AUGUST 27 EVENT

The August 27 flare released an energy $\gtrsim 10^{44}$ ergs. This probably involved a large propagating fracture in the neutron star crust with a significant rotational component [21]. This is likely to excite strong toroidal oscillations, with vibrational energy $E \approx 10^{44}$ erg. For this reason, we now consider transient spindown following a strong excitation of the crust. From eqs. (6) and (9),

$$\tau_o \frac{dP}{dt} = Z(1 + t/\tau_o)^{-2} P, \quad \text{where} \quad Z \equiv \frac{\Lambda B_\star R_\star^3}{\sqrt{2} I \, c^{3/2}} L_o^{1/2} \tau_o. \tag{14}$$

This differential equation has the solution

$$P(t) = P_o \exp\left(\frac{Z}{1 + \tau_o/t}\right). \tag{15}$$

For $t \gg \tau_o$ (see eq. [8]) this implies $(\Delta P/P) = +Z$, for $Z \ll 1$. Using eq. (5) and $\tau_o = 3E_o/L_o$ we find

$$\frac{\Delta P}{P} = \frac{3\Lambda}{(2\pi^2)^{2/3}} \frac{R_\star^2}{I} \left(\frac{\mathcal{E}}{c}\right)^{2/3} E_o^{1/3} = 1 \times 10^{-4} \, E_{44}^{1/3} \Lambda_{2/3} \, \eta^{2/3} \, I_{45}^{-1} \, R_{10}^{14/3} \, M_{1.4}^{-2/3}. \tag{16}$$

This agrees with the period shift that was measured following the 1998 August 27 event [29]. Note that the shift in P only depends upon the (known) August 27 event energy and upon well-constrained properties of the neutron star, along with a spindown coefficient close to unity ($\Lambda \sim 2/3$).[6]

---

[6] Two other explanations of this shift have been suggested. The shift could have accumulated during the Summer of 1998, if $\dot{P}$ were enhanced by a factor ~ 2 relative to its measured values in 1998 before and after the Summer (see refs. [28,29] and the final section of ref. [30]). This is possible if L_A increased by a factor ~ 4 during the Summer [eq. (9)]. No measurements of SGR 1900+14 were made between early June and late August, so we cannot be sure. Observed SGR 1900+14 $\dot{P}$ increases after 1998 make this suggestion more plausible (P. Woods, private communication). However, the simple theory given here (eq. [16]) bolsters the case that accelerated spindown occurred in the aftermath of the giant flare.
Alternatively, the period shift might be due to a sudden, superfluid *spindown glitch* triggered by the giant flare [23]. This involves the unpinning of superfluid vortices in a crust, as in radiopulsars glitches. But in magnetars, magnetically-induced crustal twists (as described above) tend to *deplete* the angular momentum of a pinned crustal superfluid, potentially leading to glitches of opposite sign, compared to radiopulsar glitches. This explanation depends upon the uncertain

Although $(\Delta P/P)$ is independent of B, the fact that all, or nearly all, excess $\dot{P}$ had decayed away by the time that the Proportional Counter Array (PCA) on the *Rossi X-ray Timing Explorer* (RXTE) began measurements on August 28 [26,29] implies $\tau_o < 24$ hours, or using eq. (8),

$$B_\star > 2 \times 10^{14} \, E_{44}^{-1/6} \, \eta^{2/3} \, R_{10}^{5/3} \, M_{1.4}^{-2/3} \text{ Gauss}. \tag{17}$$

Note the insensitivity of this result to the excitation energy. This is a new, nontrivial bound on $B_\star$, consistent with previous bounds on SGR magnetic fields [3].

ANOMALOUS X-RAY PULSAR SPINDOWN

Kaspi, Chakrabarty and Steinberger (ref. [31]) obtained phase-coherent measurements of AXP 2259+586 using RXTE. They found very smooth spindown, with $P = 6.98$ s and $\dot{P} = 4.89 \times 10^{-13}$, and timing residuals comparable or less than those found in radiopulsars. This is consistent with an *inactive* magnetar spinning down via magnetic torques [32], near or beyond its radiopulsar death line, unaided by angular momentum loss from a magnetically-powered wind. The vacuum magnetic dipole radiation (MDR) idealization may be a good first approximation in these circumstances, with $B_\star \approx 0.6 \times 10^{14}$ G for AXP 2259+586.[7] Note that $\tau_{MDR} \equiv P/2\dot{P}$ is significantly *longer* that the associated supernova remnant age, $\tau_{MDR} > t_{SNR}$. This suggests that the star was spinning down at an accelerated pace in the past, perhaps during an active phase when the star was a SGR (see §4.2 in ref. [23]).

The timing history of AXP 2259+586 during the past 20 years deviates significantly from the backward extrapolation of the Kaspi et al. spindown line (Fig. 3 in ref. [31]). However, the data can be fit if there were *two episodes of accelerated spindown, with smooth MDR spindown between episodes*. The first episode occurred around 1985, between Tenma and EXOSAT observations, with a period shift $(\Delta P/P) \approx 3 \times 10^{-6}$. The second episode occurred between 1993 and 1996 (between ASCA and RXTE), with $(\Delta P/P) \approx 2 \times 10^{-6}$.

These accelerated spindown episodes could have been due to "avalanches" of crust fractures. Such intermittent activity would not be surprising in a star that was once a SGR. The starquakes need not have produced observable X-ray bursts if the energy was released in the deep crust, without strong exterior Alfvén wave

---

physics of vortex pinning in magnetars, and in particular upon the value of the angular momentum in the pinned crustal superfluid before the giant flare, which is not yet possible to estimate reliably. Future observations of $P(t)$ during and after giant flares could determine if period shifts are truly abrupt and glitch-like, or if they occur as described by eq. (15).

[7] Vacuum MDR actually implies $B_\star \sin\alpha = 6.4 \times 10^{19} (\dot{P}P)^{1/2} I_{45}^{1/2} R_{10}^{-3}$ G. Here we quote (1/2) this value with $\sin\alpha = 1$, as conventional in radiopulsar studies. This probably underestimates the true surface dipole field, especially if a slowly-rotating star tends to spin down toward alignment and experience torque reduction, as would be the case if currents close within the light cylinder. The true dipole field of 1E2259+586 may significantly exceed 10^{14} Gauss.

pulses capable of driving nonlinear Alfvén wave cascades. The energy in an episode, if it were released suddenly under conditions in which eq. [16] applies,[8] would be $E_{crust} = 3 \times 10^{39} \, (\Delta P/3 \times 10^{-6} \, P)^3 \, \Lambda_{2/3}^{-3} \, \eta^{-2} \, I_{45}^3 \, R_{10}^{-14} \, M_{1.4}^2$ erg. Note the very steep dependence of E_{crust} on $R_\star$; this is mitigated since $I_\star^3 \propto R_\star^6$, roughly. The characteristic decay time is $\tau_o = 11 \, E_{39}^{-1/3} \, (B\star/10 \, B_Q)^{-2} \, \eta^{4/3} \, R_{10}^{10/3} \, M_{1.4}^{-4/3}$ days. Continued monitoring of $P(t)$ in AXP 2259+586 might reveal new active episodes.

STRONG-EXCITATION MODELS

Equation (15) gives a good first approximation to transient spindown following a strong excitation; however it has the property $\dot P \to 0$ at late times. We now consider more realistic models with $\dot P > 0$ in the baseline state. This could be useful for fitting $P(t)$ data from magnetar candidates with vibrating crusts.

(1) SGR-like solution For a active star, experiencing wind-aided spindown in quiescence, the persistent power in small-scale fractures is $\dot E_{fr} = \kappa E_{qs}^{4/3}$, where E_{qs} is the quiescent excitation energy of the crust (eqs. [5] and [11]). This implies $\dot E = -\kappa(E^{4/3} - E_{qs}^{4/3})$. After an excitation $E(0) = E_o$, the energy $E(t)$ decays back toward E_{qs} in a way given by

$$\tan^{-1}\left(\frac{E}{E_{qs}}\right) - \tan^{-1}\left(\frac{E_o}{E_{qs}}\right) - \frac{1}{2}\ln\left[\frac{(E+E_{qs})(E_o-E_{qs})}{(E-E_{qs})(E_o+E_{qs})}\right] + \frac{2t}{\tau_{qs}} = 0. \quad (18)$$

For $Z_{qs} = Z(E_{qs}) = \dot P_{qs}\tau_{qs}/P$ (eq. [14]), the spindown history is given by $P(t) = P_o \exp\{(Z_{qs}/\tau_{qs}) \int [E(t)/E_{qs}]^{2/3} \, dt\}$. When $\dot P_{qs} t \ll P$, this can be written $P(t) = P_o + \dot P_{qs} t + \Delta(t)$, where $\Delta(t) = \dot P_{qs} \int ([E(t)/E_{qs}]^{2/3} - 1) \, dt$.

(2) AXP-like solution If R_A does not lie within the light cylinder, then eqs. (4) and (5) do not apply. This happens in a relatively inactive star, with $L_A < (B_\star^2 R_\star^6 \Omega^4/2c^3) \sim \dot E_{MDR}$, or for crust excitation energies

$$E < E_x \equiv 4\pi \, \mathcal{E} \, R_\star^3/c \, P^3 = 8 \times 10^{38} \, P_7^{-3} \, \eta \, R_{10}^7 \, M_{1.4}^{-1} \text{ erg}, \quad (19)$$

where $P = 7P_7$ s. For $E < E_x$ the wind does not significantly blow open field lines, so $\theta_{cap} = (R_\star \Omega/c)^{1/2}$. Following the steps which led to eq. (4), but with this different polar cap radius, one finds

$$L_A = \pi^2 \, B_\star^2 \, R_\star^3 \, \Omega \, c^{-2} \, \nu^2 \, \langle \xi^2 \rangle \quad \text{for } E < E_x \text{ or } L_A < \dot E_{MDR}. \quad (20)$$

Using eq. (3), this implies $L_A = E/\tau_{in}$ for a (usually) inactive star, with

---

[8] Alternatively, if the seismic episodes in 1E2259+586 involved multiple energy releases, each with $E < E_x$ (see eq. [19] below), separated in time by $\Delta t > \tau_{in}$ (eq. [21]), then the total energy needed to drive ΔP would be $E_{crust} = 4 \times 10^{39} \, (\Delta P/3 \times 10^{-6} \, P) \Lambda_{2/3}^{-1} \, P_7^{-2} \, I_{45}$ erg according to eq. (23). This is numerically similar to the $E > E_x$ case, but the scaling parameters are different. The spindown history following each $E < E_x$ excitation would be given by eq. (22).

$$\tau_{in} = \frac{\mathcal{E}\, c^2\, P}{2\pi^3\, B_\star^2\, R_\star^3} = 3.9 \left(\frac{B_\star}{10\, B_Q}\right)^{-2} P_7\, \eta\, R_{10}\, M_{1.4}^{-1} \text{ days}. \qquad (21)$$

Thus $E(t) = (E_o - E_{qs}) \exp(-t/\tau_{in}) + E_{qs}$ following an excitation to $E_o \leq E_x$.[9] The spindown law is $I\Omega\dot{\Omega} = -[\dot{E}_{MDR} + \Lambda\, L_A(t)]$, including both wind and MDR torques. For $t \ll P/\dot{P}_{qs}$, this yields

$$P(t) = P_o + \dot{P}_{qs} t + \Delta P\, [1 - \exp(-t/\tau_{in})], \qquad (22)$$

with a net period shift

$$\frac{\Delta P}{P} = \frac{\Lambda\, P^2\, (E_o - E_{qs})}{4\pi^2\, I} = 0.8 \times 10^{-6}\, E_{39}\, \Lambda_{2/3}\, P_7^2\, I_{45}^{-1}. \qquad (23)$$

If an inactive star were excited to $E_o > E_x$, then the energy of crust vibrations would decay according to eq. (6) until it reaches $E \approx E_x$. Subsequently the decay would be exponential with time constant τ_{in}. The spindown history would follow eq. (15) before matching onto eq. (22).

CONCLUDING DISCUSSION

The agreement of eq. [16] with the measured August 27 period shift gives evidence that crust vibrations drive accelerated spindown in SGRs. Magnetic fields $> 10^{14}$ Gauss seem to be required (eq. [17]). Future observations of $P(t)$ following strong excitations could be used to determine $B_\star$ for SGRs and AXPs since the decay times [eq. (8) and (21)] scale as $\tau \propto B_\star^{-2}$.

Significant $\dot{P}$ variability was found in SGR 1806-20 [33] and recently in SGR 1900+14 (P. Woods, private communication). This could mean that the baseline excitation level E_{qs} varies with time. Triggered fractures energize the modes and also drive *modal drift*, with patterns of strain wandering over the star. This helps release strains throughout the crust, while magnetic evolution continually builds stresses [7]. This complex dynamical interplay may lead to variable E_{qs} and $\dot{P}$. Another potentially important effect occurs when the order-number of excited modes satisfy $\ell > (2\pi/\theta_{cap})$, i.e., for mode frequencies $\nu < 2$ kHz $R_{10}^{-1}\, (B/10\, B_Q)^{1/4}\, (\dot{P}/10^{-11} P)^{-1/4}$. In this case, the footpoints of the open field lines could lie at a node or at a peak of harmonic strain, and L_A, along with $\dot{P}$, can vary as the excitation pattern moves between these extremes even when E_{qs} varies little.

No clear correlations of $\dot{P}$ with SGR bursting activity has yet been found [26,29,33]. To understand this, note that when an energy E is imparted suddenly to crust excitation, it drives a net period shift $\Delta P \propto E^{1/3}$ (eq. 16). But

---

[9] This assumes quiescent, quasi-steady crust vibrations with energy $E_{qs} < E_x$. Such vibrations may power nonthermal X-ray emissions in AXPs. It is also possible that $E_{qs} = 0$ in truly quiet stars with $\dot{P}_{qs} \simeq \dot{P}_{MDR}$.

if the same amount of energy is imparted in N events each of energy E/N, then $\Delta P \propto N (E/N)^{1/3} \propto N^{2/3} E^{1/3}$; i.e., ΔP is larger for more events. If the energy injections follow a distribution law $(d\mathcal{N}/dE) \propto E^{-5/3}$, as found in self-organized critical systems, and in particular as found in SGR bursts [34], then the total released energy is $E_{tot} \propto \int E\, d\mathcal{N} \propto E_{max}^{1/3} - E_{min}^{1/3}$. This means that E_{tot} is dominated by the largest events: for a wide dynamic range, E_{tot} is insensitive to the lower cutoff. However, the total spindown driven by these events is $\Delta P \propto \int E^{1/3}\, d\mathcal{N} \propto E_{min}^{-1/3} - E_{max}^{-1/3}$. Thus small energy injections dominate the angular momentum loss.[10] Energetic SGR bursts are relatively inefficient at driving spindown for this reason, and also because energy is lost to observable X-rays (via strong Alfvén pulses undergoing nonlinear damping in the magnetosphere). The aftermaths of bright bursts, in which an avalanche of small-scale strains are released, could actually show *less* small-scale fracturing. For all these reasons, it is not surprising that $\dot{P}$ does not correlate strongly with most observed burst activity. (Very bright bursts and flares may prove an exception.)

We conclude with a few cautionary remarks. It is possible that the simple models in this paper describe some of the essential physics involved in magnetar seismic activity and angular momentum loss. However, we have considered only a few very idealized physical elements, and it is far from certain that other effects are negligible. Other modes of internal vibration can be excited in magnetars, especially in giant flares which might involve magnetic interchange instabilities in the liquid interior [3]. Also other processes may help damp magnetar vibrations [7].

Finally, other torques could affect magnetar spindown histories. Material driven from bursting SGRs by hyper-Eddington photon fluxes might form a "spin-out disk" near the co-rotation radius, with possible affects on $\dot{P}$ [33,35]. Even in the absence of crust vibrations and Alfvén-driven winds, particle flows around magnetars are driven electrodynamically.[11] If these currents flow out far enough to affect conditions at the Alfvén radius or the light cylinder, then the stellar torque is changed. Furthermore, magnetar stellar shapes are distorted by interior magnetic stresses, which might lead to *periodic* precessional spindown modulations.[12] Continued monitoring of SGRs and AXPs will be needed to disentangle these interesting effects.

---

[10] Of course, the $\Delta P \propto E^{1/3}$ law breaks down for $E < E_{qs}$ (eq. [11]), or for $E < E_x$ (eq. [19]) in a star with $E_{qs} < E_x$, but this is precisely the condition that the released energy be insufficient to change $\dot{P}$ very much.

[11] Crustal twists, as described in the Introduction, can drive currents in a magnetar's magnetosphere (§5.1 in [23]). In particular, if a bundle of field lines, describing an arch in the magnetosphere, has one footpoint twisted (with the motion driven from below by the evolving field), then a current must flow along the arch to maintain the twisted exterior field. This is because $\oint \mathbf{B} \cdot d\boldsymbol{\ell} = 4\pi I/c + (1/c)(\partial/\partial t) \int \mathbf{E} \cdot d\mathbf{A}$, so whenever $I < (c/4\pi) \oint \mathbf{B} \cdot d\boldsymbol{\ell}$, the displacement current term builds an electric field to increase I. Surface impacts of these flowing charges create hotspots at the arch's footpoints which can help account the quiescent X-rays and X-ray variability of SGRs and AXPs.

[12] See ref. [36] and §4.3 of ref. [23]. Note that the data of [31] contradict predictions of [36].

ACKNOWLEDGMENTS

This work was supported by NASA grant NAG5-8381 and Texas Advanced Research Project grant ARP-028.

REFERENCES

1. Pethick, C.J. & Ravenhall, D.G., *Ann. Rev. Nucl. Part. Sci*, **45**, 429 (1995)
2. Duncan R.C., & Thompson C. *Ap.J.*, **392**, L9 (1992)
3. Thompson, C., & Duncan, R.C., *M.N.R.A.S.*, **275**, 255 (1995)
4. Goldreich P., & Reisenegger A., *Ap.J.* **395**, 250 (1992)
5. Thompson C., & Duncan R.C., *Ap.J.* **473**, 322 (1996)
6. Heyl, J.S. & Kulkarni, S.R. *Ap.J.*, **213**, 313 (1998)
7. Duncan, R.C. *ApJ*, **498**, L45 (1998)
8. Flowers, E. & Ruderman, M. *Ap.J.*, **315**, 302 (1977)
9. Hansen, C.J. & Cioffi, D.F. *Ap.J.*, **238**, 740 (1980)
10. McDermott, P.N., Van Horn, H.M. & Hansen, C.J. *Ap.J.*, **325**, 725 (1988)
11. Pethick, C.J. & Potekhin, A.Y., *Phys. Lett.*, **B427**, 7 (1998)
12. Ruderman, M. *Ap.J.*, **382**, 576 (1991); *Ap.J.*, **382**, 587 (1991)
13. Ichimaru, S., Iyetomi, H., Mitake, S. & Itoh, N., *Ap.J.*, **265**, L83 (1983)
14. De Blasio, F.V. *Ap.J.*, **452**, 359 (1995)
15. Vasisht, G., & Gotthelf, E.V., *Ap.J.*, **486**, L129 (1997)
16. Gotthelf, E.V., Vasisht, G. & Dotani, T. *Ap.J.*, **522**, L49 (1999)
17. Strohmayer, T., et al., *ApJ*, **375**, 679 (1991)
18. Negele, J.W. & Vautherin, D. *Nuc.Phys.A* **207** 298 (1973)
19. Flowers, E. & Itoh, N. 1976, *Ap.J.*, **206**, 218
20. Thompson, C. & Blaes, O., *Phys. Rev. D*, **57**, 3219 (1998)
21. Feroci, M., Hurley, K., Duncan, R. & Thompson, C. *Ap.J.*, in press (2000)
22. Harding, A.K., Contopoulos, I. & Kazanas, D. *Ap.J.*, **525**, L125 (1999)
23. Thompson, C., et al., *ApJ* in press (2000) [astro-ph/9908086]
24. Kouveliotou, C. et al., *Nature*, **393**, 235 (1998)
25. Hurley, K. et al., *Ap.J.*, **510**, L111 (1999)
26. Kouveliotou, C. et al., *Ap.J.*, **510**, L115 (1999)
27. Murakami, T. et al., *Ap.J.*, **510**, L119 (1999)
28. Marsden, D., Rothschild, R.E. & Lingenfelter, R.E *Ap.J.*, **520**, L107 (1999)
29. Woods, P. et al., *Ap.J.*, **524**, L55 (1999)
30. Duncan, R.C. in *Gamma-Ray Bursts: Fifth Huntsville Symposium*, eds. R. Kippen et al. (New York: AIP) in press (2000) [astro-ph/0002442]
31. Kaspi, V., Chakrabarty, D. & Steinberger, J., *Ap.J.*, **525**, L33 (1999)
32. Thompson C., & Duncan R.C., *Ap.J.* **408**, 194 (1993)
33. Woods, P. et al., *Ap.J.*, **535**, L55 (2000)
34. Göğüş, E. et al., *Ap.J.*, **526**, L93 (1999); *Ap.J.*, **532**, L121 (2000)
35. Ibrahim, A.I. et al., *Ap.J.* in press (2000) [astro-ph/0007043]
36. Melatos, A. *Ap.J.*, **519**, L77 (1999)

The Afterglows of Gamma-Ray Bursts

S. R. Kulkarni*, E. Berger*, J. S. Bloom*, F. Chaffee¶,
A. Diercks*, S. G. Djorgovski*, D. A. Frail†, T. J. Galama*,
R. W. Goodrich¶ F. A. Harrison*, R. Sari* & S. A. Yost*

*California Institute of Technology, Pasadena, CA 91125, USA
†National Radio Astronomy Observatory, Socorro, NM 87801, USA
¶W. M. Keck Observatory, Kamuela, HI 96743, USA

Abstract. Gamma-ray burst astronomy has undergone a revolution in the last three years, spurred by the discovery of fading long-wavelength counterparts. We now know that at least the long duration GRBs lie at cosmological distances with estimated electromagnetic energy release of $10^{51} - 10^{53}$ erg, making these the brightest explosions in the Universe. In this article we review the current observational state, beginning with the statistics of X-ray, optical, and radio afterglow detections. We then discuss the insights these observations have given to the progenitor population, the energetics of the GRB events, and the physics of the afterglow emission. We focus particular attention on the evidence linking GRBs to the explosion of massive stars. Throughout, we identify remaining puzzles and uncertainties, and emphasize promising observational tools for addressing them. The imminent launch of *HETE-2* and the increasingly sophisticated and coordinated ground-based and space-based observations have primed this field for fantastic growth. This overview is a combined write-up of talks given at this conference and in NASA's Goddard Space Flight Center.

I INTRODUCTION

GRBs have mystified and fascinated astronomers since their discovery. Their brilliance and their short time variability clearly suggest a compact object (black hole or neutron star) origin. Three decades of high-energy observations, culminating in the definitive measurements of CGRO/BATSE, determined the spatial distribution to be isotropic yet inhomogeneous, suggestive of an extragalactic population (see [14] for a review of the situation prior to the launch of the BeppoSAX mission). Further progress had to await the availability of GRB positions adequate for identification of counterparts at other wavelengths.

In the cosmological scenario, GRBs would have energy releases comparable to that of supernovae (SNe). Based on this analogy, Paczyńsk & Rhoads [65] and Katz [44] predicted that the gamma-ray burst would be followed by long-lived but fading emission. These papers motivated systematic searches for radio afterglow, including our effort at the VLA [15]. The broad-band nature of this "afterglow" and its detectability was underscored in later work [59,78]

Ultimately, the detection of the predicted afterglow had to await localizations provided by the Italian-Dutch satellite, BeppoSAX [6]. The BeppoSAX Wide Field Camera (WFC) observes about 3% of the sky, triggering on the low-energy (2 – 30 keV) portion of the GRB spectrum, localizing events to $\sim 5 - 10$ arcminutes. X-ray afterglow was first discovered by BeppoSAX in GRB 970228, after the satellite was re-oriented (within about 8 hours) to study the error circle of a WFC detection with the 2 – 10 keV X-ray concentrators. The detection of fading X-ray emission, combined with the high sensitivity and the ability of the concentrators to refine the position to the arcminute level, led to the subsequent discovery of long lived emission at lower frequencies [10,77,16].

Optical spectroscopy of the afterglow of GRB 970508 led to the definitive demonstration of the extragalactic nature of this GRB [60]. The precise positions provided by radio and/or optical afterglow observations have allowed for the identification of host galaxies, found in almost every case. Not only has this provided further redshift determinations, but it has been useful in tying GRBs to star formation through measurements of the host star formation rate (e.g. [46,11]). HST with its exquisite resolution has been critical in localizing GRBs within their host galaxies and thereby shed light on their progenitors (e.g. [29,41,4]). Observations of the radio afterglow have directly established the relativistic nature of the GRB explosions [16] and provided evidence linking GRBs to dusty star-forming regions. Radio observations are excellent probes of the circumburst medium and the current evidence suggests that the progenitors are massive stars with copious stellar winds. The latest twist is an apparent connection of GRBs with SNe [5]. Separately, an important development is the possible association of a GRB with a nearby (40 Mpc) peculiar SN [30,47].

In this paper we review the primary advances resulting from afterglow studies. §II discusses the statistics of detections to-date, including possible causes for the lack of radio and optical afterglows from some GRBs. In §III we review constraints on the nature of the progenitor population(s), in particular evidence linking some classes of GRBs to SNe. §IV describes the status of current understanding of the physics of the afterglow emission. Here we compare observations to predictions of the basic spherically-symmetric model, and describe complications arising from deviations from spherical symmetry and non-uniform distribution of the circumburst medium. We conclude with speculations of the near and long-term advances in this field (§V).

We point out that this review has two biases. First, given the concentration of previous review articles on optical and X-ray observations, we emphasize the unique contributions of radio afterglow measurements. Second, this article is intended to also provide a summary of the efforts of the Caltech-NRAO-CARA GRB collaboration, and therefore details our work in particular. This review is in response to review talks given at the 1999 Maryland October meeting (SRK) and the 5th Huntsville GRB meeting (DAF and SRK).

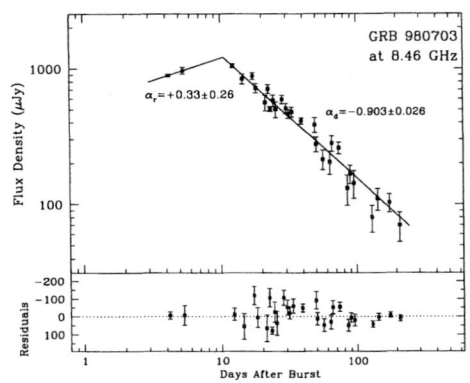 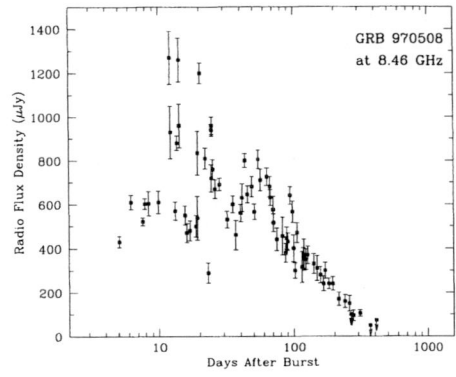

FIGURE 1. *Left: The radio light curve of GRB 980703. This is a typical afterglow, a rise to a peak followed by a power law decay. The longer lifetime of the radio afterglow allows us to see both the rise and the fall of the afterglow emission. In contrast, at optical and X-ray emission, most of the times we see only the decaying portion of the light curve. Right: The radio light curve of GRB970508 [21]. The wild fluctuations of the light curve in the first three weeks are chromatic. At later times, the fluctuations become broad-band and subdued. These fluctuations are a result of multi-path propagation of the radio waves in the Galactic interstellar medium. As the source expands (at superluminal speeds) the scintillation changes from diffractive to refractive scintillation. This is analogous to why stars twinkle but planets do not.*

II STATISTICS OF AFTERGLOW DETECTIONS

Afterglow emission was first detected from GRB 970228, both at X-ray [10] and optical frequencies [77], but not at radio wavelengths [17]. The first radio afterglow detection came following the localization of GRB 970508 [16]. Figure 1 shows two examples of radio lightcurves. The radio afterglow of GRB 970508 is famous for several reasons: it was the first radio detection, it gave the first direct demonstration of relativistic expansion, and it remains the longest-lived afterglow [21].

Afterglow emission is now routinely detected across the electromagnetic spectrum. BeppoSAX has been joined in studying the X-ray afterglows by the All Sky Monitor (ASM) aboard the X-ray Timing Explorer (XTE), the Japanese ASCA mission, and recently the Chandra X-ray observatory (CXO). A veritable armada of optical facilities (ranging from 1-m class telescopes to the 10-m Keck telescopes) routinely discover and study optical afterglows. The HST has been primarily used to make exquisite images of the host galaxies (see above) but in the near future we expect other uses such as UV spectroscopy and identification of underlying SNe. The VLA has led the detection in radio. However, other centimeter-wavelength facilities (the Australia Telescope National Facility, Westerbork Synthesis Radio Telescope, the Ryle Telescope) and millimeter wavelengths (James Clerk Maxwell Telescope, the Owens Valley Millimeter Array, IRAM and the Plateau de Bure Interferometer) are now regularly contributing to afterglow studies.

Figure 2 summarizes the statistics of afterglow detections. In almost all cases, X-ray emission has been detected, establishing the critical importance of prompt X-ray observations. Optical afterglow appears to be detected in about 2/3 of all well-localized events if sufficiently deep optical images are taken rapidly (i.e. within a day or so of the burst). Radio afterglows are detected in 40% of the cases – far more often than usually assumed. We refer the reader to the Frail et al. [22] for a comprehensive summary of the X-ray/optical/radio afterglow detection statistics. The non-detections are, as discussed below, as interesting as the detections.

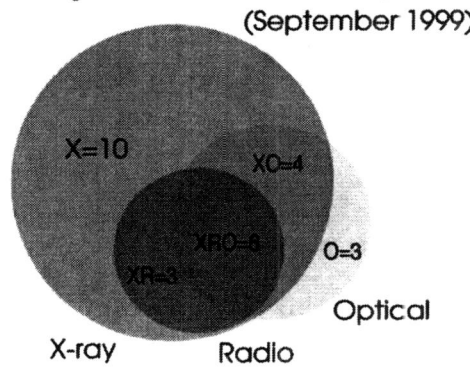

FIGURE 2. *A Venn diagram showing the detection statistics for 26 well-localized GRBs in the Northern and Southern hemispheres. Of the 23 GRBs for which X-ray afterglows have been detected to date, 10 have optical afterglows (XO + XOR) and 9 have radio afterglows (XR + XOR). In total there are 13 optical and/or radio afterglows with corresponding X-ray afterglows.*

Radio Non-detection. The failure to find radio afterglow is most likely due to lack of sensitivity. The brightest radio afterglow to date is that from GRB 991208 (Frail GCN[1] 451) with a peak flux of 2 mJy, a 60-σ detection (at centimeter wavelengths) whereas the weakest afterglow is typically around 5σ. In contrast, at optical and X-ray wavelengths, afterglow emission is routinely detected at hundreds of sigma. If the VLA were to be upgraded by a factor of 10 in sensitivity, then we

---

[1] GCN refers to the GRB Coordinates Network Circular Services. This network is maintained by S. Barthelmy at the Goddard Space Flight Center;
see http : //lheawww.gsfc.nasa.gov/docs/gamcosray/legr/bacodine/gcn_main.html

predict that radio afterglow emission would, like X-ray emission, be detected from most GRBs.

Optical Non-detection. Non-detection at optical wavelengths is more interesting, as it may result in some cases from extinction along the line of sight or within the source. Bad weather as well as rapid fading of the afterglow has certainly hindered some optical searches, which, due to notification delays, typically begin some hours after the event. Furthermore, low Galactic latitude events may be obscured, or hidden in crowded foregrounds. However, in some cases deep searches have been performed with no success. Here, non-detection likely results from extinction by dust in the burst host galaxy and/or absorption by the intergalactic medium. GRB 970828 [38] is one example, as is the more dramatic case of GRB 980329. This burst was one of the brightest events in the WFC [42]. Searches for optical afterglow emission failed to identify any counterpart. VLA observations identified an unusual radio variable in the field [76]. Soon thereafter, a red afterglow and a bright IR afterglow were identified (Klose GCN 43, Larkin et al. GCN 44). Taylor et al. [76] suggest that the GRB arose in a region with high extinction. Further optical and IR work on this interesting afterglow can be found in [34], [64], and [69].

Optically dim "red" but bright IR afterglows can also result from the GRB being located at high redshift. Intergalactic HI absorption will result in a wavelength cutoff below the Lyman limit, $< 912(1+z)$ Å, where z is the redshift of the source. This effect was originally invoked to explain the faint R-band but bright IR emission from GRB 980329 [27]. We now know, based on recent Keck observations, that the GRB host is blue, incompatible with a high-z origin. Rather, it is more tenable that the host is a typical star-forming galaxy with dusty star-forming regions, and that the GRB occurred in one such region [76]. We are presently carrying out IR spectroscopy of this host to determine the redshift and the star formation rate (SFR). While searching for "R dropouts" may in the future provide an effective method for finding high-redshift events, it is clear that cross-calibrated multi-band photometry of higher quality than currently exists will be required to make this useful.

X-ray Non-detection. The spectra of most GRB events clearly extend into the X-ray band, as established by *GINGA* observations [75]. How the X-ray emission observed during the burst connects to the X-ray afterglow is uncertain, due to sensitivity limitations of wide-field monitors. X-ray afterglow emission appears to be ubiquitous. Observations of the X-ray afterglow are important for two reasons: (i) the observations of the X-ray afterglow by sensitive imaging instruments (e.g. the concentrators aboard BeppoSAX) result in sufficiently precise (arcminute) localization and (ii) a significant (perhaps even a dominant) fraction of the explosion energy appears to be radiated in this band. Of all the SAX bursts, GRB 970111 is peculiar for the absence of X-ray afterglow (admittedly the data were obtained about 17 hours after the burst) [25]. In view of the critical role played by X-ray afterglow in localization of GRBs we regard this non-detection to be worthy of

TABLE 1. Basic Properties of Selected GRBs

GRB	α(J2000) (h m s)	δ(J2000) (° ′ ″)	R_{host} (mag)	S ×10^{-6} (erg cm^{-2})	z	References[a]
970228	05 01 47	+11 46.9	25.2[b]	1.7	0.695	Djorgovski et al. GCN 289
970508	06 53 49	+79 16.3	25.7	3.1	0.835	[60,2]
970828	18 08 32	+59 18 52	TBD	74	0.957	[12]
971214	11 56 26	+65 12.0	25.6	11	3.418	[46]
980326	08 36 34	−18 51.4	≳ 27.3	1	...	
980329	07 02 38	+38 50.7	25.4	50	...	
980519	21 22 21	+77 15.7	26.2	25	...	
980613	10 17 58	+71 27.4	24.5	1.7	1.096	Djorgovski et al. GCN 189
980703	23 59 07	+08 35.1	22.6	37	0.966	[11]
981226	23 29 37	−23 55 54	≳ 22	N.A.	...	
990123	15 25 31	+44 46 00	24.4	265	1.600	[48]
990510	13 38 07	−80 29 49	≳ 28	23	1.619	Vreeswijk et al. GCN 324
990712	22 31 53	−73 24 29	21.78	N.A.	0.430	Galama et al. GCN 388
991208	16 33 54	+46 27 21	≳ 25	100	0.706	Dodonov et al. GCN 475
991216	05 09 31	+11 17 07	24.5	256	1.020	Vreeswijk et al. GCN 496

[a] References to redshift determination.
[b] V-band magnitude from HST. All others are R magnitude in the Johnson system.

further investigation.

III THE NATURE OF THE PROGENITORS

In almost all cases, a host galaxy has been identified at the location of the fading afterglow. GRB redshifts can be obtained either via absorption spectroscopy (when the transient is bright) or by emission spectroscopy of the host galaxy. In Figure 3 and Table 1 we summarize the measured redshifts and host galaxy magnitudes. While the distance scale debate is settled (at least for the class of long duration GRBs, see below) we remain relatively ignorant of the nature of the central engine. Currently popular GRB models fall into two categories: (i) the coalescence of compact objects (neutron stars, black holes and white dwarfs [13,54,61,63]) and (ii) the collapse of the central iron core of a massive star to a spinning black hole, a "collapsar" [85,57]. We now summarize the light shed on the progenitor problem by afterglow studies.

The Location of GRBs Within Hosts. A fundamental insight into the nature of SNe came from their location with respect to other objects within the host galaxy (specifically HII regions and spiral arms) and the morphology of the host galaxy itself (elliptical versus spiral). In a similar manner, we are now making progress in understanding GRB progenitors by measuring offsets with respect to other objects in the host galaxies. The rather good coincidence of GRBs with host galaxies already suggests that they are unlikely to be a halo population (as would be expected in the coalescence scenario [3]). On the other hand, with the possible

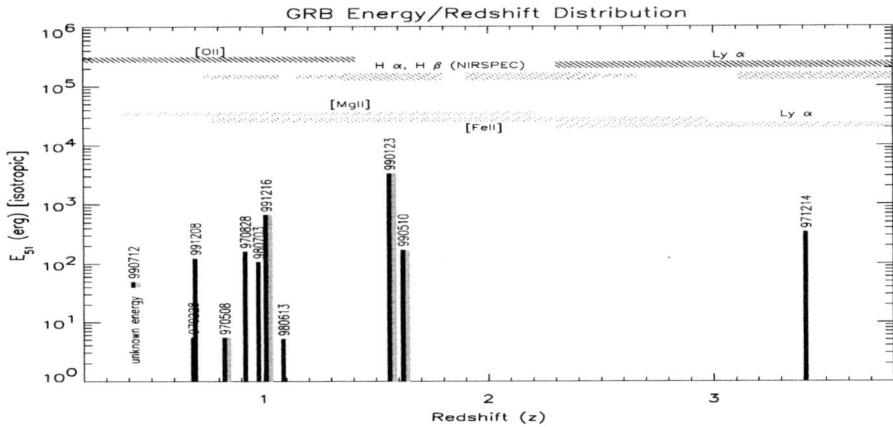

FIGURE 3. *The isotropic gamma-ray energy distribution of GRBs with confirmed redshifts. Bursts indicated in black are those with spectroscopically confirmed emission lines from the host galaxies; bursts indicated by a shaded column (e.g. 990123) are those with absorption line redshifts. The relevant key absorption or emission features are noted at the top of the figure.*

exception of GRB 970508 [66], they are clearly not associated with galactic nuclei (i.e. massive central black holes). Typical offsets of GRBs from the centroid of their host galaxies are comparable to the half-light radii of field galaxies at comparable magnitudes, suggesting that GRBs originate from stellar populations.

Host Galaxies. Demonstrating a direct link between GRBs and (massive) star formation is more difficult. On the whole, the population of identified hosts seems typical in comparison to field galaxies in the same redshift and magnitude range. The hosts have average luminosities for field galaxies, modulo corrections due to evolution. Their emission line fluxes and equivalent widths are also statistically indistinguishable from the normal field galaxy population. The observed star formation rates, derived from recombination line fluxes (mostly the [O II] 3727 Å line) and from the UV continuum flux range from less than $1\,M_\odot\,\text{yr}^{-1}$ to several tens of $M_\odot\,\text{yr}^{-1}$ – typical of normal galaxies at comparable redshifts (extinction corrections can increase these numbers by a factor of a few, but similar corrections apply to the comparison field galaxy population as well). It will probably be necessary to have a sample of several tens of GRB hosts before a correlation of GRBs with the (massive) star formation rate can be tested statistically. However, below we point to several specific examples which are suggestive of a link between GRBs and star-forming regions.

Association with Starforming Regions. There is evidence showing that GRBs arise from dusty regions within their host galaxies. In this respect, radio observations provide a unique tool for detecting events in regions of high ambient density (as was the case for GRB 980329). An even more extreme example is GRB 970828, where the host was identified based *solely* on the VLA discovery of a radio flare [12]. Interestingly enough, this is the dustiest galaxy in the sample of GRB hosts to-date.

Second, some GRBs appear to be located within identifiable star-forming regions. An example is GRB 990123 [4,28,41]. VLA observations of GRB 980703 [19] are perhaps more convincing. The radio observations can be sensibly interpreted by appealing to free-free absorption from a foreground HII region (which would dwarf the Orion complex). If this interpretation is correct then this would be strong evidence for a GRB being located within a starburst region.

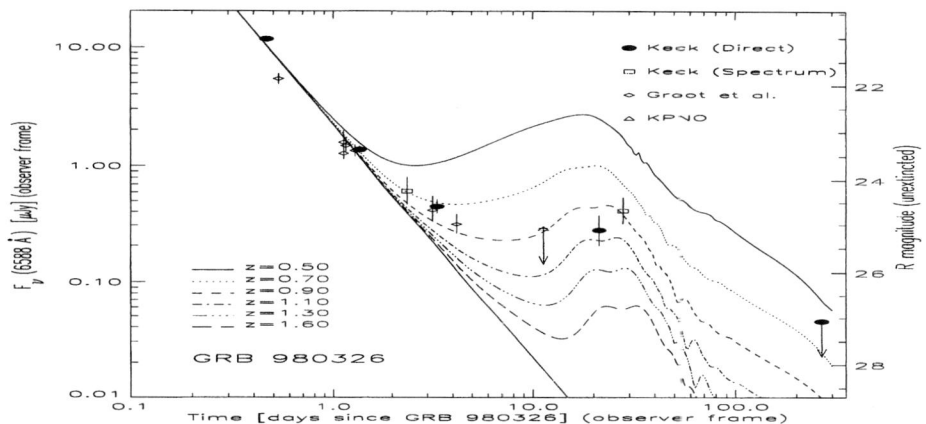

FIGURE 4. *R-band light curve of GRB 980326 and the sum of an initial power-law decay plus Ic supernova light curve for redshifts ranging from $z = 0.50$ to $z = 1.60$; from [5].*
Reprinted with permission from Nature [5], Copyright 1999 Macmillan Magazines Limited.

The GRB–SN link. If GRBs arise from the collapse of a massive star, it is an unavoidable consequence that emission from the underlying supernova should be superimposed on the afterglow. Bloom et al. [5] may have made the first detection of a possible SN component in the GRB 980326 lightcurve (Fig. 4). These authors noted that SNe, in contrast to afterglows, have distinctive temporal and spectral signatures: rising to a maximum at $\sim 20(1 + z)$ days, with little emission blueward of about 4000 Å in the restframe (and certainly blueward of 3000 Å) owing to a multitude of resonance absorption lines. This discovery has led to other possible SN detections, most notably GRB 970228 [31,68].

The suggestion of a GRB–SN connection is an intriguing one but it has yet to be placed on a firm footing. Important questions are: (i) are all long-duration

GRBs accompanied by SNe? (ii) if so, are these SNe of type Ib/c? Ground-based observations are possible in those cases where the afterglow decays rapidly (e.g. GRB 980326) or if high quality optical and IR observations exist (e.g. GRB 970228).

We need more examples to test the GRB–SN link. Future progress will depend on a combination of ground and HST observations. For relatively nearby GRBs especially those with a rapidly decaying optical afterglow it would be attractive and feasible to obtain the spectrum of the SN around the time when the flux from the SN peaks. A moderate quality spectrum with SN-like features would have the singular advantage of definitively confirming the SN interpretation (as opposed to alternatives involving re-radiation by dust [80]). However, for most GRBs, we expect HST observations to play a critical role. HST's widely recognized strengths in accurate photometry of sources embedded in galaxies [32] and photometric stability make the detection of a faint SN against the optical afterglow and the host galaxy possible.

Diversity of the Progenitor Population. As was the case with SNe, it is likely naive to think of a single progenitor population. Below, we discuss the two additional classes which show some promise: the mysterious short duration GRBs and a possible class of low luminosity GRBs associated with SNe.

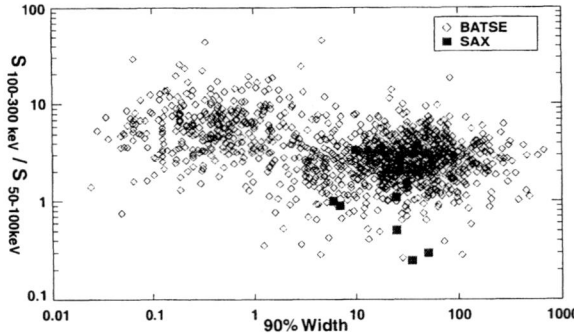

FIGURE 5. *Distribution of duration (T_{90}) vs. spectral hardness for BATSE bursts (diamonds) from the 4B catalogue. There is a clear suggestion of two groups of GRBs: short/hard and long/soft events. Events localized by BeppoSAX (solid squares) appear to belong to the long duration class.*

Short Events. It has been known for some time that the distribution of the duration of GRBs appears to be bimodal [14]; see Figure 5. Furthermore, these two groups may have different spatial distributions [45], with the short bursts being detected out to smaller limiting redshifts. However, we know very little about this class of GRBs since, as noted earlier, all bursts localized by BeppoSAX and RXTE thus far are of long duration (Figure 5). Fortunately, improvements in BeppoSAX and the imminent launch of HETE-2 provide for the first time the opportunity to follow-up

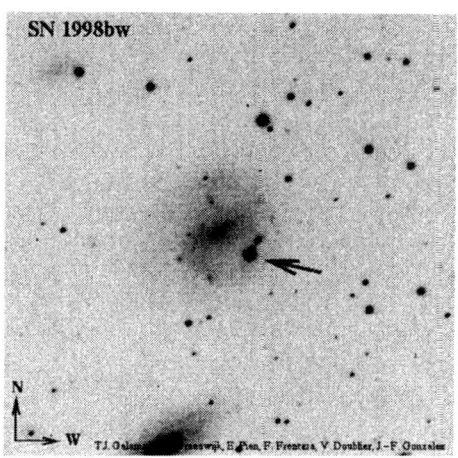

FIGURE 6. *Discovery image of SN 1998bw [30]. The SN is the bright object (marked with an arrow) SW of the nucleus. Relative to typical SNe, this SN is more energetic and appears to have synthesized ten times more Nickel.*

short GRBs.

The short duration bursts are difficult to accommodate in the collapsar model, given the long collapse time of the core. However, they find a natural explanation in the coalescence models. How would these bursts manifest themselves? Li & Paczyński [56] speculate that if the short-duration bursts result from NS–NS mergers then they may leave a bright, but short-lived ($\lesssim 1$ day) optical transient. Radio observations provide a complementary tool for determining the nature of the short duration bursts. The low ambient density would result in weak afterglows (since flux $\propto \rho^{1/2}$) which are potentially detectable. Radio observations have additional advantages of a longer lived afterglow, immunity from weather and freedom from the diurnal cycle.

Gamma-ray Bursts Associated with Supernovae. Observers and theorists alike have been intrigued by the possibility that the bright supernova, SN 1998bw, discovered by Galama et al. [30] in the error circle of GRB 980425 [67], is associated with the gamma-ray event (Figure 6). Kulkarni et al. [47] discovered that the SN had an extremely bright radio counterpart; see Figure 7. We noted that the inferred brightness temperature exceeded the inverse Compton catastrophe limit of 5×10^{11} K and to avoid rapid cooling we postulated the existence of a relativistically expanding blastwave ($\Gamma \gtrsim 2$). This relativistic shock is, of course, in addition to the usual sub-relativistic SN shock. This relativistic shock may have produced the GRB at early times. (We note here that we disagree with the much lower energy estimates of [81]; our recent calculations using the same assumptions as those made in [81] result in an energy estimate similar to that obtained earlier [47] from minimum-energy formulation). The optical modeling of the lightcurve and the spectra show that GRB 980425 was especially energetic [43,86] with an energy release of 3×10^{52} erg and Nickel production of nearly nearly a solar mass.

If GRB 980425 is associated with 1998bw, then this type of event is rare among the SAX localizations. GRB 980425 is most certainly not a typical GRB: the redshift of SN 1998bw is 0.0085 and the γ-ray energy release in GRB 980425 is at least four orders of magnitude less than in other cosmologically located GRBs. For this reason, most astronomers (especially those in the GRB field; see Wheeler's foray in experimental sociology [82]) do not believe the association between GRB 980425 and SN 1998bw. On the other hand, as evidenced by the intense interest in and modeling of the radio and optical data of SN 1998bw, this object is of considerable interest to the SN community. Indeed, we believe that the proposed GRB–SN association controversy has muddied the main issue: SN 1998bw is an interesting SN in its own right.

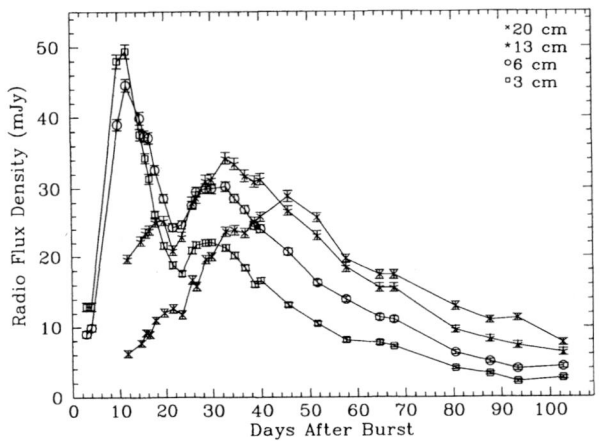

FIGURE 7. *The radio light curve of SN 1998bw at four wavelengths [47]. The peak brightness temperature of SN 1998bw at early times is 10^{13} K, well in excess of the inverse Compton limit of 5×10^{11} K, and can be best understood if the radio emission originates from a relativistic shock ($\Gamma \gtrsim 2$).* Reprinted with permission from Nature [47], Copyright 1998 Macmillan Magazines Limited.

What is the true distinguishing feature of SN 1998bw that may connect it to a GRB event? Is it the large energy release, as suggested by several authors [43,33]? We argue that in fact it is the energy *coupled into relativistic ejecta* that most closely connects SN 1998bw to a GRB. In a typical SN, about 10^{51} erg is coupled to the envelope of the star (a small fraction of the total SN energy release of 10^{53} erg). In a GRB, a similar amount of energy (10^{51}–10^{52} erg depending on the event) is coupled to a much smaller ejecta mass, resulting in relativistic outflow. For SN 1998bw, applying the minimum energy formulation to the radio observations we infer the relativistic shell to contain $\sim 10^{50}$ erg. Not only is this uncharacteristic of a typical SN (there exists no evidence for relativistic ejecta in ordinary SN), but it is not dissimilar from the energy implied for GRB outflows. One could therefore envisage a continuum of physical phenomenon between SN 1998bw and cosmological GRBs provided we use the energy in the relativistic ejecta as the basic underlying parameter and not the isotropic gamma-ray release.

IV AFTERGLOW: THE PHYSICS AND ENERGETICS OF THE FIREBALL

One can consider a GRB to be like a SN explosion with a central source releasing energy E_0 (comparable to the mechanical release of energy in an SN). This is the so-called fireball model. The difference between an SN and a GRB is primarily in ejecta mass: 1–10 $M_\odot$ for SNe whereas only 10^{-5} $M_\odot$ for GRBs. The evolution of a GRB is much faster than that of a SN due to two factors: the ejecta expand relativistically and, thanks to the smaller ejecta mass, the optical depth is considerably smaller.

As the ejecta encounter ambient gas, two shocks are produced: a short-lived reverse shock (traveling through the ejecta) and a long-lived forward shock (propagating into the swept-up ambient gas). Afterglow emission is identified with emission from the forward shock. In order to obtain significant afterglow emission, several conditions are necessary. (1) Rapid equipartition of electrons with the shocked protons (which hold most of the energy). (2) Acceleration of electrons to a power law spectrum (particle Lorentz factor distribution, $dN/d\gamma \propto \gamma^{-p}$). (3) Rapid growth of the magnetic field with energy density in the range of 10^{-2} of that of the protons. Under these circumstances, afterglow emission is dominated by synchrotron emission of the accelerated particles; see [71,79]. The weakness of this model is the assumption of growth in the magnetic field strength to the high values noted above (R. Blandford, pers. comm.).

The theoretically expected afterglow spectrum is shown in Figure 8. Three key frequencies can be identified: ν_a, the synchrotron self-absorption frequency; ν_m, the frequency of the electron with a minimum Lorentz factor (corresponding to the thermal energy behind the shock) and ν_c, the cooling frequency. Electrons which radiate above ν_c cool on timescales equal to the age of the shock. The evolution of these three frequencies is determined by the hydrodynamical evolution of the shock which in turn is affected by two principal factors: the environment of the GRB and the geometry of the explosion.

The GRB environment. The earliest afterglow models made the simplifying assumption of expansion into a constant density medium. This is an appropriate assumption should the GRB progenitor explode into a typical location of the host galaxy. However, there is increasing evidence tying GRBs to massive stars (see §III). It is well known that massive stars lose matter throughout their lifetime and thus one expects the circumburst medium to exhibit a density profile, $\rho \propto r^{-2}$ where r is the distance from the progenitor. Chevalier & Li [8] refer to these two models as the ISM (interstellar medium) and the wind model respectively. As can

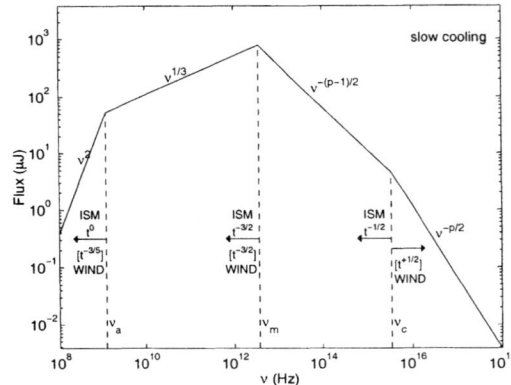

FIGURE 8. *Broad-band spectrum (f_ν) of the afterglow from a spherical fireball with constant density ("ISM" model; see text) and $\rho \propto r^{-2}$ medium ("wind" model; see text). This is representative of the observed spectrum few days after the burst. Note the distinct evolution of ν_a and ν_c in the two models.*

be seen from Figure 8 these two models give rise to rather different evolution of the three critical frequencies.

Geometry: Jets versus Spheres. The hydrodynamics is also affected by the geometry of the explosion. Many powerful astrophysical sources have jet-like structure. There is evidence (from polarization observations) indicating asymmetric expansion in SNe [82], so it is only reasonable to assume that GRB afterglows also have jet-like geometry as well. A clear determination of the geometry is essential in order to infer the true energy of the explosion. This is especially important for energetic bursts such as GRB 990123 whose isotropic energy release approaches $M_\odot c^2$.

Let the opening angle of the jet be θ_0. As long as the bulk Lorentz factor, Γ, is larger than θ_0^{-1}, the evolution of the jet is exactly the same as that of a sphere (for an observer situated on the jet axis). However, once Γ falls below θ_0^{-1} then two effects become important. First, for a well defined jet, the on-axis observer sees an edge and thus one expects to see a break in the afterglow emission. Second, the lateral expansion of the jet (due to heated and shocked particles) will start affecting the hydrodynamical explosion.

Wind or ISM? The two key diagnostics to distinguish these two models are the evolution of the cooling frequency (see Figure 8) and the early behavior of the radio emission. In the wind model, the radio emission rises rapidly (relative to the ISM model) and the synchrotron self-absorption frequency falls rapidly with time. Both these result from the fact that the ambient density decreases with radius (and hence in time) in the wind model.

Unfortunately, in general, the current data are not of sufficient quality to firmly distinguish the two models. For example in GRB 980519, the same optical and

X-ray data appear to be adequately explained by the jet+ISM model [73] and the sphere+wind model [8]. Including the radio data tips the balance, but only slightly, in favor of the wind model [23]. In our opinion, the best example for the wind model is that of GRB 980329 [20]; see Figure 9. This afterglow exhibits the two unique signatures of the wind model: high ν_a and a rapid rise. Given the importance of making the distinction between the wind and the ISM model we urge early wide band radio observations (especially at high frequencies).

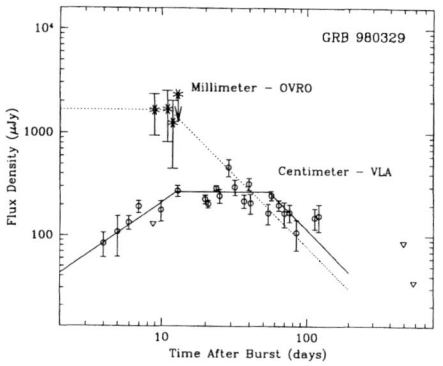
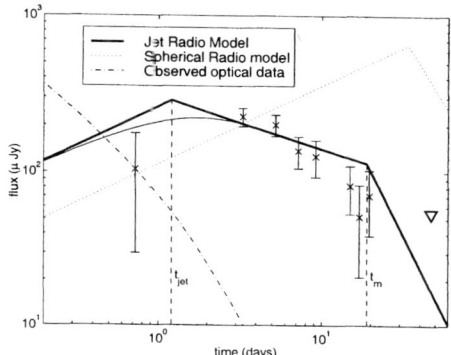

FIGURE 9. Left: Radio afterglow of GRB 980329 [20]. The rapid rise of the centimeter flux and the high absorption frequency (signified by the considerable strength of the millimeter emission) offer good support for GRB 980329 expanding into a circumburst medium with density falling as inverse square distance. The lines represents a wind model based on X-ray, optical, IR, mm and cm data. Right: Observed and model radio light curves of GRB 990510 [39]. The model predictions for the radio afterglow emission are displayed by the solid line (jet fireball model) and dotted line (spherical fireball model). The observed optical afterglow emission is displayed by the dotted-dashed line; see text for more details.

Energetics. Of all the physical parameters of the fireball, the most eagerly sought parameter is the total energy E_0. By analogy with supernovae, it is E_0 which sets the GRB phenomenon apart from other astrophysical phenomena. Classes of GRBs may eventually be distinguished and ranked by their energy budget; for example, long-duration events, short duration events and supernova-GRBs (see §III).

One approach has been to use the isotropic γ-ray energy as a measure of E_0; see Figure 3. There are three well known problems with such estimates. First, collimation of the ejecta (jets) will result in overestimation of the total energy release. For GRB 990510 where a good case for a jet has been established (Figure 9), the standard isotropic energy estimate is probably a factor of 300 more than the true energy [39]. Second, even after accounting for a possible jet geometry, the efficiency of converting the shock energy into gamma-ray emission is very uncertain. For example, some authors [51] advocate low efficiency ($\sim 1\%$) which would result in an enormous upward correction to the usual isotropic estimates. Third, the bulk Lorentz factor is extremely high during the emission of γ-rays and thus the estimates critically depend on assumption of the geometry and granularity [52] of the emitting region. In particular, if the emission is from small blobs [52] then the inferred estimates are grossly in error.

In contrast to this highly uncertain situation, afterglows offer (in principle) more robust methods to evaluate E_0. In view of the importance of determining E_0 we summarize the different methods of determining E_0 from afterglow observations. One approach is to fit a "snapshot" broad-band afterglow spectrum (from radio to X-rays) to an afterglow model; this approach was pioneered by Wijers & Galama [83]. The strength of this method is that the estimated E_0 is, in principle, robust. Specifically, the estimate does not depend on the usually unknown environmental factors (run of density). However, in practice, this method is very sensitive to the values of the critical frequencies (Figure 8) which are usually not well determined. This difficulty explains the wildly differing estimates of E_0 for GRB 970508 [83,35]. Furthermore, this method uses measurements obtained at early times (when the afterglow at high frequencies is bright) with the result that the true source geometry is hidden by relativistic beaming.

A second approach is to model the light curves of the afterglow in a given band, specifically a radio band. The advantages of this method are the photometric stability of radio interferometers and the low Lorentz factor at the epoch of the peak of the radio emission. The disadvantages are two-fold: the sensitivity to the environmental parameters (density) and the assumption of the constancy of the microphysics parameters (electron and magnetic field equipartition factors). Application of this approach to GRB 980703 has resulted in seemingly accurate measures of the fireball parameters [19].

Freedman & Waxman [24] take yet another approach, and estimate the energy release from late time X-ray observations. They show that the X-ray flux is insensitive to the GRB environment, and obtain robust estimates of the fireball energy per unit solid angle: from 3×10^{51} erg to 3×10^{53} erg.

With all the above approaches, however, the possible collimation of the ejecta in jets is still a major uncertainty. This can be addressed by observing the evolution of the afterglow as the "edge" of the jet becomes visible. In most cases no evidence for jets has been seen, with the notable exceptions of GRB 990510 and possibly GRB 990123. In addition, a variety of statistical arguments (the absence of copious numbers of "orphan afterglows") [37,36,70] suggests that, on average, the collimation cannot be extreme, and that for most bursts the opening angle is not less than 0.1 radian. Thus the total energy for most bursts may be reduced to the range of 10^{50} erg to 3×10^{51} erg, but could easily be much higher in at least some cases.

Possibly the best approach to determining the energetics, which minimizes uncertainties due both to collimation (jets) and to the environment is to model the afterglow after it becomes non-relativistic. This method builds on the well established minimum energy formulation and the self-similarity of the Sedov solution.

Not only are the ejecta truly non-relativistic, but they are also essentially spherical, as by this time jets will have had sufficient time to have undergone significant lateral expansion. Indeed, we can justifiably call this "fireball calorimetry" [21]. Applying this technique to the long-lived afterglow of GRB 970508 (Figure 1) led to the surprising result that $E_0 \sim 5 \times 10^{50}$ erg – weaker than a standard SN! This is an astonishing result. If true, this result would suggest that it is not E_0 which is the prime distinction between GRBs and SNe but the ejecta mass. However, Chevalier & Li [9] interpret the same data in the wind framework and derive much larger E_0. Clearly, we need more well studied afterglows with sufficient observations to first distinguish the circumburst environment (wind versus ISM) and then radio observations over a sufficiently long baseline to undertake calorimetry. Nonetheless, one should bear in mind that the current evidence for large energy release in GRBs is not as strong as is usually assumed.

V   EPILOGUE AND FUTURE

Clearly, the GRB field is evolving rapidly. Along what direction[s] will this field proceed in the coming years? One way to anticipate the future is by considering analogies from the past.

In §III we already discussed the parallels between the SN field and the GRB field. Here we discuss the numerous parallels with quasar astronomy. First discovered at radio wavelengths, we now study quasars across the electromagnetic spectrum. Although still identified by their gamma-ray properties, we now recognize the tremendous value of pan-chromatic GRB and afterglow studies. In both cases, there was considerable controversy about the distance scale. However, once this issue was settled, it became clear that quasars are the most energetic objects (sustained power) whereas GRBs are the most brilliant. For both, the ultimate energy appears to be related to black holes (albeit of different masses).

The raging issues in GRB astronomy today are the same that fueled quasar studies in the 60's: the spatial distribution, the extraction of energy from the central engine, the transfer of energy from stellar scales to parsec scales, and the geometry of the relativistic outflow (sphere or jet). Astronomers took decades to unify the seemingly diverse types of quasars, and to conclude that there are two types of central engines: radio loud and radio quiet. Likewise, there may well be two types of GRB engines: rapidly and slowly spinning black holes emerging respectively from collapse of a rotating core of a massive star or coalescence of compact objects and the collapse of a massive star. This picture could potentially explain both the cosmologically located GRBs and SN 1998bw. Finally, we can project that in the future, GRBs may be used to probe distant galaxies, just as quasars are used today to study the IGM.

There is a feeling in the astronomical community (outside the GRB community) that the GRB problem is "solved". The truth is that the GRB problem is now getting defined! We now summarize our view of the major issues and anticipated near term advances. In our opinion the major issues are Diversity, Progenitors and Energy Generation.

As discussed earlier, high energy observations suggest the existence of two classes: short and long duration bursts. It is possible that afterglow observations may demarcate additional classes. If so, one can contemplate that within a year (assuming abundant localizations by HETE-2) that we will have new GRB designations such as sGRBs (GRBs with late time bump indicative of an underlying SN), wGRBs (GRBs whose afterglow clearly indicates a wind circumburst medium shaped by stellar winds), iGRBs (GRBs which explode in the interstellar medium) and so on.

The broad indications are that GRBs are associated with stars and most likely massive stars. However, we know little beyond this. Comparing the unbeamed GRB event rate of 1.8×10^{-10} yr^{-1} Mpc^{-3} [74] with 3×10^{-5} Type Ibc SN yr^{-1} Mpc^{-3} and 10^{-6} yr^{-1} NS–NS merger Mpc^{-3} [53] shows that GRBs events are extremely rare; here we note that the present data do not support a collimation correction in excess of 100. It will be quite some time before we will be in a position to identify the conditions necessary for a star to die as a GRB.

It is our opinion that SN 1998bw is a major development in the field of stellar collapse. The association (or lack) with GRB 980425 unfortunately has distracted our attention of this important development. The existence of a significant amount of mildly relativistic material, $\sim 10^{50}$ erg [47], is fascinating and it is ironic that none of the models can account for this inferred value whereas most of the theoretical effort has gone into explaining the gamma-ray burst itself (especially considering the uncertain association of GRB 980425 with SN 1998bw). Clearly, SN 1998bw is a rare event but we are convinced that more such events will be found and accordingly have mounted a major campaign to identify these SNe. The robust signatures of this class are high T_B and prompt X-ray emission since these are necessary consequences of a relativistic ejecta. We note that if these future events are as bright as SN 1998bw then the energy in the relativistic ejecta can be directly measured by VLBI observations of the expanding radio shell.

It is vitally important to make quantitative progress in determining the energy release in GRBs. As discussed in §IV, firm estimates of the energy release require well sampled broad-band data at early times and densely sampled radio light curves out to late times. This will require a *coordinated* approach and necessarily involve many observatories around the world and in space. The same datasets will also help us understand a profound puzzle: if GRBs indeed arise from the death of massive stars then why do we not see signatures for a circumburst medium shaped by stellar winds in *all* long duration GRBs? Even ardent supporters of the wind model [8,9] concede that some GRBs (e.g. GRB 990123, 990510) are due to a jet expanding into a constant density medium.

We now discuss the anticipated returns. True to our tradition as observers, we order the discussion by wavelength regimes!

Radio Observations: Dusty galaxies, Circumstellar Edges and Reverse Shocks. Perhaps the most exciting use of radio afterglow is in identifying dusty star-forming host galaxies. Such host galaxies are not readily seen at optical wavelengths. Currently, such galaxies are eagerly sought and studied at sub-millimeter wavelengths. However, the sensitivity and localization of such galaxies by sub-millimeter telescopes is poor. In contrast, GRB host galaxies are identified at the sub-arcsecond level. The present radio afterglow detection rate of 40% already places an upper limit on the amount of star-formation in dusty regions, viz. this rate is not larger than that measured from optical observations. This result is entirely independent of the conclusion based on studies in the sub-millimeter regime, or the diffuse cosmic FIR background found in the COBE data. However, the result does rely on two assumptions: (i) GRBs trace star formation and (ii) the GRB explosion and its aftermath does not radically alter the ambient medium (i.e., with a prompt and complete destruction of dust grains along the line of sight)

Radio observations of SNe offer a probe of the distribution of the circumstellar matter. A spectacular example is SN 1980K whose radio flux dropped 14 yrs after the explosion [58]. A progenitor star which suffered mass loss with variation in the wind speed could explain the observations. Indeed, one *expects* significant radial structure in the circumburst medium as the progenitor evolves from a blue star to a red supergiant and thence to possibly a blue supergiant etc. If GRBs come from binary stars which undergo a phase of common envelope envolution [7] then the structure would be even more complicated. Thus radio observations have the potential (in fortunate circumstances) to give us insight into the mass loss history of the progenitor star[s].

The prompt optical emission from GRB 990123 [1] has been interpreted to arise from the reverse shock [72]. Far less discussed is the prompt radio emission – a radio flare – also seen from this burst [49]. Sari & Piran [72] suggest that the radio emission also originates from the reverse shock as the electrons cool. Observations related to the reverse shocks are important since it is only through these observations that we have a chance of studying the elusive ejecta. We now have four such examples of radio flares [50] and this represents an order of magnitude better success rate than ROTSE+LOTIS. We urge theorists to pay attention to these new findings. More to the point, radio observations appear to be fruitful for the study of reverse shocks, especially when combined with observations of the prompt optical emission. This bodes well for the coming years given the efforts underway to increase the sensitivity of ROTSE [1].

X-ray Observations: Diversity & Progenitors. GINGA identified a number of X-ray rich GRBs. BeppoSAX has found several such examples with some bursts lacking significant gamma-ray emission – the so-called X-ray flashes [40]. We know very little about these X-ray transients. Could they be GRBs in a very dense environment (with red giant progenitors)? We need to take such transients more seriously and intensively followup on such bursts.

Another interesting finding from *GINGA* was the discovery of precursor soft X-ray emission [62]. There is no simple explanation for this phenomenon in the current internal-external shock model. We suggest that the soft X-ray emission precursor is similar to the UV breakout of ordinary SNe. This hypothesis can be confirmed or rejected by obtaining the redshift to such bursts.

The X-ray rich GRB 981226 [26,18] was marked with two additional peculiarities: a precursor emission and afterglow emission which is seemingly undetectable after about 12 hours but then rises rapidly before commencing decay. Above we alluded to the fact that massive stars do not have a single phase of mass loss but instead have a veritable history of mass loss (from birth to death). The X-ray observations of GRB 981226 could be accounted for in a model in which the progenitor has first a red supergiant wind followed by a blue supergiant wind.

Optical Observations: SN link, Short bursts & Geometry. The GRB–SN connection is best probed by optical observations. The value of optical observations has already been demonstrated by the current observations of GRB 980326 and 970228. Clearly, more observations are needed to establish this link. Once this link is established then one can undertake detailed spectroscopic studies of the SN with large ground-based telescopes and photometric studies with HST.

Offsets of GRBs and the morphology of the host galaxies will continue to be of great interest. Such observations will help us differentiate whether some GRBs come from nuclear regions or always from star-forming regions. Under the current paradigm, the discovery of GRBs coincident with elliptical galaxies would be a major surprise. On the other hand, one expects short bursts to arise in the halo of their galaxies and thus in this case no coincidence is expected. We expect HETE-2 to contribute significantly to these issues. Finally, polarization measurements offer a very convenient way to probe the geometry of the emitting region as has already been demonstrated from the discovery of polarization in GRB 990510 (e.g. [55,84]).

Acknowledgments. Our research is supported by NASA and NSF. JSB holds a Fannie & John Hertz Foundation Fellowship, AD holds a Millikan Postdoctoral Fellowship in Experimental Physics, TJG holds a Fairchild Foundation Postdoctoral Fellowship in Observational Astronomy and RS holds Fairchild Foundation Senior Fellowship in Theoretical Astrophysics. The VLA is a facility of the National Science Foundation operated under cooperative agreement by Associated Universities, Inc. The W. M. Keck Observatory is operated by the California Association for Research in Astronomy, a scientific partnership among California Institute of Technology, the University of California and the National Aeronautics and Space Administration. It was made possible by the generous financial support of the W. M. Keck Foundation.

REFERENCES

1. Akerlof, C., Balsano, R., Barthelmy, S. et al., *Nature* **398**, 400 (1999).
2. Bloom, J. S., Djorgovski, S. G., Kulkarni, S. R. & Frail, D. A., *ApJ* **508**, L17 (1998).
3. Bloom, J. S., Sigurdsson, S. & Pols, O. R., *MNRAS* **305**, 763 (1999).
4. Bloom, J. S. et al., *ApJ* **518**, L1, (1999).
5. Bloom, J. S. et al., *Nature* **401**, 453 (1999).
6. Boella, G. et al., *A.&A. Suppl. Ser.* **122**, 298 (1997).

7. Brown, G. E., Lee, C. -H., Lee, H. K. & Bethe, H. A., astro-ph/9911458 (1999).
8. Chevalier, R. A. & Li, Z.-Y., *ApJ* **520**, L29, (1999).
9. Chevalier, R. A. & Li, Z. -Y., astro-ph/9908272 (1999).
10. Costa, E. et al., *Nature* **387**, 783 (1997).
11. Djorgovski, S. G., Kulkarni, S. R., Bloom, J. S., Goodrich, R., Frail, D. A., Piro, L & Palazzi, E., *ApJ* **508**, L17 (1998).
12. Djorgovski, S. G. et. al., in preparation, (2000).
13. Eichler, D., Livio, M., Piran, T., & Schramm, D. N., *Nature* **340**, 126 (1989).
14. Fishman, G. J. & Meegan, C. A., *Annu. Rev. Astron. Astrophys.* **33**, 415 (1995).
15. Frail, D. A. & Kulkarni, S. R., *Astrophys. Space Sci.* **231**, 277 (1995).
16. Frail, D. A., Kulkarni, S. R., Nicastro, S. R., Feroci, M., & Taylor, G. B., *Nature* **389**, 261 (1997).
17. Frail, D. A., Kulkarni, S. R., Shepherd, D. S., & Waxman, E., *ApJ* **502**, L119, (1998).
18. Frail, D. A., et al., *ApJ* **525**, L81 (1999).
19. Frail, D. A., Bloom, J. S., Kulkarni, S. R., Sari, R., & Taylor, G. B., in preparation, (2000).
20. Frail, D., Kulkarni, S., Sari, R., Taylor, G., Shepherd, D., Bloom, J., Young, C., Nicastro, L., & Masetti, N., *ApJ in press*, (1999).
21. Frail, D., Waxman, E. & Kulkarni, S. R., *ApJ in press*, (2000).
22. Frail, D. A., Kulkarni, S. R., Wieringa, M. H. et al., astro-ph/9912171, (1999).
23. Frail, D. A., Kulkarni, S. R., Sari, R. et al., astro-ph/9910060, (2000).
24. Freedman, D. L. & Waxman, E., astro-ph/9912214 (1999).
25. Frontera, F., Amati, L., Costa, E. et al., astro-ph/9911228, (1999).
26. Frontera, F., Antonelli, L. A., Amati, L. et al., *ApJ*, submitted (2000).
27. Fruchter, A. S., *ApJ* **512**, L1 (1999).
28. Fruchter, A. S., Thorsett, S., Metzger, M. R., *ApJ* **519**, L13 (1999).
29. Fruchter, A. S., Pian, E., Thorsett, S. E. et al., *ApJ* **516**, 683 (1999).
30. Galama, T. J. et al., *Nature* **395**, 670, (1998).
31. Galama, T. J. et al., *ApJ* in press, (1999).
32. Garnavich, P. M. et al., *ApJ* **493**, L53 (1998).
33. Germany, L., Reiss, D. J., Sadler, E. M., Schmidt, B. P & Stubbs, C. W., astro-ph/9906096 (1999).
34. Gorosabel, J., Castro-Tirado, A. J., Pedrosa, A. et al., *A.&A.* **347**, L31 (1999).
35. Granot, J., Piran, T. & Sari, R., *ApJ* **527**, 236 (1999).
36. Greiner, J., et al., *A& AS* **138**, 441 (1999)
37. Grindlay, J. E., *ApJ* **510**, 710 (1999)
38. Groot, P. J., Galama, T. J., van Paradijs, J. et al., *ApJ* **493**, L27 (1998).
39. Harrison, F. A., et al., *ApJ* **523**, L121 (1999).
40. Heise, J., talk at the 5th Hunstville GRB conference, (1999).
41. Holland, S. & Hjorth, J., *A&A* **344**, L67 (1999).
42. In't Zand, J. J. M., Amati, L., Antonelli, L. A. et al., *ApJ* **505**, 1191 (1998).
43. Iwamoto, K. et al., *Nature* **395**, 672, (1998).
44. Katz, J. I., *ApJ* **422**, 248 (1993).
45. Katz, J. I. & Canel, L. M., *ApJ* **471**, 915 (1996).

46. Kulkarni, S. R., Djorgovski, S. G., Ramaprakash, A. N. et al., *Nature* **393**, 35 (1998).
47. Kulkarni, S. R., Frail, D. A., Wieringa, M. H. et al., *Nature* **395**, 663 (1998).
48. Kulkarni, S. R., Djorgovski, S. G., Odewahn, S. C. et al., *Nature* **398**, 389 (1999).
49. Kulkarni, S. R. et al., *ApJ* **522**, L97 (1999).
50. Kulkarni, S. R. & Frail, D. A., in preparation, (2000).
51. Kumar, P., astro-ph/9907096 (1999).
52. Kumar, P. & Piran, T., astro-ph/9909014 (1999).
53. Lamb, D. Q., astro-ph/9909026 (1999).
54. Lattimer, J. M. & Schramm, D. N., *ApJ* **192**, L145 (1974).
55. Lazzati, D., Covino, S. & Ghisellini, G., astro-ph/9912247 (1999).
56. Li, L.-X. & Paczynski, B., *ApJ* **507**, L59, (1998).
57. Macfadyen, A. I. & Woosley, S. E., *ApJ* **524**, 262, (1999).
58. Montes, M. J., van Dyk, S. D., Weiler, K. W., Sramek, R. A. & Panagia, N., *ApJ* **506**, 874, (1998).
59. Mészáros, P. & Rees, M. J., *ApJ* **476**, 232 (1997).
60. Metzger, M. R., Djorgovski, S. G., Kulkarni, S. R., Steidel, C. C., Adelberger, K. L., Frail, D. A., Costa, E., & Frontera, F., *Nature* **387**, 879 (1997).
61. Mochkovitch, R., Hernanz, M., Isern, J., & Martin, X., *Nature* **361**, 236–238, (1993).
62. Murakami, T. et al., *Nature* **350**, 592 (1991).
63. Narayan, R., Paczyński, B., & Piran, T., *ApJ* **395**, L83 (1992).
64. Palazzi, E. et al., *A.&A.* **336**, L95 (1998).
65. Paczyński, B. & Rhoads, J., *ApJ* **418**, L5 (1993).
66. Pian, E., Fruchter, A. S., Bergeron, L. E. et al., *ApJ* **492**, L103 (1999).
67. Pian, E. et al., *A&AS* **138**, 463 (1999).
68. Reichart, D. E., *ApJ* **521**, L111 (1999).
69. Reichart, D. E., Lamb, D. Q., Metzger, M. R. et al., *ApJ* **517**, 692 (1999).
70. Rhoads, J. E., *ApJ* **487**, L1 (1997)
71. Sari, R., Piran, T. & Narayan, R., *ApJ* **497**, L17 (1998).
72. Sari, R. & Piran, T., *ApJ* **517**, L109 (1999).
73. Sari, R., Piran, T., & Halpern, J. P., *ApJ* **519**, L17 (1999).
74. Schmidt, M., *ApJ* **523**, L117 (1999).
75. Strohmeyer, T. E., Fenimore, E.E., Murakami, T. & Yoshida, A., *ApJ* **500**, 873 (1998).
76. Taylor, G. B., Frail, D. A., Kulkarni, S. R. et al., *ApJ* **502**, L115 (1998).
77. van Paradijs, J., Groot, P. J., Galama, T. et al., *Nature* **368**, 686 (1997).
78. Vietri, M., *ApJ* **488**, L105 (1997).
79. Waxman, E., *ApJ* **489**, L33 (1997).
80. Waxman, E., & Draine, B. T., astro-ph/9909020 (1999).
81. Waxman, E. & Loeb, A., *ApJ* **515**, 721 (1999).
82. Wheeler, J. C., astro-ph/9912403, (1999).
83. Wijers, R. A. M. J. & Galama, T. J., *ApJ* **523**, 177, (1999)
84. Wijers, R. A. M. J., Vreeswijk, P. M., Galama, T. J. et al., astro-ph/9906346 (1999).
85. Woosley, S. E., *ApJ* **405**, 273, (1993).
86. Woosley, S. E., Eastman, R. G. & Schmidt, B. P., *ApJ* **516**, 788 (1999).

Prompt GRB Optical Follow-up Experiments

H. S. Park[1], G. Williams[8], E. Ables[1], D. Band[5], S. Barthelmy[3], R. Bionta[1], T. Cline[3], N. Gehrels[3], D. Hartmann[2], K. Hurley[6], M. Kippen[4], R. Nemiroff[7], W. Pereira[7], R. Porrata[1]

[1] *Lawrence Livermore National Laboratory, Livermore CA 94550*
[2] *Dept. of Physics and Astronomy, Clemson University, Clemson, SC 29634-1911*
[3] *NASA/Goddard Space Flight Center, Greenbelt, MD 20771*
[4] *NASA/Marshall Space Flight Center, Huntsville, AL 35812*
[5] *Los Alamos National Laboratory, Los Alamos, NM 87545*
[6] *Space Sciences Laboratory, University of California, Berkeley, CA 94720-7450*
[7] *Dept. of Physics, Michigan Technological University, Houghton, MI 49931*
[8] *University of Arizona, Tucson, AZ 85721*

Abstract. Gamma Ray Bursts (GRBs) are brief, randomly located, releases of gamma-ray energy from unknown celestial sources that occur almost daily. The study of GRBs has undergone a revolution in the past three years due to an international effort of follow-up observations of coordinates provided by Beppo/SAX and IPN GRB. These follow-up observations have shown that GRBs are at cosmological distances and interact with surrounding material as described by the fireball model. However, prompt optical counterparts have only been seen in one case and are therefore very rare or much dimmer than the sensitivity of the current instruments. Unlike later time afterglows, prompt optical measurements would provide information on the GRB progenitor. LOTIS is the very first automated and dedicated telescope system that actively utilizes the GRB Coordinates Network (GCN) and it attempts to measure simultaneous optical light curve associated with GRBs. After 3 years of running, LOTIS has responded to 75 GRB triggers. The lack of any optical signal in any of the LOTIS images places numerical limits on the surrounding matter density, and other physical parameters in the environment of the GRB progenitor. This paper presents LOTIS results and describes other prompt GRB follow-up experiments including the Super-LOTIS at Kitt Peak in Arizona.

INTRODUCTION

The dramatic breakthrough in our understanding of GRBs occurred when the high resolution x-ray detector on the Beppo/SAX satellite was able to determine the position of a GRB with sufficient accuracy [1] to enable a large telescope to observe a faint, fading afterglow days later. Optical [2] and radio afterglows now have been observed for about a dozen GRBs during the last two years. These long-lasting but faint afterglows have been successfully explained in the fireball models as the result of the heating up of surrounding material by the GRB energy release [3]. From the spectra of the apparent hosts of the afterglows we now know that the GRBs are at cosmological distances and also have gained some understanding of the GRB energy output, ambient environment, and dynamics.

The goal of the LOTIS experiment is to measure prompt visible emission occurring within seconds of the gamma-ray energy release and presumably containing information about the GRB progenitor. To accomplish this we developed and have been operating an automated wide field-of-view telescope for over 3 years at Lawrence Livermore National Laboratory (LLNL) that responds to triggers distributed by the Gamma-ray burst Coordinate Distribution Network (GCN) by rapidly imaging the GRB coordinate error boxes. This instrument collected data on over 75 GRBs and in all cases found no evidence for prompt visible emission down to its sensitivity limit.

During the onset of the rainy season in Livermore, a similar competing instrument, ROTSE, at Los Alamos, NM, responded to a GCN trigger from GRB990123 whose GRB error box was later localized by the Beppo/SAX detection of a fading X-ray source which led to the later detection of optical afterglows [4]. Subsequent analysis of the ROTSE images showed a prompt optical transient at mag$\sim$9 [5]. A signal this bright would clearly have been seen by LOTIS, however to this date prompt optical signals have not been seen from any other GRB in the large LOTIS sample. This burst was a very strong event having a peak gamma-ray flux of 16.4 γ cm^{-2} s^{-1} and a total fluence of 2.0×10^{-4} erg cm^{-2}. In addition, GRB990123 exhibited an extremely hard gamma-ray spectrum and the prompt optical flux measurement is not consistent with an extrapolation of the burst spectrum to low energies [6]. With only this single case of bright prompt optical emission, it is difficult to determine which burst characteristics might predict such a flash. We attempt to compare the LOTIS limits with the observed prompt signal from GRB990123 to constrain the physical parameters of the reverse-shock model suggested by Sari & Piran [7].

LOTIS

LOTIS was constructed to rapidly cover the real-time GCN trigger error box, which was limited by the BATSE 1σ error of $2 \sim 10°$. This large error requires wide field-of-view optics to obtain statistically meaningful results. LOTIS utilizes commercially available Canon f/1.8 telephoto lenses with 200mm focal lengths yielding effective apertures of 110mm diameter. The electronic focal plane sensors are 2048 × 2048 pixel Loral 442A CCDs with 15 × 15μm pixels driven by custom read-out electronics. Each Canon telephoto lens/camera assembly has a field-of-view of 8.8 × 8.8° with a pixel scale of 15″. Four cameras are arranged in a 2 × 2 array to cover a total field-of-view of 17.4 × 17.4° overlapping 0.2° in each dimen-

 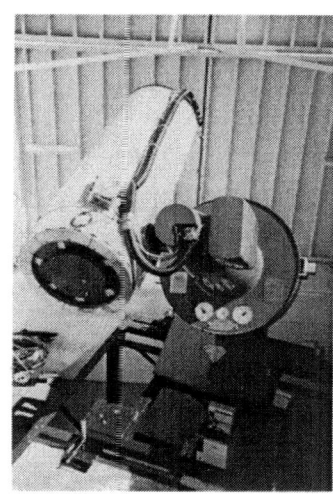

FIGURE 1. Lotis and Super-LOTIS telescopes.

sion. LOTIS is located 25 miles east of Livermore, CA. The first picture of Figure 1 shows the LOTIS system.

LOTIS RESULTS & INTERPRETATION

During more than 1330 possible nights of observations (Oct. 1996 through May 2000), LOTIS responded to 145 GCN triggers. Of these, 75 triggers were unique GRB events. The quality of the LOTIS coverage for a given event depended on: the observing conditions, the LOTIS response time, the error in the rapidly distributed GRB coordinate, the size of the final error box, and the duration of the GRB. The following table lists the 13 events for which LOTIS achieved the best overall coverage. These results have been published [8], [9], [10].

Table 1 shows our best events during this period. Since LOTIS did not see any optical activity in the visible, correlated with any of these events, we determine from the conditions in each case an upper limit on the brightness of any possible visible signal missed by LOTIS.

This set of LOTIS upper limits has been used to determine if typical GRBs, for which LOTIS has shown not to have counterparts, are similar to GRB990123, the only GRB event ever observed to have a prompt visible counterpart. Since GRB990123 also had an unusually strong gamma-ray emission, it is interesting to compare the LOTIS limits with predicted optical signals based on the measured optical intensity of GRB990123 scaled by the ratio of the x-ray intensity for each LOTIS burst to the x-ray intensity of GRB990123. Since the x-ray intensity of a burst can be quantified by its peak flux and/or its total fluence we present predic-

TABLE 1. Best LOTIS events and scaled GRB990123 limits

Event	Peak Flux	Fluence	Burst duration	LOTIS response time	LOTIS limit	Scaled mag to peak flux	Scaled mag to fluence
	γ cm^{-2}s^{-1}	erg cm^{-2}	s	s			
961017	1.98	5.07	1.2	11.0	11.5	11.3	15.9
961220	1.60	18.1	9.8	9.0	11.5	11.5	14.5
970223	16.84	968	16.3	11.5	11.0	9.0	10.2
970714	1.32	17.1	2.0	14.1	11.3	11.7	14.6
970919	0.77	22.5	20.9	11.8	11.5	12.3	14.3
971006	1.79	258	48.1	17.1	12.1	11.4	11.7
971227	2.11	9.25	6.8	10.0	12.3	11.2	15.3
990129	4.99	585	200	140.8	14.5	10.3	10.8
990308	1.26	164	50	132.1	13.5	11.8	12.2
990316	3.67	529	40	13.6	14.3	10.6	10.9
990413	2.57	68.1	15	13.0	14.0	11.0	13.1
990803	12.19	1230	28	15.0	14.5	9.3	10.0
990918	3.17	25.2	6.5	8.3	14.3	10.8	14.2

tions for both in the last 2 columns of Table 1. When we compare these results with the LOTIS limiting magnitude in the column 5, we find that the predicted prompt optical signals are in most cases brighter that the LOTIS upper limits (especially after GRB 970919 when LOTIS was upgraded). This suggests that GRB990123 was not a typical GRB.

The LOTIS results have also been used by Sari & Piran to constrain their

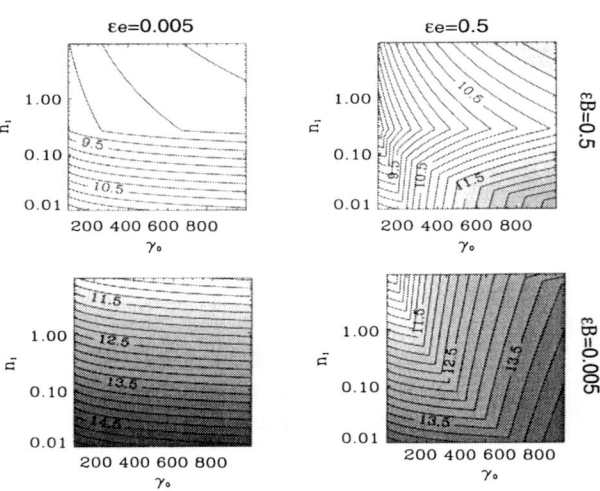

FIGURE 2. LOTIS constraints on GRB production parameters. The darker regions are preferred by LOTIS.

external-reverse-shock GRB model. This model assumes that the fraction of the GRB energy emitted in the optical band depends on the values of the cooling frequency and the characteristic synchrotron frequency. Extending this theory to the external reverse shock model for prompt optical signal, requires physical parameters of ε_e and ε_B that are the fraction of equipartition energy in the electrons and magnetic field; n, the density of the surrounding matter in cm^{-3}; and γ_o is the initial Lorentz factor. Assuming that the observable parameters of the total fluence and the observation time are fixed at S=2.33 × 10^{-7} erg cm^{-2} and t=10 s, we predict the likely gamma-ray burst production environment shown as the contour lines are shown in Figure 2. The darker regions in these plots are favored by LOTIS limits. From this analysis LOTIS results imply that the GRBs are created at low surrounding matter density, high initial Lorentz factors, low equipartition energy in the magnetic field, and high equipartition energy in the electron.

LOTIS SKY PATROL DATA

The typical rate for burst triggers is ~1/month. While not responding to burst triggers, LOTIS systematically acquires all sky data ~4 times/night. We have just begun to analyze this data set for variable objects. Figure 3 shows the light curves for two example stars over 3 months selected from the field near SS-Cyg. This data set includes most seeing conditions experienced over the 3 month period including periods of bright Moon phase. We analyzed the data utilizing the daophot photometry code to produce the light curves. Shown in the figure 3 are the light curves for a non-varying star and a cataclysmic variable, SS-Cyg. The data for the non-varying star shows that we can monitor variability with a photometric accuracy of mag~0.03 for mag~12.0 stars. The second panel shows the light curve for SS-Cyg. We can clearly monitor its outbursts. When the outburst was brighter than mag~9.5, SS-Cyg saturated the camera. Encouraged by this result, we plan to expand this analysis to all of the data.

SUPER-LOTIS

Super-LOTIS is a next-generation system designed to be sensitive enough to detect GRB prompt optical levels predicted by current theories. The telescope (second picture in Figure 1) is a Boller and Chivens 0.6 m reflective telescope of f/3.5. We automated the telescope for GRB follow-up work by adding computer controllable drives. These drives can point to any part of the sky within 30 s upon receipt of a GCN trigger. We also designed and fabricated a custom 4-element coma corrector to match the point spread function to the pixel scale at the corners of the imaging CCD. The sensor is an upgraded LOTIS CCD camera. Super-LOTIS has a 0.84 × 0.84° field-of-view (1.5″/pixel) sufficient to cover the error boxes expected from upcoming GRB satellite missions.

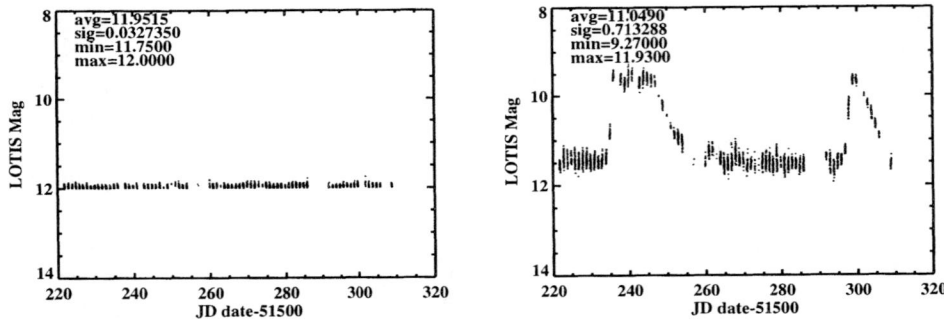

FIGURE 3. LOTIS sky patrol data: light curves for a non-varying star and a cataclysmic variable, SS-Cyg

Our data acquisition system includes custom readout electronics, a custom hardware power control unit, a weather station and a housing control unit. Extensive on-line scheduling software has been written to handle various triggers. Priority is given to the most recent trigger that has smallest error box. Our scanning strategy and automation allows us to record GRB optical activity as early as 30 s.

In April 2000, we moved Super-LOTIS to the Kitt Peak National Observatory and are ready to receive real-time GRB coordinates form the HETE-II satellite which was successfully launched in October, 2000.

OTHER PROMPT FOLLOW-UP EXPERIMENTS

TAROT: TAROT-1 has been operating since July 1999. It uses 20 cm aperture, f/3.2 telescope with a 1024×1270 pixel CCD. Its field-of-view is $2 \times 2°$ and it is located in France. Because of its small-field-of-view optics, it didn't cover many real-time BATSE triggers [11] even though its mount can respond to a burst trigger within a few seconds. They have a plan to construct TAROT-2 that has bigger optics (1.5 m aperture) and a 2000×2000 CCD with a spectrometer. This system is under development.

ROTSE: ROTSE-I is similar to the LOTIS except the CCD array which has smaller pixel size thus the total field of view is $16.4 \times 16.4°$, slightly smaller than LOTIS. The ROTSE collaboration is now constructing ROTSE-II which has 45 cm aperture with f/1.9. This system is under construction as of September, 2000.

BOOTES: BOOTES utilizes 30 cm f/3.3 telescope with a 1530×1020 pixel CCD. Its total field of view is $49' \times 33'$ and it is operating in Spain. They have responded to many BATSE triggers but again its small-field-of-view provided only limited coverage of the GRB error boxes [12].

There are other telescopes under construction including YSTAR [13] that will attempt to automatically respond to burst triggers given by GRB satellites but not many are in operation as of September, 2000.

FUTURE AND CONCLUSION

The promptly responding automatic telescopes require real-time burst coordinate triggers from GRB satellites. In June 2000, the CGRO that distributed BATSE coordinates in real-time was shut-down. Until HETE-II is fully operational, only Beppo/SAX and the IPN are distributing coordinates with a typically delay of 1∼2 days. For these events, we examine the sky-patrol archives for the earliest LOTIS image taken after the GRB of the location of the Beppo/SAX or IPN coordinates. In some cases the archived images are taken as early as 2∼4 hour after the GRB.

The trigger delay problem will be solved when the HETE-II is fully operational and starts distributing coordinates in November, 2000. Other satellites such as Integral (2002) and SWIFT (2003) will provide more frequent real-time triggers as well.

Observation and measurement of prompt optical signals from GRBs will greatly enhance our understanding of the GRB progenitor. LOTIS results show that the flux of the prompt optical emission does not scale with gamma-ray peak flux or fluence. With LOTIS and Super-LOTIS, we will be able to observe GRB optical activity from 10 s to many hours after the event down to a magnitude 14∼19. With HETE-II and other prompt GRB coordinate distributing satellites, we will be able to measure early time optical activity.

ACKNOWLEDGMENTS

This work was performed under the auspices of the U.S. Department of Energy by University of California Lawrence Livermore National Laboratory under contract No. W-7405-Eng-48. This work is also supported by the NASA contract number S-03975G.

REFERENCES

1. Costa, E., et al., *Nature*, **387**, 783 (1997).
2. van Paradijs, J., et al., *Nature*, **386**, 686 (1997).
3. Piran, T., *Phy. Rep.,*, **314**, 575 (1999).
4. Odewahn, S., et al., *GCN Circular*, 201 (1999).
5. Akerlof, C., et al., *Nature*, **398**, 400 (1999).
6. Briggs, M., et al., *ApJ*, **524**, 82 (1999).
7. Sari, R. & Piran, T., *ApJ*, **517**, L109 (1999).
8. Park, H., et al., *ApJ*, **490**, L21 (1997).
9. Williams, G., et al., *ApJ*, **519**, L25 (1999).
10. Williams, G., et al., *AIP Conf. Proc.*, **526**, 250 (2000).
11. Boer, E., et al., *AIP Conf. Proc.* **526**, 255 (2000).
12. Castro-Tirado, A., et al., *AIP Conf. Proc.* **526**, 260 (2000).
13. Byun, Y., et al., *These proceedings* (2001).

The GRB-Afterglow Transition: Black Holes, Bullets, and Beams

Ralph A.M.J. Wijers

Dept. of Physics and Astronomy, SUNY, Stony Brook, NY 11794-3800, USA.
rwijers@mail.astro.sunysb.edu

Abstract. A continuing irony of gamma-ray burst research is the fact that while the discovery of counterparts has led to much improved understanding of gamma-ray bursts, almost all this improvement is to do with the late emission from the counterparts. The origin of the gamma-ray burst itself, both in terms of its emission process and its dynamics, remain unclear, as does the eventual energy source. Here I discuss some of the progress that has been made in this area, as well as issues that remain open.

INTRODUCTION

Since their discovery in the late sixties, we have tried to understand what causes the gamma-ray emission from gamma-ray bursts. It became clear early on that unless the objects were in our immediate vicinity, there had to be relativistic outflow in them [1–3], else the gamma rays would not escape from the radiating region without thermalizing. More recently, direct observations have confirmed this by measurements of the X-ray, optical and radio afterglow of some bursts. In GRB 970508, early scintillation that later disappears indicates a mean expansion velocity over the first month close to the speed of light [4]. In GRB 990123, the very rapid robotic detection of an optical counterpart [5] provides the highest directly measured flow speed: from the brightness temperature follows a lower limit of 50 to the Lorentz factor one minute after the GRB trigger.

A persistent problem with modeling the emission from the gamma-ray burst proper is the fact that no two bursts are the same. The time histories are all different, the durations range from milliseconds to over an hour, and the spectra — while nearly always highly non-thermal — vary during the burst and between bursts. (See [6] for a review of observed properties of gamma-ray bursts.) This indicates that some aspects of the burst emission must be purely statistical, and these aspects almost certainly dominate the observational appearance of the event. How, then, do we find the underlying physical regularities ('climate') in a sea of incidental details ('weather')?

Still, the situation is not entirely hopeless. Progress is being made, in part with help of what we have learned from afterglows. The known redshifts of some GRBs have begun to set their energy scale. Also, the fact that some afterglows show signs of being collimated means that the central energy source produces a collimated outflow rather than a spherical explosion. In many central-engine models, this is not surprising. Other issues that remain at the forefront are the origin of the variability of the emission and the gamma-ray emission efficiency. Eventually, these and other questions may be resolved when earlier alerts and more sensitive robotic rapid-alert telescopes close the gap between burst emission and afterglow, to tell us directly what lies there, and whether the phenomena are continuous. I discuss some recent work that has begun to suggest answers to those questions.

JETS AND COLLIMATION

Two basic reasons to think about jets for GRB have emerged; both are mostly theoretical, at least in origin. First, the total energy emitted in gamma rays assuming isotropic emission are sometimes in excess of 10^{54} erg $\simeq M_\odot c^2$. This much energy would not be easy to get even out of a stellar-mass object, especially if one considers that the conversion efficiency of whatever the primary energy is to gamma rays is unlikely to be 100%, and could in fact be quite small [7,8]. Second, even the most optimistic energy estimates combined with the short variability time scale of bursts do seem to point to deep gravitational collapse as the ultimate energy source. Almost all astrophysical objects contain angular momentum, and some of the leading contenders for GRB sources are no exception: stellar cores, binary neutron stars and black holes. This means that after collapse, it is likely that they will rotate rapidly, in which case it is logical that the ejection from it might have a preferred direction: the rotation axis.

Such considerations could of course still be due to our limited imagination of what might be behind GRBs, but nonetheless, stellar collapse and merger models are the bulk of what is currently being considered, and in these collimation of the energy outflow is not unexpected. It may even be energetically required. If the outflow is collimated then the high Lorentz factor guarantees that the radiation is highly anisotropic as well, being emitted almost only in those directions at which the outflow is directly aimed. The outflow will expand sideways due to internal pressure, but as long as the Lorentz factor is high this expansion is limited to an extra angle of order γ^{-1}. Therefore, as long as the opening angle θ_c of the collimated flow is less than γ^{-1} we may neglect sideways expansion, and the dynamical and spectral evolution of the blast wave is indistinguishable from a spherical flow. The relativistic beaming of every surface element concentrates the radiation into an emission cone that also subtends an angle of about γ^{-1} around the direction of the flow. So as the flow slows down, and the Lorentz factor reaches $\gamma \sim \theta_c^{-1}$, both opening of the emission angle and sideways expansion begin to affect the observations at about the same time [9–11] (see also the contribution by Kumar to this volume). As a result, the theoretical predictions have not converged, and it is unclear whether the rather varied observed breaks in light curves can all be captured under the heading of beaming. For example, simultaneous presence of a cooling break may be needed for the sharp break in GRB 990510 [12]. It has also been suggested that beaming in the ultrarelativistic regime causes no break at all, and that the steep breaks have to do with transition to the nonrelativistic regime (essentially a classic Sedov-Taylor phase) [13]. However, analysis of this claim shows that it may well be due to unrealistically high radiative losses in the early phases, and particular choices for the opening angle that make it hard to separate the beaming break from the transition to the nonrelativistic regime. Clearly, current models are inadequate to resolve this: the sideways expansion is not treated correctly, but rather with ad hoc recipes that, among other things, assume that the radiation from the shock front does not depend on angle from the axis even after sideways expansion has started. Such calculations are in fact quite challenging, requiring a minimum of two-dimensional ultrarelativistic hydrodynamics.

Also, even with this idealized problem solved, one should probably consider complexities such as the fact that even initially the Lorentz factor and energy per unit mass are likely to vary from the center of the jet to the edge, giving rise to new classes of solution [14,15].

THE ORIGIN OF VARIABILITY

A number of causes for variability of the gamma-ray burst light curve have been suggested. Certainly some stochastic process must be involved, given how even bursts with roughly the same duration look very different. The original suggestion that density irregularities in the ambient medium cause it [16] have been out of favor for a while, but have recently been revived in an interesting manner [17,18]. Despite various attempts, most arguments against this model have subsequently disappeared or weakened [19,20]. The most popular suggestion in recent times has been internal shocks in the relativistic outflow from the engine [21,22], worked out in considerable detail [23]. A more recent addition to the repertoire has been the idea that the ejecta are blobby, and collide with a fairly uniform medium [24]. Again, the saving graces we have in afterglow analysis do not apply: the problem is intrinsically multidimensional. Another handicap is that the emission process for the GRB is not known, so unlike for afterglows we cannot take a simple emission model as a given and invert the light curves to get or test dynamical models empirically. This may have to await the GLAST mission, which will sample large numbers of GRB from keV to GeV energies; this may constitute a broad enough spectral range to infer the radiation mechanism unambiguously.

While most of the attempts to understand the variability start from theory and try to mimic observations, some analyses start with data, trying to reveal the underlying stochastic processes. An interesting recent development in this older tradition is the discovery that Fourier transforms of very bright bursts, or means

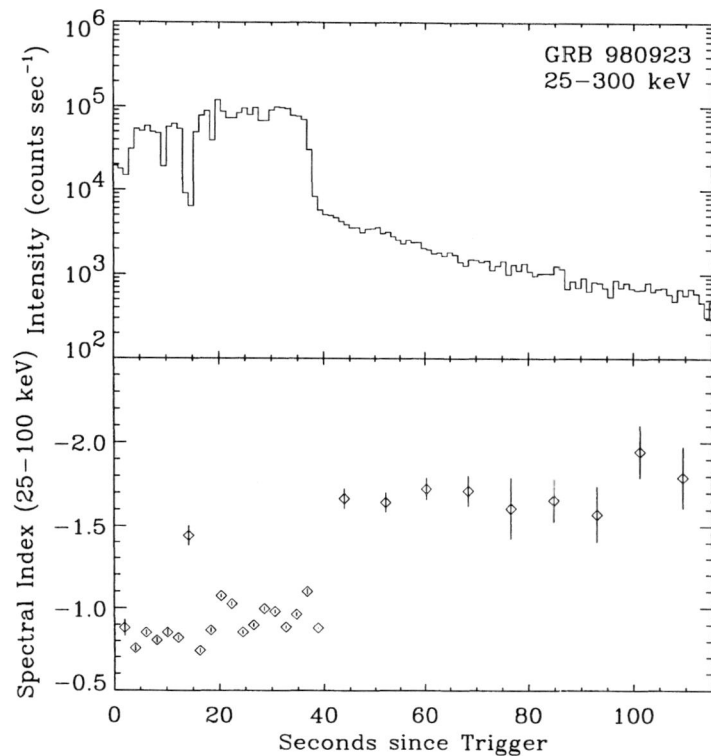

FIGURE 1. GRB 980923 and its tail, from [32,31]. The difference in temporal character between the burst and tail is clear (top), as is the fact that the tail is noticeably softer. The appearance of that same soft spectrum during a dip in the main burst at $t = 14\,\mathrm{s}$ suggests that both emission components overlap in time.

of samples of weaker ones, show a power-law spectrum with an index consistent with Kolmogorov turbulence [25]. An extension of the early variability to later fluctuations is also suggested: it appears that the known amount of turbulence in the interstellar medium suffices to induce low-level variability in the afterglow [26].

One of the pressing reasons why a resolution of the origin of the variability is urgently needed is that it has strong implication for the viability of some central-engine models (see contributions to this volume by MacFadyen and Höflich). If the variability is due to internal shocks in the outflow, one can show that the total duration of the GRB emission is similar to the time period over which the central engine operates. For the external and shotgun models, that may not be the case, and specifically the central engine could emit all its energy over an arbitrarily short period. A long central-engine life almost certainly requires a Blandford-Znajek type engine (see contribution by Lee in this volume), whereas short-duration engines could also be neutrino powered.

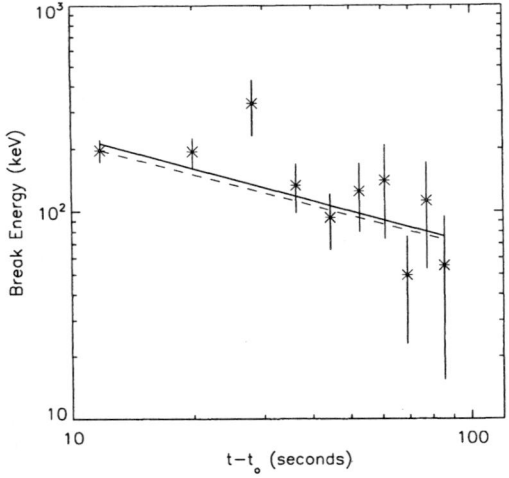

FIGURE 2. The temporal evolution of the break energy in GRB 980923 (from [31]), which is consistent with the proposition that the tail of this burst is the earliest part of a spherically symmetric, adiabatic afterglow.

THE TRANSITION: LATE BURSTS AND EARLY AFTERGLOWS

Another way to investigate the GRB, and perhaps extrapolate our increased understanding of afterglows to the earliest times, is to look at very early afterglows, or tails of emission in GRBs, to see whether one can detect signs of a transition between the two. There are a number of early recordings of such events, e.g. the HEAO-1 transient that followed GRB 780506 after a few minutes' silence [27]. More recently, it was found by very careful background subtraction that BATSE bursts have long tails of emission, lasting hundreds of seconds [28]. Others have found evidence of such tails in individual bursts [29], or in samples [30].

To examine the nature of these tails, we investigated the case of the bright BATSE burst GRB 980923 [31]. It has a 40-second highly variable outburst, followed by a lower-level smooth tail (Fig. 1). This tail has a clearly different (softer) spectrum than the main burst emission, and the appearance of this same soft spectrum during a dip in the main burst suggests that the tail does not strictly follow the main emission, but starts while the main emission is still ongoing. Furthermore, detailed spectral analysis showed that the tail emission has a spectral break in it. This break is consistent in shape with a cooling break. The break energy decreases with time as $t^{-1/2}$, as expected for a spherically symmetric, adiabatic afterglow cooling break (Fig. 2). This would also agree nicely with the fact that after about a day (1000 times later) the cooling break lies in or just below the soft X-ray band (30 times lower energy), as is seen in some afterglows.

Inspired by this find, we attempted to find more such cases. We visually selected about 40 GRB from the BATSE catalog which seemed to have a smooth

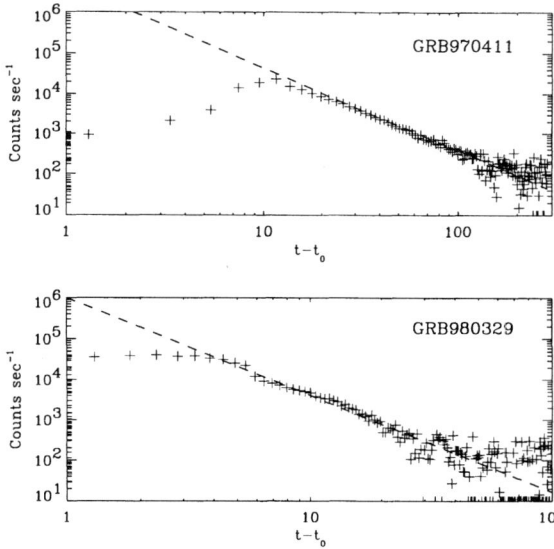

FIGURE 3. Two examples of power-law decay in a sample of bursts with smooth tails (from [31]). Note that t_0 is fitted, and not equal to the trigger time of the burst.

tail to them, and in which the signal-to-noise was high enough to do time-resolved spectroscopy to check the evolution of the tail in some detail. The outcome was very mixed: many of the tails did have power-law decline in time (Fig. 3), but only a handful matched a relation between the temporal decay rate and spectral slope that would be expected from regular afterglow models. It therefore appears that the case of GRB 980923 is not representative, and while GRB tails may be a separate component of emission, or represent a transition of behavior in the GRB, it is usually not simply the start of the afterglow. Or at least, if it is the early afterglow, its physics is different from the later afterglow.

A handful of burst tails do obey the relation that the temporal decay index is twice the (energy flux) spectral slope (Fig. 4). This behavior is characteristic of the late phase of collimated GRB outflows, once the Lorentz factor has fallen below the inverse of the opening angle. Perhaps this indicates the presence of narrow sub-jets or high-density bullets in the outflow, in the manner suggested in ref. [24]. On the other hand, the emission from a small target hit by a blast wave could be very similar, and therefore the same data could be consistent with ambient-medium irregularity [18].

CONCLUSION

We have explored some issues concerning the character of the prompt emission from gamma-ray bursts. Not much in this area is known with any certainty yet. We argue that some properties of the central engine, such as how long it has to work, how it emits matter, and whether it is collimated and therefore less energy is

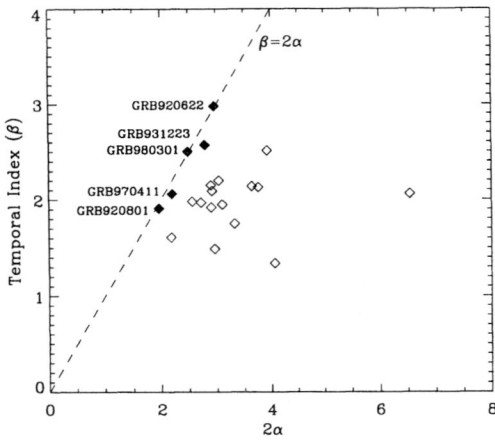

FIGURE 4. The temporal index as a function of spectral index for a sample of GRB tails (from [31]). The solid cases with names fall on the relation expected for a narrow jet or bullet. It is perhaps interesting that this group forms an upper envelope in the diagram. Especially the steepest-spectrum bursts on the right have far too slow temporal decays to fit any normal afterglow model.

required from it than we now estimate, can be inferred from prompt emission and early afterglow observations.

Near-future missions will bring new, crucial information bearing on this problem: HETE2 will alert us much more rapidly, and jointly between fast robotic cameras and major observatories we shall get the first uninterrupted light curves from gamma-ray bursts, elucidating the prompt-to-afterglow transition. GLAST will provide us with keV to GeV spectra of bursts, clarifying the emission mechanism for the prompt radiation. Swift will extend the continuous coverage of the light curves into X rays and UV, providing us with more diagnostics on the physics of the afterglow, and at earlier times. We can therefore look forward to good progress in this area in the next few years, and probably to some surprises.

REFERENCES

1. Schmidt, W. K. H., *Nature*, **271**, 525–527 (1978).
2. Cavallo, G. and Rees, M. J., *Mon. Not. R. Astron. Soc.*, **183**, 359–365 (1978).
3. Krolik, J. H. and Pier, E. A., *Astrophys. J.*, **373**, 277–284 (1991).
4. Frail, D. A., Kulkarni, S. R., Nicastro, L., Feroci, M., and Taylor, G. B., *Nature*, **389**, 261–263 (1997).
5. Akerlof, C. et al., *Nature*, **398**, 400–402 (1999).
6. Fishman, G. J. and Meegan, C. A., *Annu. Rev. Astron. Astrophys.*, **35**, 415–458 (1995).
7. Kumar, P., *Astrophys. J.*, **523**, L113–L116 (1999).
8. Kumar, P. and Piran, T., *Astrophys. J.*, **535**, 152–157 (1999).
9. Rhoads, J. E., *Astrophys. J.*, **487**, L1–L4 (1997).
10. Rhoads, J. E., *Astrophys. J.*, **525**, 737–749 (1999).
11. Mészáros, P. and Rees, M. J., *Mon. Not. R. Astron. Soc.*, **306**, L39–L43 (1999).

12. Kumar, P. and Panaitescu, A., *Astrophys. J.*, submitted (astro–ph/0003264) (2000).
13. Huang, Y. F., Gou, L. J., Dai, Z. G., and Lu, T., *Astrophys. J.*, submitted (astro–ph/9910493) (1999).
14. Wijers, R. A. M. J., Rees, M. J., and Mészáros, P., *Mon. Not. R. Astron. Soc.*, **288**, L51–L56 (1997).
15. Mészáros, P., Rees, M. J., and Wijers, R. A. M. J., *Astrophys. J.*, **499**, 301–308 (1998).
16. Rees, M. J. and Mészáros, P., *Mon. Not. R. Astron. Soc.*, **258**, L41–L43 (1992).
17. Dermer, C. D. and Mitman, K. E., *Astrophys. J.*, **513**, L5–L8 (1999).
18. Dermer, C. D., Böttcher, M., and Chiang, J., *Astrophys. J.*, **515**, L49–L52 (1999).
19. Fenimore, E. E., Ramirez-Ruiz, E., and Wu, B., *Astrophys. J.*, **518**, L73–L76 (1999).
20. Fenimore, E. E. and Ramirez-Ruiz, E., *Astrophys. J.*, submitted (astro–ph/9909299) (1999).
21. Rees, M. J. and Mészáros, P., *Astrophys. J.*, **430**, L93–L96 (1994).
22. Paczyński, B., and Xu, G., *Astrophys. J.*, **424**, 708–713 (1994).
23. Kobayashi, S., Piran, T., and Sari, R., *Astrophys. J.*, **490**, 92–98 (1997).
24. Heinz, S., and Begelman, M. C., *Astrophys. J.*, **527**, L35–L38 (1999).
25. Beloborodov, A. M., Stern, B. E., and Svensson, R., *Astrophys. J.*, **535**, 158–166 (2000).
26. Wang, X., and Loeb, A., *Astrophys. J.*, **535**, 788–797 (2000).
27. Connors, A., and Hueter, G. J., *Astrophys. J.*, **501**, 307–324 (1998).
28. Connaughton, V., In R. M. Kippen, R. S. Mallozzi, and G. J. Fishman, editors, *Gamma-Ray Bursts*, 5th Huntsville Symposium, volume 526 of *AIP Conf. Proc.*, pages 385–389, AIP:Melville, NY (2000).
29. Burenin, R. A. et al., *Astron. Astrophys.*, **344**, L53–L56 (1999).
30. Litvine, D. A., Mitrofanov, I. G., and Kosyrev, A. S., In R. M. Kippen, R. S. Mallozzi, and G. J. Fishman, editors, *Gamma-Ray Bursts*, 5th Huntsville Symposium, volume 526 of *AIP Conf. Proc.*, 390–393, AIP:Melville, NY (2000).
31. Giblin, T., van Paradijs, J., Kouveliotou, C., Connaughton, V., Wijers, R. A. M. J., and Fishman, G. J., *Astrophys. J.*, **524**, L41–L50 (1999).
32. van Paradijs, J., Kouveliotou, C., and Wijers, R. A. M. J., *Annu. Rev. Astron. Astrophys.*, **38**, 381–429 (2000).

Effects of Environment and Energy Injection on Gamma-Ray Burst Afterglows

Z. G. Dai

Department of Astronomy, Nanjing University, Nanjing 210093, China

Abstract. There is growing evidence that some long gamma-ray bursts (GRBs) arise from the core collapse of massive stars, and thus it is inevitable that the environments of these GRBs are preburst winds or dense media. We studied, for the first time, the wind model for afterglows based on the Blandford-McKee self-similar solution of a relativistic shock, and suggested that GRB 970616 is an interactor with a stellar wind. We also proposed a dense medium model for some afterglows, e.g., the steepening in the light curve of the R-band afterglow of GRB 990123 may be caused by the adiabatic shock which has evolved from an ultrarelativistic phase to a nonrelativistic phase in a dense medium. We further discussed the dense medium model in more details, and investigated the effects of synchrotron self absorption and energy injection. A shock in a dense medium becomes nonrelativistic rapidly after a short relativistic phase. The afterglow from the shock at the nonrelativistic stage decays more rapidly than at the relativistic stage. Since some models for GRB energy sources predicted that a strongly magnetic millisecond pulsar may be born during GRB formation, we discussed the effect of such a pulsar on the evolution of the nonrelativistic shock through magnetic dipole radiation. We found that in the pulsar energy injection case, the dense medium model fits very well all the observational data of GRB 980519. Recently, we combined the dense medium model with the pulsar energy injection effect to provide a good fit to the optical afterglow data of GRB 000301C.

INTRODUCTION

In the standard afterglow shock model (for a review see [27,38]), a gamma-ray burst (GRB) afterglow is usually believed to be produced by synchrotron radiation or inverse Compton scattering in an ultrarelativistic shock wave expanding in a homogeneous medium. As more and more ambient matter is swept up, the shock gradually decelerates while the emission from such a shock fades down, dominating at the beginning in X-rays and progressively at optical to radio energy band. The standard model is based on four basic assumptions: (1) the total energy of the shock is released impulsively before its formation; (2) the medium swept up by the shock is homogeneous and its density (n) is the one of the interstellar medium $\sim 1\,\mathrm{cm}^{-3}$;

(3) the electron and magnetic field energy fractions of the shocked medium and the index (p) in the accelerated electrons' power-law distribution are constant during the whole evolution stage; and (4) the shock is spherical.

Each of these assumptions has been varied to discuss why some observed afterglows deviate from that expected by the standard afterglow model. For example, the R-band light curve of GRB 970508 afterglow peaks around two days after the burst, and there is a rather rapid rise before the peak which is followed by a long power-law decay. There are two models explaining this special feature: (i) It was envisioned [28] that a postburst fireball may contain shells with a continuous distribution of Lorentz factors. As the external forward shock sweeps up ambient matter and decelerates, internal shells will catch up with the shock and supply energy into it. A detailed calculation shows that this model can explain well this special feature [26]. (ii) We considered continuous energy injection from a strongly magnetized millisecond pulsar into the shock through magnetic dipole radiation [8]. This model can also account for well the observations. It is very clear that these models don't use basic assumption (1).

There are several models in the literature that discuss the effect of inhomogeneous media on afterglows [9,23,3,4], dropping the second assumption. Generally, an $n \propto r^{-k}$ ($k > 0$) medium is expected to steepen an afterglow's temporal decay. We studied, for the first time, the wind model for afterglows based on the Blandford-McKee self-similar solution [1] of a relativistic adiabatic shock, and suggested that GRB 970616 is an interactor with a stellar wind of $n \propto r^{-2}$ [9]. It was found [3] that a Wolf-Rayet star wind likely leads to an $n \propto r^{-2}$ medium, and thus if GRB 980519 resulted from the explosion of such a massive star, subsequent evolution of a relativistic shock in this medium is consistent with the steep decay in the R-band light curve of the afterglow from this burst. Another way of dropping the second assumption is that the density of an ambient medium is invoked to be as high as $n \sim 10^6$ cm^{-3}. The temporal decay of the R-band afterglow of GRB 990123 has been detected to steepen about 2.5 days after this burst [21,2,16]. We proposed a plausible model in which a shock expanding in a dense medium has evolved from a relativistic phase to a nonrelativistic phase [11]. We found that this model fits well the observational data if the medium density is about 3×10^6 cm^{-3}. We further suggested that such a medium could be a supernova or supranova or hypernova ejecta.

In basic assumption (3), the electron and magnetic field energy fractions of the shocked medium may not be varied during whole evolution, as argued in [34], where all the observational data including both the prompt optical flash and the afterglow of GRB 990123 were analyzed.

The steepening in the light curves of the afterglows of some bursts may also be due to lateral spreading of a jet, as analyzed in [29,31] when the jet expands in a homogeneous interstellar medium (ISM). This in fact drops basic assumption (4). However, numerical studies of [24,20,36,37] show that the break of the light curve is weaker and smoother than the one analytically predicted when the light travel effects related to the lateral size of the jet and a realistic expression of the lateral expansion speed are taken into account. In the case of a jet expanding in a wind, the calculated light curve is even much weaker and smoother than the ananlytical one [22,17]. We recently calculated light curves for GRB afterglows when anisotropic jets ($dE/d\Omega \propto \theta^{-k}$) expand both in the interstellar medium and in the wind medium [5]. We found that in each type of medium, one break appears in the late-time afterglow light curve for small k but becomes weaker and smoother as k increases.

We discussed the dense medium model in more details [12], by taking into account both the synchrotron self-absorption effect in the shocked medium and the energy injection effect of [8,10]. Recently, we combined the dense medium model with the pulsar energy injection effect to provide a good fit to the optical afterglow data of GRB 000301C [13]. Here we want to give a brief review of some of our studies on GRB afterglows.

SHOCK EVOLUTION

It is well known that the evolution of a partially radiative shock depends on both the efficiency with which the shock transfers its bulk kinetic energy to electrons and magnetic fields and on the efficiency with which the electrons radiate their energy. In 1998, we proposed, for the first time, a *unified* model for dynamical evolution of a partially radiative shock [6]. This model is not only valid during the whole evolution stage including the Sedov phase for an adiabatic shock, but also can describe well an adiabatic shock as well as a highly radiatve shock. This model was later re-investigated and referred to as a *generic* one in [19]. For simplicity, we here assume that a relativistic shock expanding in a dense medium is adiabatic. This assumption is correct particularly for a low electron energy density fraction in the shocked medium [6,7]. The Blandford-McKee self-similar solution [1] gives the Lorentz factor of an adiabatic relativistic shock,

$$\gamma = \frac{1}{4}\left[\frac{17E_0(1+z)^3}{\pi n m_p c^5 t_\oplus^3}\right]^{1/8} = 1 E_{52}^{1/8} n_5^{-1/8} t_\oplus^{-3/8}[(1+z)/2]^{3/8}, \tag{1}$$

where $E_0 = E_{52} \times 10^{52}$ ergs is the total isotropic energy, $n_5 = n/10^5 \, \text{cm}^{-3}$, $t_\oplus$ is the observer's time since the gamma-ray trigger in units of 1 day, z is the the redshift of the source generating this shock, and m_p is the proton mass. We assume $\gamma = 1$ when $t_\oplus = t_b$. This implies

$$n_5 = E_{52} t_b^{-3}[(1+z)/2]^3. \tag{2}$$

For $t_\oplus > t_b$, the shock will be in a nonrelativistic phase. In the following we will discuss the spectrum and light curve during the non-relativistic phase.

As usual, only synchrotron radiation from the shock is considered. To analyze the spectrum and light curve, one needs to know three crucial frequencies: the synchrotron peak frequency (ν_m), the cooling frequency (ν_c), and the self-absorption frequency (ν_a). We assume a power law distribution of the electrons accelerated by the shock: $dn'_e/d\gamma_e \propto \gamma_e^{-p}$ for $\gamma_e \geq \gamma_{em}$, where γ_e is the electron Lorentz factor and $\gamma_{em} = 610\epsilon_e(\gamma - 1)$ is the minimum Lorentz factor. We further assume that ϵ_e and ϵ_B are the electron and magnetic energy density fractions of the shocked medium respectively. The ν_m is the characteristic synchrotron frequency of an electron with Lorentz factor of γ_{em}, while the ν_c is the characteristic synchrotron frequency of an electron which cools on the dynamical age of the shock. According to Sari et al. [32], we have derived the synchrotron peak frequency, the cooling frequency and the synchotron self-absorption frequency, measured in the observer's frame [12]. They are correct for the whole evolution stage.

Now we give the spectrum and light curve of the afterglow during the non-relativistic phase. First, for the case without energy injection, the shock velocity decays as $\propto t_\oplus^{-3/5}$ and thus we have

$$F_\nu = \begin{cases} (\nu_a/\nu_m)^{-(p-1)/2}(\nu/\nu_a)^{5/2}F_{\nu_m} \propto \nu^{5/2}t_\oplus^{11/10} & \text{if } \nu < \nu_a \\ (\nu/\nu_m)^{-(p-1)/2}F_{\nu_m} \propto \nu^{-(p-1)/2}t_\oplus^{(21-15p)/10} & \text{if } \nu_a < \nu < \nu_c \\ (\nu_c/\nu_m)^{-(p-1)/2}(\nu/\nu_c)^{-p/2}F_{\nu_m} \propto \nu^{-p/2}t_\oplus^{(4-3p)/2} & \text{if } \nu > \nu_c. \end{cases} \quad (3)$$

We easily see that for high-frequency radiation the temporal decay index $\alpha = (21-15p)/10$ for emission from slow-cooling electrons or $\alpha = (4-3p)/2$ for emission from fast-cooling electrons. If $p \approx 2.8$, then $\alpha \approx -2.1$ or -2.2. Comparing this with the relativistic result, we conclude that the afterglow decay steepens at the nonrelativistic stage.

Some models for GRB energy sources (for a brief review see [10]) predict that during the formation of an ultrarelativistic fireball required by GRB, a strongly magnetized millisecond pulsar will be born. If so, the pulsar will continuously input its rotational energy into the forward shock of the postburst fireball through magnetic dipole radiation because electromagnetic waves radiated by the pulsar will be absorbed in the shocked medium [8,10]. Since an initially ultrarelativistic shock discussed in [12] rapidly becomes nonrelativistic in a dense medium, we next investigate the evolution of a nonrelativistic adiabatic shock with energy injection from a pulsar. The total energy of the shock is the sum of the initial energy and the energy which the shock has obtained from the pulsar:

$$E_0 + \int_0^{t_\oplus} L dt_\oplus = E_{tot} \propto v^2 r^3, \quad (4)$$

where L is the stellar spindown power $\propto (1 + t_\oplus/T)^{-2}$ (T is the initial spindown time scale). The term on the right-hand side is consistent with the Sedov solution. Please note that L can be thought of as a constant for $t_\oplus < T$, while L decays as $\propto t_\oplus^{-2}$ for $t_\oplus \gg T$. Because of this feature, we easily integrate the second term on the left-hand side of equation (17). We now define a time at which the shock has obtained energy $\sim E_0$ from the pulsar, $t_c = E_0/L$, and assume $t_c \ll T$. The evolution of the afterglow from such a shock can be divided into three stages.

Stage (i): $t_\oplus \ll t_c$, viz., the second term on the left-hand side of equation (4) can be neglected. The evolution of the afterglow is the same as in the above case without any energy injection.

Stage (ii): for $T > t_\oplus \gg t_c$, the term E_0 in equation (4) can be neglected. At this stage, the shock's velocity $v \propto t_\oplus^{-2/5}$, we have derived the spectrum and light curve of the afterglow [12]

$$F_\nu = \begin{cases} (\nu_a/\nu_m)^{-(p-1)/2}(\nu/\nu_a)^{5/2} F_{\nu_m} \propto \nu^{5/2} t_\oplus^{7/5} & \text{if } \nu < \nu_a \\ (\nu/\nu_m)^{-(p-1)/2} F_{\nu_m} \propto \nu^{-(p-1)/2} t_\oplus^{(12-5p)/5} & \text{if } \nu_a < \nu < \nu_c \\ (\nu_c/\nu_m)^{-(p-1)/2}(\nu/\nu_c)^{-p/2} F_{\nu_m} \propto \nu^{-p/2} t_\oplus^{2-p} & \text{if } \nu > \nu_c. \end{cases} \quad (5)$$

It can be seen that for high-frequency radiation the temporal decay index $\alpha = (12-5p)/5 \approx -0.4$ for emission from slow-cooling electrons or $\alpha = 2 - p \approx -0.8$ for emission from fast-cooling electrons if $p \approx 2.8$. This shows that the afterglow decay may significantly flatten due to the effect of the pulsar.

Stage (iii): for $t_\oplus \gg T$, the power of the pulsar due to magnetic dipole radiation rapidly decreases as $L \propto t_\oplus^{-2}$, and the evolution of the shock is hardly affected by the stellar radiation. Thus, the evolution of the afterglow at this stage will be the same as in the above case without any energy injection.

In summary, as an adiabatic shock expands in a dense medium from an ultrarelativistic phase to a nonrelativistic phase, the decay of radiation from such a shock will steepen, subsequently may flatten if a strongly magnetic millisecond pulsar continuously inputs its rotational energy into the shock through magnetic dipole radiation, and finally the decay will steepen again due to disappearance of the stellar effect. In the next section, we will see how to explain some unusual afterglows based on the above conclusion.

SOME UNUSUAL AFTERGLOWS

GRB 980519

The optical afterglow ~ 8.5 hours after GRB 980519 decayed as $\propto t_\oplus^{-2.05\pm0.04}$ in $BVRI$ [18], while the power-law decay index of the X-ray afterglow $\alpha_X = 2.07\pm0.11$ [25], in agreement with the optical. The spectrum in optical band alone is well fitted by a power low $\nu^{-1.20\pm0.25}$, while the optical and X-ray spectra together can also be fitted by a single power law of the form $\nu^{-1.05\pm0.10}$. In addition, the radio afterglow of this burst was observed by the VLA at 8.3 GHz, and its temporal evolution $\propto t_\oplus^{0.9\pm0.3}$ between 1998 May 19.8UT and 22.3UT [14].

We now analyze the observed afterglow data of GRB 980519 based on our model. We assume that for this burst, the forward shock evolved from an ultrarelativistic phase to a nonrelativistic phase in a dense medium at ~ 8 hr after the burst. So, the detected afterglow, in fact, was the radiation from a nonrelativistic shock. This implies $\gamma \sim 1$ at $t_b \approx 1/3$ days. From equation (2), therefore, we find

$$n_5 \sim 27 E_{52}[(1+z)/2]^3. \tag{6}$$

If $p \approx 2.8$, and if the observed optical afterglow was emitted by slow-cooling electrons and the X-ray afterglow from fast-cooling electrons, then according to equation (3), the decay index $\alpha_R = (21-15p)/10 \approx -2.1$ and $\alpha_X = (4-3p)/2 \approx -2.2$, in excellent agreement with observations. Furthermore, the model spectral index at the optical to X-ray band and the decay index at the radio band, $\beta = -(p-1)/2 \approx -0.9$ and $\alpha = 1.1$, are quite consistent with the observed ones, -1.05 ± 0.10 and 0.9 ± 0.3, respectively.

We [12] took into account three observed data which correpond to the radio, R-band and X-ray frequencies respectively, and inferred intrinsic parameters of the shock and the redshift of the burst, $\epsilon_e \sim 0.16$, $\epsilon_B \sim 2.8 \times 10^{-4}$, $E_{52} \sim 0.27$, $n_5 \sim 3.4$, and $z \sim 0.55$. After considering these reasonable parameters, we numerically studied the trans-relativistic evolution of the shock [35] and found that our dense medium model can provide an excellent fit to all the observational data of the radio afterglow from GRB 980519 shown in [15].

GRB 000301C

The optical afterglow data of GRB 000301c were presented in [30]. In addition, the spectral index $\beta = 1.1 \pm 0.1$. We [13] combined the dense medium model with the pulsar energy injection effect to explain the unusual optical afterglow. For stage (ii), if $p = 3.4$, then $\alpha = (12-5p)/5 = -1.0$ and $\beta = -(p-1)/2 = -1.2$ are consistent with the GRB 000301c R-band afterglow data in initial 7.5 days after the burst. These data indicate $\alpha_1 \sim -1.1$, which implies $\alpha_{obs} \sim \beta_{obs}$ at early times. If the afterglow were radiated by fast-cooling electrons in the shocked medium, we would find $\alpha = 2(1-\beta)$, which is clearly inconsistent with the observational result. Therefore, the GRB 000301c R-band afterglow arose from those slow-cooling electrons in the shocked medium. For stage (iii), in the case of $p = 3.4$, the model's time index $\alpha = (21-15p)/10 = -3.0$ is quite consistent with the observational data of the GRB 000301c R-band afterglow at late times, $\alpha_2 = -3.01 \pm 0.53$ [30]. We also carried out simulations of the evolution of a shock with energy injection from a pulsar and the resulting emission [33]. Our numerical results indeed show one sharp break in the late-time afterglow light curve and give a good fit to the R-band afterglow data of GRB 000301c.

CONCLUSIONS

We discussed the evolution of an adiabatic shock expanding in a dense medium from an ultrarelativistic phase to a nonrelativistic phase in more details in this paper. In particular, we discussed the effects of synchrotron self absorption and energy injection on the afterglow from this shock. In a dense medium, the shock becomes nonrelativistic rapidly after a short relativistic phase. This transition time varies from several hours to a few days when the medium density is from 10^5 to a few $\times 10^6$ cm^{-3}, and the shock energy from 10^{51} to 10^{54} ergs. The afterglow from the shock at the nonrelativistic stage decays more rapidly than at the relativistic stage, while the decay index varies from -1.35 to -2.1 if the spectral index of the accelerated electron distribution, $p = 2.8$, and the radiation comes from those slow-cooling electrons. Since some models mentioned above predict that a strongly magnetic millisecond pulsar may be born during the formation of GRB, we also discuss the effect of such a pulsar on the evolution of a nonrelativistic shock through magnetic dipole radiation, in contrast to the case discussed in [8,10]. We found that after the energy which the shock obtains from the pulsar is much more than the initial energy of the shock, the afterglow decay will flatten significantly and the decay index will become -0.4. When the pulsar energy input effect disappears, the index will still be -2.1. These features are in excellent agreement with the afterglow of GRB 980519. Furthermore, our model fits very well all the observational data of GRB 980519 including all the radio data. Our model also provides a good fit to the R-band afterglow data of GRB 000301c.

ACKNOWLEDGMENTS

We would like to thank T. Lu, D. M. Wei and Y. F. Huang for helpful discussions. This work was supported by the National Natural Science Foundation of China (grant 19825109) and by the National 973 Project.

REFERENCES

1. Blandford, R. D., & McKee, C. F., Phys. Fluids **19**, 1130 (1976).
2. Castro-Tirado, A. J. et al., Science **283**, 2069 (1999).
3. Chevalier, R. A., & Li, Z. Y., ApJ **520**, L29 (1999).
4. Chevalier, R. A., & Li, Z. Y., ApJ **536**, 195 (2000).
5. Dai, Z. G., & Gou, L. J., ApJ, submitted (2000).
6. Dai, Z. G., Huang, Y. F., & Lu, T., preprint: astro-ph/9806334.
7. Dai, Z. G., Huang, Y. F., & Lu, T., ApJ **520**, 634 (1999).
8. Dai, Z. G., & Lu, T., Phys. Rev. Lett. **81**, 4301 (1998).
9. Dai, Z. G., & Lu, T., MNRAS **298**, 87 (1998).
10. Dai, Z. G., & Lu, T., A&A **333**, L87 (1998).
11. Dai, Z. G., & Lu, T., ApJ **519**, L155 (1999).
12. Dai, Z. G., & Lu, T., ApJ **537**, 803 (2000).
13. Dai, Z. G., & Lu, T., A&A, submitted (astroph/0005417).
14. Frail, D. A., Taylor, G. B., & Kulkarni, S. R., GCNC **89** (1998).
15. Frail, D. A., et al., preprint: astro-ph/9910060.
16. Fruchter, A. S. et al., preprint: astro-ph/9902236.
17. Gou, L. J. Dai, Z. G., Huang, Y. F., & Lu, T., A&A, submitted (2000).
18. Halpern, J. P., Kemp, J., Piran, T., & Bershady, M. A., ApJ **517**, L105 (1999).

19. Huang, Y. F., Dai, Z. G., & Lu, T., MNRAS **309**, 513 (1999).
20. Huang, Y. F., Gou, L. J., Dai, Z. G., & Lu, T., ApJ, in press (2000).
21. Kulkarni, S. R. et al., Nature **398**, 389 (1999).
22. Kumar, P., & Panaitescu, A., preprint: astro-ph/0003264.
23. Mészáros, P., Rees, M. J., & Wijers, R. A. M. J., ApJ, 499. 301
24. Moderski, R., Sikora, M., & Bulik, T., ApJ **529**, 151 (2000).
25. Owens, A. et al., A&A **339**, L37 (1998).
26. Panaitescu, A., Mészáros, P., & Rees, M. J., ApJ **503**, 315 (1998).
27. Piran, T., Phys. Rep. **314**, 575 (1999).
28. Rees, M. J., & Mészáros, P., ApJ **496**, L1 (1998).
29. Rhoads, J., ApJ **525**, 737 (1999).
30. Sagar, R. et al., preprint: astro-ph/0004223.
31. Sari, R., Piran, T., & Halpern, J. P., ApJ **519**, L17 (1999).
32. Sari, R., Piran, T., & Narayan, R., ApJ **497**, L17 (1999).
33. Wang, W., & Dai, Z. G., ApJ, submitted (2000).
34. Wang, X. Y., Dai, Z. G., & Lu, T., MNRAS, in press (2000).
35. Wang, X. Y., Dai, Z. G., & Lu, T., MNRAS, in press (2000).
36. Wei, D. M., in these proceedings.
37. Wei, D. M., & Lu, T., ApJ, in press (2000).
38. Wijers, R. A. M. J., in these proceedings.

The Emission Features of GRB Afterglows

D.M. Wei[1,2], and T. Lu[3]

[1] *Purple Mountain Observatory, Chinese Academy of Sciences, Nanjing, China*
[2] *National Astronomical Observatories, Chinese Academy of Sciences, China*
[3] *Department of Astronomy, Nanjing University, Nanjing, 210093, China*

Abstract. Here we talk about the effects of inverse Compton scattering (ICS) and jet on GRB afterglows. In the simplest fireball model, synchrotron radiation is believed to be the main mechanism of GRB emission, however, here we will show that under some circumstances, the inverse Compton scattering may play an important role, and can change the light curves of GRB afterglows. Beaming of relativistic ejecta in GRBs has been postulated by many authors in order to reduce the total GRB energy, thus it is very important to look for the observational evidence of beaming. Here we analyse the dynamical evolution of the jet blast wave, calculate the jet emission analytically, we find that the sharp break predicted by Rhoads will actually not exist, and for most cases the afterglow light curve will almost not be affected by sideways expansion unless the beaming angle is extremely small. We demonstrate that only when $\theta_0 < 0.1$, the afterglow light curves may be steepened by sideways expansion, and in fact there cannot be two breaks as claimed before.

THE EFFECT OF ICS ON GRB AFTERGLOW

The observed properties of GRB afterglows are in approximate accord with the fireball model (Meszaros & Rees 1997; Waxman 1997; Wijers, Rees & Meszaros 1997; Wei & Lu 1998a). In this simplest models, the surrounding medium density is assumed to be constant. However, there has been increasing evidence that at least some GRBs have massive star progenitors, as suggested by the link between some GRBs and supernovae, which means that the GRB blast wave should be expanding into the stellar wind of the progenitor star, the densiy $\rho \propto r^{-2}$ (Dai & Lu 1998; Chevalier & Li 1999a, 1999b). Although, in principle, the standard fireball model can approximately explain the afterglow light curves well, there are still some problems that cannot be explained by this simplest model. For example, it is well known that in some GRB afterglows, the light curves cannot be described by a simple power-law, but show sharp breaks (e.g. Castro-Tirado et al. 1999; Kulkarni et al. 1999; Stanek et al. 1999). These observed breaks have generally been interpreted as evidence for collimation of the GRB ejecta (Rhoads 1999), but a difficulty with this model is that the predicted break is quite smooth, while the observed breaks are rather sharp (Panaitescu & Meszaros 1999; Moderski, Sikora & Bulik 2000; Kumar & Panaitescu 2000; Wei & Lu 1999a). The transition of blast wave from relativistic to non-relativistic regime has been proposed as another mechanism for light curve breaks (Dai & Lu 1999).

In the fireball model, the effect of inverse Compton scattering is always neglected. Here we discuss the effect of ICS in a general case, the surrounding medium density can be either uniform or non-uniform ($\rho \propto r^{-2}$), and the blast wave can be in the relativistic or non-relativistic stage.

Relativistic Case

The process of inverse Compton scattering has been discussed by several authors (e.g. Waxman 1997a; Panaitescu & Kumar 2000; Sari & Esin 2000), and here we will consider the influence of ICS on GRB afterglow in details. There is a simple way to estimate the intensity of ICS, i.e. by calculating the ratio of the synchrotron radiation energy density ($u'_{\rm syn}$) to the magnetic energy density ($u'_{\rm B}$), $R = u'_{\rm syn}/u'_{\rm B}$. It is easy to show that

$$R = 10^{-24} \bar{\gamma}^2 r \Gamma^{-1} \frac{\hat{\gamma}\Gamma + 1}{\hat{\gamma} - 1} n \tag{1}$$

where $\bar{\gamma}^2$ is the average value of electron Lorentz factor square, γ_e is the minimum electron Lorentz factor, $\gamma_e = \xi_e(\Gamma - 1)\frac{m_p}{m_e}\frac{p-2}{p-1}$, ξ_e is the energy fraction occupied by electrons, and p is the index of electron distribution, for $p = 3$, we have $\gamma_e \simeq 900\xi_e(\Gamma - 1)$, and $\bar{\gamma}^2 \simeq 10\gamma_e^2$. This formula is valid for both the relativistic and non-relativistic cases, now we first consider the relativistic case.

We have shown earlier that in the relativistic case and for uniform density ($n \sim 1 \,{\rm cm}^{-3}$), the effect of ICS is usually important, the value $R = 48(\frac{\xi_e^2}{0.1})n_1^{1/2}E_{52}^{1/2}t_{\rm day}^{-1/2}$ (Wei & Lu 1998b). Now we extend to the case for non-uniform density, $\rho \propto r^{-2}$.

Chevalier & Li (1999a,b) have discussed the blast wave dynamical evolution in the wind environment, and they gave $\gamma = 4.2(\frac{1+z}{2})^{1/4}E_{52}^{1/4}A_\star^{-1/4}t_{\rm day}^{-1/4}$, $r = 2.9 \times 10^{17}(\frac{1+z}{2})^{-1/2}E_{52}^{1/2}A_\star^{-1/2}t_{\rm day}^{1/2}$ cm, where $A = \dot{M}_w/4\pi V_w = 5 \times 10^{11}A_\star \,{\rm g cm}^{-1}$, $\dot{M}_w$ is the mass loss rate, and V_w is the wind velocity, the reference value of A corresponds to $\dot{M}_w = 1 \times 10^{-5}M_\odot \,{\rm yr}^{-1}$ and $V_w = 1000 \,{\rm km s}^{-1}$. Thus we obtain

$$R = 54(\frac{\xi_e^2}{0.1})(\frac{1+z}{2})A_\star t_{\rm day}^{-1} \tag{2}$$

We see that the emission power of ICS is not smaller than that of synchrotron radiation, thus, the effect of ICS should not be neglected.

Now let us consider the effect of ICS on GRB afterglow. The GRB emission spectrum should consist of two components, below the critical frequency ν_c the spectrum is dominated by synchrotron radiation, and above ν_c the spectrum is dominated by ICS. We assume that in the comoving frame the synchrotron radiation intensity has the form $I_\nu \propto \nu^{-\alpha}$ for $\nu < \nu_m$ and $I_\nu \propto \nu^{-\beta}$ for $\nu > \nu_m$. Since in our situation the soft photons produced through synchrotron radiation are scattered by the same electrons, the Compton-scattered spectrum should have nearly the same form as that of synchrotron radiation. Therefore the total emission intensity is $I_\nu \propto \nu^{-\alpha}$ for $\nu < \nu_m$ or $\nu_c < \nu < \nu_n$, and $I_\nu \propto \nu^{-\beta}$ for $\nu_m < \nu < \nu_c$ or $\nu > \nu_n$, where $\nu_n = \gamma_e^2 \nu_m$ is the peak frequency of the ICS spectrum, and $\nu_n I_{\nu_n} = R \nu_m I_{\nu_m}$. Then, from the relation $I_{\nu_m}(\nu_c/\nu_m)^{-\beta} = I_{\nu_n}(\nu_c/\nu_n)^{-\alpha}$ we can obtain

$$\frac{\nu_c}{\nu_m} = a_1 \xi_e^{-\frac{2\alpha}{\beta-\alpha}} \left(\frac{1+z}{2}\right)^{-\frac{1+\alpha}{2(\beta-\alpha)}} A_\star^{-\frac{3-\alpha}{2(\beta-\alpha)}} E_{52}^{\frac{1-\alpha}{2(\beta-\alpha)}} t_{\rm day}^{\frac{1+\alpha}{2(\beta-\alpha)}} \qquad (3)$$

where $a_1 = 540^{-\frac{1}{\beta-\alpha}}(900 \times 4.2)^{\frac{2(1-\alpha)}{\beta-\alpha}}$. So this critical frequency is dependent on the fireball parameters, i.e. the fireball energy, surrounding gas density, energy fractions in electrons and magnetic field, and the spectrum index of synchrotron radiation.

The afterglow light curves will also be greatly modified by ICS. It has been shown that the electron Lorentz factor $\gamma_e \propto t^{-1/4}$, the typical synchrotron radiation frequency $\nu_m \propto t^{-3/2}$, and the comoving specific intensity of synchrotron radiation at peak frequency is $I_{\nu_m} \propto t^{-5/4}$. From the relation $I_{\nu_n}/I_{\nu_m} \sim R\gamma_e^{-2}$, it is easy to show that the intensity of ICS at peak frequency is $I_{\nu_n} \propto t^{-7/4}$, and $\nu_n \propto t^{-2}$, then the observed peak flux $F_{\nu_m} \propto t^{-1/2}$, and $F_{\nu_n} \propto t^{-1}$. Therefore, we conclude that if our observation is fixed at frequency ν, then the observed flux has four components: $F_\nu \propto F_{\nu_m}(\nu/\nu_m)^{-\alpha} \propto t^{-(1+3\alpha)/2}$ for $\nu < \nu_m$, $F_\nu \propto F_{\nu_m}(\nu/\nu_m)^{-\beta} \propto t^{-(1+3\beta)/2}$ for $\nu_m < \nu < \nu_c$, $F_\nu \propto F_{\nu_n}(\nu/\nu_n)^{-\alpha} \propto t^{-(1+2\alpha)}$ for $\nu_c < \nu < \nu_n$, and $F_\nu \propto F_{\nu_n}(\nu/\nu_n)^{-\beta} \propto t^{-(1+2\beta)}$ for $\nu > \nu_n$. So we can see that, if take $\alpha = 0$, $\beta = 1$, then $F_\nu \propto t^{-1/2}$ for $\nu < \nu_m$, $F_\nu \propto t^{-2}$ for $\nu_m < \nu < \nu_c$, $F_\nu \propto t^{-1}$ for $\nu_c < \nu < \nu_n$, and $F_\nu \propto t^{-3}$ for $\nu > \nu_n$.

Non-relativistic Case

For non-relativistic blast wave, $\hat{\gamma} = 5/3$, and $\Gamma \sim 1$, then $R = 2 \times 10^{-3}(\frac{\xi_e^2}{0.1})\beta^5 n t_{\rm day}$. We now consider two cases (the density $\rho \propto r^{-s}$) for both uniform density ($s = 0$) and for wind density ($s = 2$).

(1) s=0 In this case, the evolution of blast wave velocity is $\beta = (t/t_{\rm NR})^{-3/5}$, where $t_{\rm NR} \simeq 168(\frac{1+z}{2})E_{52}^{1/3}n^{-1/3}$ days. Then it is easy to show that $R(t_{\rm NR}) = 0.34(\frac{\xi_e^2}{0.1})E_{52}^{1/3}n^{2/3}$, and $R(t) = R(t_{\rm NR})(t/t_{\rm NR})^{-2}$. So we see that in general case the ICS is unimportant unless the surrounding density is very high.

If the ICS is important, then, as in the previous section, we can write the typical quantities as follows: $\nu_m(t_{\rm NR}) = 2.8 \times 10^{-5}(\frac{\xi_e^2}{0.1})(\frac{\xi_B}{0.1})^{1/2}n^{1/2}$ eV, and $\nu_m(t) = \nu_m(t_{\rm NR})(t/t_{\rm NR})^{-3}$; for $\alpha = 0$, $\beta = 1$, $\nu_c(t_{\rm NR}) = 1.5(\frac{\xi_e^2}{0.1})(\frac{\xi_B}{0.1})^{1/2}E_{52}^{-1/3}n^{-1/6}$ eV, and $\nu_c(t) = \nu_c(t_{\rm NR})(t/t_{\rm NR})^{-17/5}$; $\nu_n(t_{\rm NR}) = 0.63(\frac{\xi_e^2}{0.1})^2(\frac{\xi_B}{0.1})^{1/2}n^{1/2}$ eV, and $\nu_n(t) =$

$\nu_n(t_{NR})(t/t_{NR})^{-27/5}$. The specific intensity at peak frequency is $I_{\nu_m} \propto t^{-1/5}$, $I_{\nu_n} \propto t^{1/5}$, and the observed peak flux $F_{\nu_m} \propto t^{3/5}$, $F_{\nu_n} \propto t$. Therefore, the observed flux at fixed frequency ν is $F \propto t^{3/5-3\alpha}$ for $\nu < \nu_m$, $F \propto t^{3/5-3\beta}$ for $\nu_m < \nu < \nu_c$, $F \propto t^{-(27\alpha-5)/5}$ for $\nu_c < \nu < \nu_n$, and $F \propto t^{-(27\beta-5)/5}$ for $\nu > \nu_n$.

(2) s=2 In this wind environment, Chevalier & Li (1999b) and Wei & Lu (1999b) have given the blast wave evolution $\beta = (t/t_{NR})^{-1/3}$, where $t_{NR} \simeq 1000 E_{52} A_{\star}^{-1}$. Then we can obtain $R(t_{NR}) = 0.09(\frac{\xi_e^2}{0.1}) A_{\star}^2 E_{52}^{-1}$, and $R(t) = R(t_{NR})(t/t_{NR})^{-2}$. It is also obvious that the ICS is usually unimportant unless the mass loss rate is very high.

We can write the typical quantities as before: $\nu_m(t_{NR}) = 6 \times 10^{-6}(\frac{\xi_e^2}{0.1})(\frac{\xi_B}{0.1})^{1/2} A_{\star}^{3/2} E_{52}^{-1}$ eV, and $\nu_m(t) = \nu_m(t_{NR})(t/t_{NR})^{-7/3}$; for $\alpha = 0$, $\beta = 1$, $\nu_c(t_{NR}) = 1.25(\frac{\xi_e^2}{0.1})(\frac{\xi_B}{0.1})^{1/2} A_{\star}^{-1/2}$ eV, and $\nu_c(t) = \nu_c(t_{NR})(t/t_{NR})^{-5/3}$; $\nu_n(t_{NR}) = 0.14(\frac{\xi_e^2}{0.1})^2(\frac{\xi_B}{0.1})^{1/2} A_{\star}^{3/2} E_{52}^{-1}$ eV, and $\nu_n(t) = \nu_n(t_{NR})(t/t_{NR})^{-1/3}$. The specific intensity at peak frequency is $I_{\nu_m} \propto t^{-5/3}$, $I_{\nu_n} \propto t^{-7/3}$, and the observed peak flux $F_{\nu_m} \propto t^{-1/3}$, $F_{\nu_n} \propto t^{-1}$. Therefore, the observed flux at fixed frequency ν is $F \propto t^{-(1+7\alpha)/3}$ for $\nu < \nu_m$, $F \propto t^{-(1+7\beta)/3}$ for $\nu_m < \nu < \nu_c$, $F \propto t^{-(3+11\alpha)/3}$ for $\nu_c < \nu < \nu_n$, and $F \propto t^{-(3+11\beta)/3}$ for $\nu > \nu_n$.

THE EFFECT OF JET ON GRB AFTERGLOW

The discovery of GRB afterglow shows that GRBs are at cosmological distances. If so, the total energy for typical GRB event is about 10^{52} ergs, especially for GRB990123, the total energy of this source is $\geq 1.6 \times 10^{54}$ ergs if the emission is isotropic (Andersen et al. 1999; Kulkarni et al. 1999). This energy is so large that it gives a great challenge to the popular models. For models involving stellar mass central engines it is necessary to assume that the ejecta are beamed in order to explain such a huge energy. Here we will first give an analytical treatment of the dynamical evolution of the jet blast wave and its emission features, and will demonstrate that the sharp break will not actually exist, we may observe the steepening of the light curve only when the jet angle is extremely small, i.e. $\theta_0 < 0.1$.

Dynamical Evolution of the Jet

Now we consider an adiabatic relativistic jet expanding in surrounding medium. For energy conservation, the evolution equation is $\Gamma^2 V = \text{const}$, where Γ is the bulk Lorentz factor, and V is the jet volume, $V = 2\pi r^3 (1 - \cos\theta_j)/3 \propto r^3 \theta_j^2$ for $\theta_j \ll 1$, and $\theta_j = \theta_0 + \theta' = \theta_0 + c_s t_{co}/ct$, where θ_0 is the initial jet opening half-angle, θ' describes the lateral expansion, c_s is the expanding velocity of ejecta material in its comoving frame, and t (t_{co}) is the time measured in the burster frame (comoving frame). Then we can obtain $\Gamma \propto T^{-3/8}[1 + (\frac{T}{T_b})^{3/8}]^{-1/4}$ for $T < T_b$, and

$\Gamma \propto T^{-3/8}[1+(\frac{T}{T_b})^{1/2}]^{-1/4}$ for $T > T_b$. It is obviously that $\Gamma \propto T^{-3/8}$ for $T \ll T_b$, and $\Gamma \propto T^{-1/2}$ for $T \gg T_b$. The rapid decrease with time of Lorentz factor Γ is due to the fact that larger amounts of surrounding matter has been swept up by ejecta (Rhoads 1997, 1999). And also we can obtain the value of T_b

$$T_b \simeq \frac{r_b}{4\Gamma_b^2 c} = 70(\frac{c_s}{c/\sqrt{3}})^{-8/3}(\frac{\theta_0}{0.1})^2 E_{52}^{1/3} n_1^{-1/3} \quad (day) \tag{4}$$

We see that the break time T_b is very large for typical parameters, which means that the transition from $\Gamma \propto T^{-3/8}$ to $\Gamma \propto T^{-1/2}$ is usually very slowly and smoothly.

The Emission from Jet

Now we calculate the emission flux from the jet. Here we adopt the formulation and notations of Mao & YI (1994). The observed flux is

$$F(\nu,\theta) = \int_0^{2\pi} d\phi' \int_0^{\theta_j} sin\theta' d\theta' D^3 I'(\nu D^{-1}) \frac{r^2}{d^2} \tag{5}$$

where $D = [\Gamma(1-\beta cos\Theta)]^{-1}$ is the Doppler factor, , $\beta = (1-\Gamma^{-2})^{1/2}$, $\nu = D\nu'$, $I'(\nu')$ is the specific intensity of synchrotron radiation at ν', and d is the distance of the burst source. Here the quantities with prime are measured in the comoving frame. For simplicity we have ignored the relative time delay of radiation from different parts of the cone.

For the expanding jet, we have $r = D\Gamma\beta cT \propto D\Gamma T$ ($\beta \simeq 1$), $r' = D\beta cT \propto DT$, the magnetic field strength $B' \propto \Gamma$, the peak frequency of synchrotron radiation $\nu_m = D\nu'_m \propto D\Gamma^3$, and $I'(\nu'_m) \propto n'_e B' r' \propto D\Gamma^2 T$. Assuming that the emission spectrum $I'(\nu') \propto \nu'^{-\alpha}$, then $I'(\nu') = I'(\nu'_m)(\frac{\nu'}{\nu'_m})^{-\alpha} = I'(\nu'_m)(\frac{\nu}{\nu_m})^{-\alpha} \propto D^{1+\alpha}\Gamma^{2+3\alpha}T\nu^{-\alpha}$. Therefore we have the flux

$$F(\nu,\theta) \propto \nu^{-\alpha}\Gamma^{2(\alpha-1)}T^3 g(\theta,\Gamma,\alpha) \tag{6}$$

where

$$g(\theta,\Gamma,\alpha) = \int_0^{2\pi} d\phi' \int_0^{\theta_j} sin\theta' d\theta' (1-\beta cos\Theta)^{-(6+\alpha)} \tag{7}$$

In general, the value of g can only be calculated numerically. However here we consider the case $\theta_j \ll 1$ and $\theta \ll 1$, then $cos\Theta \approx cos\theta cos\theta'$. In this case we can calculate the value of g analytically under certain conditions. Here we define the Lorentz factor, $\Gamma_j \equiv \Gamma(r_j) = \theta_j^{-1}$, then $r_j = (\Gamma_0 \theta_j)^{2/3} r_d$, and the corresponding timescale $T_j \simeq r_j/4\Gamma_j^2 c \simeq 1.3(\frac{\theta_j}{0.1})^2 E_{52}^{1/3} n_1^{-1/3} (\frac{\theta_j}{\theta_0})^{2/3}$ day. Then the observed flux at fixed frequency is $F \propto T^{-3\alpha/2}$ for $T < T_j$, $F \propto T^{-\frac{3}{2}\alpha - \frac{3}{4}}$ for $T_j < T < T_b$, and $F \propto T^{-2\alpha-1}$ for $T_b < T$.

From above it seems that there should be two temporal index breaks in light curves. However, if we compare the values of T_j and T_b, we will find that $T_j \ll T_b$, i.e. the time interval between T_j and T_b is very large, the beaming break is much earlier than the break due to sideways expansion. We know that in order to see the steepening of the light curve, T_j or T_b must be small, so in fact we can only see one temporal break. In addition, in the Rhoads' treatment, the effect of the sideways expansion on the Γ evolution was ignored when $T < T_b$, however in fact, there is still some sideways expansion during this phase, so the evolution of Γ must be affected by sideways expansion when $\Gamma \sim \theta_0^{-1}$. Therefore, we expect that the evolution of Γ is continuous, and the transition from $\Gamma \propto T^{-3/8}$ to $\Gamma \propto T^{-1/2}$ is much smoother than previously claimed.

DISCUSSION AND CONCLUSION

The detection of GRB afterglows has greatly furthered our understanding these objects. In particular, the shape of the afterglow light curves provides important information for exploring their emission mechanism. Here we have calculated the effects of ICS on the GRB afterglows. We have shown that, when the blast wave is relativistic, the ICS emission is usually important for both uniform medium and wind environment, while when the blast wave is non-relativistic, the effect of ICS is usually unimportant unless the surrounding medium density is very high, such as $n \sim 10^6 \, \text{cm}^{-3}$ as proposed by Dai & Lu (1999). When the ICS contribution is important, then it may have great influence on the shape of afterglow light curves, i.e. it can flatten or steepen the light curves.

In addition, the GRB afterglows provide very good opportunity to study whether and how much the GRB ejecta are beamed. Rhoads predicted that the afterglow light curves should have a sharp break around T_b. However, Moderski et al. (1999) have performed numerical calculation and shown that the break of the light curve is weaker and smoother than the prediction. Here we reanalyse the dynamical evolution of the jet blast wave, calculate the emission from the jet. Our calculations show that the main reason why the results of Moderski et al. being different from that of Rhoads is that the value of T_b is very large when taking the parameters adopted by Modersli et al.. Our formular indicates that the evolution of Lorentz factor Γ with time T is continuous, changing the slope from -3/8 to -1/2 smoothly. In particular, if the value of T_b is large, then the transition is much smoothly, in this case one expects that the sharp break will not exist.

ACKNOWLEDGEMENTS

This work is supported by the National Natural Science Foundation (19703003 and 19773007) and the National Climbing Project on Fundamental Researches of China.

REFERENCES

1. Andersen, M. I., et al., 1999, *Science*, 283, 2075
2. Castro-Tirado, A.J., et al., 1999, *Science*, 283, 2069
3. Chevalier, R.A., Li, Z.Y., 1999a, *ApJ*, 520, L29
4. Chevalier, R.A., Li, Z.Y., 1999b, astro-ph/9908272
5. Dai, Z.G., Lu, T., 1998, *MNRAS*, 298, 87
6. Dai, Z.G., Lu, T., 1999, *ApJ*, 519, L155
7. Kulkarni, S.R., et al., 1999, *Nature*, 398, 389
8. Kumar, P., Panaitescu, A., 2000, astro-ph/0003264
9. Mao, S., Yi, I., 1994, *ApJ*, 424, L131
10. Mészáros, P., Rees, M.J., 1997, *ApJ*, 476, 232
11. Mészáros, P., Rees, M.J., Wijers, R., 1998, *ApJ*, 499, 301
12. Moderski, R., Sikora, M., Bulik, T., 2000, *ApJ*, 529, 151
13. Panaitescu, A., Kumar, P., 2000, astro-ph/0003246
14. Panaitescu, A., Mészáros, P., 1999, *ApJ*, 526, 707
15. Rhoads, J.E., 1997, *ApJ*, 487, L1
16. Rhoads, J.E., 1999, *ApJ*, 525, 737
17. Sari, R., Esin, A., 2000, astro-ph/0005253
18. Stanek, K.Z., et al., 1999, *ApJ*, 522, L39
19. Veillet, C., Boer, M., 2000, GCNC 611
20. Waxman, E., 1997, *ApJ*, 485, L5
21. Wei, D.M., Lu, T., 1998a, *ApJ*, 499, 754
22. Wei, D.M., Lu, T., 1998b, *ApJ*, 505, 252
23. Wei, D.M., Lu, T., 1999a, *ApJ, accepted (astro-ph/9908273)*
24. Wei, D.M., Lu, T., 1999b, astro-ph/9912063
25. Wijers, R.A.M.J., Rees, M.J., Mészáros, P., 1997, *MNRAS*, 288, L51

On the Physics of Gamma-Ray Burst Engines and Its Observational Consequences

H.-Y. Chang, I. Yi, C. Kim, and K. Kwak

Korea Institiute for Advanced Study
207-43 Cheongryanri-dong, dongdaemun-gu, Seoul
132-012, Korea

Abstract. Assuming that gamma-ray bursts (GRBs) result from emission originated by relativistically moving particles in a beam with a large bulk Lorentz factor, we demonstrate that a conical beam affects the width and the shape of the luminosity function. We investigate effects of a spatial variation of the Lorentz factor and spatial density fluctuations within the cone on the luminosity function and implications on the redshift distribution of the observed GRBs. A non-uniform conical beam yields substantially higher maximum redshifts than those inferred with uniform beam models. Consequently, when high redshift values of GRBs are interpreted, the effect of the luminosity function should be considered appropriately as well as the source evolution effect.

We study the power density spectrum (PDS) of artificial light curves of GRBs. We discuss implications on interpretations of PDS analysis results. We present an example to emphasize that, though an attempt to identify the most sensitive physical parameter has been made on the basis of the internal shock model, conclusions of this kind of approach should be derived with due care. We show that the reported slope of the averaged PDS and the distribution of individual power can be reproduced by adjusting the sampling interval in the time domain for a given decaying timescale of individual pulse in a specific form of GRB light curves. We propose that the PDS analysis could be used for more important problems such as the evolution of the GRB emission in a single burst and the classification of the origin of short and long bursts.

BEAMING AND LUMINOSITY FUNCTION

Yi [12] considered the effect of the luminosity function due to the cylindrical-beam on statistical implications of GRBs. He showed that a broad luminosity function can be a reasonable answer for a wide range of observed GRB luminosities and that it can be naturally expected if we assume beaming of GRBs. We examine effects of the cylindrical beaming and the conical beaming on the luminosity function.

We consider the beaming-induced luminosity function first [11,12], where the relativistic beam is perfectly collimated. Assumed that the beamed emission from 'standard bursts' this model defines a range of the luminosity, i.e. $[L_{\min}, L_{\max}]$, for a given photon index α, and the bulk Lorentz factor γ. An apparent luminosity function $\Phi(L)$ due to uniformly distributed beams in space is given by

$$\Phi(L) = \frac{1}{p\beta\gamma} L_{\text{int}}^{1/p} L^{-(p+1)/p} \tag{1}$$

where $\beta = (1-\gamma^{-2})^{1/2}$ and $p = \alpha + 2$.

As a more realistic model, we adopt conically-beamed emission from GRBs, which was studied by Mao and Yi [5], and extended by Chang and Yi [3]. The probability that we observe the bright bursts rapidly increases as the opening angle increases, while relatively dim bursts are not as detectable as brighter ones. Therefore, as the opening angle increases, the derived luminosity function becomes similar to that of the standard candle case, in which all bursts have the same maximum luminosity given by this luminosity function. When the opening angle is small ($\Delta\theta \ll 1/\gamma$), the luminosity function gives the same result as the cylindrical beaming case.

EFFECTS OF LUMINOSITY FUNCTIONS INDUCED BY BEAMINGS

We adopt two statistical quantities of our sample out of BATSE 4B catalog [7]. We selected bursts detected in 1024 ms trigger timescale with the peak count rates satisfying $C_{\max}/C_{\min} \geq 1$, which left us 775 bursts. We divide this sample into the long bursts (588 bursts) and the shorts bursts (149 bursts) according to the burst duration T_{90}. First, we calculate the $\langle V/V_{\max}\rangle$ as a function of threshold flux F_{th} [4,9]. Second, we calculate the fraction of GRBs. We define the fraction of bursts located at a redshift larger than z' as

$$f_{>z'} = \frac{N(z' < z < z_{\max})}{N(0 < z < z_{\max})} \tag{2}$$

where

$$N(z' < z < z_{\max}) = \int_{z'}^{z_{\max}} \frac{4\pi}{1+z} n(z) r^2(z) dr(z). \tag{3}$$

We obtain results from different combinations of luminosity functions and number density distributions. We use two number density distributions: SFR-motivated number density distribution derived by the recently observed Star Forming Rate (SFR) data [10], and the uniform number density distribution. Apparently all the luminosity functions we have studied satisfy the observed $\langle V/V_{\max}\rangle$ curve at the 3σ significance level. Although the number density distribution of GRB sources is believed to correlate with the SFR, we cannot rule out the uniform distribution of burst sources in this analysis. It is interesting to note that, if we assume $\alpha = 1.0$, a large fraction of GRBs are distributed at high redshifts, provided that GRBs are assumed to follow the SFR, regardless of specific beaming models. However, in the case of $\alpha = 2.0$, $z_{\max}$ is significantly reduced, though it still remains quite large. The estimated $z_{\max}$ for the cylindrical-beam case is as high as ~ 14 ($\alpha = 1.0$) and ~ 6 ($\alpha = 2.0$) for the long bursts and ~ 3 ($\alpha = 1.0$) and ~ 1.6 ($\alpha = 2.0$) for the short bursts where α is the photon index. When we take $z' = 3.42$ and apply the luminosity function derived for the cylindrical-beam, the expected fraction of GRBs $f_{>z'}$ is $\sim 75\%$ ($\alpha = 1.0$) and $\sim 50\%$ ($\alpha = 2.0$) for long bursts. This is too large a value given the fact that the observed GRBs with such high redshifts account for $\leq 10\%$ of the total observed bursts. It is also interesting to note that, though their $\langle V/V_{\max} \rangle$ value is close to the Euclidean value, most of the short bursts are distributed at high redshifts in the case of $\alpha = 1.0$. When we assume that the long bursts are uniformly distributed in space, the $z_{\max}$ estimate for the broad beaming cannot explain the largest observed redshift in both cases of $\alpha = 1.0$ and 2.0. We conclude that beaming-induced luminosity functions are compatible with the redshift distribution of the observed GRBs and that the apparent "Euclidean" value of $\langle V/V_{\max} \rangle$ might not be due to the near Euclidean space distribution but to the luminosity distribution.

EFFECTS OF NONUNIFORM EMISSION FROM RELATIVISTIC CONE

We also study the luminosity functions for both the varying bulk Lorentz factor and the inhomogeneous electron density. We assume the axisymmetry of the Lorentz factor profile around the cone axis, $\gamma = \gamma(\theta')$, where θ' is measured from the symmetric axis of the cone, and hence the Lorentz factor profile could mimic a simplified model for the jet-environment drag. The window function we adopt is the Gaussian function centered at the center of the cone:

$$W(\theta') = \exp[-A(\frac{\theta'}{\Delta\theta})^2], \qquad (4)$$

where A is a constant, $\Delta\theta$ is a given opening angle, and θ' varies from 0 to $\Delta\theta$. The local peak count rate P_{loc} represents a photon count rate at the source. The logarithm of the probability function of the logarithm of the local peak count rate p_L is defined by the distribution of the cone's axis. At the center of the cone, γ has the maximum value and decreases with θ'. As γ decreases more steeply with θ', the $\log p_L(\log P_{\text{loc}})$ shows a smaller peak at $\log(P_{\text{loc}}/P_{\text{loc max}}) = 0$, and a higher level of the 'tail' of the $\log p_L(\log P_{\text{loc}})$. It is because this type of γ effectively reduces the 'average' value of γ over the surface of the cone and the 'effective opening angle' simultaneously. The photon-emitting electrons are supposed to be distributed according to the γ distribution such that the local electron number density is inversely proportional to the square of the bulk Lorentz factor, γ.

It shows that the redshift of the most distant GRBs becomes larger as γ falls more steeply from the cone center. The effects of the Gaussian window on the values of

z_{max} are substantial enough for further comments. For instance, the model for a narrow window with $A = 4$ gives $z_{max} = 3.7$ (average redshift $<z> = 1.03$) while the uniform beam model (i.e. without any variation of γ) gives $z_{max} = 1.6 (<z> = 0.51)$. This simple case indicates that the effects of the luminosity function have significant implications on the cosmological spatial distribution of GRBs.

FOURIER TRANSFORM TECHNIQUE

The Fourier transform technique is widely used to study turbulent fluid and to search for the underlying process in a stochastic system as well as periodical phenomena [2]. Beloborodov et al. [1] applied the Fourier transform technique to the GRB subject. They analysed 214 light curves of long GRBs ($T_{90} > 20$ sec). And they reported that even though individual PDSs were very diverse the averaged PDS was in accord with a power law of index $-5/3$ over 2 orders of magnitude of a frequency range, and that fluctuations in the power were distributed according to the exponentional distribution. They claimed that the GRB emission was generated in a relativistic and fully developed turbulent outflow, resulting from the coalescence of two neutron stars or a neutron star and a black hole, by noting that the value of the slope was the same as the Kolmogorov spectrum of velocity fluctuation in turbulent fluid.

We study the PDS of artificially generated light curves of GRBs. Light curves of GRBs show the diverse temporal profiles. Besides differences in different bursts, pulse shapes exhibit a broad range in a form of individual pulses, in the rise and decay time scales, in a variability. Burst asymmetry on short time scales results from the tendency for most (~ 90 %) pulses to rise more quickly than they decay, the majority having rise-to-decay time scale ratios of $0.3 - 0.5$, independent of energy. Nonetheless, it is worth noting that not all of the bursts show FRED shape (Fast Rise, Exponential Decay) [6]. We describe GRB light curves as a sum of two-sided exponential functions given by:

$$f(t) = \sum_m f_m(t), \quad (5)$$

where

$$f_m(t) = \Lambda_m \exp(a_m(t - t_m)), \quad t < t_m \quad (6)$$
$$\quad\quad\quad \Lambda_m \exp(-b_m(t - t_m)), \quad t > t_m,$$

Λ_m being the height of peaks, t_m being the time of the pulse's maximum intensity, a_m^{-1} and b_m^{-1} being the rise and decay timescales, respectively. Then, the Fourier transform of the function is obtained analytically. Since the Fourier transform is a linear operator, the Fourier transform of $f(t)$ is a sum of the Fourier transforms of $f_m(t)$, which reads

$$F_m(\omega) = \frac{\Lambda_m \exp(-a_m t_m)}{i\omega + a_m}[\exp(i\omega + a_m)t_m - 1] \quad (7)$$
$$+ \frac{\Lambda_m \exp(b_m t_m)}{i\omega - b_m}[\exp(i\omega - b_m)T - \exp(i\omega - b_m)t_m],$$

where $i = \sqrt{-1}$, T is the observational duration, or the duration of the burst, and ω is the angular frequency. The PDS of $f(t)$ is defined as a square of the modulus of the Fourier transform of $f(t)$. The final resulting PDS of $f(t)$ is given by

$$P(\omega) = \sum_m \sum_n F_m(\omega) F_n^*(\omega) \quad (8)$$

$$= \sum_m \sum_n \Big[\frac{\Lambda_m \Lambda_n \exp(-(a_m t_m + a_n t_n))}{(a_m a_n + \omega^2)^2 + \omega^2 (a_m - a_n)^2} \quad (9)$$
$$\{(a_m a_n + \omega^2) g_1(\omega) + \omega(a_m - a_n) g_2(\omega)\}$$
$$+ \frac{\Lambda_m \Lambda_n \exp(b_m t_m + b_n t_n)}{(b_m b_n + \omega^2)^2 + \omega^2 (b_m - b_n)^2}$$
$$\{(b_m b_n + \omega^2) g_3(\omega) + \omega(b_m - b_n) g_4(\omega)\}$$
$$+ \frac{2\Lambda_m \Lambda_n \exp(-a_m t_m + b_n t_n)}{(\omega^2 - a_m b_n)^2 + \omega^2 (a_m + b_n)^2}$$
$$\{(\omega^2 - a_m b_n) g_5(\omega) + \omega(a_m + b_n) g_6(\omega)\}\Big],$$

where $g_k(\omega)$'s are complicated cos and sin terms which cause fluctuations on the PDS.

IMPLICATIONS OF INTERPRETATION OF PDS OF GRB

In practice, however, we sample GRB light curves every pre-determined time interval, e.g., 64 ms. The time interval defines the Nyquist frequency, which limits the region we see the information in the frequency domain [2]. Therefore, the resulting PDS cannot reveal generic features of the PDS of the original function in the time domain, unless the sampling interval is short enough. Consider the PDS of the bi-exponential function as an example of this. The PDS of the exponential function is given by

$$P(\omega) \approx \frac{1}{a^2 + \omega^2}, \quad (10)$$

where a^{-1} is the typical decaying timescale. One is likely to expect that the slope of the averaged PDS should follow the slope of -2, provided that the sampling interval is sufficiently shorter than the characteristic decaying timescale. However, if an observer takes insufficiently frequent samples, that is, the Nyquist frequency is insufficiently large, then one may see the transition region of the PDS from the flat part to the power law part with the slope of -2. Under this circumstance, the slope one may end up with is a function of the decaying timescale and the sampling interval. We demonstrate that the shortest sampling interval which is currently available may be insufficiently short indeed.

In the essentially same PDS, one may see different parts of the PDS in the frequency domain as a function of sampling interval. As the sampling interval becomes shorter, the Nyquist frequency becomes larger. Subsequently, the maximum frequency becomes larger for more frequent sampling. Different parts of the PDS appears to follow a slightly different slope. The observed slope of the averaged PDS is indeed a function of the rising and decaying timescales, and the sampling interval. The distribution of individual powers appears to follow the exponential distribution.

DISCUSSIONS

We have demonstrated that the beaming-induced luminosity function may explain the statistics of the observed GRB data. Our study is based on assumptions that all GRBs have the same intrinsic luminosity and they are all beamed. Because the $\langle V/V_{\max}\rangle$ test is not sufficiently sensitive to the observed data, different luminosity functions are essentially indistinguishable. If there is an intrinsic luminosity function and the degree of beaming of GRBs is moderate, we may obtain results similar to what we have presented in this study. If all GRBs are highly beamed and their sources follow a SFR-like distribution, the maximum detectable redshift becomes very large, even in the case that the observed $\langle V/V_{\max}\rangle$ is sufficiently close to the Euclidean value, i.e., 0.5. Therefore, it is difficult to rule out a possibility that the apparent Euclidean value of $\langle V/V_{\max}\rangle$ may be due to the luminosity function, for instance, induced by beaming. In other words, the so-called Euclidean value may have nothing to do with the Euclidean distribution. The obtained $z_{\max}$'s in the cylindrical-beam case are $\sim 10, 14, 3$ ($\alpha = 1.0$) and $\sim 4.5, 5.5, 1.6$ ($\alpha = 2.0$) for the total selected sample, and long and short subgroups, respectively. The obtained $z_{\max}$'s in the broader beam case (i.e. $\Delta\theta = 3°.0$) are $\sim 4, 6, 2$ ($\alpha = 1.0$) and $\sim 2.2, 2.6, 1.2$ ($\alpha = 2.0$) for the total selected sample, and long and short subgroups, respectively. Without any detailed information on the redshift distribution of GRBs, the maximum redshift gives a rather strong constraint on the ratio between the Lorentz factor and the opening angle $\Delta\theta$. For the conic beam case, the luminosity function derived for the narrow opening angle seems to fit the $\langle V/V_{\max}\rangle$ curve better.

In the case of the cylindrically beamed luminosity function, we determined the intrinsic luminosity L_{int} by the $\langle V/V_{\max}\rangle$ test for each subsample. In all cases, the intrinsic luminosity is much smaller than that of non-beamed luminosity functions, i.e. $L \sim 10^{52-53}$ erg/sec. The maximum luminosity obtained from our luminosity functions, in both the cylindrical-beam and the conic-beam cases, is

$L_\mathrm{max} \sim 10^{50-51}$ erg/sec. This is compatible with the required value for isotropic radiation. The ratio of the maximum luminosity for the long bursts to that for the short bursts $L_\mathrm{max,long}/L_\mathrm{max,short}$ is ~ 7 for the narrow beam and ~ 3 for the broad beam, respectively. This implies that these two subgroups may have two different intrinsic luminosity functions, and hence probably two different origins. The present work has shown that the simple beaming model and their resulting apparent luminosity function have significant effects in interpreting the observed data.

We have also demonstrated that the Fourier transform technique can be used in investigations of the behavior of the 'central engine'. Furthermore, unless one resolves the issue as to whether the currently available time interval is short enough, in comparison with the rising and decaying timescales, efforts to identify a controlling parameter on the behavior of the PDS should be carried out with due care [8]. As we have demonstrated a conclusion from such kind of analysis is not unique.

It is a challenging question of whether one may determine the decaying timescale for a given observational sampling interval and the observed slope. So far such a data analysis method has not been introduced to individual GRB light curve. Provided that one implements a sophisticated algorithm to accommodate the diversity of the light curves with further efforts, this method could be used for more important problems such as the evolution of the GRB emission in a single burst and the classification of the origin of short and long bursts. The two classes may have intrinsically different flare time scales which could be identified in an analysis similar to the present one. One would like to apply this method to long bursts individually, and to see whether there is a signature of a possible evolution in the GRB emission mechanism. One may also explore the emission timescale of long bursts and short bursts to account for two different origins of them, for instance, as suggested by Yi and Blackman [13].

REFERENCES

1. Beloborodov, A., Stern, B., & Svensson, R., *Ap. J. Lett.* **508**, L25 (1998).
2. Bracewell, R., *The Fourier Transform and Its Applications* New York:McGraw-Hill, 1965.
3. Chang, H.-Y. & Yi, I., astro-ph/0005302 (2000).
4. Che, H., Yang, Y. & Nemiroff, R. J., *Ap. J.* **516**, 559 (1999).
5. Mao, S. & Yi, I., *Ap. J. Lett.* **424**, L131 (1994).
6. Norris, J. P. et al., *Ap. J.* **459**, 393 (1996).
7. Paciesas, W. S. et al., *Ap. J. Supp.* **122**, 465 (1999).
8. Panaitescu, A., Spada, M., & Mészáros, P. *Ap. J. Lett.* **522**, L105 (1999)
9. Schmidt, M., Higdon, J. C., & Hueter, G., *Ap. J.* **329**, L85 (1998).
10. Steidel, C. et al., *Ap. J.* **519**, 1 (1999).
11. Yi, I., *Phy. Rev. D* **48**, 4518 (1993).
12. Yi, I., *Ap. J.* **431**, 543 (1994).
13. Yi, I. & Blackman, E. G., *Ap. J. Lett.* **494**, L163 (1998).

SUPERNOVAE AND JETS

Jet Induced Supernovae: Hydrodynamics and Observational Consequences

A. Khokhlov [1], P. Höflich [2]

[1] *Naval Research Lab, Washington DC, USA*
[2] *Department of Astronomy, University of Texas, Austin, TX 78681, USA*

Abstract. Core collapse supernovae (SN) are the final stages of stellar evolution in massive stars during which the central region collapses, forms a neutron star (NS), and the outer layers are ejected. Recent explosion scenarios assumed that the ejection is due to energy deposition by neutrinos into the envelope but detailed models do not produce powerful explosions. There is mounting evidence for an asphericity in the SN which is difficult to explain within this picture. This evidence includes the observed high polarization, pulsar kicks, high velocity iron-group and intermediate-mass elements material observed in remnants, etc.

The discovery of highly magnetars revived the idea that the basic mechanism for the ejection of the envelope is related to a highly focused MHD-jet formed at the NS. Our 3-D hydro simulations of the jet propagation through the star confirmed that the mechanism can explain the asphericities.

In this paper, detailed 3-D models for jet induced explosions of "classical" core collapse supernovae are presented. We demonstrate the influence of the jet properties and of the underlaying progenitor structure on the final density and chemical structure. Finally, we discuss the observational consequences, predictions and tests of this scenario.

INTRODUCTION

Supernovae (SN) are among the most spectacular events because they reach the same brightness as an entire galaxy. This makes them good candidates to determine extragalactic distances and to measure the basic cosmological parameters. Moreover, they are thought to be the major contributors to the chemical enrichment of the interstellar matter with heavy elements. Energy injection by SN into the interstellar medium, triggers star formation and feedback in galaxy formation, and is regarded as a key for our understanding of the formation and evolution of galaxies.

Core collapse supernovae are thought to be the final stages of the evolution of massive stars which live only 10^6 to 2×10^8 years. Such supernovae could be the brightest objects in the distant past when stars first began to form. A detailed understanding of core collapse is essential to probe the very early phases of the Universe right after the initial star forming period which occurs at redshifts $z \geq 3...5$. Understanding the mechanism of core collapse supernovae explosions is a problem that has challenged researchers for decades (Hoyle & Fowler 1960). In the general scenario for the explosion, the central region of a massive star collapses and forms a neutron star. Eventually, parts of the potential energy will cause the ejection of the envelope. This general scenario has been confirmed by a wealth of observations including the direct detection of neutrinos in SN1987A and neutron stars in young supernovae remnants.

In recent years, there has been a mounting evidence that the explosions of massive stars (core collapse supernovae) are highly aspherical. (1) The spectra of core-collapse supernovae (e.g., SN87A, SN1993J, SN1994I, SN1999em) are significantly polarized indicating asymmetric envelopes (Méndez et al. 1988; Höflich 1991; Jeffrey 1991; Wang et al. 1996; Wang et al. 2000). The degree of polarization tends to vary inversely with the mass of the hydrogen envelope, being maximum for Type Ib/c events with no hydrogen (Wang et. al. 1999, Wang et al. 2000). For supernovae with a good time and wavelength coverage the orientation of the polarization vector tends to stay constant both in time and in the wavelength. This suggests that there is a global symmetry axis in the ejecta (Wang et al. 2000). (2) Observations of SN 1987A showed that radioactive material was brought to the hydrogen rich layers of the ejecta very quickly during the explosion (Lucy 1988; Tueller et al. 1991). (3) The remnant of the Cas A supernova shows rapidly moving oxygen-rich matter outside the nominal boundary of the remnant and evidence for two oppositely directed jets of high-velocity material (Fesen & Gunderson 1997). (4). Recent X-ray observations with the CHANDRA satellite have shown an unusual distribution of iron and silicon group elements with large scale asymmetries in Cas A (Huges et al. 2000). (5) After the explosion, neutron stars are observed with high velocities, up to 1000 km/s (Strom et al. 1995).

There is a general agreement that the explosion of a massive star is caused by the collapse of its central parts into a neutron star or, for massive progenitors, into a black hole. The mechanism of the energy deposition into the envelope is still debated. The process likely involves the bounce and the formation of the prompt shock (e.g. Van Riper 1978, Hillebrandt 1982), radiation of the energy in the form of neutrino (e.g. Bowers & Wilson 1982) and the interaction of the neutrino with the material of the envelope and various types of convective motions (e.g. Herant et al. 1994, Burrows et al. 1995, Müller & Janka 1997, Janka & Müller 1996), rotation (e.g. LeBlanc & Wilson 1970, Saenz & Shapiro S.L. 1981, Mönchmeyer et al. 1991) and magnetic fields (e.g. LeBlanc & Wilson 1970, Bisnovati-Kogan 1971).

Spherically symmetric explosion models rely on the neutrino deposition mechanism. The results depend critically on the progenitor structure, equation of state, neutrino physics, and implementation of the neutrino transport. Currently, results are inconclusive even when using sophisticated Boltzman solvers for the neutrino transport. For example, Mezzacappa et al. (2000) find an explosion whereas

Yamada et al. (1999) and Rammp & Janka (2000) do not. Even if successful, these models cannot explain the observed asymmetries. Within the spherical core-collapse picture, additional mechanisms must be invoked which operate within the envelope itself.

Two such mechanisms have been studied in some detail. One is the Rayleigh-Taylor instability which causes mixing of the layers of different composition when the outgoing shock front passes through (Müller et al. 1989, Benz & Thielemann 1990, Fryxell et al. 1991). This effect can explain mixing of the carbon, oxygen and helium-rich layers required for the SN1987A, but none of the simulations were able to account for the high velocity of Ni observed in SN1987A (Kifonidis et al. 2000). Rayleigh-Taylor mixing provides a rather small-scale structures and can hardly account for the observed polarization which requires a global asymmetry of the expanding envelope (Höflich 1991). Another mechanism involves an explosion inside a rapidly and differentially rotating supernova progenitor (Steinmetz & Höflich 1992). With this mechanism it was possible to account for the polarization in SN1987A which originated from a blue supergiant. This mechanism may have difficulty accounting for the early polarization in some Type II supernovae (Wang et al. 2000) whose light curves indicate a red-giant progenitors. A strong differential rotation in red supergiants can hardly be expected due to their convective envelopes (Steinmetz & Höflich, 1991).

Attempts have also been made to include multi-dimensional effects into a model of collapse itself. The collapsing core becomes unstable due to the gradients of both electron mole fraction Y_e and entropy. The developing convection then affects both the neutrino flux and the energy deposition behind the stalled shock. Numerous studies have demonstrated the presence of this effect. It is still debated whether convection combined with the neutrino transport provides the solution to the supernova problem (Rammp et al. 1998 and references therein). In the current calculations, the size and scale of the convective motions seem to be too small to explain the observed asymmetries. The angular variability of the neutrino flux caused by the convection has been invoked to explain the neutron star kicks (Burrows et al. 1995, Janka & Müller 1994). Calculations give kick velocity up to $\simeq 100$ km/s whereas NS with velocities of several 100 km/s are common.

Rotation of the collapsing core may also be important. It tends to facilitate the explosion because the centrifugal barrier reduces the effective potential for the material moving in the equatorial plane and introduces an axial symmetry in the fluid motions. Simulations made so far indicate that the rotation alone has no or has only a weak effect on the explosion (e.g. Mönchmeyer et al. 1991, Zwerger & Müller 1997). In the latter case it induces a rather weak asymmetry of the explosion with more energy going along the rotational axis.

It has long been suggested that the magnetic field can play an important role in the explosion (LeBlanc & Wilson 1970; Ostriker & Gunn 1971, Bisnovati-Kogan 1971, Symbalisty 1984). LeBlanc and Wilson simulations showed the amplification of the magnetic field due to rotation and the formation of two oppositely directed, high-density, supersonic jets of material emanating from the collapsed core. Their simulations assumed a rather high initial magnetic field $\sim 10^{11}$ Gauss and produced a very strong final fields of the order of $\sim 10^{15}$ Gauss which seemed to be unreasonable at the time. The recent discovery of pulsars with very high magnetic fields (Kouveliotou et al. 1998, Duncan & Thomson 1992) revives the interest in the role of rotating magnetized neutron stars in the explosion mechanism. It is not clear whether a high initial magnetic field required for the LeBlanc & Wilson mechanism is realistic. On the other hand it may not be needed. Recently Mayer and Wilson (2000, private communication) have suggested that the field amplification and the generation of the jet may continue during the first few seconds of the cooling of the neutron star when the neutron star shrinks rapidly. The current picture of the core collapse process is unsettled. A quantitative model of the core collapse must eventually include all the elements mentioned above.

Due to the difficulty of modeling core collapse from first principles, a very different line of attack on the explosion problem has been used extensively and proved to be successful in understanding of the supernova problem, SN1987A in particular (Arnett et al. 1990, Hillebrand & Höflich 1991). The difference of characteristic time scales of the core (a second or less) and the envelope (hours to days) allows one to divide the explosion problem into two largely independent parts - the core collapse and the ejection of the envelope. By assuming the characteristics of the energy deposition into the envelope during the core collapse, the response of the envelope can be calculated. Thus, one can study the observational consequences of the explosion and deduce characteristics of the core collapse and the progenitor structure. This approach has been extensively applied in the framework of the 1D spherically symmetric formulation. The major factors influencing the outcome have been found to be the explosion energy and the progenitor structure. The same approach can be applied in multi-dimensions to investigate the effects of asymmetric explosions. In this paper we study the effects and observational consequences of an asymmetric, jet-like deposition of energy inside the envelope of a core-collapse supernova.

NUMERICAL METHODS AND MODEL SETUP

3-D Hydrodynamics: The explosion and jet propagation are calculated by a full 3-D code within a cubic domain of size D. The stellar material is described by the time-dependent, compressible, Euler equations for inviscid flow with an ideal gas equation with $\gamma = 5/3$ plus a component due to radiation pressure with $\gamma = 4/3$. The Euler equations are integrated using an explicit, second-order accurate, Godunov type, adaptive-mesh-refinement, massively parallel, Fully-Threaded Tree (FTT) program, ALLA (Khokhlov 1998). Euler fluxes are evaluated by solving a Riemann problem at cell interfaces. FTT discretization of the computational domain allowes the mesh to be dynamically refined or coarsened at the level of individual cells. For more details, see Khokhlov (1998) and Khokhlov et al. (1999ab).

1-D Radiation-Hydrodynamics: About 1000 seconds after the core collapse and in case of the explosion of red supergiants, the propagation of the shock front becomes almost spherical (see below). To be able to follow the developement up to the phase of homologous expansion ($\approx 3 - 5$ days), the 3-D structure is remapped on a 1-D grid, and the further evolution is calculated using a one-dimensional radiation-hydro code (e.g. Höflich et al. 1998) that solves the hydrodynamical equations explicitly in the comoving frame by the piecewise parabolic method (Colella and Woodward 1984).

Radiation Transport: Detailed polarization and flux spectra for asymmetric explosions are calculated using our Monte Carlo code including detailed equations of state. For details, see Höflich (1995), Höflich et al. (1995) & Wang et al. (1998).

The Setup: The computational domain is a cube of size L with a spherical star of radius R_{star} and mass M_{star} placed in the center. The innermost part with mass $M_{core} \simeq 1.6 M_\odot$ and radius $R_{core} = 4.5 \times 10^8$ cm, consisting of Fe and Si, is assumed to have collapsed on a timescale much faster than the outer, lower-density material. It is removed and replaced by a point gravitational source with mass M_{core} representing the newly formed neutron star. The remaining mass of the envelope M_{env} is mapped onto the computational domain. At two polar locations where the jets are initiated at R_{core}, we impose an inflow with velocity v_j ρ_j. At R_{core}, the jet density and pressure are the same as those of the background material. For the first 0.5 s, the jet velocity at R_{core} is kept constant at v_j. After 0.5 s, the velocity of the jets at R_{core} was gradually decreased to zero at approximately 2 s. The total energy of the jets is E_j. These parameters are consistent within, but somewhat less than, those of the LeBlanc-Wilson model.

RESULTS

Jet Propagation:

As a baseline case, we consider a jet-induced explosion in a helium star. Jet propagation inside the star is shown in Fig. 1. As the jets move outwards, they remain collimated and do not develop much internal structure. A bow shock forms at the head of the jet and spreads in all directions, roughly cylindrically around each jet. The jet-engine has been switched off after about 2.5 seconds the material of the bow shock continues to propagate through the star. The stellar material is shocked by the bow shock. Mach shocks travels two wards the equator resulting in a redistribution of the energy. The opening angle of the jet depends on the ratio between the velocity of the bow shock to the speed of sound. For a given star, this angle determines the efficiency of the deposition of the jet energy into the stellar envelope. Here, the efficiency of the energy deposition is about 40 %, and the final asymmetry of the envelope is about two.

Influence of the Jet Properties

Fig. 2 shows two examples of an explosion with with a low and a very high jet velocity compared to the baseline case (Fig. 1). (Fig.1). Fig.2 demonstrates the influence of the jet velocity on the opening angle of the jet and, consequently, on the efficiency of the energy deposition. For the low velocity jet, the jet engine is switched off long before the jet penetrates the stellar envelope. Almost all of the

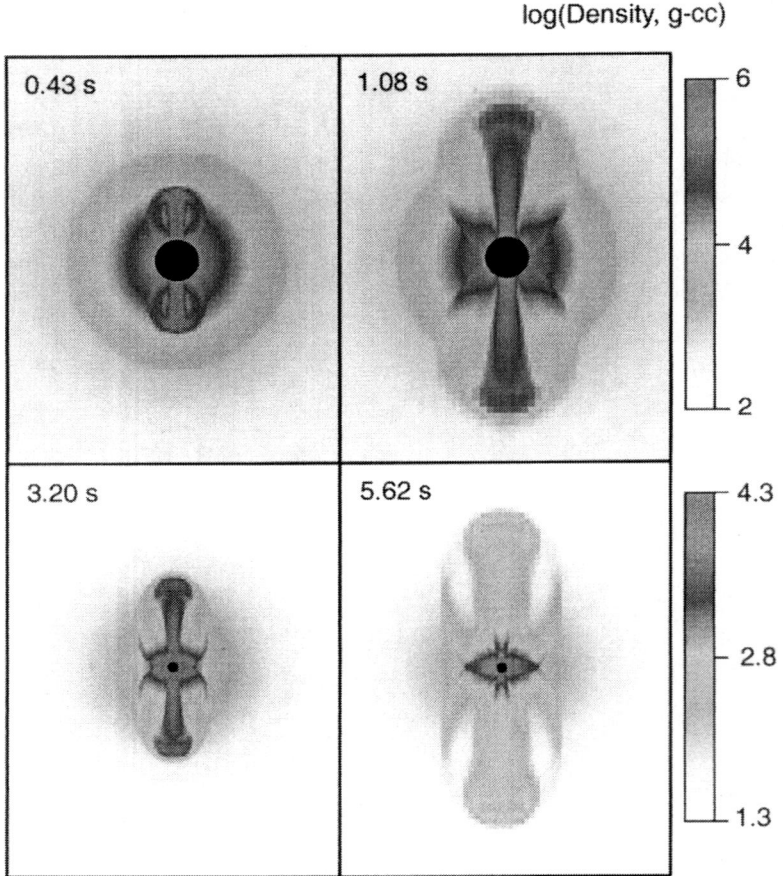

FIGURE 1. Logarithm of the density structure as a function of time for a helium core. The total mass of the ejecta is 2.6 $M_\odot$. The initial radius, velocity and density of the jet were taken to 1200 km 32,000 km/sec and $6.5E5 g/cm^3$, respectively. The shown domains 7.9, 9.0, 36 and 45 $\times 10^9 cm$. The total energy is about 9E50 erg. After about 4.5 seconds, the jet penetrates the star. The energy deposited in the stellar envelope by the jet is about 4E50 erg, and the final asymmetry is of the order of two.

energy of the jet goes into the stellar explosion. On a contrary, the fast jet (61,000 km/sec) triggers only a weak explosion of 0.9 foe although its total energy was $\approx 10 foe$.

Influence of the Progenitor

For a very extended star, as in case of 'normal' Type II Supernovae, the bow shock of a low velocity jet stalls within the envelope, and the entire jet energy is used to trigger the ejection of the stellar envelope. In our example (Fig. 3), the jet material penetrates the helium core at about 100 seconds. After about 250 seconds the material of the jet stalls within the hydrogen rich envelope and after passing about 5 solar masses in the radial mass scale of the spherical progenitor. At this time, the isobars are almost spherical, and an almost spherical shock front travels outwards. Consequently, Strong asphericities are limited to the inner regions. After about 385 seconds, we stopped the 3-D run and remaped the outer layers into 1-D structure, and fullowed the further evolution in 1-D. After about 1.8E4 seconds, the shock front reaches the surface. After about 3 days the envelope expands homologously. The region where the jet material stalled, expands at velocities of about 4500 km/sec.

Fallback: Jet-induced supernovae have very different characteristics with respect to fallback of material and the innermost structure. In 1-D calculations and for stars with Main Sequence Masses of less than 20 $M_\odot$ and explosion energies in excess of 1 foe, the fallback of material remains less than 1.E-2 to 1.E-3 $M_\odot$ and an inner, low density cavity is formed with an outer edge of ^{56}Ni. For explosion

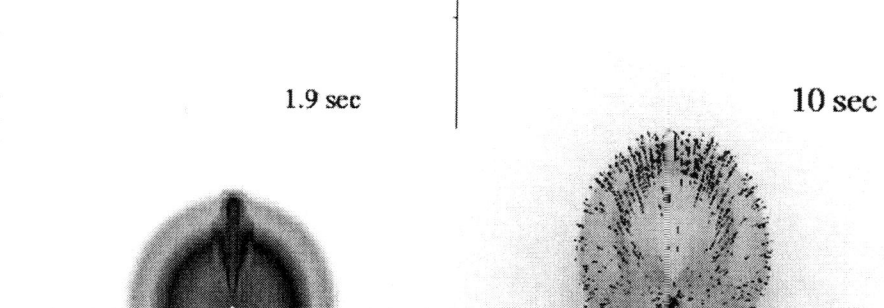

FIGURE 2. Same as Fig. 1 ($0.5 \leq log(\rho) \leq 5.7$) but for a jet velocity of 61,000 km/sec and a total energy of 10 foe at $\approx 1.9 sec$ (left), and 11,000 km/sec and a total energy of 0.6 foe (right). The size of the presented domains are 5 (left) and 2 $10^{10} cm$ (right), respectively. For the high velocity jet, most of the energy is carried away by the jet. Only 0.9 foe are deposited in the expanding envelope. In case of a low velocity jet, the bow-shock still propagates through the star after the jet is swiched off (at $\approx 3 sec$), and the entire jet energy is deposited in the expanding envelope.

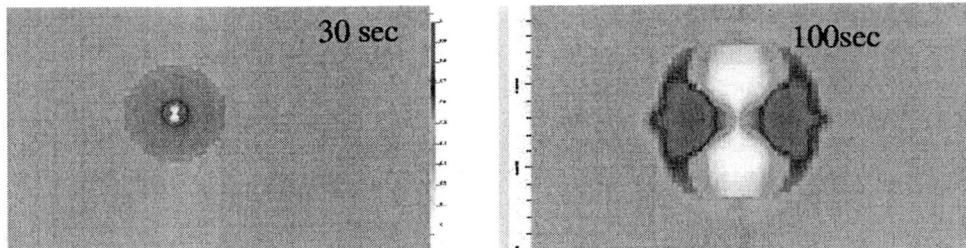

FIGURE 3. Same as Fig. 1 but helium abundance (between 0 to 1) for the explosion of a red supergiant with 207 $R_\odot$ and 7.6 $M_\odot$. The jet velocity of 11,000 km/sec and a total energy of 2 foe has been taken. A domain of about $1.4 \times 10^{12} cm$ is shown. After about 30 seconds, the material of the bow shock penetrates the Helium core, and, at about 250 seconds, the jet material 'stalls' in the hydrogen rich layers. Subsequently a almost spherical shock front propagates through the star. In the final configuration, asymmetries are restricted to the layers within the 2 to 3 solar masses of the h-rich envelope. After homologous expansion, the region corresponding to the stalled shock expands with about 4000 km/sec. All of the jet energy is deposited in the expanding envelope.

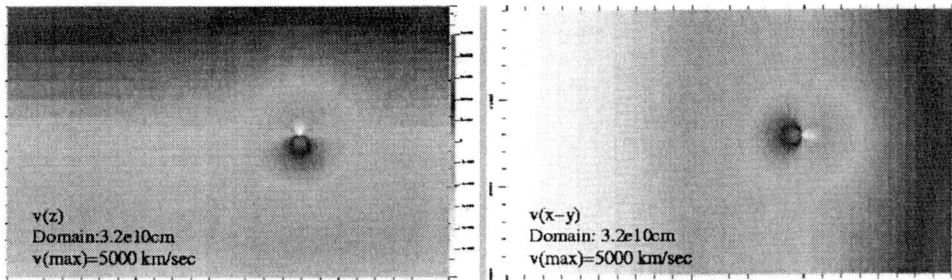

FIGURE 4. Same as Fig. 3 but the velocity distribution in the xy- and yz plane for the very inner regions at about 250 sec. Note the qualitative difference between 1-D and multidimensional results. In 1-D, a large, almost empty cavity is found with expansion velocities of a about 4000 km/sec for corresponding explosions. In multidimensional simulations shown here, this cavity is all but absent. Still, after in multidimensional simulations material can be found up with low velocities. Even infall can persist over a rather extended period of time.

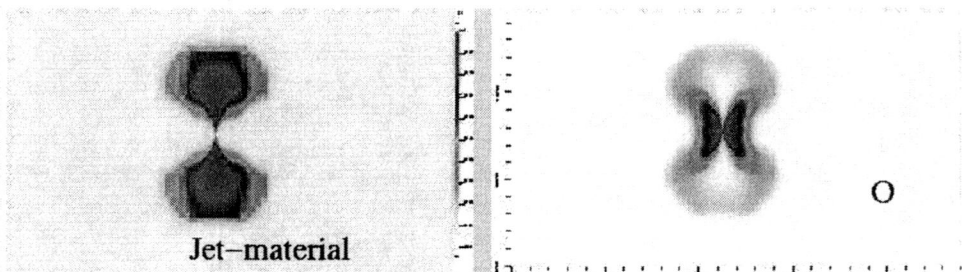

FIGURE 5. Same as Fig. 3 but the distribution of O and the jet material

energies between 1 and 2 foe, the outer edge of the cavity expands typically with velocities of about 700 to 1500 km/sec (e.g. Woosley 1997, Höflich et al. 2000). In contrast, we find strong, continuous fallback of $\approx 0.2 M_\odot$ in the the 3-D hydro models, and no lower limit for the velocity of the expanding material (Fig. 4). This significant amount of fallback must have important consequences for the secondary formation of a black hole. The exact amount and time scales for the final accretion on the neutron star will depend sensitively on the rotation and momentum transport.

Chemical Structure: The final chemical profiles of elements formed during the stellar evolution such as He, C, O and Si are 'butterfly- shaped' whereas the jet material fills an inner, conic structure (Figs. 3 and 5).

The composition of the jets must reflect the composition of the innermost parts of the star, and should contain heavy and intermediate-mass elements, freshly synthesized material such as ^{56}Ni and, maybe, r-process elements because, in our examples, the entropy at the bow shock region of the jet was as high as a few hundred. In any case, during the explosion, the jets bring heavy and intermediate mass elements into the outer H-rich layers.

CONCLUSIONS

We have numerically studied the explosion of Core Collapse supernovae caused by supersonic jets generated in the center of the supernova as a result of the core collapse into a neutron star. We simulated the process of the jet propagation through the star, and the redistribution of elements. A strong explosion and a high efficiency for the conversion of the jet energy requires low jet velocities or a low, initial collimation of the jet. With increasing extension of the envelope, the conversion factor increases. Typically, we would expect higher kinetic energies in SNe II compared to SNe Ib/c if a significant amount of explosion energy is carried away by jets. Within the framework of jet-induced SN, the lack of this evidence suggests that the jets have low velocities.

For the compact progenitors of SNe Ib/c, the final departures of the iso-density contours from sphericity are typical a factor of two. This will produce a linear polarization of about 2 to 3 % (Fig. 6) consistent with the values observed for Type Ib/c supernovae. In case of a red supergiant, i.e. SNe II, the asphericity is restricted to the inner few solar masses. In the latter case, the iso-densities show an axis ratios of up to ≈ 1.4 at the innermost, hydrogen-rich layers. The outer layers remain spherical. This has strong consequences for the observations, in particular, for polarization measurements. In general, the polarization should be larger in SNe Ib/c compared to classical SNe II which is consistent with the observations by Wang et al. (2000). Early on, we expect no or little polarization in supernovae with a massive, hydrogen rich envelope which will increase with time to about 1 % (Höflich 1991), depending on the inclination the supernovae is observed. This is also consistent both with the long-term time evolution of SN1987A (e.g. Jefferies

1991) and, in particular, the plateau supernova 1999em which has been observed recently with VLT and Keck (Wang et al. 2000; Leonard et al. 2000).

The He, C, O and Si rich layers of the progenitor show characteristic, butterfly-shape structures. This overall morphology and pattern should be observable in supernovae remnants, e.g. with the Chanda observatory despite some modifications and instabilities when the expanding medium interacts with the interstellar material.

During the explosion, the jets bring heavy and intermediate mass elements into the outer layers including ^{56}Ni. Due to the high entropies of the jet material close to the center, this may be a possible site for r-process elements. Spatial distribution of the jet material will influence the properties of a supernova. In our model for a SN II, the jet material stalled within the expanding envelope corresponding to a velocity of $\approx 4500 km/sec$ during the phase of homologous expansion. In SN1987A, a bump in spectral lines of various elements has been interpreted by material excited by a clump of radioactive ^{56}Ni (Lucy 1988). Within our framework, this bump may be a measure of region where the jet stalled. This could also explain the early appearance of X-rays in SN1987A which requires strong mixing of radioactive material into the hydrogen-rich layers (see above). We note that, if this interpretation is correct, the 'mystery spot' (Nisenson et al. 1988) would be unrelated. In contrast to 1-D simulations, we find in our models strong, continuous fallback over an extended period of time, and a lack of an inner, almost empty cavity. This significant amount of fallback and the consequences for the secondary formation of a black hole shall be noted. Moreover, fallback and the low velocity material may alter the escape probablity for γ-rays produced by radioactive decay of ^{56}Ni. In general, the lower escape probability is unimportant for the determination of the total ^{56}Ni production by the late LCs because full thermalization can be assumed in core collapse SN during the first few years. However, in extreme cases such as SN98BW (e.g. Schaefer et al. 1999), only a small fraction of gamma's are trapped.

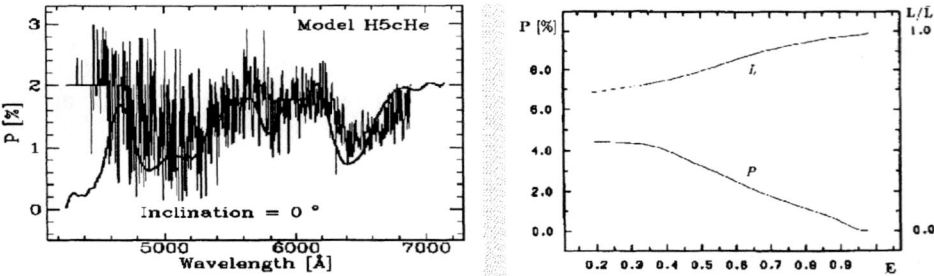

FIGURE 6. Polarization spectrum for SN1993J for an axis ratio of 1/2 for an oblate ellipsoide in comparison with observations by Trammell et al. (1993) are given in the left plot. On the right, the dependence of the continuum polarization (right) and directional dependence of the luminosity is shown as a function axis ratios for oblate ellipsoids seen from the equator (from Höflich, 1991 & Höflich et al. 1995b).

Effects of multi-dimensionallity will strongly alter the energy input by radioactive material and disallow a reliable estimate for the total ^{56}Ni mass.

Finally, we want to emphasize the limits of this study and some of the open questions which will be addressed in future. We have assumed that jets are formed in the course of the formation of a neutron star, and have addressed observational consequences and constrains. However, we have not calculated the jet formation, we do not know if they really form, and, if they form, whether they form in all core-collapse supernovae. We cannot claim that the jets are the only mechanism that can explain asphericity in supernovae although we are not aware of the others. Qualitatively, the observational properties of core collapse supernovae are consistent with jet-induced supernovae and support strongly that the explosion mechanism is highly aspherical but no detailed comparison with an individual object has been performed.

ACKNOWLEDGMENTS

We want to thank our collegues for helpful discussions, in particular, E.S. Oran, L. Wang, J.C. Wheeler, Inzu Yi A., C. Mayers, J.C. Wilson, A. Chieffi, M. Limongi, and O. Straniero. This work is supported in part by NASA Grant LSTA-98-022.

REFERENCES

1. Arnett W.D., Bahcall J.N., Kirshner, R.P., Woosley, S.E. 1990, ARAA 27, 62
2. Benz W., Thielemann F.-K. 1990, ApJ 348, 17
3. Bisnovatyi-Kogan 1971, Soviet Astronomy AJ, 14, 652
4. Bowers R.L., Wilson J.R. 1982 ApJS 50, 115
5. Colella, P.; Woodward, P.R. 1984, J.Comp.Phys. 54, 174
6. Duncan, R. C. & Thompson, C. 1992, ApJ, 392, L9
7. Fesen, R. A. & Gunderson, K. S. 1996, ApJ, 470, 967
8. Fryxell B., Arnett D., Müller E. 1991, ApJ 367, 619
9. Herant M., Benz W., Hix W.R., Fryer C.L., Colgate S.A. 1994, ApJ 435, 339
10. Hillebrandt W., Höflich 1991, Nuclear Physics B 19, 113
11. Hillebrandt W. 1982, ApJ 103, 147
12. Höflich P., Straniero O., Limongi M. Dominguez I. Chieffi A. 2000, 7th TexMex-Conference, eds. W. Lee & S. Torres-Peimbert, UNAM-Publ., in press & astro-ph/005037
13. Höflich, P., Wheeler, J. C., and Thielemann, F.K. 1998, ApJ 495, 617
14. Höflich, P. 1995, ApJ 443, 89
15. Höflich P., Wheeler, J.C., Hines, D., Trammell S. 1995b, ApJ 459, 307
16. Höflich, P. 1991 A&A 246, 481
17. Hoyle F., Fowler W. A. 1960, ApJ 132, 565
18. Huges J.P., Rakowski C.E., Burrows D.N., Slane P.O. 2000, AJ, in press & astro-ph/9910474

19. Janka H.T. & Müller E., 1994 A&A 290, 496
20. Jeffrey D.J., 1991, ApJ, 375, 264
21. Kifonidis K., Plewa T., Janka H.T., Müller E. 2000, ApJ 531, 123
22. Khokhlov, A.M., Oran, E.S. 2000, Combustion & Flame, in press.
23. Khokhlov A., Höflich P., Oran E.S., Wheeler J.C., P. Wang L., 1999, ApJ 524, L107
24. Khokhlov, A.M., Oran, E.S., Chtchelkanova, A.Yu., Wheeler, J.C. 1999a, Combustion & Flame, 117, 99
25. Khokhlov, A.M., Oran, E.S., Thomas, G.O., 1999b, Combustion & Flame, 117, 323
26. Khokhlov, A.M. 1998, J.Comput.Phys., 143, 519
27. Kouveliotou, C., Strohmayer, T., Hurley, K., Van Paradijs, J., Finger, M. H., Dieters, S., Woods, P., Thompson, C. & Duncan, R. C. 1998, ApJ, 510, 115
28. Lucy L.B. 1988, Proc. of the 4th George Mason conference, ed. by M. Kafatos, Cambridge University Press, p. 323
29. LeBlanc, J. M. & Wilson, J. R. 1970, ApJ, 161, 541
30. Leonard D.C., Filippenko, A.V., Barth A.J., Matheson T. 2000, ApJ 536, 239
31. Mezzacappa A., Liebendoerfer M., Bronson Messer O.E. Hix R., Thielemann F-K, Bruenn S.W. 2000, Phy.Rev.Let., accepted
32. Mönchmeyer R., Schaefer G., Mueller E., Kates R.E. 1991 A&A 246, 417
33. Müller E., Janka H.T. 1997, A&A 317, 140
34. Müller E., Hillebrandt W., Orio M., Höflich P., Mönchmeyer R., Fryxell B.A. 1989, A&A 220, 167
35. Nisenson P., Papaliolios C., Karovska M., Noyes R. 1988, ApJ 324, 35
36. Ostriker, J. P. & Gunn, J. E. 1971, ApJ, 164, L95
37. Rammp M., Müller E., Ruffert M. 1998, A&A 332, 969
38. Rammp M. and Janka, H.-T. 2000, ApJ 593, L33
39. Schaefer B. 2000, ApJ 533, 21
40. Saenz R.A., Shapiro S.L. 1981, ApJ 244, 1033
41. Steinmetz M., Höflich P. 1992, A&A 257, 641
42. Symbalisty E.M.D. 1984, ApJ 285, 729
43. Straniero, O. Chieffi, A. and Limongi M. 1999 ApJ, in press
44. Strom R., Johnston H.M., Verbunt F., Aschenbach B. 1995, Nature, 373, 587
45. Symbalisty E.M.D. 1984, ApJ 285, 729
46. Trammell S., Hines D., Wheeler J.C. 1993, ApJ 414, 21
47. Tueller J., Barthelmy S., Gehrels N., Leventhal M., MacCallum C.J., Teegarden B.J. 1991, in: Supernovae, ed. S.E. Woosley, Springer Press, p. 278
48. Van Riper K.A. 1978, ApJ 221, 304
49. Wang L., Howell A., Höflich P., Wheeler J.C. 2000, ApJ, in press
50. Wang, L., Wheeler, J. C., Li, Z. W., & Clocchiatti, A. 1996, ApJ, 467, 435
51. Wang, L., Wheeler, J.C., Höflich, P. 1997, ApJ, 476, 27
52. Yamada S., Janka H.T., Suzuki H. 1999, A&A 344, 533
53. Zwerger T., Müller E. 1997, A&A 320, 209

The Collapsar Model for Gamma-Ray Bursts

A. I. MacFadyen

Astronomy Department
University of California
Santa Cruz, CA 95064
USA[1]

Abstract. Cosmic gamma-ray bursts (GRBs) have been an intriguing mystery for over thirty years. Recent observation and theoretical progress has led to important progress in the last three years. I describe here some of the observational background in the field of GRB research with an emphasis on how collapsars fit into the picture. Collapsars are massive rotating stars which collapse to form rapidly accreting black holes. These accretion powered transients are capable of producing classical gamma-ray bursts of long duration class. They may also make asymmetric stellar explosions which should be observable as supernovae with large explosion energies. Here, I review some of the GRB observations as they relate to the collapsar model and mergers of compact ninaries.

INTRODUCTION

The "delayed" explosion mechanism is the currently favored means thought to produce explosions in massive stars via neutrino energy deposition exterior to a contracting proto-neutron star. Massive stars ($M_{ms} \gtrsim 25$ M$_\odot$) may not always explode successfully via this mechanism [1,2]. The collapsar model postulates the failure of neutrino energy deposition to explode the core of some massive rotating stars and follows the continued evolution after the stellar cores collapse to black holes and accrete the surrounding stellar mantle. For the most massive stars ($M_{ms} \gtrsim 35$ M$_\odot$) with sufficient angular momentum, a collapsar – a rapidly accreting ($\dot{M} \approx 0.1$ M$_\odot$ s^{-1}) stellar mass black hole – forms promptly at the center of a collapsing star [15]. These are referred to as Type-I collapsars. Less rapidly accreting black holes can also form over longer time periods due to the fallback of stellar gas which failed to escape during the initial weak supernova explosions. This class constitutes Type II collapsars and probably occurs for main sequence masses, $M_{ms} \gtrsim 20$ M$_\odot$. Stars

---

[1] This research has been supported by NASA (NAG5-2843, MIT SC A292701, and NAG5-8128), the NSF (AST-97-31569), the US DOE ASCI Program (W-7405-ENG-48)

with masses below this usually explode as normal supernovae and leave behind neutron star remnants. Collapsars power jetted explosions by tapping a fraction of the binding energy released by the accreting star through magnetohydrodynamical processes or neutrino annihilation, or possibly by extracting some of the black hole spin energy. The vast majority of stellar explosions do not make collapsars, only those which make black holes and have sufficient angular momentum. Not all collapsars make GRBs. Only those which happen in sufficiently small (in radius) stars and manage to accelerate a fraction of the explosion energy to sufficiently high Lorentz factor. Other collapsar explosions may be responsible for hyper-energetic and asymmetric supernovae like SN1998bw.

In this paper, I will review gamma ray burst properties as they relate to the various models for the central engine. I will also contrast the expected observational properties with mergers of compact objects.

GAMMA-RAY BURST (GRB) OBSERVATIONS

Discovery

Cosmic gamma-ray bursts were discovered in 1969 in data from the *Vela 4a* and *4b* military satellites taken on July 2, 1967. While the sun could not be excluded as a source for this burst due to poor timing resolution, subsequent *Vela* missions established the "cosmic origin" of the bursts and they were announced by Klebesadel, Strong and Olsen in the Astrophysical Journal Letters [3]. The *Vela* satellites were first launched by the U.S. Air Force in 1963 to monitor for the recently signed Limited Nuclear Test Ban Treaty (October 10, 1963) which prohibited nuclear explosions in the atmosphere, in space or underwater. The satellites were equipped with x-ray detectors to detect the x-ray flash from a nuclear explosion containing the bulk of the high energy photons. Gamma-ray detectors were included to confirm the explosion especially in the case that the x-ray flash was obscured.

For many years after their discovery there was a general belief that GRBs were associated with neutron stars in the Galaxy. This association was supported by reported detections of line features in GRB spectra by the *Ginga* and *Konus* instrument teams. Though GRBs appeared to be distributed isotropically on the sky, many researchers expected a concentration of GRBs in the Galactic plane to appear at faint levels as improved detectors went into operation (see [4] and [6] for reviews of the history of GRB research). The Burst and Transient Experiment (BATSE) on the *Compton Gamma-Ray Observatory (CGRO)* surprised many by showing that even the faint GRBs were isotropic on the sky. Moreover, BATSE revealed a paucity of faint bursts as seen in the deviation from a Euclidean slope of $-3/2$ in the $\log N(> P) - \log(P)$ distribution [7], where $N(> P)$ is the number of bursts with peak flux greater than P. GRBs were not spatially correlated with any know astrophysical structure, for example the Milky Way, Andromeda, galaxy clusters.

The observed isotropy and deficiency of faint bursts are well explained if GRBs originate at cosmological distances [8]. Observations by *BeppoSAX*, the Italian–Dutch x-ray satellite launched in 1996, led to the identification of x-ray afterglow, the first non-gamma-ray detection of GRBs. The higher spatial resolution of the x-ray instrument led to the discovery of lower energy counterparts to GRBs and a consequent revolution in the field with the first x-ray counterpart detection for GRB 970228 [9]. The case for cosmological location of some GRBs was firmly established in May 1997 by the identification of redshifted absorption lines of ionized iron and magnesium (Fe II and Mg II) in the spectrum of the optical afterglow emission of GRB970508 with $z > 0.835$ [11]. Many important discoveries quickly followed the detection of lower-energy afterglows from GRBs.

GRB Properties

Gamma-ray bursts are short flashes of high energy ($\gtrsim 100$ keV) radiation. They show a broad range of behavior in duration and intensity and diverse light curve morphologies. GRB durations range over five orders of magnitude from tens of milliseconds to thousands of seconds. When plotted in terms of the duration for the central 90% of emission, T_{90}, two distinct groups centered around 0.5 s and 35 s are discernible [10]. There is a correlation of these two groups with spectral hardness ratios such that the shorter bursts have harder spectra than the longer group. The division of classical GRBs into the two duration classes, "short hard" bursts and "long soft" bursts, has important implications for engine models and points to the possibility of multiple burst engines. Here, we are primarily concerned with the "long soft" bursts since the short time scales < 2 s of the "short hard" bursts are not natural for the cores of massive stars which have collapse times ~ 10 s. Unless otherwise stated, the term "GRB" will henceforth refer to the "long soft" bursts. It is important to note that the GRBs for which *BeppoSAX* positions have been determined, and subsequently redshifts via optical spectroscopy have been determined, are all of the "long soft" class.

While the statistical properties of GRBs are often discussed, it is important to keep in mind their great diversity. GRB 940217, for example, is an extremely long duration burst which emitted GeV photons up to 1.5 hours after the initial outburst. Some bursts show little or no time structure on time-scales shorter than the burst duration, others, show distinct well-separated episodes of emission. Still others are complex and chaotic with fine structure on time scales much shorter than the burst duration. Fourier and wavelet analysis of burst light curves show no preferred frequencies or strictly periodic emission. Time structure as short as 200 μs has been detected for GRB 910711.

Afterglow Detection

The discovery of x-ray emission by the Italian-Dutch x-ray satellite *BeppoSAX* led to arcminute localization of GRBs and the consequent detection of afterglow emission in optical, radio, UV, and IR wavelengths. The detection of lower energy emission has revolutionized the field of GRB research and led to fundamental discoveries including: 1) the confirmation of the fireball model, 2) evidence for relativistic expansion (from radio scintillation measurements and GRB 990123 optical flash brightness temperature), 3) proof that *at least some* GRBs are at cosmological distance, 4) indications that GRBs are associated with star forming regions in galaxies, 5) the association of an unusual TypeIb/c supernova, SN1998bw, with a weak GRB, and 6) possible signature of supernova emission in the light curves of several GRBs. These discoveries offer important clues to the nature of GRB central engines. The last four points constitute strong observational support for the collapsar model.

OBSERVATIONAL CONSTRAINTS

Energy

Redshift measurements for GRBs allows for the determination of isotropic equivalent energis, E_{iso}, whixh span more than three orders of magnitude, even excluding GRB 980425 which may be from a distinct class. The most energetic bursts, such as GRB 990123 [12], require that GRB engine models be capable of producing in excess of $2 \times 10^{54} f_\Omega f_\gamma^{-1}$ erg, where $f_\Omega \equiv \Omega/4\pi$ is the fraction of the sky irradiated by the burst and f_γ is the fraction of explosion energy converted into gamma-rays. This number approaches or exceeds the rest mass energy equivalent of a neutron star, 2.5×10^{54} erg. Even using beaming factors of $f_\Omega \approx 0.01$ and assuming efficient conversion of engine energy into observed gamma rays ($f_\gamma \approx 0.1$), the energy of such energetic bursts imply conversion of a significant fraction of a solar rest mass into gamma rays. Models with a limited amount of accretable material, e.g. merging neutron stars and black holes with $M_{torus} \approx 0.01 - 0.5$ M$_\odot$, must resort to very efficient beaming, and/or electromagnetic extraction of rotational energy from the rapidly rotating black hole, to provide the energies observed for the most energetic bursts. Lee *et al.* [13] have shown that the Blandford–Znajek mechanism [14] is capable of extracting $\sim 9\%$ of the black hole rest mass, though typical values may be an order of magnitude lower. In any case, efficient extraction of rotational energy from a rapidly rotating 3 M$_\odot$ black hole with $f_\Omega f_\gamma^{-1} \approx 0.1$ can in principle explain the energies of the most luminous bursts observed so far.

Explaining the most luminous bursts is less of a problem for models involving the collapse of massive stars because the reservoir of accretable gas is potentially much larger, up to ~ 10 M$_\odot$ (though the explosion of the parent star limits this). The rest mass energy of 10 M$_\odot$ is 1.8×10^{55} erg; clearly, any process capable of converting even a small percentage of this rest mass into gamma rays could account for the most luminous bursts, especially since beaming of ~ 0.01 may be natural for these models [15]. Extraction of black hole rotational energy could also work in these models, though it may not be required. In fact, massive stars can build up sizeable magnetic fields in the convective shells of the pre-collapsar star and any existing field will be amplified during collapse due to flux conservation. The fields can be further amplified by differential rotation in the accretion disk and will naturally thread the ergosphere of the black hole.

Since beaming can change the energy requirements of an observed burst by orders of magnitude, it is important to be clear about what is meant by a burst's energy. We shall refer in this work to the *isotropic equivalent energy* (E_{iso}) as the apparent energy of a burst assuming that the source emitted isotropically. The actual energy required of the GRB engine is $f_\gamma^{-1} f_\Omega E_{iso}$ where $0 \leq f_\Omega \leq 1$. f_Ω is expected to be a function of wavelength and time since radiation detected at different epochs in the fireball evolution is produced by electrons with different bulk Lorentz factor Γ with a corresponding range in beaming angle $\theta \approx \Gamma^{-1}$. E_{iso} is a useful quantity since the beaming of particular bursts and their afterglows is folded into the f_Ω specified for the observation.

Duration

GRBs range in duration from roughly 30 ms to over 1000 s. While it is difficult to precisely measure burst duration due to trigger effects, the BATSE team has defined a duration measure, T_{90}, to be the time during which the central 90% of the burst fluence is detected [4]. In the internal shock scenario [5] the GRB engine must last as long as the burst itself. "Long" GRBs are observed to last for tens to hundreds of seconds, with some bursts lasting more than 1000 s. This is many of orders of magnitude longer than the characteristic orbital time scales of a stellar mass compact object (~ 1 ms) and, in many cases, long even compared to the collapse time of massive stellar cores (~ 10 s). Such long time scales might be possible from accretion disks if the viscous time $t_{visc} > t_{dyn}$ were very long, due to very low disk viscosity, and determined the duration of the burst. Disks with low viscosity imply many orbits for the disk gas, but the magnetic fields necessary for the extraction of black hole rotational energy should build up and provide a turbulent viscosity via the magneto-rotational instability. Long time scales can be explained naturally by fallback onto a compact object formed in a weak supernova explosion. Long bursts can also be explained coming from engines of arbitrarily short duration by light travel time effects in the external shock model [16].

Rate

BATSE detects about one burst per day, corresponding to ~ 1000 GRBS per year since BATSE views about one third of the sky. Because GRBs have a broad intrinsic luminosity distribution, i.e. they are not standard candles, it is not possible to infer their event rate simply from the log N – log S distribution. The exact rate of GRBs in the universe depends on their spatial (redshift) and luminosity distribution. In the simplest "no evolution" cosmological model this corresponds to approximately one burst per L^* galaxy per 10^7 years. However, if the burst rate follows the cosmological star formation rate, as most viable models predict, then the bursts are further, brighter and rarer with a current rate of $\sim 10^{-8}$ per year per galaxy [17] since the star formation rate, and hence the GRB rate, was ~ 10 times higher at $z \approx 1$ than at present [18]. Viable GRB engines must make GRBs at an event rate consistent with the observed rate, taking into account any beaming factor natural to the model or required to explain the observed energetics. Even with substantial beaming the observed rates are a small fraction of the supernova rate of a few per L^* galaxy per 100 years [19].

The association of GRB 980425 with SN 1998bw implies an isotropic equivalent energy of $\sim 10^{48}$ erg, about four orders of magnitude lower than other GRBs with determined redshifts. Since the spatial volume probed by a flux-limited sample scales with luminosity as $V \propto L^{3/2}$, low-luminosity events like GRB 980425 are detected in a volume $\sim 10^6$ times smaller than cosmological GRBs. This component of low-luminosity bursts constitutes a subset of GRBs with a Euclidean slope which can constitute at most $\sim 10\%$ of the total GRB population detected by BATSE [20]. This rate of $\lesssim 100$ low-luminosity GRBs detected per year implies $\lesssim 10^{-4}$ such events per galaxy per year, less than the observed rate of Type Ib/c supernovae of a few 10^{-3} per galaxy per year [19].

Viable models must provide a sufficient rate of events to accommodate the observed rate of $\sim 10^{-7}$ yr^{-1} per galaxy times a beaming factor, f_Ω^{-1}, which may be large (~ 100). A number of groups have estimated the rate for merging compact objects using Monte Carlo population synthesis calculations [21–23]. These calculations evolve a population of binary progenitor systems from birth, through orbital decay, common envelope evolution, supernova explosions and kicks, as appropriate, to the eventual merger. Due to the many uncertainties inherent in such calculations, the resulting merger rates are uncertain to at least three orders of magnitude [23].

Merger Locations

Due to the space velocities imparted to NS–NS and BH–NS binaries during the two supernova explosions the systems experience and the long time scale for orbital decay to drive the merger, some fraction of merging NS–NS and NS–BH systems are expected to be located outside of the star forming regions in which they formed. This expectation is testable given the growing number of well localized bursts. Up to the present it appears that all well localized bursts are located exactly on top of star forming regions in actively star forming galaxies [24]. However the numbers are still too small to draw definite conclusions. Only ~ 10 bursts are localized and perhaps only one or two (15%) would be expected far from a star forming region [25].

Recent population synthesis studies taking into account asymmetric kicks and a range of galactic gravitational potentials make predictions for the locations of NS–NS and BH–NS mergers relative to their galaxies of origin [22,25,23]. Bloom et al. [25] find the observed localizations of GRBs with afterglows is consistent with merging NS–NS for massive galaxies, but that 15% of the mergers should occur far removed ($R \gtrsim 30$ kpc, or about 4 arcsec at $z = 1$) from dwarf galaxy hosts. Fryer et al. [23] find that half of NS–NS mergers will occur more than 60 kpc from the galactic center for a large galaxy and more than 5 Mpc from the galactic center for a small galaxy (though it is often difficult to determine the true center of faint high redshift galaxies). They also find that up to ~ 40 % of the NS–NS and BH–NS mergers should be more than 1 Mpc away from the center of galaxies with one fourth the mass of the Milky Way, $\sim 5\%$ for galaxies as massive as the Milky Way. The close association observed for GRBs with afterglows with star-forming regions is beginning to make this difficult for merging NS–NS and BH–NS binaries.

COLLAPSAR PROPERTIES

Collapsars should occur directly in star forming regions. The time from birth to death of the massive star progenitors ($\sim 10^7$ yr) is short compared to the time it would take for the stars to travel appreciable distances ($\sim 10^8$ yr). Therefore, collapsars are expected to explode at their birth sites in the star forming regions of galaxies. Identification of a GRB clearly outside of a star forming region would constitute support for another model. There may be a bias, however, to locating GRBs within star forming regions if the star forming environment is favorable to afterglow production and subsequent localization [26].

Collapsars naturally make long duration (~ 10 s) bursts which may be hard to make with compact object mergers unless the disk viscosity is very low. This is because, even with high viscosity, the lifetime of the collapsar engine is set by the collapse time of the stellar envelope feeding the accretion disk and not by the viscous time scale for an isolated disk. The dynamical time scale for a helium star is $\tau \approx 446 \text{ s}/\rho^{1/2} \approx 10$ s which roughly gives its collapse time. It is also the time scale for explosion, which can choke the source of accretion material into the black hole. The star can explode by the disk wind or by lateral expansion of a polar jet. The longer duration bursts could still be accounted for by larger stars with longer dynamical time scales, by relatively cold jets which fail to explode the star at all angles (though the disk wind could still explode the star so low viscosity might also be necessary for this to happen), or by fallback in a weak neutrino powered supernova.

Collapsar Environment and GRB Diversity

The huge diversity in observed GRB light curves is an important clue to the nature of the central engine and its environment. Since a system of two merging compact objects in a baryon free region of space is fairly simple, not too much diversity would be expected in the GRBs made by these objects. Massive stars, however, offer a variety ways to add complexity to the burst. There is the interaction of the jet with the star and with the surrounding clumpy circumstellar medium. This environment contains winds of varying densities and topologies from the pre-collapse mass shedding episodes. In addition the star forming regions themselves are clumpy and inhomogeneous. The GRB jet can interact in a variable way with the stellar and circumstellar gas so that heterogeneity in the GRBs made from massive stars is to be expected. Eta carinae and the pistol nebulae are examples of the messy circumburst medium to be expected surrounding GRBs formed in massive stars.

Evidence for Massive Star Progenitors

Since massive stars are short lived (10^7 yr), they are expected to die near the star forming regions where they were born. This is in contrast to merging compact objects which can take 10^9 years to merge and some are expected to merge far from star forming regions. *BeppoSAX* localizations of GRBs with bright afterglow has allowed for follow up observations with Keck and HST to identify the host environment. All current optically identified GRBs appear on or near vigorous star forming regions [24]. This is evident from the host morphologies and star formation rates as deduced by blue colors and star formation indicators such as the strength of Hα and the flux at 1500 Angstroms. Indirect evidence includes the possibility that location in a dusty environment can cause the lack of an optical transient detection (e.g., GRB 970828).

Indications of Supernova Emission in GRB Light Curves

The fireball model predicts afterglow at longer wavelengths as the relativistic fireball is decelerated by the circum-burster medium. The simplest models predict power law decays. Recent observations of deviations from power law decay of optical afterglow transients — brightening about three weeks after the GRB which is fit by the relativistic blast wave power law plus a Type Ib/c supernova (SN 1998bw) light curve — have been interpreted as evidence for supernovae explosions occurring simultaneously with GRB 980326 [27] and GRB 970228 [28,29]. The supernova interpretation is strengthened by the red color of the afterglow relative to the color expected for the GRB blast wave component of the optical transient. An alternative explanation is that dust surrounding the GRB scatters the optical afterglow from the burst which arrives with a delay of about one month [30]. Spectra of

GRB optical transients taken when the brightening is prominent should be able to distinguish between these possibilities.

Several days after the GRB the optical transient fades and the supernovae brightens to a point roughly two weeks after the burst where the SN can be seen as a bump in the optical light curve.

Host Galaxies

Deep HST images of the GRB 990123 field shows that the GRB optical transient is located near star forming regions in the host galaxy. A redshift of z=1.6004 was determined from metallic absorption lines in the spectrum of the optical transient. The host galaxy is fairly typical of galaxies at this redshift and has the appearance of a merger remnant. The unobscured star formation rate is estimated at $6\,M_\odot\,\mathrm{yr}^{-1}$. The optical transient does not seem to be associated with the galactic nucleus and favors an association with star forming regions. The fact that this and other OTs lie very near to star forming regions favors the death of massive stars over the coalescence of massive star remnants.

CONCLUSIONS

Long duration, classical gamma-ray bursts and energetic asymmetric supernova explosions can result from rotating massive stars ($M_{ZAMS} \gtrsim 25 M_\odot$) whose iron cores collapse to black holes. The explosion is powered by rapid accretion ($\dot{M}_\odot \approx 10^{-4} - 10^{-1} M_\odot\,\mathrm{s}^{-1}$) into the stellar mass black hole formed by the collapsed iron core. The model explored in most detail in (14A in [15]) demonstrates gas accreting through a disk into a newborn black hole at an average rate of $0.07\,M_\odot\,\mathrm{s}^{-1}$, with extended peaks of twice that value. This is a sufficiently high rate and occurs for a sufficiently long time to power the most common GRBs found at cosmological distances. Results have been directly calculated for the most tractable case when energy transport is by neutrino annihilation, but MHD processes are equally viable. In fact, MHD processes may be necessary for the most energetic bursts (e.g., GRB 990123, GRB 971214) unless strong beaming is invoked. Models presented here do, in fact, indicate that beaming to 1% of the sky or less is natural for collapsars. The star acts to compress and focus the expanding fireball as it propagates along the polar axis of the collapsar.

Variability of the accretion rate with time scales similar to the viscous time of the accretion disk, $\sim 50\,\mathrm{ms}$ may imprint short time scale variability in the observed GRB. Future high-resolution calculations will study hydrodynamical instabilities resulting from the interaction of the jet with the star at larger radii which may provide longer (~ 1 s) time scale variations in the Lorentz factor of the jet. The collapsing star forms a disk whose geometry is conducive to jet formation. When energy is deposited near the black hole along the rotation axis, the geometry of the accretion disk naturally channels the expanding fireball into narrowly collimated jets.

The passage of the jet through the star can make it explode. This is especially true in the case of "hot" jets. The degree to which the explosion resembles a Type Ib/c supernova depends on the amount of radioactive ^{56}Ni that is produced and how it is distributed in the exploding star. It is possible that some collapsars will make GRBs without bright accompanying supernovae. A "cold" jet might interact so little with the star through which it passes that there would be insufficient ^{56}Ni in the explosion to power a bright supernova. Disk winds powered by viscous dissipation of energy in a collapsar accretion disk may eject sufficient ^{56}Ni into the stellar envelope to produce a bright supernova. The star is subsequently exploded by either the disk wind itself or the lateral expansion of a "hot" jet. We predicted that GRBs of a certain class should be accompanied by supernovae. Supernova detections were subsequently reported in three GRBS.

REFERENCES

1. Burrows, A.: 1998, in *Proceedings of the 9th Workshop on Nuclear Astrophysics*, p. 76
2. Fryer, C. L. 1999, ApJ, in press, astro-ph/9902315
3. Klebesadel, R. W., Strong, I. B., & Olson, R. A.: 1973, *Astrophys. J., Lett.* **182**, L85
4. Fishman, G. J. & Meegan, C. A.: 1995, *ARA&A* **33**, 415
5. Rees, M. J. & Mészáros, P.: 1994, *Astrophys. J., Lett.* **430**, L93
6. van Paradijs, J., Kouveliotou, C., & Wijers, R. A. M. J.: 2000, *ARA&A*, in press
7. Paciesas, W. S., et al. : 1999, *Astrophys. J., Suppl. Ser.* **122**, 465
8. Paczyński, B.: 1986, *Astrophys. J., Lett.* **308**, L43
9. Costa, E., et al. *Nature* **387**, 783
10. Fishman, G. J. et al. : 1994, *Astrophys. J., Suppl. Ser.* **92**, 229
11. Metzger, M. R., Djorgovski, S. G., Steidel, C. C., Kulkarni, S. R., Adelberger, K. L., & Frail, D. A.: 1997, *IAU Circ.* 6655
12. Andersen, M. I., et al. *Science* **283**, 2075
13. Lee, H. K., Wijers, R. A. M. J., & Brown, G. E.: 2000, *Physics Reports* **325**, 83
14. Blandford, R. A. & Znajek, R. L.: 1977, *Mon. Not. R. Astron. Soc.* **179**, 433
15. MacFadyen, A. I., & Woosley, S. E. 1999, ApJ, 524, 262
16. Dermer, C. D., Böttcher, M., & Chiang, J.: 1999, *Astrophys. J., Lett.* **515**, L49
17. Wijers, R. A. M. J., Bloom, J. S., Bagla, J. S., & Natarajan, P.: 1998, *Mon. Not. R. Astron. Soc.* **294**, L13
18. Madau, P., et al. : 1996, *Mon. Not. R. Astron. Soc.* **283**, 1388
19. Cappellaro, E., Evans, R., & Turatto, M.: 1999, *Astron. Astrophys.* **351**, 459
20. Kommers, J. M. et al. : 2000, *Astrophys. J.* **533**, 696
21. Bethe, H. A. & Brown, G. E.: 1998, *Astrophys. J.* **506**, 780
22. Portegies Zwart, S. F. & Yungelson, L. R.: 1998, *Astron. Astrophys.* **332**, 173
23. Fryer, C. L., Woosley, S. E., & Hartmann, D. H.: 1999, *Astrophys. J.* **526**, 152

24. Fruchter, A. S. et al. : 1999, *Astrophys. J.* **516**, 683
25. Bloom, J. S., Sigurdsson, S., & Pols, O. R.: 1999, *Mon. Not. R. Astron. Soc.* **305**, 763
26. Mészáros, P. & Rees, M. J.: 1997, *Astrophys. J.* **476**, 232
27. Bloom, J. S. et al. : 1999, *Nature* **401**, 453
28. Reichart, D. E.: 1999, *Astrophys. J., Lett.* **521**, L111
29. Galama, T., et al. : 1999, *Astrophys. J., Lett.* submitted, astro-ph/9907264
30. Esin, A. A. & Blandford, R.: 2000, *Astrophys. J., Lett.*, submitted, astro-ph/0003415
31. Aloy, M.A., Müller, E., Ibanez, J.M., Marti, J.M., MacFadyen, A.I., 1999, astro-ph/9911098.
32. MacFadyen, A.I., Woosley, S.E., & Heger, A. 1999, astro-ph/9910034
33. Iwamoto, K., et al. 1998, Nature, 395, 672
34. E. Müller, these proceedings.
35. Woosley, S. E., Eastman, R. G., Schmidt, B. P. 1999, ApJ. 516, 788,

Bi-polar Supernova Explosions

Lifan Wang, Peter Höflich, and J. Craig Wheeler

Department of Astronomy and McDonald Observatory
The University of Texas at Austin
Austin, TX 78712
lifan, pah, wheel@astro.as.utexas.edu

Abstract.
Core-collapse SNe are found to be significantly polarized. Our preliminary results are: (a) Core-collapse supernonae are normally polarized at 1% level; (b) the polarization of core-collapse SNe increases with the mass of the stellar envelope; (c) their polarizations increase as deeper into the ejecta. We discuss here the optical spectropolarimetry of several core-collapse SNe, SN 1996cb (Type IIB), SN 1997X (Type Ic), and SN 1998S (Type IIn). The data show polarization evolution of several spectral features at levels from 0.5% to above 4%. These data suggest that the the distribution of ejected matter is highly aspherical. In the case of SN 1998S, the minimum major to minor axis ratio must be larger than 2.5 to 1 if the polarization is 3% from an oblate spheroidal ejecta seen edge on. A well-defined symmetry axis can be deduced from spectropolarimetry for the peculiar Type IIn SN 1998S. The observed degree of polarization of the Type Ic SN 1997X is above 4%. The data reveal a trend that the degree of polarization increases with decreasing envelope mass and with the depth within the ejecta. The high axial ratio of the ejecta is difficult to explain in terms of the conventional neutrino driven core-collapse models for Type II explosions. Highly asymmetric explosion mechanisms such as the formation of bipolar jets during core-collapse may be a necessary ingredient for models of all core-collapse SNe.

I INTRODUCTION

Only a few SNe have been observed with polarimetry. SN 1987A represented a breakthrough (Cropper et al. 1988; Méndez et al. 1988; Jeffery 1991; Höflich 1991; Wang & Wheeler 1996). SN 1993J also provided a wealth of data (Trammell, Hines, and Wheeler; Doroshenko, Efimov, & Shakhovskoi 1995; Tran et al 1996). More recently we have begun a program to obtain routine spectropolarimetry on all accessible SNe. We have used mostly the 2.1 and 2.7 meter telescopes at McDonald Observatory. Wang et al. (1996) compared broad band polarimetry data obtained at the McDonald Observatory with SN polarimetry data published before 1996 and found that all Type II in their sample are polarized at about the 1 percent level and that Type Ia are much less polarized, less than 0.2-0.3%. One normal Type Ia, SN

1996X are polarized at levels around 0.2% near optical maximum (Wang, Wheeler, & Höflich 1997). A subluminous Type Ia SN 1999by is polarized at about 0.7% (Howell et al 2000). It is not clear if the relatively high polarization of SN 1999by is related to its peculiar photometric behavior. The data confirm that core-collapse events are significantly polarized. The degree of polarization seems to increase with decreasing masses of the stellar envelope, with the highest polarization observed for Type IIB and Type Ib/c events (Wang, Wheeler, Höflich 1999).

II OBSERVATIONS

All of the data on SN 1996cb, SN 1997X, and SN 1998S were obtained at the McDonald Observatory using the 2.7 meter telescope or the 2.1 meter telescope. The instrumental polarization of the polarimeter is less than 0.1%, much lower than the polarization of the SNe. We have observed a given SN with the same configuration for 2-4 nights to get enough photons to reduce statistical noise. Polarized and unpolarized standard stars were observed in all of our observing runs. Multiple sets of data were generally taken with each individual exposure normally 20 to 30 minutes at each waveplate position angle. We have also observed a couple of flux standard stars each observing run to allow for flux calibrations of the SN spectra.

A SN 1996cb

SNe IIB have lost most of their hydrogen envelopes and have only a tenuous hydrogen layer surrounding them before the explosion. Early spectroscopic data are discussed in a recent paper by Qiu et al. (1999). The SN reached optical maximum around 1996 Jan. 2. Despite some minor differences pointed out by Qiu et al. (1999), the spectroscopic and photometric behavior of SN 1996cb and SN 1993J are remarkably similar. Both SN 1996cb and SN 1993J showed evidence of clumpiness in the ejecta (Wang & Hu 1994; Qiu et al. 1999), and both SNe were positively detected in the radio (Van Dyk et al 1996).

Polarimetry data of SN 1993J were obtained around optical maximum. Trammell et al. (1993) reported data taken at the McDonald Observatory on 1993 April 20. These data were analyzed in detail by Höflich et al. (1996) where asymmetric helium cores were assumed to produce the observations. More spectropolarimetry of SN 1993J was reported in Tran et al. (1997). The degree of continuum polarization is an evolving function of time and is around 1.7-2% around 1993 April 20 - suggesting a highly distorted envelope with major to minor axis ratio $\gtrsim 1.4$ (Höflich, 1991). The SN must be viewed nearly edge on for an oblate spheroid or pole on for an prolate spheroid according to models of Höflich et al. (1996).

The real surprise is that the two SNe are not only spectroscopically similar but also spectropolarimetrically similar. Fig. 1 compares the polarimetry of SN 1996cb obtained on 1996 January 5 with that of SN 1993J taken on 1993 April 26. The

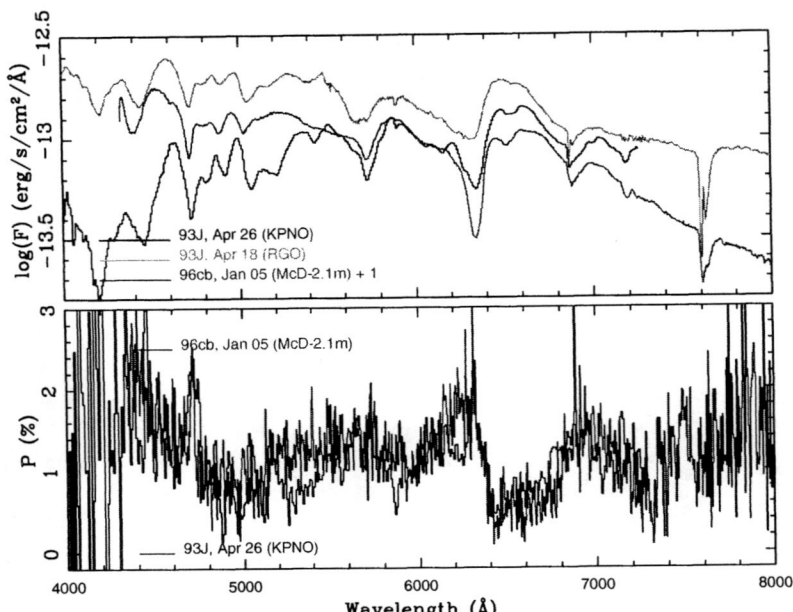

FIGURE 1. The total flux and polarization spectra of SN 1996cb are compared to those of SN 1993J.

SN 1993J data were obtained at the Kitt Peak National Observatory. Sharp polarization changes were seen across the Hα and He 5876 lines. The variation of the polarization is as large as 1.5% in both SNe. In addition, data taken after optical maxima show that the polarization of both SN 1993J and SN 1996cb tends to grow with time suggesting that the asymmetries are larger deeper within the ejecta.

B SN 1998S

1 Spectropolarimetry Data at the McDonald Observatory

Our spectropolarimetry were obtained using the 2.1 meter telescope of the McDonald Observatory with the Imaging Grism Polarimeter (IGP) on two observing runs each lasting 4 nights. Two sets of data were acquired in the first run, both on the night of 1998 March 30, about 10 days after optical maximum.

The March 30 and May 1 data were taken at about 10 and 41 days after optical maximum of the SN, respectively. The latter data were taken just before the light curve began to decline rapidly from the plateau. The SN showed dramatic evolution during this period where both the spectral and polarimetric behavior had changed dramatically. The March 30 data show an abnormally strong Si II 6355 Å line and a weak Hα line on top of a blue continuum. The velocity of the absorption troughs of the P-Cygni features is typically 3200 km s^{-1}. The emergence of the strong lines from Fe II, Sc II, Si II indicates that the photosphere has receded into slower moving envelope material. The presence of narrow P-Cygni features indicates that there is slow moving matter, perhaps of circumstellar origin, outside the rapidly expanding ejecta.

Some important features can be summarized as follows:

(a) The degree of polarization evolves with time. Polarization variations are observed across strong emission/absorption lines such as Hα, Si II, He I, Fe I, and Fe II. Unlike the pre-maximum data (Leonard et al. 1999), the continuum polarization is no longer flat in our post-maximum data. The degree of polarization is the highest in the blue ($\sim 1.6\%$) and is the lowest at around 5500 Å ($\sim 0.4\%$); (b) The position angles of the polarization vector evolved significantly during the two epochs of our observations compared to the earlier data from Keck II; (c) A sharp change of polarization position angle by 90 degrees is observed across the Hα line in the May 1 spectrum; (d) the polarization position angle changed by nearly 90 degrees from red to blue in the May 1 spectroscopic data.

2   How Large is the Intrinsic Polarization?

The maximum degree of interstellar polarization is found to be $9 \times$ E(B-V) from observations of interstellar polarization in the Galaxy (Serkowski et al 1975). The maximum achievable polarization corresponds to the configuration in which all the dust particles are aligned to the optimum angle for polarization. The circle in Fig. 2 shows the limit of allowable interstellar polarization according to the Serkowski relation. It is, however, uncertain because of the uncertainty of the derived extinction.

We will assess the interstellar polarization via a new approach by relaxing the constraints on the polarization properties of emission lines and assume only that at each specific epoch, the polarization produced by the SN ejecta has a single polarization position angle independent of wavelength. This can be achieved, for instance, if the polarization is produced purely by electron/dust scattering in the vicinity of the SN with no large scale chemical inhomogeneities.

The Q-U plot in Fig. 2 is consistent with our assumptions. The polarization at different wavelengths falls roughly on a straight line on the Q-U plot which is typical of the combination of two polarization vectors of which one is nearly constant (like interstellar polarization) while the other highly variable. Under this assumption, the interstellar polarization must lie somewhere along the line drawn by the observed Q and U vectors. The allowable areas are marked in Fig. 2 as A and B.

We believe that the shorter wavelength suffers severe de-polarization due to numerous Fe II, and Sc II lines whereas there are few strong lines at wavelengths

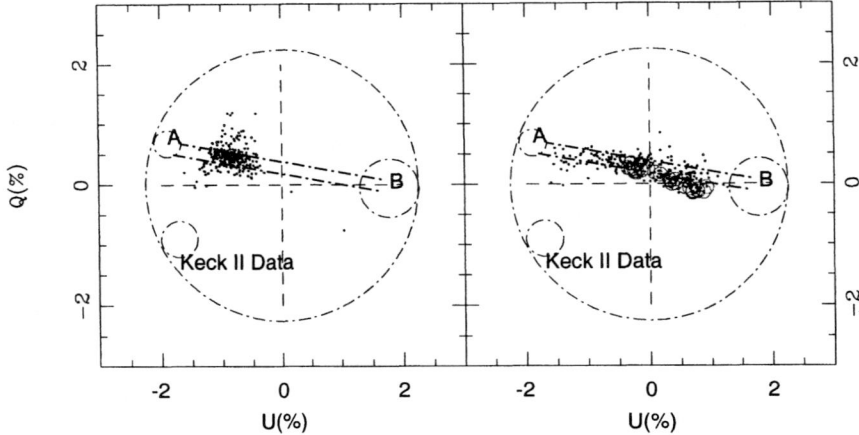

FIGURE 2. The Spectropolarimetry of SN 1998S on the Q-U plane for 1998 March 30 (left panel) and 1998 May 1 (right panel). The outer circles are the limit of the interstellar polarization from the extinction. The premaximum data from Leonard et al. (1999) is labeled by Keck data. The circles labeled A and B represent the two choices for the ISM, of which we prefer A (see text). The small colored circles in the May 1 data are from 6397Å to 6802Å in the vicinity of the Hα line.

longer than Hα and therefore the longer wavelength side is more likely to have higher polarization. This limits the possible interstellar component to area A. In the subsequent discussion we will take $Q \sim 0.7\%$ and $U \sim -1.9\%$ for the interstellar polarization.

This determination of the interstellar polarization leads to intrinsic polarization of the SN of around 1% for the March data and over 3% a month later. To produce such a high degree of polarization for density profiles which resemble those of SNe II, the major/minor axis ratios must be larger than 1.2 to 1 and 2.5 to 1 for the early and late observations, respectively, if the ejecta distribution is an oblate spheroid viewed edge on. Any other viewing angle Θ will require larger axis ratios.

Another Type IIn, SN 1994Y, was also observed in our program close to optical maximum and showed similar behavior to SN 1998S (Wang et al. 1996). The degree of polarization variation across the Hα line is larger than 1.5% in that case, and by using the same method for separating the interstellar polarization, the intrinsic polarization can be as large as 3%. This is consistent with SN 1998S. These are the only two Type IIn SNe with polarimetry observations so far.

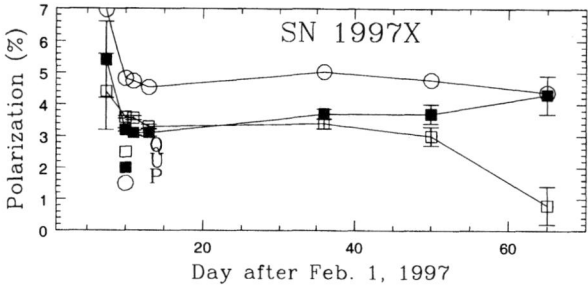

FIGURE 3. The evolution of the continuum polarization (Q, U, and total polarization P), for SN 1997X obtained by averaging from 5000Å to 6000Å. The error bars reflect also the actual variations across the spectral range. Note the large change with time which must be intrinsic to the SN and suggests a polarization of order 4%.

C Type Ic SN 1997X

SN 1997X is one of the best observed SNe in terms of polarimetry. The observed total degree of polarization is as high as 7%. A significant fraction of the polarization may be of interstellar origin. The polarized spectra do show clear spectral features across the strong He I 5876, 6678 lines (Wang, Wheeler, Höflich 1997). Although it is hard to separate the intrinsic and interstellar polarization, it is also clear that a large faction of the of the polarization must be associated with the SN ejecta. Fig. 3 shows the time evolution of the continuum polarization. A polarization level of at least 4% can be attributed to an asymmetric ejecta with axis ratios larger than 3 to 1 for oblate spheroids.

III DISCUSSION

These polarization observations reveal that core-collapse SNe are generally polarized and are highly aspherical. The degree of polarization evolves with time and generally increases after optical maximum. The polarization is also anti-correlated with the mass of the remaining hydrogen shell. The trend is that highest polarizations are observed for SNe that have lost most of their hydrogen envelope before explosion. This trend is illustrated by the sequence Type II SN 1987A, Type IIn SN 1994Y and SN 1998S, Type IIB SN 1993J and 1996cb, and Type Ic SN 1997X. From their relatively large variations of the polarization position angles across spectral lines, Type IIB events seem to possess not only large global asymmetry but also strong large scale chemical or ionization clumps.

Both the temporal evolution and the dependency on ejecta mass of the polarization are consistent with aspherical explosions. The asphericity is expected to be the largest near the center of the ejecta which is normally revealed only by observations past optical maximum. The asymmetry is also expected to have the largest effect on the ejecta geometry for bare-core progenitors such as those for Type Ib/c since a massive hydrogen envelope tends to smear out the asphericity.

These observations argues strongly *against* the conventional picture for Type II SNe explosions. It is hard to imagine that neutrino-driven explosions can produce such large asymmetries throughout the ejecta. We note that the neutrino-driven explosion models have so far failed to produce robust explosions. The recent jet-driven explosion models by Khokhlov et al. (1999) and MacFadyen & Woosley (1999) provide a more promising approach to accounting for the polarization.

REFERENCES

1. Cropper, M., Bailey, J. A., McCowage, J., Cannon, R. D., Couch, W. J., Walsh, J. R., Straede, J. O., & Freeman F. 1988, MNRAS, 231, 695
2. Doroshenko, V. T., Efimov, Y. S., & Shakhovskoi, N. M. 1995, Astrophys. Lett., 21, 513
3. Gerady, C. L., Fesen, R. A., Höflich, P., & Wheeler, J. C. 1999, ApJ, in press
4. Höflich, P. 1991, A&A 246 481
5. Höflich, P., Wheeler, J. C., Hines, D. C., & Trammell, S. R. 1996, ApJ, 459, 307
6. Howell, D. A. et al 1999, in preparation
7. Leonard, D. C., Filippenko, A. V., Bath, A. J., & Matheson, T. 1999, submitted to ApJ (astro-ph/9908040).
8. Leonard, D. C., Filippenko, A. V. & Chornock, R. T. 1999, IAUC, 7305
9. McFadyen, A. I., & Woosley, S. E. 1999, ApJ, 524, 262
10. Mendez, M., Clocchiatti, A., Benvenuto, O. G., Feinstein, C., & Marraco, H. G. 1988, ApJ, 334, 295
11. Qiu, Y., Li, W., Qiao, Q., & Hu, J. 1999, AJ, 117, 736
12. Serkowski, K., Mathewson, D. S., & Ford, V. L. 1975, ApJ, 196, 261
13. Trammell, S. R., Hines, D. C., & Wheeler, J. C. 1993, ApJ, 414, L21
14. Tran, H.D., Filippenko, A.V., Schmidt, G.D., Bjorkman, K.S., Jannuzi, B.T., & Smith, P.S. 1997, PASP, 109, 489
15. Van Dyk, S. D., Sramek, R. A., Montes, M. J., Weiler, K. W., & Panagia, N. 1996, IAUC 6528
16. Wang, L., & Hu, J. Y. 1994, Nature, 369, 380
17. Wang, L., Wheeler, J. C., & Höflich, P. 1997, ApJ, 476, L27
18. Wang, L., Wheeler, J. C., & Höflich, P. 1999, in SN 1987A: Ten Years After, ed. M. M. Phillips & N. Suntzeff (Provo: ASP), in press
19. Wang, L., Wheeler, J. C., Li, Z., & Clocchiatti, A. 1996, ApJ, 467, 435

Cosmic Ray Acceleration at Supernova Remnants

Hyesung Kang [*]

[*]*Pusan National University, Pusan, 609-735, Korea*

Abstract. We believe supernova remnants (SNRs) probably supply all but the highest energy cosmic rays (CRs) below the Ankle energy. Diffusive shock acceleration (DSA) is now widely accepted as the model to explain the production of CRs in a wide range of astrophysical environments. Owing to complex nonlinear physics involved in the model numerical simulations have been quite useful and successful in understanding the details of the acceleration process and dynamical feedback of the CRs to the underlying plasma. Despite the initial success of the theory in explaining the energetics and the spectrum of CRs accelerated by SNRs, there still remain some unresolved issues such as particle injection out of the thermal plasma at shocks, CR diffusion due to the self-generated MHD waves and yet-to-be-detected gamma- ray emission due to the ionic CRs. One of the greater technical difficulties in full DSA theory for SNRs comes from the likelihood that the CR scattering lengths range over several orders of magnitude from the thickness of the gas subshock to much greater scales. In order to overcome this numerical challenge, we have developed a novel computer code which uses multi-level adaptive mesh refinement and sub-zone shock tracking to provide enhanced spatial resolution around shocks at modest cost compared to the coarse grid and vastly improved cost effectiveness compared to a uniform, highly refined grid.

INTRODUCTION

The origin of cosmic rays (CRs) had been by and large an unsolved mystery until 1980s, but during last two decades we have put together many pieces of this important astrophysical puzzle. We now believe most of galactic cosmic rays, at least up to 10^{14} eV of the particle energy, are accelerated by the supernova (SN) blast waves within our Galaxy via Fermi first order process (Blandford & Eichler 1987, Jones et al. 1998). For a wide range of particle energy, the energy spectrum of the observed CRs can be represented by a broken power law, $N(E)dE \propto E^{-\alpha}dE$ with indices $\alpha \sim -2.5$ (Hillas 1984). Two distinct breaking energies where the power-law changes the slope are the *Knee* energy at around $10^{15.5}$ eV and the *Ankle* energy at $10^{18.5}$ eV. We believe the particles above the Ankle energy are likely to come from outside of our Galaxy, although their origins still remain on-going controversies. Several models have been proposed to explain the particles between the Knee and the Ankle energies such as blast-waves from SN Type II/Ib which propagate into a wind bubble generated by strong stellar winds (Biermann 1993), galactic wind terminal shocks (Jokipii & Morfill 1987), and neutron stars (Hillas 1984). But here more detailed quantitative studies are necessary to come to any agreements.

DIFFUSIVE SHOCK ACCELERATION

Theoretical Aspects

There are several physical reasons why we believe SNRs are the primary acceleration sites of galactic cosmic rays.

1. First of all, we find from the direct measurements at the interplanetary shocks that the particles do get accelerated at the shock via Fermi First-order process (Ellison et al 1990). Also plasma simulations (Quest 1988) show that ions can be scattered back and forth across the shock by self–generated waves, and that these scattered ions can provide a seed population of cosmic rays.

2. Secondly, SNRs are the most energetic phenomena occurring in our Galaxy. According to numerical studies, about 30 % of SN explosion energy can be transferred into cosmic rays, although the exact fraction depends on the detailed model parameters. Considering that one SN goes off about every 30 years in our Galaxy, the energy injection rate into CRs from SNR is about 10^{42} erg/s. This can replenish the energy of CRs escaping from our Galaxy, keeping the CR density at the observed level (Blandford & Eichler 1987).

3. According to the standard acceleration model where we assume a spherically symmetric shock propagating into the uniform ISM and a mean magnetic field parallel to the shock normal and particle diffusion by randomly fluctuating field, the maximum energy of the particles that can be accelerated by the standard SNR is about 10^{14} eV (Lagage & Cesarsky 1983). But we note this is a zero-th order "spherical cow" approximation, so more detailed study with realistic models is necessary to come to a definite conclusion.

4. Also according to numerical studies, the accelerated particle spectrum at the SNR is a power-law of index of -2.1 at the source (Berezhko et al 1994). After considering the propagation through ISM and escape process, the expected spectrum at the Earth steepens to $E^{-2.7}$ which is close to what we observe.

Observational Evidences

There are also plenty of observational evidences supporting the idea that the CR acceleration takes place at SNRs. Most successful observations have been made for CR Electrons. Many SNRs are observed by radio Synchrotron radiation due to relativistic electrons gyrating around the magnetic field lines. In some remnants this synchrotron radiation extends to X-ray (e.g. SN1006, Cas A, and IC443). The X-ray energy spectrum of SN1006 turns out to be a power-law, a typical signature of synchrotron emission (Koyama et al 1995). The maximum electron energy emitting this radiation was estimated to be 100 TeV, which is consistent with what the standard model predicts. Relativistic electrons can be detected also in gamma rays by nonthermal bremsstrahlung and by Inverse Compton (IC) scattering of cosmic microwave background radiation. The TeV gamma rays from SN1006 detected by Cangaroo experiment is believed to be due to the Inverse Compton Scattered CBR from 40 TeV electrons (Tanimori et al 1998).

We have not been so lucky, however, in detecting source CR protons at SNRs. Relativistic protons collide with the ISM and emit gamma rays via pion decay (i.e. $p + p \to \pi^0 \to \gamma\ ray$). According to theoretical estimates, the detection of pion decay gamma rays from nearby SNRs may be difficult with current detectors but not impossible (Drury et al. 1994, Berezhko & Volk 2000). Also There is an intrinsic difficulty in distinguishing pion decay gamma rays from electron IC gamma rays, since often IC gamma rays dominate in some remnants as in SN1006. So far most of gamma ray observations, however, gave only upper limits, but no positive detections yet (e.g. SN1006: Tanimori et al 1998). Although theoretical predictions and data are marginally consistent, this failure for proton γ ray detection calls for further improvements on theoretical modeling and numerical calculations. Especially more realistic MHD simulations of SNRs of various types including interactions with non-uniform circumstellar medium are much desired. Also we need to quantify in more details the following aspects of the acceleration theory: non-linear interactions among particles, waves and the underlying plasma, that is, the injection process, self-generation of Alfvén waves, particle scatterings, CR pressure feedback and etc.

Basic Physics

I review briefly the basics physics of the *Diffusive Shock Acceleration* (DSA) theory. In this model, the particles are randomly scattered by inhomogeneities in the magnetic field, that is, MHD waves which are moving with the background flow converging near the shock front. So the shock effectively provides converging mirrors and the particles are scattered between them gaining energy at every shock-crossing (Bell 1978). According to plasma simulations, some supra-thermal particles in the high energy tail of the Maxwellian velocity distribution can re-cross the shock upstream and provide a seed population of the CRs, and the streaming motion of the CR particles against the background fluid can induce MHD waves which then scatter the particles (Lee 1983).

In the standard theory (Drury 1983), the mean magnetic field is parallel to the shock normal. which is called "parallel" shock. The suprathermal particles can be injected into CRs efficiently and MHD wave are efficiently generated by the CR particles at parallel shocks (Malkov 1998). On the other hand, in "perpendicular" shocks where the mean magnetic field is perpendicular to the shock normal, both self-generation of waves and particle injection may become inefficient, but the acceleration process can be very fast and effective due to reflections upstream of the shock. In SNRs the relative orientation of magnetic field and shock front can be very diverse, even in one single object. For example, if a spherically symmetric shock front expands in a region of homogeneous magnetic field (SN Ia), the directions of the shock normal and the magnetic field change over the shock surface from parallel to perpendicular. Therefore, regions of SNRs where quasi–parallel shocks exist, are likely to be where the most effective injection occurs, while effective acceleration, i.e. short acceleration time scales and hard spectra, may be realized in other parts of a SNR, where an oblique geometry of magnetic field and shock normal is found.

In the kinetic version of diffusive shock acceleration theory, we solve the diffusion convection equation for CR proton distribution function along with the usual gas dynamics equations including the contribution of CR pressure. The diffusion–convection equation which describes the time evolution of the particle distribution $f(p, x, t)$ (e.g. Skilling 1975) takes the form:

$$\frac{df}{dt} = \frac{1}{3}\frac{\partial u}{\partial x} p \frac{\partial f}{\partial p} + \frac{\partial}{\partial x}\left(\kappa(x,p)\frac{\partial}{\partial x}f\right) + Q. \tag{1}$$

The diffusion coefficient κ and the injection rate Q from thermal particles to CRs are the primary free parameters in this model. Thus most of uncertainties in the DSA theory lie in the magnetic field configuration and MHD wave spectrum which determines the particle injection process at the shocks and the diffusion model.

Injection Process

One of the key questions to the efficiency of the DSA is the injection process out of thermal particles into CRs. The particles with a gyro radius greater than the wavelength of resonant waves can have an effective velocity with respect to the wave frame. Some of these particles that are in the appropriate part of the phase space would be able to cross the shock from downstream to upstream. Thus some supra-thermal particles in the high energy tail of the Maxwellian velocity distribution may gain energies and have velocities that allow them to re-cross the shock. These particles in turn provide a seed particle beam which generates Alfven waves responsible for the particle scattering. In previous studies of the DSA simulations, the particle injection rate was treated as a free parameter which ranges from 10^{-4} to 10^{-2} of incoming gas particles (Kang & Jones 1991, Berezhko et al. 1994).

A self-consistent injection model based on the interactions of the suprathermal particles with self-generated MHD waves has been developed recently by Malkov

(1998). By adopting this analytic solution, a numerical treatment of the plasma-physical injection model at a strong quasi–parallel shock has been devised and incorporated into the combined gas dynamics and the CR diffusion–convection code (Gieselr et al 1999). According to their simulations, the injection process is self–regulated in such a way that after a quick initial adjustment the injection rate reaches and stays at a nearly stable value where about 10^{-3} of the incoming thermal particles are injected into the CRs. This numerical method provides a self-consistent injection model without relying on a free parameter. On the other hand, for electrons whose gyro-radius is much smaller than ionic gyro-radius, some additional process is needed to bridge the gap between the thermal electron population and the relativistic region. Recently energy transfer from waves amplified by ions reflected off the shock has been suggested as a pre-acceleration (electron injection) mechanism (Levinson 1996; McClements et al. 1997). It still remains mostly uncertain, however, how the particles are injected and accelerated at the perpendicular shock.

Particle Diffusion

The diffusive acceleration depends on the existence of resonant scatterings by self-generated Alfvén waves both upstream and downstream of the shock. So the diffusion coefficient, $\kappa(p) = 1/3\ \lambda v$, is determined by the scattering length λ which is in turn determined by the intensity of resonantly interacting Alfvén waves. Thus, a fully self-consistent theory must include the evolution of those waves through nonlinear interactions with particles. In standard calculations, however, the Bohm diffusion model, $\kappa_B = 1/3 r_g v$, where the particles scatter within one gyration radius (r_g), is often adopted for parallel shocks, assuming the self-generated waves provide completely random scatterings. For the blastwaves from Type Ia SNe propagating into uniform ISM and magnetic field, the shock is mostly quasi-parallel. So the Bohm type diffusion can be a reasonable approximation, since self-generated Alfvén wave are effectively excited at parallel shocks. But for the Type II SN into the stellar bubble, the mean field configuration is mostly azimuthal and so perpendicular to the shock normal. But the particle diffusion at quasi-perpendicular shocks remains uncertain and needs further studies. Even at parallel shocks a self-consistent treatment of wave spectrum may be necessary to calculate the diffusion coefficient, because the resonant scatterings of the highest energy particles may become weak and the Bohm diffusion may become invalid at such energy (Malkov 1998).

Numerical Methods

In order to make a direct comparison with gamma ray observations, it is necessary to refine current numerical calculations of the particle acceleration at SNRs which predicts the particle spectrum of CR protons and pion decay gamma ray

TABLE 1. Types of SN considered in DAS theory

SN Type	Background Density	Composition	Magnetic Field	E_{max}
Ia	uniform ISM	ISM (H & He)	radial	the Knee
Ib or II	wind bubble	Wind (Heavy ions)	azimuthal	the Ankle (?)

radiation flux. The main difficulty in numerical technique to solve nonlinear diffusive shock acceleration is the fact the diffusion-convection equation includes a extremely wide range of length scales that needs to be resolved (i.e. the diffusion length $l_d = \kappa(p)/u_s$). This can depend strongly on the particle momentum in realistic diffusion models including a Bohm type diffusion model. So if we consider the acceleration from supra-thermal particles to the Knee energy, for example, the ratio of largest length scale to the smallest length scale to be resolved is as high as 10^9. If conventional hydrodynamics codes with uniform grid zones are used, such simulations will be prohibitively expensive. Berezhko et al (1994) introduced a new numerical method that can handle such a strongly momentum dependent diffusion model by adopting a set of grids expanding exponentially from the shock. But the code can be applied to a single, spherically expanding shock. We have developed a new hydro/CR code that can track exactly a shock wave as a discontinuous jump by sub-zone *shock tracking* (LeVeque & Shyue 1995), and use multi-level *adaptive mesh refinement* (Berger & LeVeque 1997) around shocks (Kang & Jones 1999). In the shock tracking scheme, the shock front is moved to a new location using the Riemann solutions and the shock remains as an exact discontinuity without smearing. By tracking the shock location exactly, we can refine only the regions around the shock to an arbitrary level of refinements. Since the injection and acceleration of suprathermal particles are confined in a few diffusion length around shocks, our AMR method can provide enhanced spatial resolution required for lower energy particles, while higher energy particles with larger diffusion lengths can be treated at coarser grids.

SUMMARY

The standard diffusive shock acceleration theory have considered the particle acceleration in two types of remnants (see Table 1). In case of Type Ia propagating into uniform ISM, the shock accelerates the material in the ISM and the shock is mostly quasi-parallel. The estimated maximum acceleration energy of CR protons is 10^{14} eV (Berezhko et al. 1994). For Type II/Ib, the shock propagates into the stellar wind bubble first and then into the swept shell of ISM (Berezhko & Volk 2000). The accelerated material includes the metal enriched stellar wind. The magnetic field is likely to be azimuthal due to stellar rotation and wind, so the shock is perpendicular. The particle injection and acceleration is rather uncertain in perpendicular shocks. However, there is a study by Biermann (1993) where this type of SN remnants may possibly accelerate the heavy ions up to the Ankle energy with energy independent diffusion due to macroscopic turbulences. Numerical simulations of both Type Ia and II/Ib suggest that about 30 % of the explosion energy of these SNRs can be transferred into CR energy via first order Fermi process, assuming that $10^{-4} - 10^{-3}$ of incoming thermal particles are injected into CRs and the Bohm diffusion is valid. Although some pieces are still missing, we now are on the verge of completing the puzzle for the origin of Galactic cosmic rays.

ACKNOWLEDGMENTS

This work was supported by Korea Research Foundation Grant (KRF99-015-DI0114).

REFERENCES

1. Bell, A. R., *MNRAS*, **182**, 147 (1978)
2. Berezhko E.G., Yelshin V.K., & Ksenofontov L.T., *Astropart. Phys.* **2**, 2 (1994)
3. Berezhko E.G., & Völk H.J., *Astro. & Astrophs.* **357**, 283 (2000).
4. Berger, J. S., & LeVeque, R. J. SIAM J. Numer. Anal. (1997)
5. Biermann, P. B. *Astron. & Astrophs.* **271**, 649 (1993).
6. Blandford R. D., & Eichler D., *Physics Reports* **154**, 1 (1987).
7. Drury L. O'C., *Rep. Prog. Phys.* **46**, 973 (1983).
8. Drury, L. O'C., Aharonian, F., & Volk, H. J., *Astron. & Astrophs.* **287**, 959 (1994).
9. Ellison, D. C., Möbius, E., & Paschmann, G. *Astrophys. J.* **352**, 376 (1990)
10. Hillas, A. M., *Annual Rev. of Astron. & Astrophys.*, **22**, 425 (1984)
11. Gieseler U.D.J., Jones T.W., & Kang H., *Proc. 26th Int. Cosmic Ray Conf.* (Salt Lake City), **4**, 419 (1999).
12. Jokipii, J.R., & Morfill, G. *Astrophys. J.* **312**, 170 (1987).
13. Jones, T. W., *Astrophys. J.* **413**, 619 (1993).
14. Jones, T. W., et al. *Pub. of Astro. Soc. of the Pacific* **110**, 125 (1998).
15. Kang H., & Jones T.W., *MNRAS* **249**, 439 (1991).
16. Kang H., & Jones T.W., *Astrophys. J.* **447**, 944 (1995).
17. Kang H., & Jones T.W., *Proc. 26th Int. Cosmic Ray Conf.* (Salt Lake City), **4**, 455 (1999).
18. Koyama, K. et al. *Nature* **378**, 255 (1995).
19. Lagage, P.O., & Cesarsky, C.J. *Astron. & Astrophs.* **118** 223 (1983).
20. Lee, M.A. *J. Geophys. Res.* **87**, 5063 (1982).
21. LeVeque, R. J., & Shyue, K. M. SIAM J. Scien. Comput. 16, 348 (1995)
22. Levinson A., *MNRAS* **278**, 1018 (1996)
23. Malkov M.A., *Phys. Rev. E* **58**, 4911 (1998).
24. McClements K.G., Dendy R.O., Bingham R., Kirk J.G, Drury L. O'C., *MNRAS* **291**, 241 (1997)
25. Quest K.B.,*J. Geophys. Res.* **93**, 9649 (1988).
26. Skilling J., *MNRAS* **172**, 557 (1975).
27. Tanimori, T., et al, *Astrophys. J. Letts.* **497**, L25 (1998)

Pulsars, Magnetars & Asymmetric Supernovae

J. Craig Wheeler[*], Insu Yi[†], Peter Höflich[*] & L. Wang[*]

[*]*Department of Astronomy, University of Texas*
[†]*Korea Institute for Advanced Study*

Abstract. The possible physical processes, associated timescales, and energetics are outlined that could lead to the production of pulsars, jets, and asymmetric supernovae, in routine circumstances and to a magnetar in more extreme circumstances in the collapse of the bare core of a massive star. A LeBlanc-Wilson MHD jet would be formed promptly, but requires 5 to 10 s to reach the surface of the progenitor of a Type Ib/c supernova. During this time, the newly-born neutron star could contract, spin up, and wind up field lines or turn on an $\alpha - \Omega$ dynamo. The infalling matter could be subject to strong torques. In addition, the light cylinder will contract from a radius large compared to the Alfvén radius to a size comparable to that of the neutron star. This will disrupt the structure of any organized dipole field and promote the generation of ultrarelativistic MHD waves at high density and large mmplitude electromagnetic waves at low density. The generation of these waves would be delayed by the cooling time of the neutron star about 1 to 10 seconds, but the propagation time is short so the MHD waves could arrive at the surface at about the same time as the original jet. These processes could account for the ubiquitous bipolar asymmetries revealed by supernova spectropolarimetry.

INTRODUCTION

The problem of core-collapse has been with us for over 40 years (Hoyle & Fowler 1960). Immediately after the discovery of pulsars, it was reasonable to explore the issue of whether or not the rotation and magnetic fields associated with pulsars could be a significant factor in the explosion mechanism (Ostriker & Gunn 1971; Bisnovatyi-Kogan 1971). With typical dipole fields of 10^{12} Gauss and rotation periods of several to several tens of milliseconds, a strong robust explosion seemed unlikely. Several factors have led to a need to re-examine this conclusion. The principle one is the accumulating evidence that core collapse supernovae are distinctly asymmetric. Aside from its famous rings, HST observations of SN 1987A resolving the debris show that the ejecta are asymmetric with an axis that roughly aligns with the small axis of the rings (Pun et al. 1997). New CXO observations of Cas A show that the jet and counter jet and associated structure are observable in the X-ray (Hughes et al. 2000; Hwang et al. 2000) as well as the optical (Fesen & Gunderson 1996; and references therein). The most direct evidence bearing on this topic is from supernova spectropolarimetry which shows that substantial asymmetry is ubiquitous in core-collapse supernovae and that a significant portion show strong evidence for a single, wavelength-independent axis of symmetry (Wang et al. 1996; Wang et al. 2000; see contribution by Wang in these proceedings). Many, even perhaps most, core-collapse supernovae are bi-polar.

The strength of the asymmetry observed with polarimetry is higher (several %) in supernovae of Type Ib and Ic that represent exploding bare non-degenerate cores. The degree of asymmetry also rises as a function of time for Type II supernovae that have retained their hydrogen envelopes (from $\lesssim 1\%$ to $\gtrsim 1\%$) as the ejecta expand and one looks more deeply into the core material. Both of these trends suggest that it is the inner machine, the core collapse mechanism itself, that is responsible for the asymmetry. The observed polarization requires significant asymmetry; axis ratios exceeding 2 to 1. To impose the observed strong asymmetry in the final homologous expansion, an axial flow must be established and maintained for at least several dynamical time scales. This is the operational definition of a jet.

Asymmetries associated with neutrino emission are unlikely to produce the global effects reflected in the spectropolarimetry. Neutrino asymmetries will yield a short-lived, essentially impulsive effect (Shimizu, et al. 1994; Burrows & Hayes 1996; Fryer & Heger 1999; Lai et al. 2000). Expansion and transverse pressure gradients will wipe out transient asymmetries before homologous expansion is achieved. Sufficient neutrino impulse might be delivered to the neutron star to yield a substantial runaway velocity, but it is very difficult to see how this impulse can be communicated in a substantial and permanent way to the final ejecta trajectories.

Highly-resolved, fully three dimensional, adaptive grid numerical calculations (Khokhlov 1998) have, however, established that non-relativistic axial jets of energy of order 10^{51} ergs originating within the collapsed core can initiate a bi-polar asymmetric supernova explosion that is consistent with the spectropolarimetry (Khokhlov et al. 1999). Some imbalance in axial jets can also account for pulsar runaway velocities. While a combination of neutrino-induced and jet-induced explosion may prove necessary for complete understanding of core-collapse explosions, jets as computed by Khokhlov et al. (see also Höflich in these proceedings) are sufficient. (See the contribution by MacFayden in these proceedings for a discussion of jets in context of black hole formation.)

The question then arises as to the mechanism of the production of the jets in routine core collapse events. A significant role for rotation and magnetic fields is the obvious candidate. Any such mechanism that purports to account for routine pulsar formation may involve large transient magnetic fields, but must ultimately be consistent with the distribution of deduced dipole strengths of "normal" pulsars of $\sim 10^{12}$–10^{13} G. In a minority of cases, the final dipole field might be consistent with a value of $\sim 10^{15}$ G, yielding a "magnetar" (Duncan & Thompson 1992; see also contribution by Duncan in these proceedings). Growing evidence for neutron stars of that field strength has been obtained by RXTE and other facilities (Kouveliotou et al. 1999, Ibrahim et al. 2000). The nature of the birth event of a magnetar is a separate problem that is significant in its own right. The ensemble of normal pulsar and magnetar births must also be consistent with observed nucleosynthetic abundances, a potentially crucial constraint on jet-induced supernova models.

Possible physical mechanisms for inducing axial jets, asymmetric supernovae, and related phenomena driven by magneto-rotational effects has recently been considered by Wheeler et al. (2000). The means of amplifying magnetic fields by differential rotation in the neutron star and possibly by $\alpha - \Omega$ dynamos was discussed. Attention was focused on the effect of the resulting net dipole field on the torquing of the infalling plasma and newly formed neutron star and on the creation of strong Poynting flux (in the form, initially, of ultrarelativistic MHD waves) at the time-variable speed of light circle in analogy to pulsar radiation mechanisms, albeit buried deeply in the core-collapse supernova ambience.

The topics of core-collapse supernovae and γ-ray bursts overlapped, at least in principle, with the discovery of SN 1998bw. Although the association is still controversial, this odd supernova is likely to have been connected to GRB 980425 (Galama et al. 1998; see also contributions by Kulkarni and by Wijers in these proceedings.). Some models of SN 1998bw invoked especially large kinetic energies, in excess of 10^{52} ergs in spherically-symmetric models, to account for the bright light curve and high velocities (Iwamoto, et al. 1998; Woosely, Eastman, & Schmidt 1998), while others took note of the measured polarization to suggest that strongly asymmetric models could account for the observations with more "normal" energies (Höflich, Wheeler, & Wang 1999). Large kinetic energy has also been attributed to several other supernovae (again based on spherically symmetric models), especially SN 1997cy (Germany et al. 1999) and SN 1997ef (Branch 2000; Iwamoto et al. 2000). Supernova-like excesses of light have been detected in the afterglow of two γ-ray bursts, GRB 970228 (Reichart 1999; Galama et al. 1999) and GRB 980326 (Bloom et al. 1999) about two weeks after the γ-ray bursts. The excess light in the afterglow of both GRB 970228 and GRB 980326 has been modeled by the addition of light from an event like SN 1998bw. While there may be other explanations for these excesses, the connection with supernovae must be pursued and may be related to the production of jets.

Evidence for exceedingly energetic supernovae cannot be easily disentangled from suggestions that they are asymmetric, a condition that affects energy estimates. Events like SN 1997cy, SN 1997ef and SN 1998bw will help to sort out the physics of explosive events, whether such events are more closely related to "ordinary" supernovae or "hypernovae," whether either of these classes leaves behind neutron stars as "ordinary" pulsars or highly magnetized "magnetars" or whether the remnant is a black hole and whether any of these events are associated with classic cosmic gamma-ray bursts as suggested by the supernova-like brightening of the afterglow of GRB 970228 and GRB 980326.

Our working hypothesis is thus that jets leading to asymmetric supernova explosions are associated with routine core collapse and pulsar formation in roughly 90% of the cases. Perhaps 10% of the core collapse events would be associated with magnetars. Magnetar formation might be associated with exceptionally strong explosions and asymmetries, but this is not necessarily the case. What is clear is that the physics of rotating, magnetic collapse needs to be studied anew.

The basic arguments of Wheeler et al. (2000) for the nature of jet formation collapse will be summarized in §2. The final dipole component and production of magnetars are discussed briefly in §§3, 4. A summary is given in §5.

THE FORMATION OF JETS IN CORE COLLAPSE

To understand routine, pulsar-forming, strongly asymmetric supernovae, there must be a deeper understanding of the generation of jets during the process of core collapse to form neutron stars. Wheeler et al. (2000) have explored the possible physical processes, associated timescales, and energetics that could lead to the production of pulsars, jets, and asymmetric supernovae and to a 10^{15} G magnetar in more extreme circumstances in the collapse of a massive stellar core

When the neutron star first forms there is likely to be linear amplification of the magnetic field. The energy in differential rotation of the proto-neutron star is sufficient, in principle, to power a significant jet. Wheeler et al. (2000) noted that when the neutron star contracts and speeds up two significant things may happen. One is that the rotational energy increases. The energy becomes significantly larger than required to produce a supernova and sufficient, in principle, to drive a cosmic γ-ray burst if the collimation is tight enough. In addition, the light cylinder may contract from a radius large compared to the Alfvén radius to a radius comparable to that of the neutron star. This could disrupt the structure of an organized dipole field and promote the generation of intense MHD waves. The frequency of the MHD waves, $\sim \Omega_{NS} \sim 10^4$ Hz, would always be less than the plasma frequency, so these waves would be reflected internally and perhaps also channeled up the rotation axis. Collimation of the flow of energy may thus be expected.

The original rotational energy of the proto-neutron star is,

$$E_{\rm rot, PNS} \simeq \frac{1}{2} I_{PNS} \Omega_{PNS}^2 \simeq 9 \times 10^{50} \text{ erg} \left(\frac{M_{NS}}{1.5 M_\odot}\right) \left(\frac{\Omega_{PNS}}{250 \text{ s}^{-1}}\right)^2 \left(\frac{R_{PNS}}{50 \text{ km}}\right)^2, \quad (1)$$

and the rotational energy after the deleptonizatin and contraction is,

$$E_{rot,NS} \simeq \frac{1}{2} I_{NS} \Omega_{NS}^2 \simeq 6 \times 10^{52} \text{ erg} \left(\frac{M}{1.5 M_\odot}\right) \left(\frac{\Omega_{NS}}{10^4 \text{ s}^{-1}}\right)^2 \left(\frac{R_{NS}}{10^6 \text{ cm}}\right)^2. \quad (2)$$

It is not clear how much of the latter energy can be dissipated in the magnetic field, but there is substantial energy for a supernova for any final rotation period less than about 7 msec. Strictly speaking only the energy associated with the differential rotation can be tapped, but the neutron star is always in strong differential rotation compared to the progenitor star. There may be constant torques on the neutron star until the explosion succeeds or fails, so this issue is complex and involves the magnetic fields we will now explore.

Generation of the Toroidal Field

If the pre-collapse progenitor core has a field strength comparable to that of a magnetized white dwarf, $\sim 10^6$ G, then a field of $\sim 10^{12}$ G could arise from flux-freezing. This field can be amplified by differential rotation in the neutron star (Meier et al. 1976; Kluźniak & Ruderman 1998). The proto-neutron star is likely to be differentially rotating with angular velocity Ω increasing outward since the proto-neutron star will be "stiffer" than the iron core which collapsed. This is certainly true for a collapsing highly degenerate white dwarf progenitor (Ruderman, Tao & Kluźniak 2000). In this process, differential rotation could wrap the poloidal seed field into a strong toroidal field which then emerges from the neutron star through buoyancy. For this mechanism, the field grows linearly with time. After n_ϕ revolutions of the neutron star, the initial seed (poloidal) field is wrapped and amplified to produce a toroidal field,

$$B_\phi \simeq 2\pi n_\phi B_p, \qquad (3)$$

where B_p is the initial seed poloidal field. Buoyancy will operate to expel the field if the amplified final field B_f satisfies,

$$\frac{B_f^2}{8\pi} \simeq f_B \rho c_s^2, \qquad (4)$$

where $f_B \simeq 0.01$ is the fractional difference in density between the rising flux tube elements and the stellar material. Wheeler et al. (2000) considered the buoyant field from near the center of the proto-neutron star and hence sound speed $c_s \simeq c/3$ and density $\rho \simeq 10^{13}$ g cm^{-3}. For these values, one finds,

$$B_f \simeq 2 \times 10^{16} \text{ G } f_{B-2}^{1/2} \rho_{13}^{1/2}, \qquad (5)$$

where $f_{B-2} = f_B/0.01$ and $\rho_{13} = \rho/10^{13}$ g cm^{-3}. Assuming that the magnetic flux tube (a torus) occupies a volume of $V_B/V_{PNS} \simeq 0.1$ where $V_{PNS} \simeq \frac{4\pi}{3} R_{PNS}^3$, the energy contained in the magnetic flux tubes at the buoyancy limit prior to deleptonization is estimated to be,

$$E_B \simeq 0.1 \times V_{PNS} \times \frac{B_f^2}{4\pi} \qquad (6)$$
$$\simeq 1.6 \times 10^{51} \text{ erg}.$$

This magnetic energy would be ejected from the neutron star in the rising magnetic flux tubes. This energy is comparable to the proto-neutron star rotation energy. For the adopted parameters, the proto-neutron star would have to lose substantial rotational energy to the magnetic field before the field would float on dynamical time scales. This magnetic energy is still likely to escape from the proto-neutron star, but the details may be complex and involve the subsequent contraction of the neutron star.

The number of revolutions to reach the buoyancy limit of $B_f \sim 2 \times 10^{16}$ G is,

$$n_f = \frac{B_f}{B_0}\frac{1}{2\pi} \simeq 3000 B_{0,12}^{-1} f_{B-2}^{1/2} \rho_{13}^{1/2}. \tag{7}$$

For $B_0 \sim 10^{12}$ G and $\rho \sim 10^{13}$ g cm^{-3}, the amplification time scale before the field is expelled by buoyancy is thus,

$$t_f \simeq n_f P_{PNS} \sim 300 \text{ s}\left(\frac{P}{100 \text{ ms}}\right) \sim 3 \text{ s}\left(\frac{P}{1 \text{ ms}}\right). \tag{8}$$

Thus linear wrapping is slow at first, but is likely to accelerate with the contraction and spin up of the neutron star. If the neutron star attains a period of 1 ms, t_f could be comparable to the cooling, contraction time of ~ 1 s. In general, the epoch of maximum expulsion of the toroidal magnetic field from deep within the neutron star is a few seconds after collapse. An important point is that this process of wrapping and expelling field does not necessarily halt after one characteristic time. The toroidal field will build up to the buoyancy limit and then begin to be expelled. Field of this strength, of order 10^{16} G in the example just illustrated, will continue to pump out as the differential rotational energy is dissipated.

The situation just outlined for generating and expelling the field applied, by assumption, to the field deep within the neutron star. The magnetic field will, however, be subject to wrapping by differential rotation throughout the collapsing medium since the field is ultimately anchored in the outer, slowly rotating progenitor. Note that the number of wrappings, n_f, to reach the buoyancy limit depends on the density. For a density of $\rho \simeq 10^{11}$ g cm^{-3}, characteristic of the neutrino sphere, the buoyancy limit is about 10^{15} G, which is achieved after only about 300 wrappings in a time of order 1 s for a period of 3 msec.

Another key component may be the strongly shearing layer between the neutron star and the infalling matter. This shear will be exacerbated as the neutron star cools and contracts. If the typical density in the infalling matter is $\rho \simeq 10^8$ g cm^{-3}, then the critical buoyant field is still in the range of 10^{14} to 10^{15} G, depending on the buoyancy parameter, f_B. The number of wrappings to reach the buoyant limit is thus reduced to a few tens and the time to tenths of a second. Although the field generated here will be less than expected deep within the neutron star, the fact that it may become buoyant quickly may cause this process to play a major role in the subsequent energetics of the neutron star and dynamics of the jet.

This discussion illustrates that the buoyancy time and subsequent, longer term dissipation of the rotational energy into magnetic field can be short enough to be a significant effect on the radial contraction time of the neutron star. There will be a gradient of buoyancy field that breaks out of the neutron star as a function of time and location, the nature of which needs to be explored. The manner in which the buoyant magnetic field affects the differential rotation that is needed to generate the field must be examined in more detail. One of the controlling parameters there will be $\beta = P_{gas}/P_{mag}$. The field is not expected to be dynamically important deep within the neutron star ($\beta \gg 1$), but it will affect the differential rotation on secular time scales. As the field emerges from the neutron star into lower density matter, there will be regime of low β, where dynamical effects are important.

The shear, field amplification, and field expulsion will go in the direction of eliminating the shear, but the infall of more slowly rotating matter will always maintain a shear with negative gradient in Ω at the neutron star boundary. The creation and expulsion of field from this layer should continue until the explosion of the star cuts off the infall. A strong toroidal field could inhibit radial motion and the poloidal component will have significant dynamical affects within the Alfvén surface, as described below.

The Dipole Component

There is likely to be a dipole component to the field generated in collapse and its affects could be considerable. An important aspect of the toroidal field is that it will be substantially unaffected at the speed of light cylinder since plasma can slide along the field and will not be forced to move at the speed of light. This is true at the equator, but the consequences of the presence of the speed of light circle may be more substantial up the axis, depending on the aspect angle of the field. This is worth deeper consideration.

In any case, it is clear that any dipole component will have very different dynamical implications than the toroidal component and that the dipole component will interact in a major way with the time-dependent speed of light cylinder since it forces plasma to co-rotate with the neutron star in the azimuthal direction.

The radius of the light cylinder is,

$$R_{LC} = \frac{c}{\Omega} = \frac{cP}{2\pi} \simeq 50 \text{ km} \left(\frac{P}{\text{ms}}\right), \tag{9}$$

where Ω is the rotational frequency and P the rotational period of the neutron star. The Alfvén radius at which magnetic pressure is balanced by the ram pressure, e.g.,

$$\frac{1}{2}\rho v^2 \simeq \frac{1}{8\pi}B^2, \tag{10}$$

is, with $B \simeq B_{NS}\left(\frac{R}{R_{NS}}\right)^{-3}$,

$$R_A \simeq 3.0 R_{NS} B_{14}^{1/3} \rho_8^{-1/6} v_8^{-1/3}. \tag{11}$$

When the proto-neutron star first forms, the speed of light cylinder is likely to be very large, of order the size of the original iron core and not significant. For an initial field of order 10^{12} G, the Alfvén radius may be comparable to or even less than the radius of the neutron star, but as the field amplifies, the Alfvén radius moves outward. The Alfvén radius can grow to be of order the radius of the standing shock (∼ 200 km in typical numerical calculations). The dipole component of the field can thus extend to the Alfvén radius and affect the dynamics of the flow through torques. As the neutron star cools, contracts, and spins up, however, the light cylinder contracts and can become less than the Alfvén radius. Under these conditions, the dipole field (and perhaps a component of the toroidal field) will be disrupted and substantial energy generated in the form of Poynting flux.

Torques can be exerted on the plasma within the Alfvén radius, The spin down time for the neutron star will be long, but the time to deflect the infalling matter will be short. The latter is given approximately by,

$$t_{tor,1} \simeq \frac{\rho R_A v}{B^2} \simeq \frac{\rho R_A v}{4\pi \rho v^2} \simeq \frac{1}{4\pi} \frac{R_A}{v}, \qquad (12)$$

$$\simeq 0.01 \text{s} \left(\frac{R_{PNS}}{50 \text{ km}}\right) B_{PNS,14}^{1/3} \rho_8^{-1/6} v_8^{-4/3}.$$

This time is sufficiently short that the torque could substantially alter the flow of matter, preventing the radial infall interior to the shock that is common to all spherically symmetric models of core collapse. Multi-dimensional models that produce non-radial circulation flows in the matter beneath the standing shock would also be substantially affected for conditions similar to those reflected in equation (12). This time is also substantially shorter than the time scale estimated earlier for a buoyant field to be generated in the shear layer at the surface of the neutron star, ∼ 0.1 s. The torque on the infalling matter could thus alter the flow into this boundary layer.

If the condition occurs that the light cylinder retreats to within the Alfvén radius, then energy will be dissipated at the light cylinder. The power generated at the light cylinder by the disruption of the dipole component of the field is about:

$$L_{MHD} \simeq 4\pi R_{LC}^2 \times \frac{c}{4\pi} |\vec{E} \times \vec{B}| \simeq \frac{\mu_{NS}^2}{R_{LC}^4} c \simeq \frac{R_{NS}^6 B_{NS}^2 \Omega_{NS}^4}{c^3}, \qquad (13)$$

or about

$$L_{MHD} \simeq 4 \times 10^{52} \text{ erg s}^{-1} \left(\frac{R_{NS}}{10 \text{ km}}\right)^6 \left(\frac{B_{NS}}{10^{16} \text{ G}}\right)^2 \left(\frac{P_{NS}}{\text{ms}}\right)^{-4}. \qquad (14)$$

This mechanism could thus be important or not, depending on the field strength and distribution. We note that, in principle, the Alfvén radius will be a complicated function of angle, perhaps ∼ 10^{14} G at the equator and ∼ 10^{16} G along poles. In particular, the magnetic field will halt, trap, and redirect the infall at the position of local Alfvén surface. A key question is whether or not a magnetic "wall" will prevent excessive inflow, and neutronization in the sort of jet model we are describing here.

The Poynting flux generated at the light cylinder will be of very high amplitude and perhaps of relativistic speeds. As an indication of the latter, the Alfvén velocity can be written:

$$\Gamma v_A \simeq \left(\frac{B^2}{4\pi\rho}\right)^{1/2} \simeq 3 \times 10^{11} \text{ cm s}^{-1} B_{16} \rho_8^{-1/2}, \tag{15}$$

where Γ is the Lorentz factor of the Alfvén waves. Whether the MHD waves generated in this ambience are, strictly speaking, Alfvén waves is not clear since they may not be describable as a linear perturbation on the field structure, but this suggests that they could propagate at relativistic speeds under certain circumstances.

THE FINAL DIPOLE COMPONENT

In the preceeding sections we have been discussing toroidal and dipole fields of order 10^{14} to 10^{16} G arising in the formation of a neutron star in a supernova. There is clearly an issue of the final effective dipole strength of the neutron stars left behind in the explosion. We must have a final dipole field distribution of the neutron stars consistent with those of observed pulsars, $\sim 10^{12}$ to 10^{13} G, ignoring magnetars. It seems plausible that the large fields generated in the collapse are dissipated. The very strong fields that may be generated in the depths of the neutron star will take longer to float out, but the time could still be short compared to the lifetime of the supernovae, never mind the pulsar. Stronger fields in the interior could also leave smaller surface dipole fields. Much of the field may be amplified and dissipated at the shearing boundary between the infalling matter and the neutron star. This layer and its field will almost surely substantially dissipate after the explosion.

MAGNETARS

All of the issues raised here will be examined in the context of magnetars. However the large effective dipole fields arise in magnetars, there is a strong presumption that they are created in the formation of the neutron star. The formation of a magnetar may require special circumstances, or they may arise as the tail of the normal pulsar formation process. One possibility is that they represent that fraction of core-collapse events that, perhaps by accident of progenitor conditions, do attain an especially rapid differential rotation and amplify the field not just by field line wrapping, but by an $\alpha - \Omega$ dynamo (Duncan & Thompson 1992). This could lead to exponential field growth, and total field strength of order 10^{17} G with dipole component of order 10^{15} G. These events could lead to stronger jets and stronger explosions. They might account for the growing sample of supernovae (SN 199ef, SN 1997cy, SN 1998bw) with large expansion line widths and hence presumed large total kinetic energy.

In the context of the physics outlined here, there are a host of issues to be explored in the magnetar regime. In addition to the possibility of higher energy, the stronger dipole fields of a magnetar could create higher torques in the inflow. The implied faster rotation implies an especially small light cylinder and hence greater potential for disrupting poloidal field and creating Poynting flux. There is an interesting question of whether a magnetar will deliver more power to a jet or give a jet of higher velocity. Ironically, jet-induced supernova models suggest that less energy is imparted to the overlying matter as the jet speed is increased (Höflich in these proceedings). Thus it is not completely clear that a magnetar will automatically lead to a "hypernova" of exceptionally high energy.

CONCLUSIONS

We have summarized the ubiquitous evidence that normal core collapse to produce neutrons stars produces strongly asymmetric debris in homologous expansion. The most obvious means to accomplish this is by jet-induced explosions. We have discussed the means by which magneto-rotational effects could torque the infalling matter and generate strong magnetic fields through linear wrapping in the differential motion of the cooling, deleptonizing, contracting neutron star. The dipole component of the field would exert substantial torques and could power strong ultrarelativistic MHD waves from the vicinity of the contracting light cylinder.

A jet emerging from deep in the collapsing iron core of a star will emerge from the helium core in 5 to 10 seconds. Wheeler et al. (2000; see also Nakamura 1998) argued that a jet launched after the deleptonization contraction of the neutron star could arrive at the surface of the helium star at about the same time. Wilson (private communication) has pointed out that the 5 to 10 seconds associated with the loss of neutrinos in the deleptonization phase represents the characteristic time scale for the luminosity, but that the luminosity is enhanced at later phases by the contraction which results in a higher temperature at the neutrinosphere. The time scale for the contraction of the radius and hence the spin up of the neutron star is actually considerably shorter, of order one second (Wilson & Mayle 1993). This means that any stronger jet launched during the contraction will catch up to and dominate any magnetic "bubble" or jet launched promptly during the intial collapse.

This discussion has summarized the generation of the toroidal field by field line wrapping and the possible consequences of the dipole component of that field. Wheeler et al. (2000) briefly mentioned the possible role of the primary toroidal field. In the future, more emphasis should be placed on this toroidal field. This component has the capacity to directly generate axial jets within core-collapse conditions by analogy with magneto-centrifugal models of jets in AGN (Koide, Shibata & Kudoh 1997; Ouyed, Pudritz & Stone 1997; Meier et al. 1997; Romanova et al. 1998; Ustyugova et al. 1999; Meier, 1999; Lery & Frank 2000; Koide, Meier, Shibata & Kudoh 2000; and references therein). The magneto-centrifugal models in the literature usually focus on black hole conditions where the field is anchored in the disk or at infinity. The physical conditions associated with core collapse may prove especially robust since the field is anchored in substantial, corporeal objects, the outer, still collapsing layers of the progenitor star and the collapsed object — a neutron star. The production of a strong toroidal field, substantially stronger than the 10^{12} G field of a pulsar, is nearly inevitable, and strong axial jets driven by that field equally so. This mechanism in the context of core collapse may prove to have its closest astrophysical analog in the production of supersonic magneto-centrifugal jets in young stellar objects where also there is an outer envelope of infalling matter and an inner, differentially rotating object (Konigle & Pudritz 2000; Lery & Frank 2000).

ACKNOWLEDGMENTS

We are grateful to Alexei Khokhlov, Elaine Oran and Almadena Chtchelkanova for helping us to understand how jets work in stars, to Rob Duncan for discussions of magnetars and to Dave Meier and Jim Wilson for helping us to understand how jets might form in core collapse. This research was supported in part by NSF Grant 95-28110, NASA Grant NAG 5-2888, a grant from the Texas Advanced Research Program, KRF 1998-001-D00365 (to IY) and the Ewha University Faculty Research Fund (to IY).

REFERENCES

1. Bisnovatyi-Kogan, G. S. 1971, Soviet Astronomy AJ, 14, 652
2. Blandford, R. D. & Payne, D. G. 1982, MNRAS, 199, 833
3. Blandford, R. D. & Znajek, R. 1977, MNRAS, 179, 433
4. Bloom, J. S. et al. 1999, Nature, submitted, astro-ph/9905301
5. Branch, D. 2000, in The Largest Explosions Since the Big Bang: Supernovae and Gamma-Ray Bursts, eds. M. Livio, K. Sahu & N. Panagia, in press
6. Burrows, A. & Hayes, J. 1996, Phys Rev Lett, 76, 352
7. Duncan, R. C. & Thompson, C. 1992, ApJ, 392, L9
8. Fesen, R. A. & Gunderson, K. S. 1996, ApJ, 470, 967
9. Fryer, C. L. & Heger, A., 1999, ApJ submitted, astro-ph/9907433
10. Galama, T. J. et al. 1998, Nature, 395, 670
11. Galama, T. J. et al. 1999, ApJ, submitted, astro-ph/9907264
12. Germany, L. M., Reiss, D. J., Sadler, E. S., Schmidt, B. P. & Stubbs, C. W. 1999, ApJ, submitted, astro-ph/9906096
13. Höflich, P., Wheeler, J. C., & Wang, L. 1999, ApJ, 521, 179
14. Hoyle, F. & Fowler, W. A. 1960, ApJ, 132,565
15. Hughes, J. P., Rakowski, C. E., Burrows, D. N. & Slane, P. O. 2000, ApJ, 528, L109
16. Hwang, U. Holt, S. S. & Petre, R. 2000, ApJ, in press (astro-ph/0005560)
17. Ibrahim, A. I., Strohmayer, T. E., Woods, P. M., Kouveliotou, C., Thompson, C., Duncan, R. C., Dieters, S., van Paradijs, J. & Finger, M. astro-ph/0007043
18. Iwamoto K. et al., 1998 Nature, 395, 672
19. Iwamoto K. Nakamura, T., Nomoto, K., Mazzali, P. A., Danziger, I. J., Garnavich, P., Kirshner, R. P., Jha, S., Balam, D. & Thorstensen, J. 2000, ApJ, 534, 660
20. Khokhlov A.M., 1998, J. Comput. Phys., 143, 519

21. Khokhlov A.M., Höflich P. A., Oran E. S., Wheeler J.C. Wang, L, & Chtchelkanova, A. Yu. 1999, ApJ, 524, L107
22. Koide, S., Shibata, K. & Kudoh, T. 1999, ApJ, 522, 727
23. Koide, S., Meier, D. L., Shibata, K. & Kudoh, T. 2000, ApJ, 536, 668
24. Konigl, A. & Pudritz, R. E. 2000, in Protostars and Planets IV, eds. V. Mannings, A. Boss, & S. Russell (Arizona: University of Arizona Press)(astro-ph/9903168)
25. Kouveliotou, C. et al. 1999, ApJ, 510, L115
26. Lai, D., Chernoff, D. F. & Cordes, J. M. 2000, ApJ, in press (astro-ph/0007272)
27. Lery, T. & Frank, A. 2000, ApJ, 533, L897
28. LeBlanc, J. M. & Wilson, J. R. 1970, ApJ, 161, 541
29. Meier, D. 1999, ApJ, 522, 753
30. Meier, D., Edgington, S., Godon, P., Payne, D. G. & Lind. K. R. 1997, Nature, 388, 350
31. Meier, D., Epstein, R. I., Arnett, W. D. & Schramm, D. N. 1976, ApJ, 204, 869
32. Nakamura, T. 1998, Prog. Theor. Phys. 100, 921
33. Ostriker, J. P. & Gunn, J. E. 1971, ApJ, 164, L95
34. Ouyed, R., Pudritz, R. E. & Stone, J. M. 1997, Nature, 385, 409
35. Pun, C. S. J., Kirshner, R. P., Garnavich, P. M. & Challis. P. 1997, BAAS, 191.9901
36. Reichart, D. E. 1999, ApJ, submitted, astro-ph/9906079
37. Romanova, M. M., Ustyuogova, G. V., Koldoba, A. V., Chechetkin, V. M. & Lovelace, R. V. E. 1998, ApJ, 500, 703
38. Ruderman, M. A., Tao, L. & Kluźniak, W. 2000, ApJ, submitted (astro-ph/0003462)
39. Shimizu, T., Yamada, S. & Sato, K. 1994, ApJ, 432, L119
40. Ustyugova, G. V., Koldoba, A. V., Romanova, M. M., Chechetkin, V. M. & Lovelace, R. V. E. 1999, ApJ, 516, 221
41. Wang, L., Howell, D. A., Höflich, P. & Wheeler, J. C. 2000, ApJ, in press
42. Wang, L., Wheeler, J. C., Li, Z. W., & Clocchiatti, A. 1996, ApJ, 467, 435
43. Wheeler, J. C., Yi, I., Höflich, P. & Wang, L. 2000, ApJ, in press (astro-ph/9912080)
44. Wilson, J. R. & Mayle, R. 1993, Phys. Rep. 227, 97
45. Woosley S., Eastman R., Schmidt M. 1998, ApJ, 516, 788

Modern Supernova Search

Myung Gyoon Lee

Astronomy Program, SEES, Seoul National University.
Seoul 151-742, Korea
Email: mglee@astrog.snu.ac.kr

Abstract. Supernovae play a critical role in observational cosmology as well as in astrophysics of stars and galaxies. Recent era has seen dramatic progress in the research of supernovae. Several programs to search systematically supernovae in nearby to distant galaxies have been very successful. Recent progresses in the modern supernova search are reviewed.

INTRODUCTION

Ancient supernova (SN) searches must have been based on random visual observations and discovered eight SNe all of which are in our Galaxy. It is almost 400 years since the last galactic supernova SN 1604 was observed. Korea has a rich record of old observations of transient objects including galactic supernovae which were called as 'guest stars' before (e.g. see Chu 1968). Fig. 1 illustrates a light curve of SN 1604 based on the old Korean and European observations (Clark & Stephensen 1977). A light curve of a Type Ia Supernova 1991T (Lira et al. 1998) is also overlayed in Fig. 1, arbitrarily shifted to match SN 1604 around the peak. The light curve of SN 1604 is approximately matched by that of the Type Ia, showing how remarkable the old visual photometry of SN 1604 was even at that time.

In 1885 the first extragalactic supernova (SN 1885) was discovered in M31 after 281 years passed since the last galactic supernova was discovered. However, it is only in 1934 that the concept of supernovae was first introduced by Baade & Zwicky (1934) and the first systematic supernova (SN) search was made by Zwicky (1938). Since then many supernova searches were conducted.

There are several reasons for doing SN search. First, it is a lot of fun to discover new objects in the sky (which is a very important motivation for amateur astronomers). Secondly, SN searches provide diverse informations of the astrophysics of stars and galaxies. Thirdly, SN searches can be also used for observational cosmology, especially to determine the cosmological parameters (see a recent review by Leibundghut (2000)).

However it is only in 1990's that systematic SN searches using modern instruments were conducted, and it is about time for SN search to bloom today. As many as about 1750 extragalactic SNe were discovered as of July, 2000 (up to SN 2000cz).

RECENT SUPERNOVA SEARCHES

Modern SN searches are characterized by either fast automatic search with small telescopes covering a wide field for nearby SNe and other transient objects such as gamma ray bursts (GRBs), or deep search in a small field with large (4m class) telescopes well scheduled for maximizing the SN discovery rate.

Recent professional SN searches can be divided into three kinds: the high-z SN searches, the low-z SN searches, and the cluster SN searches. Among the high-z SN searches are the High-z SN search team (HzSS; Garnavich etal. 1998) and the SN Cosmology Project (SCP; Perlmutter et al. 1999). Among the low-z SN searches are the Lick Observatory SN Search (LOSS; Filippenko et al. 2000), the Beijing Astronomical Observatory SN Survey (BAOSS; Li et al. 1996, 2000), the EROS Nearby SN Search (EROSNSS; Hardin etal. 2000), the Nearby Galaxies SN Search (NGSS; Strolger et al. 2000), the QUEST (Schafer et al. 1999, Schafer 2000), the European SN Cosmology Consortium (ESCC; Hardin et al. 1999), and the Super Livermore Optical Transient Imaging System (Super-LOTIS; Williams et al. 2000, Park 2000). Among the cluster SN searches are the Mount Stromlo Abell Cluster SN Search (MSACSS; Reiss etal. 1998), the Wise Observatory Optical Transients Search (WOOTS; Gal-Yam & Maoz 1998), and the Seoul National University SN Search (SNUSS; Lee et al. 1999a,b). Amateur astronomers also have played impor-

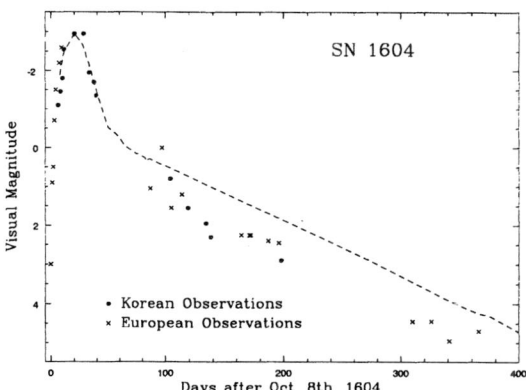

FIGURE 1. A light curve of SN 1604 based on the Korean (filled circles) and European (crosses) records (Clark & Stephenson 1977). The dashed line represents a B-band light curve of Type Ia SN 1991T (Lira et al. 1998). Note how remarkable the old visual photometry of SN 1604 was even at that time.

TABLE 1. A List of Recent SN Searches

Team	N(1998)[a]	N(1999)[a]	N(2000)[a]	N(total)[a]	Remarks
				High-z SS	
SCP	23	43	0	66	4m
HzSS	20	33	0	53	4m
				Low-z SS	
LOSS	19	40	15	74	0.7m, automatic
EROSNSS	13	22	14	49	1m
QUEST	0	0	33	33	1m Schmidt, scanning, $2.3° \times 15°$ per hour
NGSS	0	14	8	22	1m, $59' \times 59'$
BAOSS	8	5	0	13	0.6m, China
ESCC	0	12	0	12	1m-4m
				Cluster SS	
MSACSS	22	2	0	24	1m, Australia
WOOTS	5	12	0	17	1m, Israel
SNUSS	0	1	0	1	1.8m, test, Korea

[a] Numbers of SNe discovered for 1998–July 29, 2000.

tant roles in SN discovery (e.g. Evans (2000), the UK nova/SN patrol (Armstrong 2000), the Tenagra Observatories (Schwartz 2000), and the Pucket Observatory (Pucket 2000)).

Table 1 lists the number of SNe discovered by some of these SN search teams for the last three years (1998 - July 29, 2000). It shows that the recent SN searches have been very successful. Outstanding examples are SCP and HzSS for high-z SNe, and LOSS, EROSNSS and QUEST for low-z SNe.

Fig. 2 shows an example of SNe in the Abell clusters discovered by the SNUSS (Lee et al. 1999a, b), SN 1999dm in Abell 2065.

Nearby clusters of galaxies are ideal places for SN search, because there are

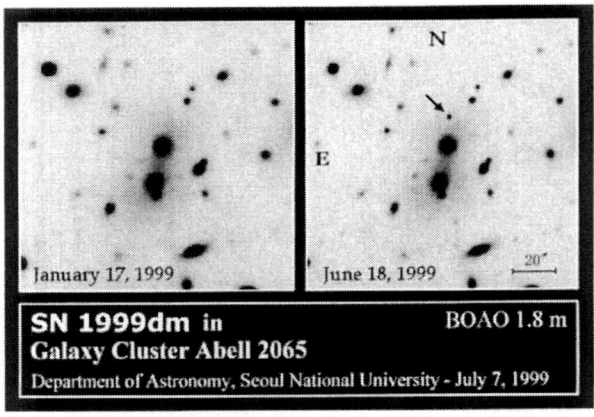

FIGURE 2. Finding charts for SN1999dm in Abell 2065 discovered by the SNUSS.

included many galaxies in a single field covered by a telescope. It is expected that many more SNe in the clusters of galaxies will be discovered soon. Saini et al. (2000) suggested an interesting idea of searching for SNe in the gravitationally lensed arclets in distant clusters, because those SNe are expected to be magnified by up to 3-4 magnitudes.

SUPERNOVA STATISTICS

A recent SN catalog was given by Barbon et al. (1999), and updated SN catalogs are provided in CBAT (http://cfa-www.harvard.edu/iau/lists/Supernovae.html) and http://merlino.pd.astro.it/ supern/. Various informations of SNe are also available in the International Supernova Network (http://www.supernovae.net/isn.htm). The recent SN catalogs show that there are 1747 SNe discovered until July 29, 2000 (up to SN 2000cz).

Fig. 3 displays the number distribution of SNe discovered for 1885 – July 29, 2000. It shows that the number of discovered SNe started increasing steeply in 1990's. 100 SNe per year was broken in 1997 and 200 SNe per year was broken in 1999. It is expected that this trend of increasing may continue for a while in the beginning of 21th century.

The SN types are known for about 1080 SNe: 540 Ia+Iap (50%), 291 II+IIp (27%), 95 I+Ip (10%), 71 Ib+Ic+Ib/c (6%), 48 IIn+IIb (4%), 36 IIn+IIb (3%). Most of host galaxies are spiral galaxies. Only I+Ip and Ia+Iap are discovered in elliptical galaxies, while no I+Ip and Ia+Iap are discovered in irregular galaxies.

There are several galaxies with multiple SNe. Two of them (NGC 5236 and NGC 6946) have six SNe each, and three (NGC 2276, NGC 3690, NGC 4321) have four SNe each. The faintest SN is SN 1999FF with $I \sim 27$ mag in the Hubble Deep

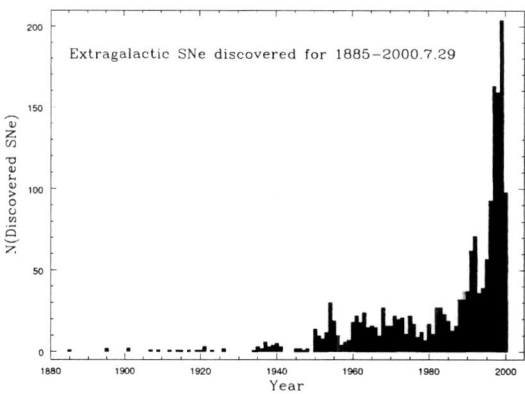

FIGURE 3. Number distribution of the supernovae discovered for 1885 – July 29, 2000.

Field (Gilliland et al. 1999) and the most distant SNe is at z=1.2 (SN 1999fv) or 1.3 (SN 1997ff).

HIGHLIGHTS OF RECENT SUPERNOVA SEARCHES

There are many important new results emerged from recent SN searches, a few of which are listed below (but see also the remaining problems pointed out by Suntzeff (2000)).

(1) The most exciting news among the recent results in the SN searches may be a discovery by two high-z SN search teams that the universe may be accelerating, with $\Omega_\Lambda \sim 0.7$ and $\Omega_M \sim 0.3$ (Perlmutter et al. 1999; Garnavich et al. 1998). Fig. 4 (Riess 2000) illustrates the Hubble diagram of distant SNe Ia discovered by the SCP and the HzSS as well as nearby SNe Ia discovered by the Cálan/Tololo Survey (Hamuy et al. 1996). Although the present data provide reasonable evidence for the acceleration universe, better data for the high-z SNe are still needed to derive a definitive conclusion on the linear systematic effect due to evolution or grey extinction.

(2) Two SNe (SN 1997ff, SN 1997fg) were discovered in the Hubble Deep Field North from the images taken with an interval of two years (Gilliland et al. 1999).

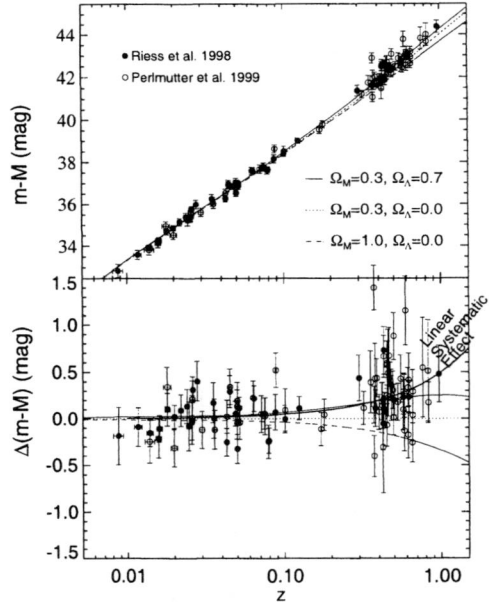

FIGURE 4. Upper panel: the Hubble diagram of SNe Ia at low to high-z. Lower panel: the difference between data and models from the $\Omega_M = 0.3$ and $\Omega_\Lambda = 0.7$ prediction, showing also the effect of a systematic error which grows linearly with redshift.
Reprinted from Riess 2000, PASP 112, 1284, Copyright 2000, Astronomical Society of the Pacific; reproduced with permission of the Editors.

The redshift of SN 1997ff determined photometrically is z = 1.32 ($I = 27.0$ mag). This is the most distant SN among the known SNe, if confirmed spectroscopically.

(3) The Hubble constant was determined with improved accuracy using the SN Ia in the galaxies to which Cepheid distances were measured with the Hubble Space Telescope (HST). The HST H_0 team presented a value of $H_0 = 68\pm7$ km s^{-1} Mpc^{-1} (Gibson et al. 2000), and the HST SN team presented a value of $H_0 = 59 \pm 6$ km s^{-1} Mpc^{-1} (Parodi e al. 2000). There is still some difference between the two estimates, but they are much closer than before.

(4) Recently it has been suggested that SN Ib/c may be linked with the Gamma-Ray Burst (GRB): SN 1998bw/GRB980425 and SN 1997cy/GRB970514, but it needs further studies to confirm it (Galama et al. 1998, Iwamoto et al. 1998, Paczyński 1999, Brown et al. 2000).

FUTURE

It is only a decade that modern supernova searches started. It is very impressive and encouraging that recent supernova searches made a great progress in such a short period of time. However, even more impressive and revolutionary results may come from the supernova searches in the 21th century. There are two directions in the future for the supernova search: all-sky search and deep-sky search. All-sky search is to search for nearby SNe in the entire sky using fast wide field equipments, and deep-sky search is to search for distant SNe in a small field using ground-based large telescopes or space telescopes (e.g. see a supernova pencil beam survey suggested by Wang (2000)).

One space mission to be exclusively used for supernova search is being planned by Perlmutter et al. (2000): SN Acceleration Probe Satellite (SNAP). SNAP will use a wide-field 2m class telescope, covering one degree^2 field of view, with a primary aim of pinning down the cosmological parameters with Type Ia SNe. It is expected that about 2000 SNe Ia at $z < 1.7$ will be discovered every year for three years using the SNAP. With these data, the linear effect as shown in the lower panel of Fig. 4 can be tested. More distant supernovae at $z < 5 - 8$ in a small field will be searched in the space using the Next Generation Space Telescope (NGST) with an 8m aperture, which is scheduled to be launched in 2009. Even more distant supernovae at $z < 10$, if any, will be discovered using the 100m-class ground-based optical telescopes. For example, a conceptual study of the OverWhelmingly Large Telescope (OWL) with the 100m aperture is being done now (Gilmozzi & Dierickx 2000). With the results to come, the theoretical predictions of the cosmic SN rate history (Kobayashi et al. 2000, Dahlén, & Fransson 2000) can be tested. A new golden age era is coming soon in the supernova search.

REFERENCES

1. Armstrong, M. 2000, IAU Circ. 7470

2. Baade, W., & Zwicky, F. 1934, Proc. Natl. Acad. Sci., 20, 254
3. Barbon, R., Buondi, V., Cappellaro, E., & Turatto, M. 1999, A&AS, 139, 531
4. Brown, G. E., et al. 2000, New Astronomy, 5, 191
5. Clark, D. H., & Stephenson, F. R. 1977, The Historical Supernovae, Oxford: Pergamon Press
6. Dahlén, T., & Fransson, C. 2000, A&A, in press
7. Evans, R. 2000, IAU Circ. 7438
8. Filippenko, A. 2000, BAAS, in press
9. Galama, T. et al. 1998, Nature, 395, 670
10. Galama, T. et al. 2000, ApJ, 536, 185
11. Gal-Yam, A., & Maoz, D. 1998, IAU Circ. 7055
12. Garnavich, P. et al. 1998, ApJ, 493, L53
13. Gilson, B. K. et al. 2000, ApJ, 529, 723
14. Gilliland, R. L., Nugent, P. E., & Phillips, M. M. 1999, ApJ, 521, 30
15. Gilmozzi, R., & Diericks, P. 2000, The Messenger, 100, 1
16. Hamuy, M. et al. 1996, AJ, 112, 2438
17. Hardin, D. et al. 1999, IAU Circ. 7182
18. Hardin, D. et al. 2000, astro-ph/0006424
19. Iwamoto, K. et al. 1998, Nature, 395, 672
20. Kobayashi, C., Tsujimoto, T., & Nomoto, K. 2000, ApJ, in press
21. Lee, M. G. et al. 1999a, IAU Circ. 7237
22. Lee, M. G. et al. 1999b, Bulletin of the Korean Ast. Soc., 24, 74
23. Leibundghut, B. 2000, A&AR, in press (astro-ph/0003326)
24. Li, W. D. et al. 1996, IAU Circ. 6379
25. Li, W., Filippenko, A. V., & Riess, A. G. 2000, astro-ph/0006291
26. Lira, P. et al. 1998, AJ, 115, 234
27. Paczyński, B. 1999, astro-ph/9909048
28. Park, H. S. 2000, in this conference
29. Parodi, B. R., Saha, A., Sandaga, A., & Tammann, G. 2000, astro-ph/0004063
30. Perlmutter, S. et al. 1999, ApJ, 517, 565
31. Perlmutter, S. et al. 2000, in preparation (SNAP proposal)
32. Pucket, T. 2000, IAU Circ. 7446
33. Reiss, D. J., Germany, L. M., Schmidt, B. P., & Stubbs, C. W. 1998, AJ, 115, 26
34. Riess, A. G. et al. 1998, AJ, 116, 1009
35. Riess, A. G. 2000, PASP, in press
36. Saini, T. D., Raychaudhury, S., & Shchekinov, Y. A. 2000, A&Ap, in press
37. Schaefer, B. E. et al. 1999, ApJ, 524, L103
38. Schaefer, B. E. 2000, IAU Circ. 7391
39. Schwartz, M. 2000, IAU Circ. 7456
40. Strolger, L. G. et al. 2000, IAU Circ. 7404
41. Suntzeff, N. B. 2000, astro-ph/0001248
42. Wang, Y. 2000, ApJ, 531, 676
43. Williams, G. G., Park, H. S., Hartmann, D. H., & Porrata, R. A. 2000, BAAS, in press
44. Zwicky, F. 1938, PASP, 50, 215

ASTROPHYSICS

Is the Cas A Point Source Discovered by Chandra a Black Hole or a Neutron Star?

Hideyuki Umeda[1], Ken'ichi Nomoto[1], Sachiko Tsuruta[2] and Shin Mineshige[3]

[1] *Department of Astronomy, University of Tokyo, Bunkyo-ku, Tokyo 113-0033, Japan*
[2] *Department of Physics, Montana State University, Bozeman, Montana 59717, USA*
[3] *Kyoto University, Sakyo-ku, Kyoto 606-8502, Japan*

Abstract. Recently the *Chandra* First Light Observation discovered a point-like source in Cassiopeia A (Cas A) supernova remnant. This detection was subsequently confirmed by the analyses of the archival data from both ROSAT and *Einstein* observations. Here we compare the results from these observations with the scenarios involving both black holes (BH) and neutron stars (NS). If this point source is a BH we offer as a promising model a disk-corona type model with low accretion rate where a soft photon source at ~ 0.1 keV is Comptonized by higher energy electrons in the corona. If it is a NS the dominant radiation observed by *Chandra* most likely originates from smaller, hotter regions of the stellar surface, but we argue that it is still worthwhile to compare the cooler component from the rest of the surface with cooling theories. We emphasize that identifying this point source provides very important information on the theories of supernova explosion, progenitor scenario, compact remnant formation, accretion to compact objects, and neutron star thermal evolution.

INTRODUCTION

Cassiopeia A (Cas A) is an interesting supernova (SN) remnant in various aspects. The remnant is very young, about 320 years old. This ring-shaped (e.g., [1]) remnant is associated with jet like structures [2]. The observed abundances of heavy elements are in good agreement with the yields of a massive star (e.g., [3]). The overabundance of nitrogen found in some knots ([2]) implies that the progenitor was a massive Wolf-Rayet star (WN type) which has lost most of its H-rich envelope during the pre-SN evolution. The SN was suggested to be faint [3], which implies that the progenitor was not a red-supergiant possibly due to loss of its H-rich envelope.

Recently the ACIS on board the *Chandra* X-ray satellite observed Cas A and found a point-like source [4]. Subsequently, [5] reported that the ROSAT/HRI image of Cas A taken during 1995-1996 also shows the point-like source at the similar location. Subsequently, [7] and [8] reported the results of their detailed analyses of the Cas A point-source data from the *Chandra* observation. These authors convincingly argue that the observed point source should, indeed, be a compact remnant of the SN explosion. The single power-law fit to the *Chandra* data by [7] yields the higher photon index Γ and lower luminosity L than those observed from typical young pulsars. The spectrum can be equally well fit by thermal models. The best fit for a one component blackbody model yields the temperature $T^\infty = 6 - 8$ MK, the effective radius $R_e = 0.20 - 0.45$ km, and the bolometric luminosity $L^\infty = (1.4 - 1.9) \times 10^{33}$ erg s^{-1}. (In this paper the temperature T^∞ and luminosity L^∞ refer to the values to be observed at infinity.) [8] obtained similar results. The size is too small for a 10 km radius neutron star (NS), but it is consistent if the dominant emission comes from localized hot spots. [7] find that the spectrum is equally well fit by a two temperature thermal model with hydrogen polar caps and the rest of the cooling NS surface composed of Fe. These authors also analyzed the archival data from ROSAT and *Einstein*, and report that the results are consistent with the *Chandra* results within the 1 σ level. Their data analyses of the point source showed no statistically significant variability (both long and short time scale) over the *Einstein* - *Chandra* period. [8] carried out detailed timing analysis and report that the 3 σ upper limit on the sinusoidal pulsed fraction is $< 25\%$ for period $P > 100$ ms, $< 35\%$ for $P > 5$ ms, and $< 50\%$ for $P > 1$ ms.

We emphasize here that the detection of the point source itself is extremely important, whether it turns out to be a neutron star (NS) or a black hole (BH). In this paper, therefore, we will consider both cases. Although the currently available data are not sufficient to distinguish between these options, the most recently completed long *Chandra* observation by [9] and already planned long XMM observations should be able to do so. Therefore, we consider that it is extremely important and timely now, to discuss the implications and offer some predictions for each case.

ACCRETING BLACK HOLE

If the Cas A progenitor is more massive than $\sim 25 M_\odot$ a BH may be formed in the explosion (e.g., [10]). After formation the inner part of the ejected matter may fall back onto the BH due to the presence of a deep gravitational potential well or a reverse shock. The property of an accreting BH depends strongly on whether or not an accretion disk is formed. Here we present plausible BH scenarios based on the disk accretion model under the following observational constraints (see, e.g., [7]): (1) the single power-law X-ray luminosity of intermediate brightness, $L_x(0.1-5.0$ keV$) = (2 - 60) \times 10^{34}$ erg s^{-1} for distance $d = 3.4$ kpc, which is much lower than the Eddington luminosity, $L_{\rm EDD} \sim 7.5 \times 10^{38}(M_{\rm BH}/3M_\odot)$ erg s^{-1} for hydrogen-free matter, (2) no significant variability being detected between the *Einstein* and

Chandra observations, (3) large $F_\mathrm{x}/F_\mathrm{opt}$ ($\gtrsim 100$), and (4) large power-law photon index, $\Gamma \sim 2.6 - 4.1$.

In our model, we assume that the fallback material has specific angular momentum greater than $\sim (GM_\mathrm{BH}r_\mathrm{S})^{1/2}$ (where r_S is the Schwarzschild radius) and thus a fallback disk is formed. There is no efficient mechanism for angular-momentum removal since the Cas A compact remnant is unlikely to have a binary companion (§4). Then the disk evolution most likely obeys the self-similar solution in which the total angular momentum within the disk is kept constant [11, 12]. This solution predicts that disk luminosity decays in a power-law fashion after the disk is formed [13] as $l \equiv L/L_\mathrm{EDD} \sim 10(M_\mathrm{fallback}/0.1M_\odot)(\alpha/0.1)^{-1.3}(t/320\mathrm{yr})^{-1.3}(M_\mathrm{BH}/3M_\odot)^{-1.15}$, where M_fallback is the amount of fallback material and α is the viscosity parameter. We should allow a factor of $0.1 - 10$ changes depending on the distribution of matter and angular momentum. In order for the black-hole accretion scenario to be consistent with the observed $l \sim 10^{-4}$ at 320yr, the amount of the fallback material should indeed be very small, $M_\mathrm{fallback} \sim 10^{-6}M_\odot$. Although the accretion models predict luminosity decrease during the last 20 years from the *Einstein* (in 1979) to the *Chandra* (in 1999) observations, it is small, only about $10\% - (320/300)^{-1.3} \sim 0.90$. Since the *Einstein* observations include larger error bars, more than several tens of %, the luminosity drop of this level cannot be detected, which is consistent with the lack of observed long range large scale variability.

The luminosity of $\sim 10^{33}$ erg s^{-1} is typical to Galactic BH candidates (GBHC) during quiescence. However, the constraint (3), the large $F_\mathrm{x}/F_\mathrm{opt}$ ratio, rules out models that invoke formation of a fallback disk whose properties are similar to those in quiescent GBHC [8]. In the case of usual GBHC, hydrogen-rich matter is continuously added to the disk from the binary companion. According to the disk-instability model for outbursts of GBHC [14], a part of the transferred material is accumulated in the outer parts of the disk, which inevitably produces large optical flux in the quiescent GBHC. Also the constraint (4), a large photon index Γ, is in conflict with the ADAF (advection-dominated) model for the quiescent GBHC [15]. For any ADAF models in which soft photons are provided only by internal synchrotron emission and no external soft photons are available, the power-law photon indices should be as small as $\Gamma \sim 1.7$ [16]. These are the reasons why [8] did not favor an accreting BH model for the Cas A source.

Here we propose a different promising BH model, a *disk-corona* type model, for which the above analogy to the GBHC *is not valid*. First, we consider the constraint (3), the large $F_\mathrm{x}/F_\mathrm{opt}$ ratio. In our model for Cas A, there is no binary companion which supplies mass at 320 years (§4). This means that the outer disk boundary is not extended enough to emit significant optical fluxes. The disk is stable due to the smaller disk size and the different composition of the disk material (mostly heavy elements with possibly a little He but no hydrogen; §4), i.e., the thermally unstable outer zones are absent. In the absence of an instability, the mass-flow rate in the disk is close to be constant [12]. Then according to the standard disk model, the effective temperature is $T(r) \sim 4000(M_\mathrm{BH}/3M_\odot)^{1/4}(r/10^{10}\mathrm{cm})^{-3/4}(l/10^{-4})^{1/4}$ K. For the disk size as small as $r_\mathrm{d} \lesssim 10^{10}$ cm and $l \sim 10^{-4}$, the constraint $L_\mathrm{opt} < 10^{32}$

erg s^{-1} is satisfied.

Next consider the constraint (4), the large Γ. In order to reproduce large photon indices by Compton scattering, we require that energy input rate into soft photons exceeds that into electrons. It is important to note that GBHC generally exhibit two states, soft and hard, and a large Γ is a characteristics of the soft-state emission which exhibits soft blackbody spectra with $kT \sim 1$ keV. The radiation from thermal photons with ~ 1 keV times the area of the emission region around a typical black hole of 3 - 10 $M_\odot$ produces higher luminosity, $L_x \sim 10^{36-38}$ erg s^{-1}, than observed from Cas A. However, we emphasize that unlike GBHC no further mass input is available in our model. Then the accretion rate monotonically decreases, and so does the maximum blackbody temperature, as $T_{max} \sim 0.1(M/3M_\odot)^{1/4}(l/10^{-4})^{1/4}$ keV. Therefore, we get $T_{max} \sim 0.1$ keV for $l \sim 10^{-4}$, instead of ~ 1 keV. A large Γ is then naturally obtained in our model, since there is copious supply of soft-photons at ~ 0.1 keV into electron clouds in the corona from the underlying cool disk [17]. In other words, the important model parameter is ℓ_{soft}/ℓ_{hard} (ratio of compactness parameter of soft photons to that of hard electrons), where the compactness parameter is proportional to the energy output rate divided by the size of the region. For $\ell_{soft} > \ell_{hard}$, we have a large spectral index ($\Gamma > 2$) because of efficient Compton cooling of hard electrons as shown in [17]. The spectral slope is rather insensitive to $\dot{M}$ and M. The conclusion is that with the low accretion rate and lower soft photon temperature our Compton model with a disk-corona configuration naturally yields large Γ with the observed luminosity.

COOLING NEUTRON STAR

Here let us assume that the observed Cas A point source is a NS. [7] and [8] convincingly argue that the dominant radiation observed by *Chandra* is most likely coming from polar hot spots or equatorial ring if it is a NS. Our main purpose in this section is to argue that it is still worthwhile to compare with theoretical models the observed upper limit to the cooling NS component (i.e., the radiation from the whole stellar surface excluding the hotter, localized areas).

[7] offered, as a possible model, a two-component thermal model where the temperature and radius of the polar caps with hydrogen are 2.8 MK and ~ 1 km, respectively, while the rest of the surface of the 10 km NS consisting of Fe is at 1.7 MK. In this model, the hotter polar caps are the result of higher conductivity of hydrogen as compared with Fe, the temperature difference between the polar caps and the rest of the surface should be small, less than a factor of 2, and hence non-standard cooling should be excluded.

Here we offer a promising alternative NS model. SN remnants are usually classified into two categories: shell-type and filled-center (plerions). Cas A is considered to be a prototype of the former, where radio pulsars are normally not found. Recently [18] emphasized the evidence for the presence of an active NS in at least some of the shell-type SNRs although radio pulsars were not found. Also, there is some evidence for significant magnetospheric activities (which can be responsible for polar cap heating) in some NS where no radio pulsar has been found. An example is Geminga (e.g., see [19]). Therefore, the apparent absence of a radio pulsar and/or a plerion should not be used as evidence against polar cap heating. [8] offers accretion as a possible cause for polar cap heating when the field strength is significant. If it is weak, their accreting NS model offers the hotter component as originating from the equatorial hot ring. In either case, with an additional heat source for the hotter component, larger temperature difference between the hotter and cooler components is expected, and hence there is no conflict with the possibility of faster non-standard cooling.

We adopt the conservative upper limit to the cooler component given by *Chandara* [7], $L^\infty < 3 \times 10^{34}$ erg s^{-1}. The neutron star thermal evolution is calculated with a general relativistic evolutionary code without making the isothermal approximation [20, 21, 22]. Our results are summarized in Figure 1.

The observed upper limit for Cas A is consistent with the 'standard' cooling. However, it is still only an upper limit, and if the actual luminosity of the cooler component turns out to be $\sim 10^{33}$ erg s^{-1} or less, the result will be extremely interesting. This is because then the observed value will be certainly below the standard cooling curve, and hence that will be considered the evidence for non-standard cooling scenarios such as those involving pion and/or kaon condensates, or the direct URCA process (e.g., [21, 22]). When the particles in the stellar core are in the superfluid state with substantial superfluid energy gaps, neutrino emissivity l_ν is significantly suppressed (e.g., see [19]). In order to examine this effect of superfluidity, we calculated pion cooling for a representative superfluid model with an intermediate degree of suppression, called the E1 − 0.5 model (see [21]). The result is shown as the thin solid curve in Figure 1.

CONSTRAINTS FROM PROGENITOR SCENARIOS

Here we discuss whether the formation of a NS or BH is consistent with the current models of stellar evolution and supernovae and whether the evolutionary scenarios constrain the radiation processes from the compact source. The overabundance of nitrogen in Cas A implies that the progenitor was a massive WN star which lost most of its hydrogen envelope before the SN explosion. Here we describe two possible evolutionary paths to form such a pre-SN WN star.

One path is the mass loss of a very massive *single* star. A star with the zero-age main-sequence mass M_{MS} larger than ~ 40 M$_\odot$ can lose its hydrogen-rich envelope via mass loss due to strong winds and become a Wolf-Rayet star (e.g., [23]). Recent theoretical models and population synthesis studies suggest that stars with $M_{MS} \gtrsim 25$ M$_\odot$ are more likely to form BHs than NSs (e.g., [10]). This implies that the WN star progenitor is massive enough to form a BH. The explosion can be energetic enough to prevent too much matter fall back to be consistent with the small fall back mass inferred in §2.

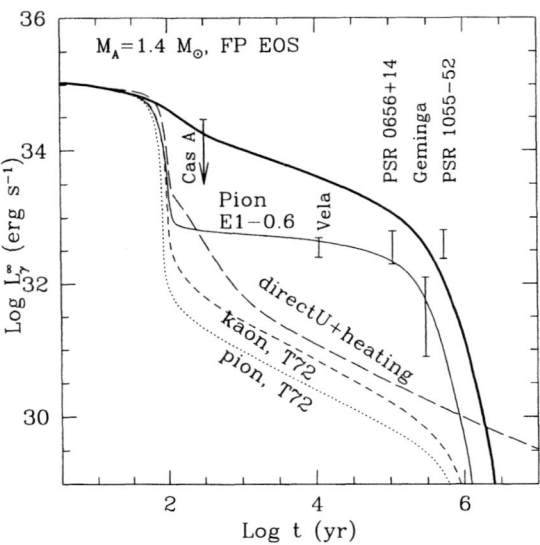

FIGURE 1. Various neutron star thermal evolution curves are compared with the observational data of the Cas A point source and several cooling NS candidates. The heavy solid curve refers to 'standard-cooling', the dashed and dotted curves show the 'non-standard' kaon and pion cooling scenarios when the superfluid effect is negligible, while the thin sold curve refers to the pion cooling with significant superfluid effect. The long dashed curve shows the effect of strong heating on the direct URCA cooling, the fastest 'non-standard' scenario. These curves are obtained for 1.4 $M_\odot$ neutron stars with the intermediate FP equation of state [24]. Cooling due to various more straightforward neutrino mechanisms such as the modified URCA is called 'standard cooling', while extremely fast cooling caused by some other more unconventional mechanisms is called 'non-standard cooling'. The Cas A data should be considered as the upper limit to the radiation from the whole stellar surface.

The other evolutionary path to form a pre-SN WN star is mass loss due to binary interaction. If the progenitor is in a close binary system with a less massive companion star, the star loses most of its H-rich envelope through Roche lobe overflow. In this case, the WN progenitor can form from a star of $M_{\rm MS} \lesssim 40$ $M_\odot$. Its SN explosion of type Ib/c would leave either a BH (if $M_{\rm MS} \sim 25 - 40$ $M_\odot$) or a NS (if $M_{\rm MS} \lesssim 25$ $M_\odot$). If the compact remnant in Cas A turns out to be a NS, therefore, the progenitor must have been in a close binary system.

In the binary scenario, the companion to the Cas A progenitor cannot be more massive than a red dwarf, as constrained from the R & I band magnitude limit [25]. When the companion star is such a small mass star, i.e., the mass ratio between the stars is large, the mass transfer is inevitably non-conservative (e.g., [26]), and the companion star will spiral-in into the envelope of the Cas A progenitor. In order for most of the H-rich envelope to be removed, the envelope should have been a red-giant size so that the orbital energy released during the spiral-in exceeds the binding energy of the envelope. After losing its envelope due to frictional heating, the star became a WN star.

If we take the model of $M_{\rm MS} = 25 M_\odot$, as an example, the star at the WN stage has 8 $M_\odot$. Since the explosion ejects Si and Fe from the deep layers [4], the mass of the compact remnants could not exceed $2 - 3 M_\odot$. Then the binary system is very likely to be disrupted at the explosion. Then the compact star in the Cas A remnant does not have a companion star, and so no mass transfer can be postulated. The implication is also that the accretion onto the compact remnant can occur only as a result of fallback of the ejected matter, and so the composition of the fall back matter is mostly heavy elements with possibly a small fraction of helium but no hydrogen.

In either the single or binary scenario, the WN star blows a fast wind which collides with the red-giant wind material to form a dense shell [27]. If the red-giant wind formed a ring-like shell (due possibly to the spiral-in of the companion), the collision between the supernova ejecta and the shell could explain the observed ring-like structure of Cas A.

DISCUSSION AND CONCLUSION

We agree with [8] that for Cas A point source the usual ADAF model for a quiescent GBHC hardly reconcile with observation. However, we emphasized in §2 that there does exist a very promising BH disk accretion model. In this model, the fallback material is like the soft state of a GBHC with a disk-corona configuration, not like a quiescent GBHC with ADAF. With the low accretion rate and Comptonization of cooler soft photons (~ 0.1 keV or less), we naturally obtain large photon index of $\Gamma \sim 2.6 - 4.1$ and lower luminosity of $L \sim 10^{34} - 10^{35}$ erg s^{-1}, as observed from the Cas A point source.

Accreting NS models are also possible (see [8]). However, we can still, without difficulty, distinguish between the BH and NS accretion models because the characteristic properties of the observed X-ray spectra in these two cases are quite different (e.g., see [28]). For instance, the radiation from an accreting NS is dominated by thermal emission from the stellar surface [29], which is absent if a BH is involved.

If the point source is a NS, the dominant radiation observed by *Chandra* most likely corresponds to the radiation from a localized small area. The detailed studies of theoretical light curves expected from anisotropic cooling of a NS have been carried out by, e.g., [30] and [19], with the latter including hot spots. The results show that pulsation depends on the relative angles between the rotation axis, magnetic axis and the line of sight. Depending on the combinations of these angles, pulsations from zero to up to about 30% are predicted, and so the observed constraints on the pulsed fraction are still consistent with a NS model.

Although the current data of Cas A point-source can be consistent with both BH and NS scenarios, future observations by the *Chandra*, XMM, and other satellite missions should be able to distinguish between these cases. If distinct periodicity is found the point source definitely should be a NS. The existence of the NS itself will significantly constrain the progenitor scenario for Cas A. Better spectral information should be able to distinguish between the BH and NS as the compact remnant. If the source is found to be a BH, the implication is significant in the sense that this will offer the first observational evidence for BH formation through a SN explosion and greatly constrain the BH progenitor mass by combining with the abundance analysis of Cas A [3].

In conclusion we emphasize that the Cas A point source can potentially provide great impacts on the theories of supernova explosion, progenitor scenario, compact remnant formation, accretion to compact objects, and NS thermal evolution.

ACKNOWLEDGMENTS

We thank Drs. G. Pavlov, M. Rees, H. Tananbaum, B. Aschenbach, and J. Trümper for valuable discussions. This work has been supported in part by the grant-in-Aid for Scientific Research (0980203, 09640325), COE research (07CE2002) of the Ministry of Education, Science, Culture and Sports in Japan, and a NASA grant NAG5-3159.

REFERENCES

1. Holt, S., Gotthelf, E.V., Tsunemi, H., & Negoro, H. 1994, PASJ, 46, L151
2. Fesen, R.A., Becker, R.H., & Blair, R.H. 1987, ApJ, 313, 378
3. Ashworth, W.B. 1980, J. Hist. Astron., 11, 1
4. Hughes, J.P., Rakowski, C.E., Burrows, D.N., & Slane, P.O. 2000, ApJ, 528, L109
5. Tananbaum, H. et al. 1999, IAUC No. 7246
6. Aschenbach, B. 1999, IAUC No. 7249
7. Pavlov, G.G., Zavlin, V.E., Aschenbach, B., & Trümper, J..E. 2000, ApJ, 531, L53
8. Chakrabarty, D., Pivovaroff, M.J., Hernquist, L.E., Heyl, J.S., & Narayan, R. 2000, ApJ, submitted (astro-ph/0001026)
9. Holt, S. et al. 2000, http://asc.harvard.edu/targets/summary_observed_daily.html
10. Ergma, E., & van den Heuvel, E.P.J. 1998, A&A, 331, L29
11. Pringle, J.E. 1974, Ph.D. Thesis, University of Cambridge
12. Mineshige, S., Nomoto, K., & Shigeyama, T. 1993, A&A, 267, 95
13. Mineshige, S., Nomura, H., Hirose, M., Nomoto, K., & Suzuki, T. 1997, ApJ, 489, 22
14. Mineshige, S., & Wheeler, J.C. 1989, ApJ, 343, 241
15. Narayan, R., McClintock, J.E., & Yi, I. 1996, ApJ, 457, 821
16. Tanaka, Y., & Lewin, W. H. G., 1995, in X-ray binaries, ed. W.H.G. Lewin, J. van Paradijs, E.P.J. van den Heuvel (Cambridge: U.P., Cambridge), 126

17. Mineshige, S., Kusunose, M., & Matsumoto, R.. 1995, ApJ, 445, L43
18. Pacini, F. 2000, in IAU Symp. 195, Highly Energetic Physical Processes and Mechanisms for Emission from Astrophysical Plasmas, eds. P. Martens, S. Tsuruta, & M. Weber, PASP, in press
19. Tsuruta, S. 1998, Phys.. Rep. 292, 1
20. Nomoto, K., & Tsuruta, S. 1987, ApJ, 312, 711
21. Umeda, H., Nomoto, K., Tsuruta, S., Muto, T., & Tatsumi, T. 1994, ApJ, 431, 309
22. Umeda, H., Tsuruta, S., & Nomoto, K. 1994, ApJ, 433, 256
23. Schaller, G., Schaerer, D., Meynet, G., & Maeder, A. 1992, A&AS, 96, 269
24. van den Bergh, S., & Pritchet, C.J. 1986, ApJ, 307, 723
25. Nomoto, K., Iwamoto, K., & Suzuki, T. 1995, Phys. Rep.. 256, 173
26. Chevalier, R.A., & Liang, E. 1989, ApJ, 346, 847
27. Tanaka, Y. 2000, in IAU Symp. 195, Highly Energetic Physical Processes and Mechanisms for Emission from Astrophysical Plasmas, eds. P. Martens, S. Tsuruta, & M. Weber, PASP, in press
28. Friedman, B., & Pandharipande, V.R. 1981, Nucl. Phys. A, 361, 502
29. Rutledge, R.E., Bildsten, L., Brown, E.F., Pavlov, G.G., & Zavlin, V.E. 2000, ApJ, 529, 985
30. Shibanov, Yu.A., Zavlin, V.E., Pavlov, G., Qin, L., & Tsuruta, S. 1995, Proc. 17th Texas Symp., eds. H. Böhringer, et al., N.Y. Acad. Sci., 291.

Accretion Disk Emission and Raman Scattering in Symbiotic Stars

Hee-Won Lee

Department of Astronomy, Yonsei University.
Seoul, Korea
Email: hwlee@galaxy.yonsei.ac.kr

Abstract. Symbiotic stars, known as binary systems of a giant with heavy mass loss and a white dwarf accompanied by an emission nebula, are believed to form an accretion disk around the white dwarf component by attracting the slow but heavy stellar wind around the giant companion. About a half of symbiotic stars exhibit mysterious broad emission features around 6830 Å and 7088 Å, which have been identified by Schmid (1989) as the Raman scattered features of the O VI 1032 Å and 1038 Å doublet by atomic hydrogen. The scattering incoherency results in very broad profiles and strong polarization. Spectroscopic and polarimetric observations show that the Raman scattered features exhibit double or triple peak profiles and a polarization flip in the red wing part. In the accretion disk emission model, it is expected that the Raman features are polarized perpendicular to the binary axis and show double peak structures in the profile, because the neutral scatterers located near the giant component views the accretion disk in the edge-on direction. Assuming the presence of scattering regions outflowing in the polar directions, we may explain the additional red wing or red peak structure, which is polarized parallel to the binary axis. Adopting the asymmetric accretion disk emission model, it is predicted that the blue peak strength relative to the red peak is larger in the 6830 Å Raman feature than in the 7088 Å Raman feature. It is concluded that Raman scattering is an important tool to investigate the physical conditions and geometrical configuration of the accretion disk in a symbiotic star.

INTRODUCTION

Symbiotic stars are widely believed to be interacting binary systems consisting of a mass-losing evolved cool giant and a hot white dwarf surrounded by a nebula showing strong emission lines in the UV, optical and IR regions. The mass loss rate is estimated to be $10^{-4} - 10^{-7}$ $M_\odot$ yr^{-1} (e.g. Kenyon 1986). Recent observations show that the morphology of the nebulae associated with symbiotic systems is dominantly bipolar, from which Soker (1998) proposed that the binarity in the central star system may play an important role in the nebular shape formation (see also Corradi & Schwarz 1995, Soker & Rappaport 2000, Livio, Salzman & Shaviv 1979). At this time it is not clear whether the mass loss from the giant companion in the form of a heavy stellar wind can be accreted to the white dwarf companion. Mastrodemos & Morris (1998) performed SPH computations to show that a permanent and stable accretion disk can be formed in a large parameter space in binary systems of a giant and a white dwarf. However, definite spectroscopic evidence for this scenario has not been provided, yet.

About a half of symbiotic stars show the so-called symbiotic emission bands around 6830 Å and 7088 Å which were identified by Schmid (1989), who proposed that they are Raman scattered features of the O VI 1032 1038 doublet by atomic hydrogen. The quantum mechanical process of Raman scattering may be described as the annihilation of an incident O VI line photon and the creation of a Raman-scattered photon accompanied by an excitation of the scattering hydrogen atom initially in the ground $1s$ state to one of np states and a subsequent de-excitation to the excited $2s$ state.

Using the Kramers-Heisenberg formula the Raman and Rayleigh scattering cross sections are computed to be $\sigma_{Ram}^{1032} = 7.48\sigma_T, \sigma_{Ray}^{1032} = 34.\sigma_T$ for O VI 1032 photons and $\sigma_{Ram}^{1038} = 2.45\sigma_T, \sigma_{Ray}^{1038} = 6.6\sigma_T$ for O VI 1038 photons, where $\sigma_T = 0.66 \times 10^{-24}$ cm^2 is the Thomson scattering cross section (e.g. Schmid 1989. Lee & Lee 1997a, Karzas & Latter 1961, Saslow & Mills 1969). Observational confirmation of Raman scattering includes simultaneous observations of the far UV O VI 1032, 1038 lines and the 6830 Å and 7088 Å features in a number of symbiotic stars (e.g. Espey et al. 1995, Birriel, Espey, & Schulte-Ladbeck 1998, Birriel, Espey, & Schulte-Ladbeck 2000, Schmid 1998).

Being slightly less energetic than Lyβ, O VI line photons are Raman-scattered to produce less energetic photons than Hα that occurs around 6563 Å. The energy conservation gives the wavelength of the scattered wave λ_o, which is

$$\lambda_o = (\lambda_i^{-1} - \lambda_{Ly\alpha}^{-1})^{-1}, \qquad (1)$$

where λ_i is the wavelength of the incident wave and $\lambda_{Ly\alpha}$ is the wavelength of the Lyα transition. This relation of scattering incoherency leads to an enhancement of the Doppler width by a factor of 7, because $\Delta\lambda_o/\lambda_o = (\lambda_o/\lambda_i)\Delta\lambda_i/\lambda_i$, and as a result, these Raman features get very broad. Nussbaumer, Schmid, & Vogel (1989) discussed the importance of Raman scattering in astrophysics. They proposed that Raman scattered Lyβ will contribute to the broad wings of Hα, and Lee (2000) computed the Hα wing profiles which are in excellent agreement with the observed data compiled by van Winckel, Duerbeck, & Schwarz (1993) and Ivison, Bode, & Meaburn (1994).

Harries & Howarth (1996) showed that the Raman features are not only very broad but also exhibit various structures including multiple peak profiles with strong polarization often accompanied by the position angle flip in the red wing part (see also Schmid & Schild 1994). Schmid (1996) proposed to attribute the multiple peak profiles and polarization structures to the slow spherical stellar wind around the giant companion mainly consisting of neutral hydrogen. Fig. 1 shows the conceptual diagram for this model. Several researchers adopted a Monte Carlo method to compute the profiles and polarization of the Raman scattered features using the spherical stellar wind model (see also Lee & Lee 1997b, Harries & Howarth 1997).

However, it seems that special velocity laws are required to produce multiple peak profiles with the polarization flip in the red wing part, which makes the spherical stellar wind model appear unnatural. Furthermore, the Raman-scattered features have a width about 20 Å, which amounts to a kinematic velocity scale of ~ 100 km s^{-1}, whereas the typical terminal velocity of the stellar wind around a giant star does not usually exceed 20 km s^{-1} and therefore it is difficult to produce broad profiles seen in the Raman scattered features. In this work, we introduce an accretion disk emission model and argue that this model is more adequate to interpret the spectropolarimetric features of the Raman scattered line.

ACCRETION DISK EMISSION MODEL

Lee & Park (1999) proposed an accretion disk emission model, according to which around the white dwarf component there exists an accretion disk that coincides with the UV emission region. The main scattering region is assumed to be located near the giant component. In this case, the scatterers view the accretion disk in the edge-on direction, because of the angular momentum associated with the accretion

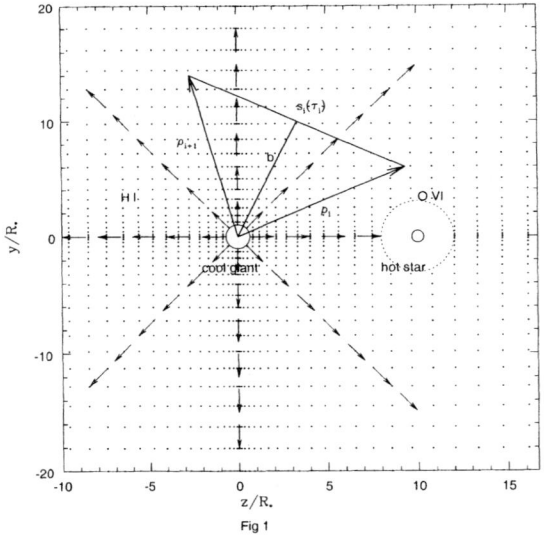

Fig 1

FIGURE 1. Scattering geometry for a spherical stellar wind of a symbiotic system. The hot star is represented by a circle on the right side, and the bigger circle in the center is the cool giant. A spherically symmetric stellar wind is shown by radial (small) arrows.

disk is expected to align with that associated with the binary orbital motion of the symbiotic system. Since the profile of the Raman scattered feature is determined from the relative motion of the source and scatterers, we expect that a double-peak profile is a natural consequence of this model.

In this case, the observed velocity width ~ 100 km s^{-1} is attributed to the kinematics of the accretion disk. From this we may deduce that the accretion disk emission region has a dimension of ~ 0.1 AU assuming a Keplerian motion around the white dwarf component. Furthermore, in the observed profiles the blue peak is weaker than the red part, which implies that the accretion disk may be asymmetrical or may possess a hot spot in the red emission region that enhances the red peak part. Fig. 2 shows a schematic diagram illustrating an asymmetrical accretion disk emission model. It is expected that the binary orbital motion severely affects the accretion flow and as a result an asymmetrical accretion disk seems an unavoidable consequence. This picture is also qualitatively consistent with the SPH simulations performed by Mastrodemos & Morris (1997).

Furthermore, in order to explain the polarization flip in the red wing part and triple-peak profiles in a number of symbiotic systems we need to invoke another scattering component of which the location is perpendicular to the scattering region near the giant with respect to the source. This scattering component should also move away from the emission source, and therefore it may be naturally identified with the polar outflow. From the observed profiles and spectropolarimetric data it is expected that the polar outflow has a typical velocity scale ~ 50 km s^{-1}, which is also often found in planetary nebulae. Depending on the velocity of the polar outflow and H I column density, the resultant profile may be either a double-peak one or a triple-peak one. In the latter case, the reddest part is polarized in the direction perpendicular to the main part of the profile, and in the former case we may expect a polarization flip in the red wing part. This is consistent with the spectropolarimetric and high resolution spectroscopic observations.

PREDICTIONS AND FUTURE WORKS

It is important to distinguish the accretion disk emission model from the spherical stellar wind model, because this distinction will point out the origin of the multiple peak structures in the profiles of the Raman-scattered features. In the accretion disk emission model, the red emission region is expected to be more optically thick than the blue emission region. The emergent flux ratio $r = I_{1032}/I_{1038}$ of O VI 1032 and O VI 1038 depends on the optical depth of the emission region, because they are collisionally excited and expected to suffer a number of resonant scatterings and absorption. In the optically thin case, the flux ratio is equal to the ratio of the oscillator strengths for the line transitions, which is 2. However, in the optically thick case the ratio approaches the thermal limit of 1.

In the escape probability formalism, the escape probability $P_{esc} = (1 - e^{-\tau_l})/\tau_l$ for a line optical depth τ_l and therefore, the emergent flux ratio will be given by

$$\frac{I_{1032}}{I_{1038}} \simeq \frac{1-e^{-2\tau_{1032}}}{1-e^{-\tau_{1032}}}. \tag{2}$$

Therefore, the flux ratio of O VI 1032 to O VI 1038 in the red emission region is expected to be smaller than that in the blue emission region.

From this consideration we expect that the blue part is more suppressed compared with the red part in the Raman-scattered 7088 Å feature than in the 6830 Å feature. The Raman-scattered 7088 Å feature is weaker than the 6830 Å feature by a factor of ~ 5, mainly due to smaller scattering cross section. Thus far, most observations have been devoted to the interpretation of the profile and polarization of the 6830 feature. This approach is valid if the O VI emission region is localized and characterized by a single physical condition, in which case the flux ratio of O VI 1032 photons and O VI 1038 is fixed and no profile difference is expected.

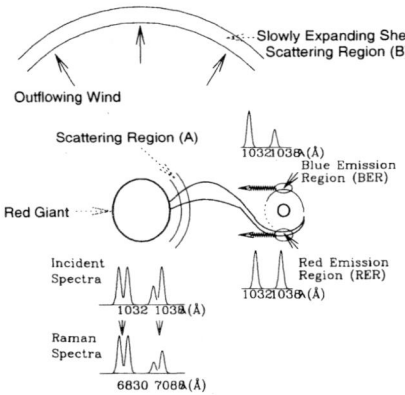

FIGURE 2. A conceptual diagram of an accretion disk emission region and the scattering geometry for a typical symbiotic star. There are two emission regions around the hot star denoted by 'RER' and 'BER', which provides red-shifted photons and blue-shifted photons to the direction of the giant, respectively. The scatterers near the giant see a double peak incident source. If 'RER' has a large O VI line center optical depth, then the strength of O VI 1032 is comparable to that of O VI 1038. On the other hand, if 'BER' is much more optically thin, then O VI 1032 is twice stronger than O VI 1038. Therefore, a much stronger blue component with respect to the red part in the Raman scattered 6830 feature is expected than in the 7088 Å band.

Therefore, in order to test the accretion disk emission model, the high quality of the weaker Raman 7088 feature need to be secured for comparison with the counterpart around 6830 Å. Observational data found in the literature do show that there is a tendency of a weaker blue peak in the 7088 Å feature, which is consistent with the accretion disk emission model (e.g. Harries & Howarth 1996, Schmid et al. 2000). However, we should wait for good quality data to be more conclusive.

The profile comparison study can be complemented further by high resolution far ultraviolet spectroscopic observations that are available from missions like the *Far Ultraviolet Spectroscopic Explorer* (FUSE) from which detailed profiles of O VI lines may be obtained. Schmid et al. (1999) used the ORFEUS data to find that the O VI 1032, 1038 lines in the symbiotic star RR Tel show single-peak profiles, which is in high contrast with the double-peak profiles of the Raman-scattered lines.

The profile and strength of the Raman-scattered features will be sensitively affected by the dust content and clumpiness (e.g. Neufeld 1991). It is known that strong Raman scattered features are observed more frequently in D-type symbiotic stars than S-type stars, where D-type symbiotics show an IR excess that may be attributed to the reprocess of UV radiation by dust. Therefore, further theoretical studies on the radiative transfer of resonantly scattered line photons in a dusty medium will be useful for more proper interpretation of the spectroscopic data for the Raman scattered lines.

Raman scattering also plays an important role in forming broad wings around Balmer emission lines, because Lyman series emission lines will be converted to Balmer emission lines with the broadening enhancement by the factor of the ratio of the wavelengths of the incident and outgoing waves. This was pointed out by Nussbaumer et al. (1989) and the wing profiles were computed and compared with the observed data by Lee (2000) (see also Selvelli et al. 2000). In particular, these broad wings are present in young planetary nebulae including IC 4997 and M2-9 (Lee & Hyung 2000, Balick 1989) and also seem to be present in post AGB stars (van de Steene 2000). A strong constraint for Raman scattering to operate is provided by the special condition of the co-existence of a hot UV emission region and a cold neutral region around it. This condition is expected to be met in objects in their late stage of evolution, suffering severe mass loss and evolving into the white dwarf stage. It is concluded that Raman scattering spectroscopy and polarimetry will play an important role to investigate symbiotic stars and objects in the similar stage of evolution.

REFERENCES

1. Balick, B., 1989, AJ, 97, 476
2. Birriel, J. J., Espey, B. R. & Schulte-Ladbeck, R. E., 1998, ApJ, 507, L75
3. Birriel, J. J., Espey, B. R. & Schulte-Ladbeck, R. E., 2000, preprint, (astro-ph/0008152)
4. Corradi, R. L. & Schwarz, H. E., 1995, A& A, 293, 871

5. Espey, B. R., Schulte-Ladbeck, R. E., Kriss, G. A., Hamann, F., Schmid, H. M., Johnson, J. J., 1995, ApJ, 454, L61
6. Harries, T. J. & Howarth, I. D., 1996, A& AS, 119, 61
7. Ivison, R. J., Bode, M. F.,& Meaburn J. 1994, A& AS, 103, 201
8. Karzas, W. J., & Latter, R., 1961, ApJS, 6, 167
9. Kenyon, S. J., 1986, The Symbiotic Stars, Cambridge University Press, Cambridge
10. Lee, H. -W., 2000, ApJ, 541, L25
11. Lee, H. -W. & Hyung, S., 2000, ApJ, 530, L49
12. Lee, H. -W. & Lee, K. W., 1997a, MNRAS, 287, 211
13. Lee, H. -W. & Park, M. -G., 1999, ApJ, 515, L89
14. Lee, K. W. & Lee, H. -W., 1997b, MNRAS, 292, 573
15. Livio, M., Salzman, J., & Shaviv, G., 1979, MNRAS, 188, 1
16. Mastrodemos, N. & Morris, M, 1998, ApJ, 497, 303
17. Neufeld, D. A., 1991, ApJ, 370, L85
18. Nussbaumer, H., Schmid, H. M., & Vogel, M., 1989, A& A, 211, L27
19. Saslow, W. M., & Mills, D. L., 1969, Physical Review, 187, 1025
20. Selvelli, P. L., & Bonifacio, P., submitted to A& A Letters, (astro-ph/0008060)
21. Schmid, H. M., 1996, MNRAS, 282, 511
22. Schmid, H. M., & Schild, H. 1994, A& A, 281, 145
23. Schmid, H. M., Corradi, R., Krautter, J., Schild, H., 2000, A& A, 355, 261
24. Schmid, H. M., et al., 1999, A& A, 348, 950
25. Soker, N., 1998, ApJ, 496, 833
26. Soker, N., & Rappaport, S., 2000, ApJ, 538, 241
27. Van de Steene, G. C., Wood, P. R., & van Hoof, P. A. M., 2000, in Asymmetrical Planetary Nebulae II: From Origins to Microstructures, ed. J. H. Kastner, N. Soker, & S. Rappaport (San Francisco: ASP)
28. Van Winckel, H., Duerbeck, H. W., & Schwarz,H. E. 1993, A& AS, 102, 401

Flares in X-ray Transients

Soon-Wook Kim[†,1]

[†]*School of Earth and Environmental Sciences, Astronomy Program*
Seoul National University, Seoul 151-742, Korea
skim@astro6.snu.ac.kr

Abstract. I discuss the outburst phenomena in binary X-ray transients, emphasizing the phenomena of the secondary flare, or reflare, based on implicit, time-dependent, irradiated, accretion disk instability models. I examine the role of irradiation, generated from the central compact region and corona, or disk atmosphere, to the disk structure and outburst evolution in black hole X-ray transients. The cause of the primary rise in X-ray nova outburst is plausibly due to the disk instability associated with the ionization of hydrogen and helium, similar to the rise observed in dwarf nova outbursts. The reflare results principally from the so-called "stagnation" phenomena, coupled with the time-dependent effects of indirect irradiation reflected by the corona above the disk. The direct radiation from the central region around the compact object also plays an important role to the reflare. The time-dependent effect of the screening, or shadowing, of the direct irradiation on the disk is included in the modeling. The direct irradiation, unlike in the reflare, makes the disk dimmer in the X-ray quiet state due to shadowing and more depletion of the disk mass. The reflares observed in X-ray transients are short-lived and, furthermore, more pronounced in the radio and optical than in X-rays, suggesting that the phenomena plausibly is local and temporal, associated with the degree of ionization in the outer portion of the disk as modeled here.

INTRODUCTION

There are several reliable black hole candidates, nominated mainly by the measurement of the mass function, among tens of X-ray transients discovered for more than last two decades by the characteristic signatures in the X-ray outbursts.

Among the commonly observed outburst features are: (1) primary maximum followed by a fast rise, a week or two, (2) a reflare, tens of days after the maximum, and (3) a secondary maximum in the late decay of the primary outburst, a few hundred days later.

There are, however, a few atypical phenomena observed in outburst evolution such as precursor activities in the rise, superluminal jets and chaotic fluctuation,

---

[1)] The BK21 Postdoctoral Research Fellow

both in the early decay, and the so-called mini-outbursts before the transients make transition to the quiescence (see [1] for details).

It is now widely accepted that the primary rise and maximum are due to the disk instability associated with the ionization of hydrogen and helium (see [2] for a review and details). The disk instability models of others, however, had difficulty in explaining the phenomena of reflares observed in black hole X-ray novae before our works [2–4], although they can reproduce the appropriate exponential or linear decays. I here present some aspects of the irradiation-induced disk instability model for the reflare, collected from works recently published [2–5].

TIME-DEPENDENT IRRADIATED DISK MODEL

Basic Formulation

In the model I am presenting [2], the analytic expression for total irradiation flux is described as a sum of direct irradiation from the innermost disk and indirect irradiation reflected by a corona, or disk atmosphere, above the disk (e.g., [6]):

$$F_{\mathrm{irr}}(r,t) = \frac{\eta L_{\mathrm{acc}}(t)}{\pi r^2} \left\{ \frac{C_{\mathrm{X}}}{4} + (1-A) \cdot \frac{1}{1+\left(\frac{dh}{dr}\right)^2} \cdot \frac{1}{R}\left[r\left(\frac{dh}{dr}\right) - h\right] \right\}. \tag{1}$$

where C_{X} is a constant, η is the efficiency of the accretion luminosity L_{acc}, A is the X-ray albedo, r and h are disk radius and height, and $R = (r^2 + h^2)^{1/2}$. I further approximate $R \approx r$, and adopt the standard result $h \ll r$ (see equation (3) in [2]), consistent with the numerical results presented here.

The resultant vertically averaged energy equation is then given in terms of the viscous heating Q_{H}^+, irradiation input Q_{irr}^+, radiative cooling Q_{C}^-, and the thermal diffusion term Q_ν:

$$\frac{C_{\mathrm{p}}\Sigma}{2}\left(\frac{\partial}{\partial t} + v_r\frac{\partial}{\partial r}\right)T_{\mathrm{c}} = Q_{\mathrm{H}}^+ - Q_{\mathrm{irr}}^+ + Q_{\mathrm{C}}^- + Q_\nu, \tag{2}$$

where C_{p} is the heat capacity and Σ is the surface density of the disk, and v_r is the radial velocity of the accretion flow.

A model presented here is for A0620-00 (X-ray Nova Mon 1975), a prototype of X-ray novae, with the disk size of 10^{11}cm around a $4M_\odot$ black hole. The detailed numerical parameters can be found in [2]. The model, however, is not a unique choice. A variety of features for the effect of irradiation in the X-ray transients in general will appear elsewhere in the future.

Disk Mass and Inner Accretion Flow

The irradiation yields to a factor two lower inner accretion rate, $\dot{M}_{in}$, due to a higher depletion of the disk mass during outburst (see Figure 1). In the case of A0620-00, the numerical result is well consistent with the limits on $\dot{M}_{in}$, $\lesssim 10^{12} g s^{-1}$, deduced by [7] (MRW) and [8] (MHR) assume the standard, steady state, thin disk model [9]. The irradiated disk instability model also can reproduce *permanent*, fully ionized inner disk ($\gtrsim 10^4 K$) and hot inner disk temperature, $\lesssim 10^7 K$, in outbursts, enough to produce ultra-soft X-rays ($\lesssim 1 keV$) as observed [3].

The model gives a strong correlation between the optical and X-rays in the rise: the optical flux from the outer disk should rise before any activity from the inner disk, as observed in some X-ray novae. As shown in the upper panel of Figure 1, the precursor activity is produced before the primary rise (for some aspects of the precursor activities, see [3,4]). In addition, 40-60 days after the maximum, there is a small but unnegligible amount of the mass enhancement due to *local* instability in the outer portion of the disk (see below).

The disk instability models, however, have not been successful to account for the observed recurrence time and hard X-rays, and hence other physics such as the advection-dominated accretion flow (e.g., [10]), or some physics related to the accretion disk corona (e.g., [11]) might be included to explain other aspects of observations in X-ray transients I do not address here.

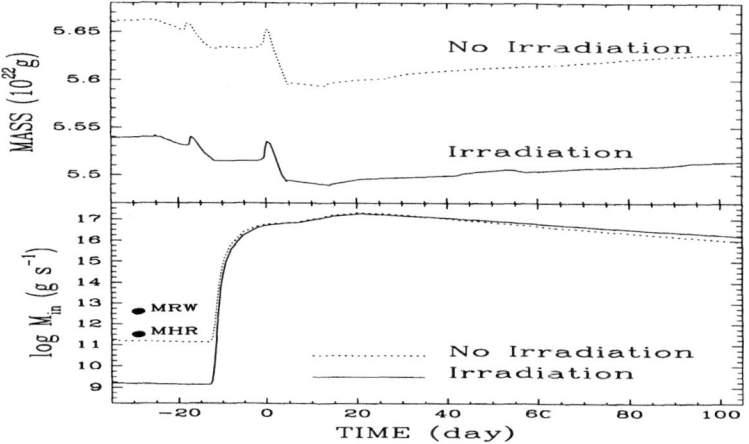

FIGURE 1. Time evolution of the disk mass and inner accretion rate in quiescence and outburst. The model optical maximum is defined as day zero. Note that the two filled circles represent the *upper* limits for the inner accretion rate, $\dot{M}_{in}$ deduced from observations.

Shadow Effect

Unlike the standard, steady state model, in which the disk height has a uniform relationship to the disk radius (e.g., $h \propto r^{9/8}$ [9]), the time-dependent disk instability model *intrinsically* produce non-linear evolution of the disk height. The natural outcome is the fact that the inner portions of the disk block the direct radiation generated from the central region, and hence prevent the irradiation of outer portions of the disk.

A systematics of the shadowing process is illustrated in Figure 2. The inner, permanently ionized, hot disk ($\lesssim 10 r_{ms}$ for a $4 M_\odot$ black hole) is always exposed to the direct irradiation (r_{ms} is the marginally stable orbit). In quiescence, the shadowing is in effect only in the middle portions of the disk, while both inner and outer disks are exposed to the mild irradiation from the central source, an order of $\lesssim$ a few$\times 10^3$K. As the ionization process is initiated at ~ 25 days prior to the optical maximum, the shadowing process acts *locally* on and off outside the disk radii $\sim 2 \times 10^8$cm due to the non-linear behaviors of local heating and cooling waves, intrinsic properties of the thermal disk instability model. After maximum light, the outer disk is essentially totally blocked from the direct irradiation by the inner hot disk except a stage of the reflare at days 50–60 (see below).

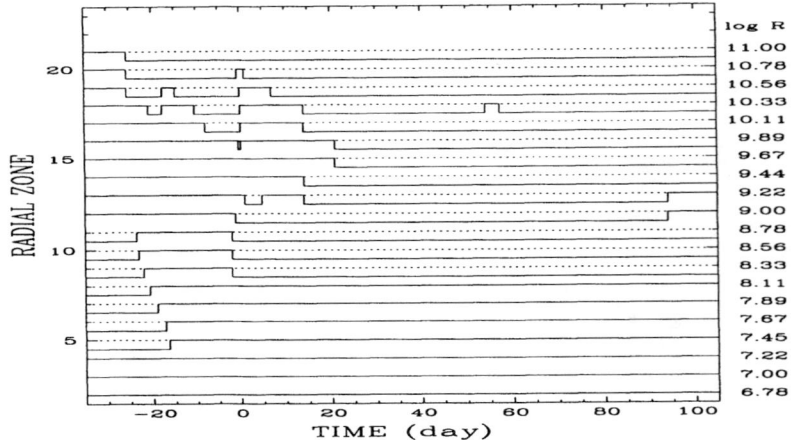

FIGURE 2. Shadowing Process for the direct irradiation. The solid lines which are one-half integer smaller than the zone number, presented together with the dotted lines, signify where a given zone is in the shadow of some interior zone and hence not receiving direct irradiation. Note the exposure of an outer zone in the reflare at $\sim$ day 55 after the maximum (day zero).

Physics of Reflare

Figures 3 and 4 display an irradiated disk model in terms of the time-dependent evolution of the disk effective temperature, T_{eff}, and height in outburst, from

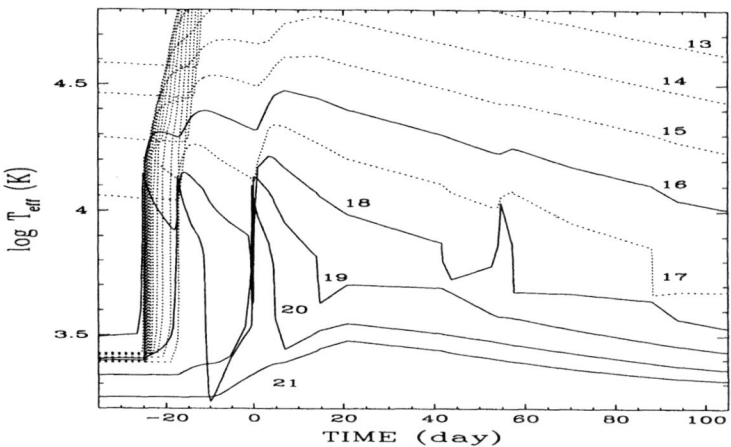

FIGURE 3. Time-dependent evolution of the disk effective temperature.

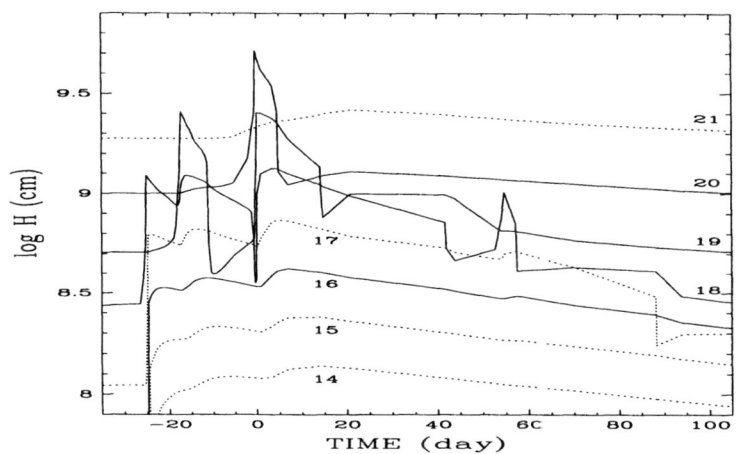

FIGURE 4. Time-dependent evolution of the disk height.

the rise to early decay up to 105 days after the maximum (a version for the disk midplane, or central, temperature, T_c, also appears in [2]). Without irradiation, the cooling wave propagates from the outer to inner zone with the sharp edge due to the rapid transition from the hot, fully ionized state to cool, neutral state, as shown in Figures 6 and 8 in [2], a typical result in the disk instability models.

With both direct and indirect irradiation, the evolution is totally non-linear: the transition is not direct from the hot to cool branch. There is, instead, a metastable branch called "stagnation" regime between the hot and cool state in the thermal hysteresis curve (so-called "thermal limit cycle" curve), due to the property of opacity [2,12,13]. A local disk zone lingers at this metastable regime when the disk temperature lies in $logT_c \sim 3.5-4.2$, or $logT_{eff} \sim 3.4-3.8$. There are two regimes in the stagnation state: molecular and partially ionized H^--dominated domains due to the nature of low temperature opacity (for details, see [2]).

Unlike the case of dwarf nova outbursts, where the stagnation is a natural product of the disk instability for the lower temperature disks (for details, see [12,14]), it is obviously due to the effects of irradiation, since it disappears without irradiation. The precursor activities in the rise, very reminiscent to those in Nova Sco 1994 (GRO J1655-40) [15], will be discussed elsewhere in the future (also see [4]).

After the optical outburst maximum at day zero, the cooling wave is initiated. The resulting outward diffusion results in heating the outer disk zones. As the outer zones enter the stagnation regime, the consequent higher temperature due to the indirect irradiation, independent of the shadowing, affects the degree of ionization in the inner zones. The regions inside the stagnation can not simply drop into the cool state due to the outward diffusion of accreted matter that is not directly irradiated. As a result, these regions linger in a near constant temperature inside the stagnation regime. The repetition of the stagnated, weak ionization processes produce a slower propagation of the cooling waves in the stagnated outer disk zones and a small increase in temperature. At about days 40−50, as matters keep accreted to the inner zone, enough to make the region (zone 18 in Figure 3) to enter the full ionization regime ($> 10^{3.8}K$), there is a sudden ionization and a transient heating wave propagates inward (days 40−60). In brief, one portion of the cooling disk undergoes a heating instability that takes it back to the hot, ionized state, and the process repeats until the final cooling wave arrives. The phenomena, however, is only local, and hence results in a short-lived "flare", only for a week or two. In the model optical light curve, there is a small increase of ~ 0.5mag [2].

Note that, in some X-ray novae (e.g., A0620-00, GS2000+25 and GS1124-683), the reflares have been observed in the optical, UV or soft X-rays, but rarely seen in hard X-rays (e.g., [1]), meaning phenomena in a lower energy regime. In the model, the *outer* disk portions causes a transient heating waves propagates inward with a small increase in the mass accretion and, in turn, in the optical as a reflare. The time-scale, amount of flux increase and low energy-dominated nature of the reflare are very consistent to the observed features in the phenomena called "reflare".

The last comment. In the UV observation of the early decay in the X-ray Nova Per 1992 (GRO J0422+32) [16], a question for the timescale related to a sudden upturn and down, $\lesssim 10-15$ days, has been raised: "the timescale involved may be hard to explain in the context of the accretion-disk instability models". With the stagnation, an intrinsic characteristic of the disk instability, and the irradiation as an external effect adopted, we *did*.

ACKNOWLEDGMENTS

The work presented here is a collaboration with J. Craig Wheeler (University of Texas at Austin) and Shin Mineshige (Kyoto University). S.-W.K. is grateful to Insu Yi at KIAS for fruitful discussion and support, and to Heon-Young Chang and Chunglee Kim at KIAS for their helps, during the proceedings. S.-W.K. is supported by the BK21 Project of the Korean Government.

REFERENCES

1. Chen, W., Shrader, C. R., and Livio, M., *Astrophy. J.* **491**, 312 (1997).
2. Kim, S.-W., Wheeler, J. C., and Mineshige, S., *Publ. Astron. Soc. Japan* **51**, 393 (1999).
3. Kim, S.-W., Wheeler, J. C., and Mineshige, S., *Physics of Accretion Disks*, Amsterdam: Overseas Publishers Association, pp. 171-174 (1996).
4. Kim, S.-W., Wheeler, J. C., and Mineshige, S., *Cataclysmic Variables and Related Objects* (IAU Colloquium 158), Dordrecht: Kluwer Academic Publishers, pp. 139-140 (1996).
5. Kim, S.-W., Wheeler, J. C., and Mineshige, S., *The Evolution of X-ray Binaries* (AIP Conference Proceedings 308), New York: AIP, pp. 213-216 (1994).
6. Fukue, J., *Publ. Astron. Sco. Japan* **44**, 663 (1992).
7. Marsh, T. R., Robinson, E. L., and Wood, J. H., *Mon. Notices. Royal. Astron. Soc.* **266**, 137 (1994).
8. McClintock, J. E., Horne, K., and Remillard, R. A., *Astrophy. J.* **442**, 358 (1995).
9. Shakura, N. I., and Synyaev, R. A., *Astron. Astrophy.* **24**, 337 (1973).
10. Narayan, R., in this proceedings.
11. Kusunose, M., in this proceedings.
12. Kim, S.-W., Wheeler, J. C., and Mineshige, S., *Astrophy. J.* **384**, 269 (1992) and **339**, 330 (1992: Erratum).
13. Mineshige, S., *Astron. Astrophy.* **190**, 72 (1988).
14. Kim, S., *Disk Instabilities in Close Binary System: 25 Years of the Disk-Instability Model*, Tokyo: Universal Academy Press, Inc., pp. 199-202 (1999).
15. Bailyn, C. D., Orosz, J. A., Girard, T. M., Jogee, S., della Valle, M., Begam, M. C., Fruchter, A. S., González, R., Ianna, R. A., Layder, A. C., Martins, D. H., and Smith, M., *Nature* **374**, 701 (1995).
16. Shrader, C. R., Wagner, R. M., Hjellming, R. M., and Han, X. H., *Astrophy. J.* **434**, 698.

Submillimeter-wave and Millimeter-wave Observations of the Interaction of Supernova Remnants with Molecular Clouds

Ken'ichi Tatematsu,* Yuji Arikawa,* Yutaro Sekimoto,* and Mt. Fuji Submillimeter-wave Telescope Team

Nobeyama Radio Observatory, Nobeyama, Minamisaku, Nagano 384-1305, Japan

Abstract. We report on submillimeter-wave and millimeter-wave observations toward supernova remnants (SNRs) by using the James Clerk Maxwell Telescope, the Mt. Fuji submillimeter-wave telescope, and the Nobeyama 45-m radio telescope. For the supernova remnant W28, which is an EGRET gamma-ray source, we have convincingly detected the broad CO, HCO^+, HCN, and SiO emission lines from the shock-accelerated gas due to the SNR-cloud interaction. The previously-reported 1720-MHz OH maser spot is found to be located at the shock front. By using the Mt. Fuji submillimeter-wave telescope, we observed the 492-GHz CI (neutral atomic carbon), 345-GHz CO (3−2) and 330-GHz ^{13}CO (3−2) emission toward SNRs, W44 and W51B/C. We found that the CI/CO and CI/^{13}CO intensity ratio tends to be high in the SNR-cloud interaction region in W51B/C SNR. This fact might suggest the CI relative abundance is enhanced by the interaction, not only in the SNR IC 443, but also in W51B/C SNR.

INTRODUCTION

Stars form in molecular clouds in the interstellar medium. Massive ($\gtrsim 8\ M_\odot$) stars end their lives with supernova explosions, and leave supernova remnants (SNRs). Since the total energy of the SNR is enormous (10^{51} erg), it will affect the interstellat medium largely. It is very interesting to study how the interstellar medium, in particular the molecular cloud, is affected (heated, compressed, or disrupted) by SNRs. Since the discovery of the shock-accelerated CO gas in the SNR IC 443 [1], this SNR had been long the sole observational example of broad molecular line emission due to the SNR-cloud interaction until recently. Several groups including us have observed SNRs to search for the SNR-cloud interaction. In G109.1−1.0, HB 21, and γ Cyg SNR, we have obtained the morphological evidence of the interaction by observing a millimeter-wave CO emission line (J = 1−0), but the broad CO emission suggesting the shock-accelerated gas was not detected [2] [3]

[4] [5]. Recently, several SNRs were found to show the shock-accelerated molecular gas with more sensitive observations [6] [7] [8] [9].

With the advent of submillimeter-wave telescopes operating on regular bases, it has become possible to search for the shock-accelerated molecular gas much more easily than ever. Because submillimeter molecular lines are more sensitive to warm ($\gtrsim 30$ K), dense ($\gtrsim 10^5$ cm^{-3}) gas than millimeter molecular lines, we can search for the shocked gas more efficiently. Furthermore, by comparing submillimeter and millimeter observations, we can determine clearly whether the emission represents shock-compressed dense, warm gas or not.

The submillimeter regime has two ground-level fine-structure lines of neutral atomic carbon (CI). The CI lines provide us with fruitful information regarding the phase change of carbon, which is the fourth most abundant element in the universe. Previously, it was claimed that the CI abundance is enhanced near the shock front propagating into molecular clouds associated with the SNR IC 443 [10] [11]. We have investigate whether such enhancement is unique or more common.

In this paper, we report on our observations toward SNRs.

OBSERVATIONS

Observations were made with three telescopes.

Observations toward W28 have been carried out with the James Clerk Maxwell Telescope (JCMT) in J = 3–2 CO line at 345 GHz and with the Nobeyama 45 m radio telescope in the J = 1–0 lines of CO, HCO$^+$, and HCN, and the J = 2–1 line of SiO. JCMT has a 15-m main dish, and the half-power beam width at 345 GHz is 13 arcsec. The Nobeyama 45 m telescope has a 15-18 arcsec beam at 100 GHz frequency band.

The Mt. Fuji Telescope [12] operates since 1998 at the summit of Mt. Fuji. It can observe 345, 500, and 800 GHz frequency bands. The observations to be reported here were made in the 492 GHz CI, 345 GHz J = 3–2 CO, and 330 GHz J = 3–2 ^{13}CO lines.

RESULTS FOR W28

In W28, we have detected the broad molecular line emission very clearly. It is very likely that this broad emission represents shock accelerated molecular gas due to the interaction with the SNR W28. Figure 1 shows examples of spectra obtained in J =3–2 and J = 1–0 CO at four positions. The J = 3–2 CO intensity tends to be stronger in the broad emission than J = 1–0 CO, whereas J = 1–0 tends to be stronger in the narrow emission line at 7 km s^{-1} (unshocked, quiescent gas). Line excitation calculations show that the broad emission is emitted from warm ($\gtrsim 60$ K), dense ($\gtrsim 10^4$ cm^{-3}) gas, while the narrow emission comes from cold ($\lesssim 20$ K), less-dense ($\lesssim 10^3$ cm^{-3}) gas.

Figure 2 shows the position-velocity diagram obtained toward W28 in J = 3−2 CO. The structure near the shock front is clearly shown. The OH maser spot [15] is located just at the shock front. It becomes clearer that the 1720-MHz OH maser is a good probe of the SNR-cloud interaction (see also [8])

Figure 3 shows the spatial distribution of the J = 3−2 CO emission. The molecular gas is strongly correlated with the radio continuum "ridge", suggesting compression of magnetic field due to the SNR-cloud interaction. Further details of observations in J = 3−2 and J = 1−0 CO are given in [9] [16].

We have also observed the J = 1−0 lines of HCO^+, and HCN, and the J = 2−1 line of SiO, which are thought to trace dense shocked gas. Figure 4 shows the position-velocity diagrams. Note that these emission lines are observed only in the broad emission.

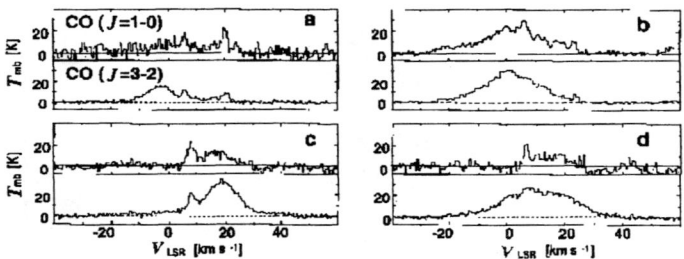

FIGURE 1. Examples of spectra obtained in J =3−2 and J = 1−0 CO at four positions toward SNR W28 (Reprinted with permission from [9], © Astronomical Society of Japan.)

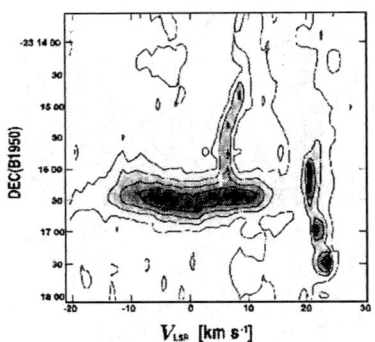

FIGURE 2. Position-velocity diagram obtained along RA (1950) = $17^h58^m08^s.2$ toward W28 in J = 3−2 CO [9]. The narrow velocity component at 7 km/s represents the unshocked gas, while the broad velocity component near DEC = -23 16' 20" represents the shocked gas. The white cross represents an OH maser spot [15].

SUBMILLIMETER CI SURVEY TOWARD SNRS

To understand the physics and chemistry in the region where the SNR interacts with the molecular cloud, by using the Mt. Fuji submillimeter-wave telescope, we

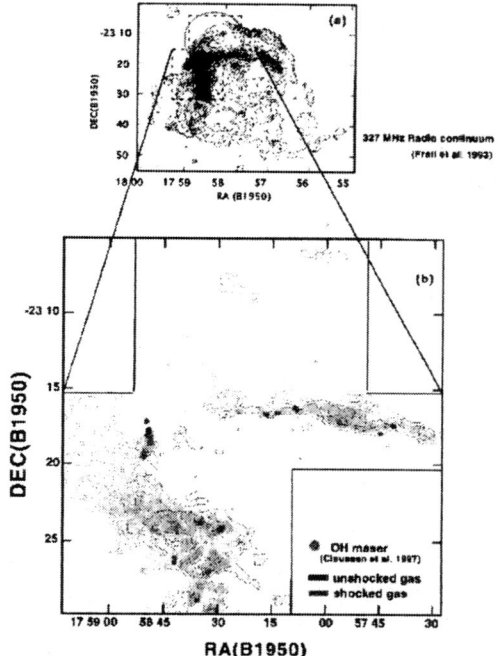

FIGURE 3. (upper) Radio contunuum map of W28 at 327 MHz [13] and the EGRET gamma-ray error circle (red oval) [14]. (lower) The grey-scale map represent the $J = 3-2$ intensity of the gas associated with W28 [9]. Small filled circles are OH maser spots [15].

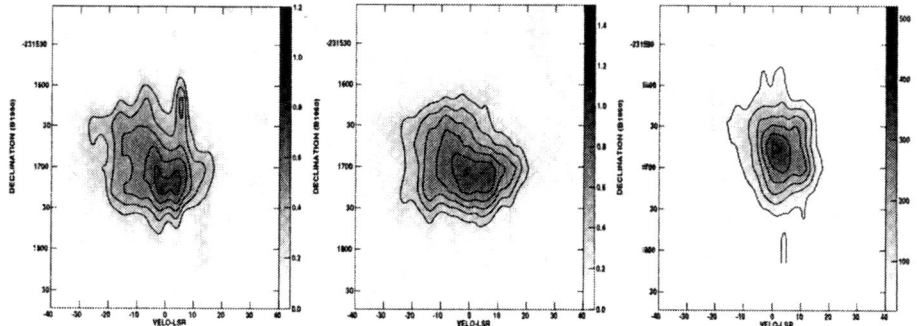

FIGURE 4. Position-velocity diagram obtained toward W28 in $J = 1-0$ HCO^+ (left), $J = 1-0$ HCN (middle), and $J = 2-1$ SiO (right). HCO^+ and HCN observations are made along RA (1950) = $17^h58^m08^s.2$ (the same as Figure 2, but narrower diclination range), while SiO observations were made along a strip passing through the $J = 3-2$ CO intensity peak at RA (1950) = $17^h57^m55^s.6$.

are surveying the CI emission toward SNRs [16]. Target SNRs are those which are known to intarct with molecular clouds by ourselves or in literature (molecular-line observations, OH maser observations, etc). In this paper, we report on two observations from this survey.

W44

For W44, some observational studies have been done, and it is now believed that W44 interacts with molecular clouds [17] [18] [8]. Figure 5 shows the CI and J = 3−2 CO maps obtained toward W44. The global intensity distribution is very similar in these two lines. We previously suggested that the CI/J = 3−2 CO intensity ratio is enhanced toward the OH maser in W44 [16]. Because J = 3−2 CO is likely to be optically thick, we have observed the optically thinner J = 3−2 ^{13}CO line toward W44. However, the noise level is not enough to detect the line toward the SNR-cloud interaction region in W44. Therefore, we cannot conclude whether the CI abundance is enhanced or not for "this" SNR.

W51B/C SNR

The shock-accelerated atomic and molecular gas was observed toward the shell-like SNR W51C [19] [6]. The 1720-MHz OH maser emission was also observed near W51B [20]. W51A contains HII regions, while W51B seems to contain HII regions and a supernova remnant. Because the relationship between W51B (OHmaser) and W51C is not clear (single SNR or two?), we just call W51B/C SNR.

Figure 6 shows the CI and J = 3−2 CO maps obtained toward W51. The global intensity distribution is very similar in these two lines, again. Figure 7 shows the CI, J = 3−2 CO, and J = 3−2 ^{13}CO spectra obtained at the OH maser in W51B/C SNR (left) and in W51A HII region (right).

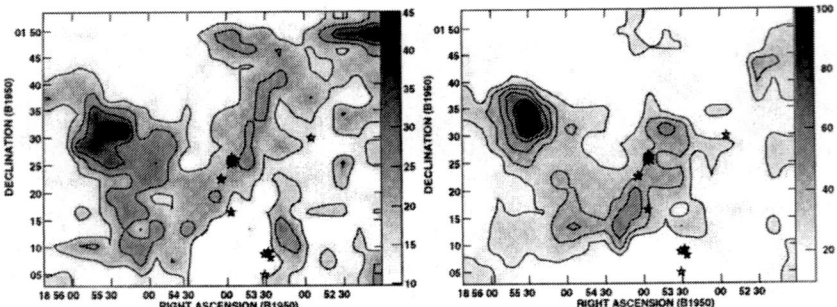

FIGURE 5. CI (left) and J = 3−2 CO (right) maps obtained toward W44. Stars mark OH maser spots [15].

Toward W51A, Tmb (CI)/Tmb (^{13}CO) is about 10. The line profile of CI is very similar to that of ^{13}CO. On the other hand, in W51B/C (OH maser) the peak velocities of these lines differ from each other. At V(LSR) < 67 km/s, Tmb(CI)/Tmb(CO) is close to unity, and exceptionally high compared with other Galactic molecular clouds (e.g. [21]). At V(LSR) = 66-68 km/s (near the CI peak velocity), Tmb(CI)/Tmb(^{13}CO) is $\gtrsim$ 20. It might suggest that the CI/CO abundance ratio tends to be high in W51B/C (OH maser) in this velocity range. The radial velocity of the OH maser spots in W51B/C SNR are 68.9 and 71.9 km/s [20]. It is known that the velocity of the OH maser is close to the systemic velocity or the velocity of quiescent gas [15] [16]. Note that in IC 443, the CI abundance enhancement was observed near the line center (systemic velocity) rather than in line

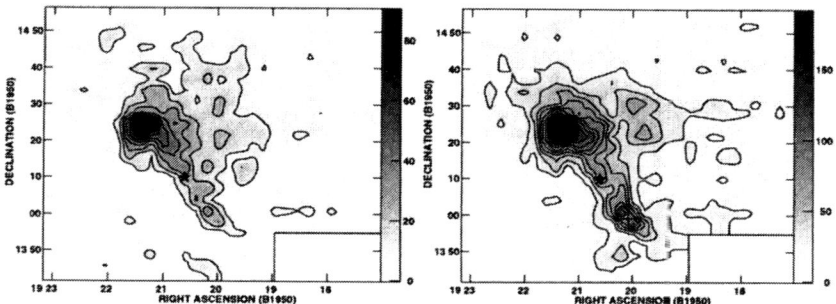

FIGURE 6. CI (left) and J = 3−2 CO (right) maps obtained toward W51 region. Stars mark OH maser spots [20].

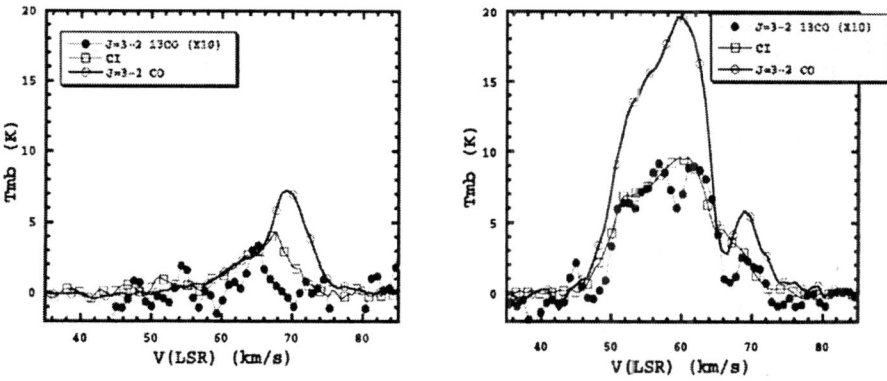

FIGURE 7. CI, J = 3−2 CO, and J = 3−2 ^{13}CO spectra obtained at OH maser spot in W51B/C SNR (left) and in W51A HII region (right).

wing [10] [11]. Regarding the difference in the peak velocity in Figure 7 (left), the strong J = 3−2 CO intensity at 70 km/s can be due to the externally UV-radiated clouds in W51B HII regions [22].

SUMMARY

Submillimeter-wave and millimeter-wave observations toward supernova remnants were reported. In W28, the broad CO, HCO$^+$, HCN, and SiO emission from the shock-accelerated gas by the SNR-cloud interaction was observed. By using the Mt. Fuji submillimeter-wave telescope, we surveyed the 492-GHz CI (neutral atomic carbon), 345-GHz CO (3−2) and 330-GHz ^{13}CO (3−2) emission toward SNRs. We found that the CI/CO and CI/^{13}CO intensity ratio tends to be high in the SNR-cloud interaction region in W51B/C SNR.

REFERENCES

1. DeNoyer, L.K., and Frerking, M.A., *Astrophys. J.* **246**, L37 (1981).
2. Tatematsu, K., et al., *Astron. Astrophys.* **184**, 279 (1987).
3. Tatematsu, K., et al., *Astrophys. J.* **351**, 157 (1987).
4. Tatematsu, K., et al., *Astron. Astrophys.* **237**, 189 (1990b).
5. Fukui, Y., and Tatematsu, K., *IAU Colloquium 101, Supernova Remnants and Interstellar Medium*, Cambridge: Cambridge University Press, 261 (1988).
6. Koo, B.-C., and Moon, D.-S. *Astrophys. J.* **485**, 263 (1997b).
7. Wilner, D.J., Reynolds, S.P., and Moffett, D.A., *Astron. J.* **115**, 247 (1998).
8. Frail, D.A., and Mitchell, G.F. *Astrophys. J.* **508**, 690 (1998).
9. Arikawa, Y., Tatematsu, K., Sekimoto, Y., and Takahashi, T., *Publ. Astron. Soc. Japan* **51**, L7 (1999).
10. Phillips, T.G., Keene, J., and van Dishoeck, E.F., *The Physics and Chemistry in Interstellar Molecular Clouds*, Berlin: Springer, 96 (1993).
11. White, G.J., *Astron. Astrophys.* **283**, L25 (1994).
12. Sekimoto, Y., et al., *Review of Scientific Instruments*, in press (2000).
13. Frail, D.A., Kulkarni, S.R., and Vasisht, G., *Nature* **365**, 136 (1993).
14. Esposito, J.A., et al. *Astrophys. J.* **461**, 820 (1996).
15. Claussen, M.J., et al. *Astrophys. J.* **489**, 143 (1997).
16. Arikawa, Y. *PhD Thesis*, The Gradute University for Advanced Studies, Japan (2000).
17. Wootten, H.A. *Astrophys. J.* **216**, 440 (1977).
18. Seta, M., et al. *Astrophys. J.* **505**, 286 (1998).
19. Koo, B.-C., and Moon, D.-S. *Astrophys. J.* **475**, 194 (1997a).
20. Green, A.J., et al. *Astron. J.* **114**, 2058 (1997).
21. Ikeda, M., et al. *Astrophys. J.* **527**, L59 (1999).
22. Koo, B.-C. *Astrophys. J.* **518**, 760 (1999).

Intense Laser Astrophysics

Hideaki Takabe

Institute of Laser Energetics, Osaka University,
Suita, Osaka 565-0871 JAPAN

Abstract. A new approach to study astrophysics is introduced. Intense lasers can be used to study thermodynamic properties of plasmas near or interior of stars. By focusing on self-similarity of dynamics of high-Mach number shock and matter interaction, radiation hydrodynamics, non-LTE atomic process, and so on, we can study integrated physical phenomena in Universe with use of intense lasers. With use of ultra-intense lasers, we can generate electron-positron fireball in a tiny scale but with high energy density. In the present paper, history from laser fusion to the present subject is described, and physical properties of laser produced plasma are briefly explained for non-specialists. Three view points necessary to connect laser plasmas with Astrophysical plasmas, sameness, similarity, and resemblance are introduced by briefly explaining examples.

INTRODUCTION AND HISTORICAL BACKGROUND

Just 40 years have passed since the invention of laser. It is usually said that the laser is one of the greatest inventions in the 20th Century. In some case, an invention will lead to an unexpected development with pressure from society's needs or with progress of science. I find that the imagination of human beings is wonderful by the fact that people had the dream to achieve the laser fusion by increasing the output of the laser by 100 thousand times. This was just at the time (1962) when a pulsed laser could be made in the laboratory. This story encourages us when we have some unusual idea. I am, however, not so naive as to believe that only such a dream of a scientist has led to the construction of the present-day huge laser systems. That achievement occurred because of the strong or even hysteric request from the public driven by the Oil-Shock in 1973, and also because of the scientific and political competition between nations during the Cold War. These unusual conditions helped to make the laser systems bigger and bigger.

When the short pulse and intense lasers come into being familiar and the word "laser-plasma" was born, John Dawson published a paper on the properties and application of the laser-plasma [1]. In that paper there is a suggestion that intense lasers can be used as a simulator of astrophysics and that experiments related to the collision-less shocks generated by the explosion of a supernova or solar-flare phenomena can be studied in laboratory. This paper, however, did not trigger astrophysics experiments. At that time nobody knew what sort of plasma the laser-plasma is, and it was enough work to generate, diagnose, and analyze the laser-plasma. In addition, when matters are irradiated by lasers, an anomalous absorption was measured and a variety of nonlinear phenomena such as the parametric instabilities were observed. Therefore, it was not the time to apply the lasers to fusion or study of astrophysics. First it was necessary to understand the physics of laser-plasma.

After the Oil-Shock, the construction of big laser systems made it possible to start to understand implosion physics and laser fusion energy research progressed dramatically. This was the time of the mid-1980's. Around that time, B. Ripin et al. in Naval Research Laboratory carried out a series of experiments on blast waves generated by the laser system "Pharos". By taking detailed data with an interference technique, they checked the self-similarity of the wave front propagation, and observed the hydrodynamic instability of the blast wave front and the time evolution of the turbulent spectrum in the later stage of the instability [2]. In addition, they also carried out an experiment on a blast wave in a gas with an externally imposed magnetic field and discussed the phenomena of reconnection of the magnetic field [3]. The same sort of experiment was also carried out in the former Soviet union, and they also mentioned that the experiment has been done to study the interaction between the blast wave generated by a supernova explosion and the cosmic background magnetic field [4]. There were, however, no continuing activities after these publications, although they titled their activity as science for astrophysics. One of the reasons is that their work was related to the SDI (Strategic Defense Initiative) program and the magnetic field modeled the earth's magnetic field to see the behavior of expanding plasma in the earth's magnetic field when nuclear weapons are exploded. This was a reasonable activity at that time from the view point of the government which funded the SDI program, for example, in US.

On the other hand, the supernova 1987A (SN1987A) observed in the south hemisphere on February 23, 1987 taught us that hydrodynamic instability is a big issue not only in the field of laser fusion research [5]. The explosions of supernovae are among the most spectacular physical events in the Universe. Elements heavier than helium are synthesized in massive stars and finally scattered out into space through the supernova explosion. Such ejected materials became the source to form the earth after billions of years. This story encourages us to study physics. To analyze the X-ray and Gamma-ray data from SN1987A the possibility of hydrodynamic instability and mixing in the explosion has been intensively discussed. This spectacular fact of a deep relation of hydrodynamic instability to physics of supernova explosion taught us that the laser fusion scientist is working for not just physics in 1 mm space, but revealing the deep thought of the nature seen in much broader fields. This gave us courage and taught us the importance of having widely opened eyes.

When laser fusion research began, the textbooks of astrophysics gave us guidance on radiation transport, equation of state, opacity and so on. Among them, theoretical papers dealing with high-density plasmas can be found only related to astrophysics and I remember I studied many of them when I was a student. Also from this fact, it is clear that lasers are very useful as simulator of physical phenomena in the Universe, especially astrophysics. With a call from Prof. K. Nomoto (Univ. Tokyo) just after the explosion of SN1987A, I began to be deeply involved in the society of astrophysics in Japan. By listening to the talks by astrophysicists and through discussion with them, I began to think strongly that the connection is not only through the hydrodynamic instability, but most of the subjects are similar, and the difference is just the scales of time and space. From this point of view, the idea of using an intense laser for studying astrophysics appeared naturally through the discussions.

Laser fusion research had progressed very rapidly after the US de-classification of the concept of implosion in 1972 which overlapped with the time of the Oil-Shock. During the middle of 1980's, high neutron yield or high-density experiments were performed with Gekko-XII [6] and NOVA lasers [7], and there was no trend to consider that such big lasers should be used not only for fusion program but also for purely academic purpose (at least, there is no room in my mind and brain). At the same time, in the form of the international joint experiment with Max-Planck Institute for Quantum Optics (MPQ) of Germany, X-ray generation, confinement, and transport in a gold cavity were studied with Gekko-XII laser [8]. The style of their research was impressive. At first, they study the phenomena of radiation thermal wave propagation theoretically with use of self-similar solutions, and then carried out a series of experiments with a small-scale German laser at Garching. With these accomplishments and results, they submitted a proposal for Japan-Germany joint experiments and carried out the joint work in the manner of gentlemen [9]. It is certain that such accomplishment of the joint work helped for US to reexamine the classification policy regarding laser fusion in 1993. After the declassification, we found that LLNL had carried out the same type of experiment and analyzed the data with the German-style self-similar solution [10].

During the latter half of the 1980's, experiments with Gekko-XII were mainly focused on the achievement of high density compression with deuterized plastic shell targets, which are thought to model the future DT solid target, in order to prove the principle of the laser implosion fusion scheme. Actually, the type of implosion has shifted from the exploding pusher type implosion by fast electrons to the ablative implosion. This means the implosion can be controlled and driven by the ablation pressure sustained by the classical heating. In addition, during these period the laser beam smoothing techniques like the random phase plate (RPP) appeared and highly precision uniformity began to be available [11]. As the result, a density of 600 times solid density has been achieved and the fusion core condition required for density and temperature are separately demonstrated, taking account of the temperature realized in high neutron yield experiments.

What has been learned from the high-density compression experiments, however, is that the structured core necessary for high gain is not achieved and the spark and main fuel structures seen in one dimensional implosion calculations are not observed experimentally because of hydrodynamic instabilities. Implosion experiments have still continued; however in 1990's the main activity shifted to a precise model experiment of hydrodynamic instabilities. At first, detailed experiment and analysis of linear and nonlinear Rayleigh-Taylor instability at the ablation front have been carried out at LLNL for the case of X-ray drive. Through the analysis of implosion experiments, the importance of turbulent mixing in the final compression phase has became widely known and many scientists became eager to see the development of academic research regarding hydrodynamic instability of a compressible fluid, its nonlinear evolution, and turbulent mixing. This is the time when laser fusion scientists met one another in "the International Workshop on the Physics of Compressible Turbulent Mixing". It also was an occasion to change our viewpoint so that Rayleigh-Taylor and Richtmyer-Meshkov instabilities are common terminology in a variety of new academic fields, although they are difficulty for the laser fusion society. More details on the academic background and applications are given, for example, in my review article [12].

The opacity of high temperature and high density plasmas is very complicated, when the plasma ions are high-Z and partially inonized. In this case, unresolved line radiation contributes to the mean value of opacity and the atomic modeling of such high-Z, partially ionized plasmas becomes important. When the author started to model the opacity of gold plasmas produced by laser radiation, I could not believe that the opacity modeling is a precision science. The appearance of OPAL code, which is the most reliable code developed at LLNL [13], impressed me very much. In producing data of OPAL, they solve atomic structure based on para-potential method to obtain all possible atomic configurations and calculate the lines by line-by-line with many hours CPU time of Cray YMP. When I attended the 3-rd International Opacity Workshop held at MPQ (Max-Planck Institute for Quantum Optics during March 7-11, 1994, I had an impression "Opacity is now precision science". The ability to make such opacity code is a fundamental measure of the scientific level of an institute whether it is located on a high level or not. I think of developing such sophisticated code by ourselve. It has been already demonstrated that OPAL gives appropriate answers in many topics in Astrophysics [14]. For example, Dr. M. Kato obtain a good light curve of Nova explosion and after by using the mean opacity of OPAL [15]. In her case, the partially ionized iron, which is small abundant, in the atmosphere of the surface of Novae is found to contribute significantly the absorption of the radiation steming from the thermonuclear explosion in the bottom of the surface. The precise opacity data is now indispensable in modeling astrophysical phenomena.

In the spring in 1995 when a big progress of experiment to understand the hydrodynamic instabilities has been accomplished with NOVA laser in LLNL, an international conference was held in Osaka [16] and Bruce Remington attended the conference from LLNL. His expertise is hydrodynamic instability in laser fusion and his attendance was due to distinguished achievement with experiment [17]. In the conference I asked Prof. K. Nomoto to give an invited talk on hydrodynamic instabilities in supernova explosions relating to the topics of SN1987A [16]. Through the conference, Bruce Remington became involved in the hydrodynamic instability of supernovae and began a model experiment of supernovae just after his return to LLNL. He told me that before attending the conference, his boss told him that the hydrodynamic experiment for laser fusion is well carried out and enough, so find a new theme. It might have driven him to rush into the initiation of the model experiment of astrophysics, establishment of a network with astrophysicists, and organizing the first international workshop on laboratory astrophysics with intense lasers [18]. And, he has come to the level to establish the Center of Laboratory Astrophysics in LLNL. A new movement, in general, is triggered with the combination of a talented man/woman and a chance. It must not be, however, forgotten that there is an inevitable background.

In US, NIF is under construction. I have a report describing the details of the NIF [19]. In the preface of the report, four items are refereed as the mission of NIF project. That is, "national security", "science and technology", "energy resources", and "industrial competitiveness". The scientific research always has two faces and this tendency was enhanced during the era of the Cold War. The Cold War is over and a new world political system is going to be constructed. However, the world contains still many elements of un-stability and uncertainty. In such situation, it is not so easy to rapidly change the direction and strategy of research. With the end of the Cold War, LLNL seems to reexamine its future plan in order to become the first runner of the world in the field of fundamental science [20]. It is unavoidable to regard that the first priority of the NIF is the national security at the present time. Through the present era of the globalization of economics and politics and rapid change of social system, it is difficult to be sure that the above order of the four items remain the same until the completion of the NIF. As the accelerator has grown after the Second World War as a big device for fundamental science to study elementary particle physics and nuclear physics, this time the laser has a potentiality to be regarded as a big device to study a new fundamental science in the near future. In this course, the relation with astrophysics, which is an everlasting academic subject for the human beings, is very important.

The possibility of the usage of the NIF for fundamental science is summarized by Dick Lee of LLNL by mainly focusing on atomic physics and radiation transport [21]. At the present time, it is announced that 15% of the NIF experiments will be used for the fundamental science. This percentage, of course, will change according to the needs of the era and I believe that this percentage is sure to increase [22]. The computer is a well-known example of the dual use of science. Physics Today journal published a special issue titled "50 years of computers and physicists" [23]. In the issue, the cradle of the scientific computer was described in the article titled "From Mars to Minerva: The origins of scientific computing in the AEC Labs" (by R. W. Sedel). It is the fact that the origin of computer is "Mars" (God of War). It is clear evidence of the wisdom of the human beings to have changed it to "Minerva" (God of Intelligence). The same should be done concerning all science and technology.

CHARACTERISTICS OF LASER PRODUCED PLASMAS

When an intense laser is irradiated on a solid material as shown in Fig. 1-(a), the laser energy is absorbed at very beginning by the multi-photon process and free electrons are produced. If the free electrons have enough kinetic energy, they enhance the ionization of the material through the collisional ionization process. Once the partially ionized plasma is produced, the free electrons predominantly contribute to the absorption of the still-coming laser energy. This absorption process is called "inverse-Bremmstrahlung" absorption. This is the inverse process of radiation emission stemming from the collisions between electrons and ions. The free electrons obtain the energy of laser photons in colliding with ions. The free electrons quiver in the electric field of laser light, and their motion is almost harmonic oscillation in the plane of the electric field. If we can neglect the collisional process, the motion is reversible. When the plasma temperature is low and the density is high, however, the collisional process plays important role to make the quivering motion random. This is the thermalization of the electrons or the heating process of plasma by the inverse-Bremmstrahlung mechanism. More detailed description is given, for example, in the text by Kruer [24]. If the laser intensity is not so high (say, less than 10^{13}W/cm2), the absorption process and the following physical processes become very complicated. However, such a case is recently studied relating to the industrial application of lasers such as material processing with laser ablation [25].

For the case when the material is of low-Z, the absorbed laser energy is transferred to the over-dense region by the electron heat conduction as shown in Fig. 1-(b). The heat conduction is nonlinear because the electron mean-free-path is a strong function of the temperature. And a heat wave is formed with its velocity decreasing rapidly as a function of time [26]. When the sound velocity of the heated region becomes comparable to the velocity of the heat wave, a shock wave is generated and propagates in front of the heat wave. On the other hand, the heated material expands into the vacuum with the velocity roughly equal to the sound velocity. The rear end of the expansion wave runs after the heat wave. Almost the same time as the shock wave is formed, the heat wave merges with the expansion wave to form the ablative heat wave. After this time, the structure seen in Fig. 1-(c) is maintained under the constant irradiation of laser. For the case of the laser intensity of about 10^{14}W/cm2, the time scale for changing from Fig. 1-(b) to Fig. 1-(c) is several pico-seconds. Therefore, it is a good approximation to assume that the structure of Fig. 1-(c) will be kept through the laser irradiation as long as the pulse duration is much longer than this time scale [27].

The rear of the shocked, high-dense region in Fig. 1-(c) is called "ablation front". The region from the ablation front to the point of laser absorption is called "conduction region" or "deflagration region". The deflagration is the name for a slow combustion wave front [28]. If we regard the laser heating as the energy increase due to chemical reaction in the combustion, we can also use this word for the case of laser heating.

In the case when the solid material is of high-Z (e.g., gold), most of the absorbed energy is converted to soft X-rays in the conduction region [29]. Then, the heat wave and the ablative heat wave in Fig. 1 are sustained by the X-ray radiation. The X-rays escape from the plasma toward the right in Fig. 1, and almost 70-80 % of the absorbed energy is released into vacuum as X-ray energy. The spectrum of the X-rays is near Planckian. Since the thermal X-rays are generated mainly in the conduction region, the temperature of the radiation is almost same as the temperature of this region, which is usually a few hundred eV. By irradiating the lasers inside the cylindrical cavity of gold, we can confine the X-rays in the cavity to produce an Planckian X-ray source [8]. This cavity is called "houlraum" and has been widely used in the scheme of radiation driven inertial confinement fusion (ICF) [7]. The cavity X-ray has also been widely used to drive strong shock waves to measure the equation of state, to heat a sample material for the opacity experiment and so on [30]. It is noted that even in the case when the laser is irradiated on low-Z materials, several to ten percents of the absorbed energy escapes from the plasma in the form of X-rays. In the case of the low-Z plasmas, however, the most of the absorbed energy is finally converted to the kinetic energy of the expending plasmas.

The intense lasers can be used to study the high energy density (HED) physics

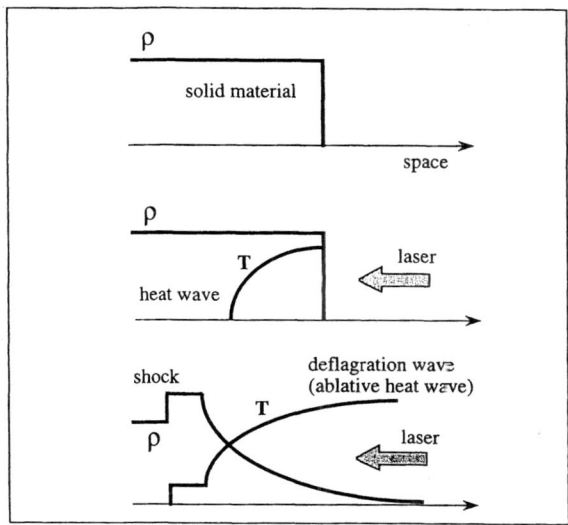

FIGURE 1. Schematics of time evolution of plasma formation when an intense laser irradiated from the right. Time evolves from the top (a), middle (b), to the bottom (c).

[30]. The high density plasmas with the temperature up to a few keV can be produced with the size of sub-mm. The pressure generated at the ablation front is called "ablation pressure" and has been studied theoretically and experimentally from the early time of laser plasma research. When lasers with the intensity of 1013-1015 W/cm2 is irradiated on a solid material, the ablation pressure in the range of a few tens to a hundred Mega-bar (1012 cgs-unit) is generated. This extremely high pressure can be used to generate a strong shock wave in gas or solids.

We can enumerate another ways, such as a shock tube and a high explosive to generate high pressures. However, the typical pressure generated by the shock tube is several tens atmosphere. With the high explosive the pressure reaches a few hundreds of kilo-bar, while the ionization of dense matters can not be expected.

The X-rays generated from the high-Z plasmas can be used to heat uniformly a small sample material. Such X-rays have been also used for diagnostic purpose. There are two ways to measure the state and dynamics of plasmas. One is the "passive" way in which the x-ray or optical signals coming from the plasmas are measured. The other is the "active" way in which X-rays generated by another lasers are used to diagnose the laser-produced plasmas or are used to generate uniform HED plasmas.

The x-ray back-lighting and side-lighting techniques are a standard method to measure the density (optical thickness) structure of laser produced plasmas. In Fig. 2, a schematic configuration of the x-ray side-lighting technique is shown. Figure 2 indicates the configuration to obtain the images of hydrodynamic instability grow-

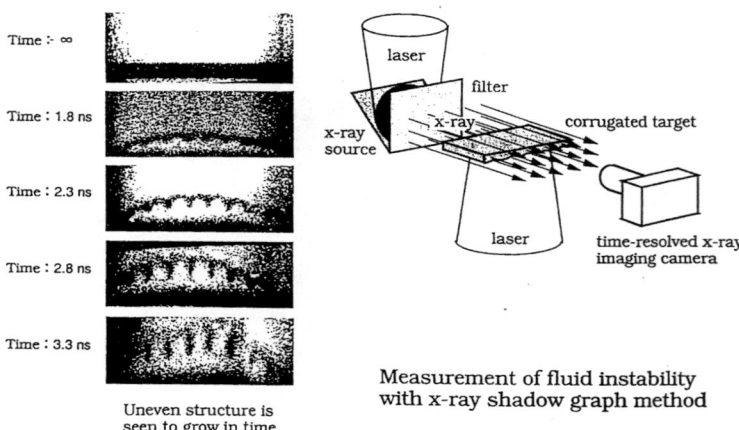

FIGURE 2. Back-lighting technique to measure the time evolution of plasma structure. A typical example is shown on the left for Rayleigh-Taylor instability growth at laser ablation front.

ing in a plastic foil being accelerated by laser which is irradiated from the bottom side. A hard x-ray is generated by another laser beam irradiated on a X-ray source material. The transmitted X-ray is measured with the time-resolved X-ray imaging camera as shown in Fig. 2. A typical example of the X-ray transmitted image is shown around the bottom in Fig. 2. The target foil is located initially near the bottom in the image and, then, accelerated upward. The target surface was corrugated to provide an initial perturbation for R-T instability. It is seen in the shadow image in Fig. 2 that the perturbation has grown into the nonlinear phase, where the initial sinusoidal perturbation has become that with the bubble-spike structure. The perturbation has already penetrated into the rear side of the target, where perturbed structure is also seen. Details of the linear growth rate of R-T instability was measured by analyzing the time evolution of the intensity of the transmitted x-ray through the target foil [31]. More detailed report on the diagnostic technique and the experimental result of the instability growth is given, for example, by Remington et. al. [32] and Azechi et. al. [33].

SAMENESS, SIMILARITY AND RESEMBLANCE

During 40 years from the invention of laser, the power of intense lasers has progressed dramatically from kW to PW (peta-watt). The diagram of annual growth of the maximum available laser power is plotted on the right in Fig. 3. This figure indicates that the power increased by factor ten every three years in average as estimated from the dotted line. In the field of accelerator, the left figure of Fig. 3 is called "Livingstone Chart" [34]. As is clear from this figure, the beam energy of accelerators has increased for the last 60 years by a factor 10 every six years. It is clear that the high power laser has grown as a new tool to investigate an advanced physics in extreme conditions. This rapid growth of the laser technology is partially driven by the need for laser fusion research and also by the invention of CPA (Chirped Pulse Amplification) technology [35] I have strong convince that such dramatic progress of laser technology open a new field of science called "Intense Laser Science".

Among this new science, the first one would be "Intense Laser Astrophysics (ILA)". If the accelerator is an experimental tool to study elementary processes up to three minutes from the Big Bang [36], we can say that the laser is an experimental tool to study the physics of dynamical phenomena of Universe after three minutes except for the physics relate to gravitational interactions.

There are three view points relating the laser-produced plasmas to astrophysics. They are

(a) Sameness of physics

(b) Similarity of physical dynamics

(c) Resemblance of physics

At first, it is easy to understand the content of the term (a). This means to produce plasma with the same temperature and density as the surface or interior of stars and to study, for example, the equation of state (EOS) and emissivity and opacity of X-rays. In addition, if we are able to use a huge laser system like NIF, we may also be able to study the cross sections and their density dependence of thermonuclear fusion reactions important inside stars.

The term (b) represents the study of physical evolution in time and space of compressible hydrodynamics and atomic processes. For example, we reproduce the physics of supernova remnants with the diameter of a few tens ly (light year) and the age of a few thousand years by reducing the scale of time and space by a factor of 1020 in the blast wave driven by laser light. This concept is same as the design of aircraft by putting a small-scale model in a wind tunnel. In this case, the non-dimensional parameters, Reynolds number and Mach number, should be kept same as those for a real aircraft in flight. In the case of the wind tunnel, the scaling factor is about 103. The difference from the case of the wind tunnel is extremely large difference of the scaling factor. Another examples of the term (b) are very strong shock and matter interactions [37], hydrodynamic instabilities of the case where

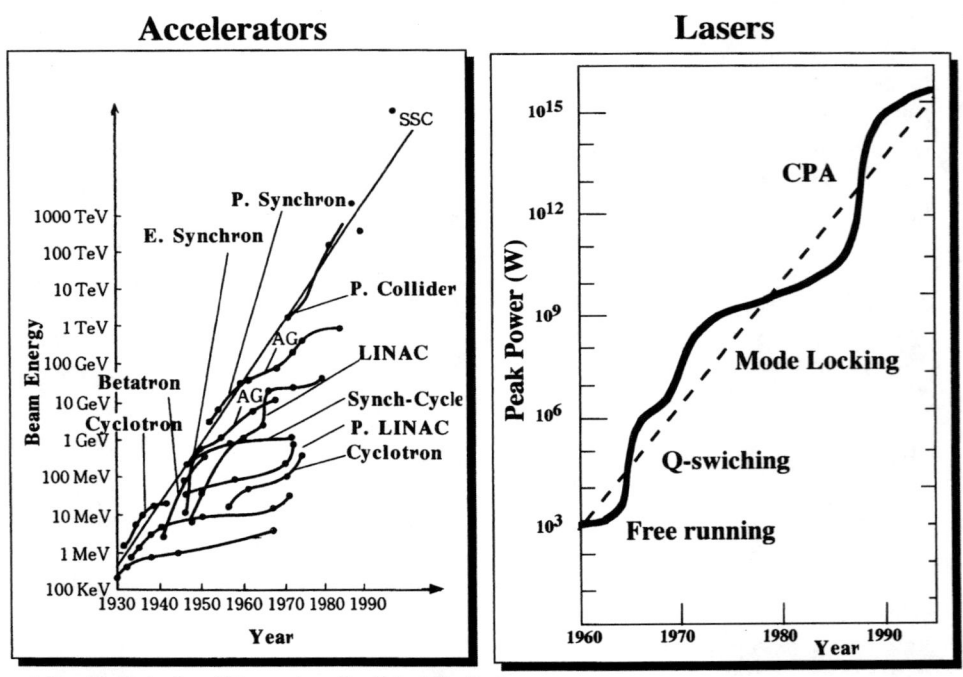

After Y. Totsuka, "Elementary Particle Physics"

FIGURE 3. Livingston chart of accelerator and corresponding chart for laser power.

energy transport essentially modifies the growth of the instabilities [38], non local thermodynamic equilibrium phenomena [39], radiation hydrodynamic phenomena [40] and so on.

The term (c) represents the case where although scaling law is not found yet, physics and phenomena are very resemble each other. For example, it indicates the creation of anti-matter by use of ultra-intense lasers [41] and electron-positron plasma formation. There are fill of electron-positron plasmas near Black Hole and also near AGN (Active Galactic Nuclei) [42]. We plan to do an experiment where the generated electron-positron plasma collide with imposed magnetic field or matter. Such experiment may be an model experiment of expanding fire-ball believed to be important as the energy source of Gamma-ray bursts [43]. At the same time, it is also interesting to study photo-nuclear reaction process [41]. The demonstration of possibility of X-ray laser astro-objects is also included in the term (c). It is reported that photo-ionized plasma is observed near the X-ray compact object, Cygnus X-3 [44]. The plasma with temperature roughly equal to 5-10 eV is under a strong irradiation of hard X-ray coming from a companion Black Hole and this low temperature plasma is photo-ionized to emit many strong line emission from metal such as Si, Mg, . As described in the previous Section, by use of gold cavity we can generate almost Planckian X-ray with temperature up to about 200 eV. By providing a low temperature plasma at the same time, we can observe the atomic state distribution of photo-ionized plasma under the irradiation of this X-ray source. Systematic study of the photo-ionized plasma will lead us to clarify a condition of photo-pumping X-ray lasing in Universe.

CONCLUSION

At the present time, B. Remington at LLNL is promoting ILA activity in US with many collaborators [45]. In Japan, a theoretical work has been done [46] and we have carried out the first experiment of ILA in 1999 on the subject of model experiment of ejecta-ring collision [47] and astrophysical jets [48]. In order to discuss and exchange information, an international conference is held every 2 years (International Conference on Laboratory Astrophysics with Intense Lasers). The last one was the third conference and held at Houston [49]. The bunch of papers reported in the second conference at Tucson is published as special issue of Astrophysical Journal Supplement recently [45]. The author recommend to read recent review papers written by B. Remington, P. Drake, D. Arnett, and H. Takabe [50]

Finally, I would like to discuss a bit about a relation with HED (High Energy Desity) physics. Physics of laser fusion is based on HED physics whose key words are shown in the bottom of Fig. 4. At the same time, the physics of astrophysics is also based on HED physics. Understanding of HED physics through Intense Laser Astrophysics activity helps not only a progress of astrophysics, but also a progress of laser fusion research. If we express the conventional astrophysics as "Astrophysics in Heaven", then the Intense Laser Astrophysics is "Astrophysics on Earth". These two link strongly through model experiments and development and improvement of computer codes with the results of the model experiments. This means ILA also leads a progress of large-scale computing, because astrophysics is essentially an integrated physics and needs a development of an integrated code and large scale computing. In order to link "Astrophysics in Heaven" and "Astrophysics on Earth", I would like to say that a good collaboration between laser plasma scientists and astrophysicists is very essential. The key word for ILA is, therefore, "Heaven, Earth, and Man".

The astrophysics with intense laser is not only a new trend of research, but also should offer students and young researchers a good opportunity to satisfy their scientific curiosity. If the astrophysics with lasers grows as one academic field, I believe that it will also provide a good place to educate experts for long-term fusion

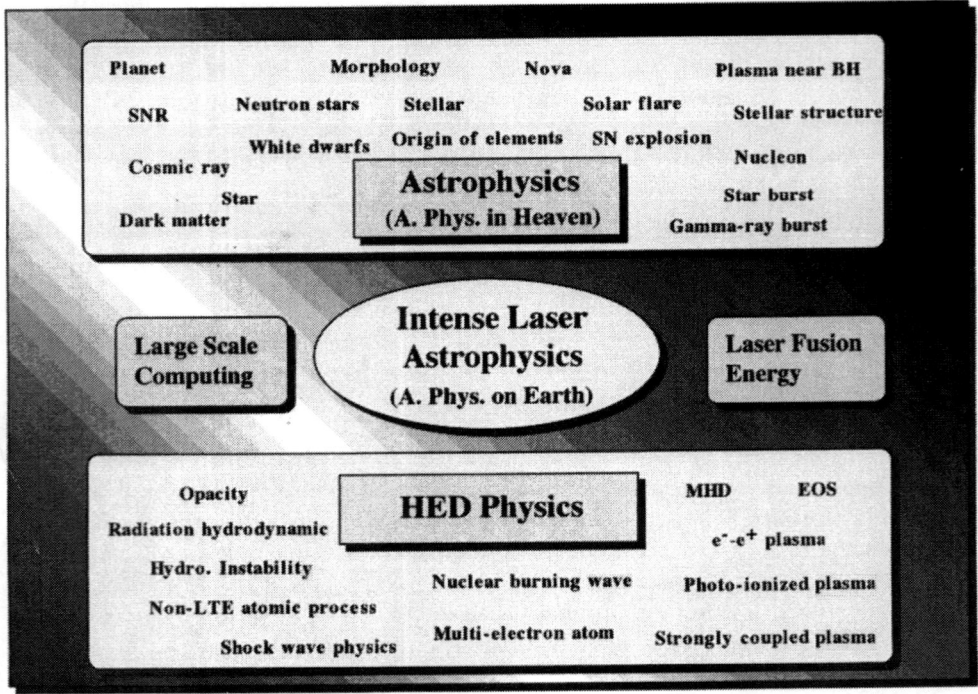

FIGURE 4. Intense laser astrophysics is to study HED physics with intense laser through model experiment of astrophysics. Many subjects in astrophysics can be clarified through better understanding of HED physics. Such study is a traction of large-scale computing and laser fusion research.

energy research.

REFERENCES

1. J. M. Dawson, "On the Production of Plasma by Giant Pulse Lasers", *Phys. Fluids* **7**, 981 (1964).
2. J. Grun et al, "Experimental Studies of Very High Mach Number Hydrodynamics", *NRL Mem. Rep.* 6790-94-7366, 1994; J. Grun et al., *Phys. Rev. Lett.* **66**, 2738 (1991).
3. B. H. Ripin et al."Sub-Alfvenic Plasma Expansion", *Physic Fluids* **B5**, 3491 (1993).
4. V. M. Antonov et al, "Laser Interaction with Matter", ed. S. Rose, IOP Conf. Ser. No. 140 (IOP, 1995), p. 167.
5. For example; W. D. Arnnet et al., "Supernova 1987A", *Annu. Rev. Astron. Astrophys.* 1989, **27**: 627-700.;
   W. Hillebrandt and P. Hoflich, "The Supernova 1987A in the Large Magellanic Cloud", *Rep. Prog. Phys.* **52**, 1421-73 (1989).;
   H. Takabe, "Inertial Confinement Fusion and Supernova Explosion", *Japanese J. Plasmas & Fusion Res.*,**69**, 1285 (1993) (in Japanese).
6. H. Takabe et al, *Phys. Fluids* **31**, 2884 (1988); H. Azechi et al., Laser Part. Beams **9**, 193 (1991).
7. For example, J. D. Lindl, "Inertial Confinement Fusion"(AIP Press and Springer, 1999).
8. R. Sigel et al., Phys. Rev. **A 38**, 5779 (1988); H. Nishimura et al., *Phys. Rev.* **A 44**, 8323 (1991).
9. The Max-Planck Prize is offered to this joint research (1990).
10. B. G. Levi, Physics Today, "Veil of Secracy is Lifted from Parts of Livermore's Laser Fusion Program", September issue, p. 17 (1994).
11. This research has got the Prize of "Excellence in Plasma Physics" of APS in 1993
12. H. Takabe, "Inertial confinement Fusion and Hydrodynamic Instabilities", *Japanese J. Plasmas & Fusion Res.* **73**, 147 (1997).
13. Web-site: http://www.phys.llnl.gov/V-Div/OPAL/
14. F. J. Rogers and C. A. Iglesias, *Science* **263**, 50(1994).
15. M. Kato and I. Hachisu, *Astrophys. J.* **437**, 802 (1994).
16. "12th Int. Conf. on Laser Interaction and Related Plasma Phenomena", Osaka, April 24-28, 1995. Proceedings; AIP Conf. Proc. 369 (AIP, 1996).
17. Their accomplishment on hydrodynamic instability is prized with "Excellence in Plasma Physics" of APS in 1995.
18. "International Workshop on Laboratory Astrophysics with Intense Lasers", Websirte; http://lasers.llnl.gov/lasers/target/astro
19. ICF Quarterly Report, Vol. 7, No.3 (April-June, 1997) Report is available with Website; http://lasers.llnl.gov/lasers/pubs/icfq.html
20. *Science* **275**, 1252 (1997).
21. Web site; http://www.llnl.gov/science_on_lasers/ or R. W Lee, "Science on the NIF" in "Energy & Technology Review", LLNL, December 1994, pp. 43-54.

22. In the workshop on "Frontier Science on the NIF -Episode-I-"(Oct. 4-6, 1999, Pleasanton, CA), the delegate of DOE announced that 15 % of the NIF experiments will be used for fundamental science conducted by the Science User Group of NIF which is chaired by Prof. R. Petrasso of MIT. The percentage for the fundamental science was 10 % previously.
23. *Physics Today*, October issue, 1996.
24. W. L. Kruer, "The Physics of Laser Plasma Interactions" (Addison-Wesley Pub., 1988).
25. S. I. Anisimov and V, A, Khokhlov, "Instabilities in Laser-Matter Interaction"(CRC Press., 1995).
26. Ya. B. Zel'dovich and Yu. P. Raizer, "Physics of Shock Waves and High-Temperature Hydrodynamic Phenomena" Vols. 1 and 2 (Academic, 1966). Chap. X.
27. H. Takabe et al, *J. Phys. Soc. Jpn*, **45**, 2001(1978).
28. L. D. Landau and E. M. Lifshitz, "Fluid Mechanics" (Pergamon Press, 1959) Chap. XIV.
29. H. Takabe and K. Nishikawa, *J. Quant. Spectrosco. Radiat. Transfer* **51**, 379-395 (1994).
30. M. D. Rosen, *Phys. Plasmas* **3**,1803 (1996).
31. K. S. Budil et al., *Phys. Rev. Lett.*, **76**, 4536 (1996).
32. B. Remington et al, *Phys. Plasmas* **2**, 241 81995).
33. H. Azechi et al, *Phys. Plasmas* **4**, 4079 (1997).
34. Y. Totsuka, "Elementary Particle Physics" (Iwanami Publ., Tokyo) in Japanese.
35. G. A. Mourou et al, Physics Today, January issue, 22 (1998): C. J. Joshi and P. B. Corkun, *Physics Today* **48**, No.1 (1995), p. 36.
36. S. Weinberg, "The First Three Minutes" (Basic Books Inc., New York, 1988).
37. J. M. Stone and M. L. Norman, *Astrophys. J.* **390**, L17-L19(1992).
38. T. Shigeyama, *Publ. Astron. Soc. Jpn*, **47**, 581-588(1995).
39. H. Tsunemi et al., in "Thermonuclear Supernovae", NATO ASI Ser.(c) Vol. 486 (Kluwer Academic Pub. Dordrecht, 1977) p. 561.
40. J. Stone et al., *Astrophys. J. Supp.*, **127**, 497-502 (2000).
41. T. Cowan et al., *Phys. Rev. Lett.*,**84**, 903 (2000).
42. E. Liang, *Nature*,**381**, 49 (1996).
43. For Example; N. Gehrels and I. Paul, *Phys. Today*, **51**, No.2 (1998) p.26.
44. K. Kawashima and S. Kitamoto, *Publ. Astron. Soc. Jpn*,**48**, L113-L116(1996).
45. B. Remington et al., Phys. Plasmas **4**, 1994 (1997): Many papers are found in "Special issue in Astrophys J. Supplement"; Vol. 127, No. 2, Part 1, April (2000): http://www.journals.uchicago.edu/ApJ/journal/contents/ApJS/v127n2.html
46. H. Takabe, in "Superstrong Fields in Plasmas" edited by M Lontano et al., AIP Conf. Proceedings 426 (AIP, Woodbury, 1998), p. 560.
47. Y. G. Kang, H. Nishimura, H. Takabe et al, "Laboratory Simulation of the Collision of the Supernova 1987A with its Ring Nebula", submitted to Astrophys. J. (1999).
48. R. Kodama et al, in preparation.
49. Web-site: http://spacsun.rice.edu/laser2000/
50. B. Remington et al., *Science*,**284**, 1488-1493 (1999): *Phys. Plasmas*, **7**, 1641-1652 (2000).

AUTHOR INDEX

A

Ables, E., 261
Agol, E., 125
Arikawa, Y., 382

B

Band, D., 261
Barthelmy, S., 261
Bastrukov, S., 197
Berger, E., 240
Bionta, R., 261
Blackman, E. G., 100
Bloom, J. S., 240
Brown, G. E., 68

C

Chaffee, F., 240
Chang, H.-Y., 291
Cline, T., 261

D

Dai, Z. G., 276
Diercks, A., 240
Djorgovski, S. G., 240
Dotani, T., 78
Duncan, R. C., 228

F

Frail, D. A., 240

G

Galama, T. J., 240
Gehrels, N., 261
Goodrich, R. W., 240

H

Harrison, F. A., 240
Hartmann, D., 261
Höflich, P., 132, 301, 324, 338
Hong, D. K., 184
Hurley, K., 261

J

Jones, T. W, 85
Jung, H., 172

K

Kang, H., 331
Khokhlov, A., 301
Kim, C., 291
Kim, S.-W., 375
Kippen, M., 261
Krolik, J. H., 10
Kulkarni, S. R., 240
Kusunose, M., 119
Kwak, K., 291

L

Lattimer, J. M., 205
Lee, C. H., 3
Lee, C.-H., 178
Lee, H. K., 3, 23, 184
Lee, H. M., 48
Lee, H.-W., 368
Lee, M. G., 350
Li, H., 119
Lu, J., 93
Lu, T., 284

M

MacFadyen, A. I., 313
Min, D.-P., 172
Mineshige, S., 359

N

Nagase, F., 56
Nemiroff, R., 261
Nomoto, K., 359
Nowak, M. A., 184

P

Park, H. S., 261
Park, M.-G., 112
Park, T.-S., 172
Peng, Q., 93
Pereira, W., 261
Podgainy, D., 197
Porrata, R., 261
Portegies Zwart, S., 39

R

Rho, M., 160, 184
Ryu, D., 85

S

Sari, R., 240
Schukraft, J., 147
Sekimoto, Y., 382

T

Takabe, H., 389
Takahara, F., 119
Tatematsu, K., 382

Totani, T., 221
Tregillis, I. L., 85
Tsuruta, S., 359

U

Umeda, H., 359

V

van Putten, H. P. M. M., 30

W

Wang, J., 93
Wang, L., 324, 338
Weber, F., 197
Wei, D. M., 284
Wheeler, J. C., 324, 338
Wijers, R. A. M. J., 268
Williams, G., 261

Y

Yang, J., 197
Yi, I., 291, 338
Yost, S. A., 240
Yuan, F., 93